D1357522

Human Gene Evolution

The HUMAN MOLECULAR GENETICS series

Series Advisors

D.N. Cooper, *Institute of Medical Genetics, University of Wales College of Medicine, Cardiff, UK*

S.E. Humphries, *Division of Cardiovascular Genetics, University College London Medical School, London, UK*

T. Strachan, *Department of Human Genetics, University of Newcastle upon Tyne, Newcastle upon Tyne, UK*

Human Gene Mutation
From Genotype to Phenotype
Functional Analysis of the Human Genome
Molecular Genetics of Cancer
Environmental Mutagenesis
HLA and MHC: Genes, Molecules and Function
Human Genome Evolution
Gene Therapy
Molecular Endocrinology
Venous Thrombosis: from Genes to Clinical Medicine
Protein Dysfunction in Human Genetic Disease
Molecular Genetics of Early Human Development
Neurofibromatosis Type 1: from Genotype to Phenotype
Analysis of Triplet Repeat Disorders
Molecular Genetics of Hypertension
Human Gene Evolution

Forthcoming title
B Cells

Human Gene Evolution

David N. Cooper

Institute of Medical Genetics, University of Wales College of Medicine, Cardiff, UK.

βIOS
SCIENTIFIC
PUBLISHERS

© BIOS Scientific Publishers Limited, 1999

First published in 1999

A CIP catalogue record for this book is available from the British Library.

ISBN 1 859961 51 7

BIOS Scientific Publishers Ltd
9 Newtec Place, Magdalen Road, Oxford OX4 1RE, UK
Tel. +44 (0)1865 726286. Fax +44 (0)1865 246823
World Wide Web home page: http://www.bios.co.uk/

Published in the United States, its dependent territories and Canada by Academic Press, Inc., A Harcourt Science and Technology Company, 525 B Street, San Diego, CA 92101–4495. www.academicpress.com

TO PAUL, CATRIN AND DUNCAN

O sweet spontaneous
earth how often
has the naughty thumb
of science prodded
thy
beauty
thou answereth
them only with
spring.

e e cummings *Tulips and Chimneys* (1924)

Production Editor: Andrea Bosher.
Typeset by Saxon Graphics Ltd, Derby, UK.
Printed by Biddles Ltd, Guildford, UK.

Contents

Preface

It seemed worthwhile to try how far the principle of evolution would throw
light on some of the more complex problems in the natural history of man.
Charles Darwin *The Descent of Man* (1871)

Ever since Darwin published his *The Origin of Species* in 1859, the theory of evolution has been the single most important unifying idea in biology. Although 140
years have separated the comparison of pigeon breeds from the comparison of
mammalian gene sequences, the underlying evolutionary questions posed are not
dissimilar. In the intervening years, evolutionary theory has permeated all the
natural sciences. Indeed, the study of human evolution has now become so broad
that it spans biochemistry, anatomy, physiology, psychology and social behavior,
linguistics, epidemiology, demography, paleontology and of course genetics. It has
also provided the impetus to forge new alliances between some of these disciplines
whose practitioners have traditionally shared little common ground.

Knowledge of our origins, and of our relationship to the rest of the natural
world, has the potential to enrich a wide range of human activities. Indeed, if 'the
proper study of mankind is man', then it is arguable that the study of human evolution should be regarded as fundamental to our continuing quest for self-knowl-
edge. As McConkey and Goodman (1997) put it:

> It is our conviction that comparative analysis of human and ape genomes is far
> more than an excursion into natural history at the molecular level. Until we
> have a detailed understanding of the genetic differences between ourselves
> and our closest evolutionary relatives, we cannot really know what we are.

The dramatic progress made in human molecular genetics over the last two
decades has produced a wealth of information on gene structure, mapping and
expression as well as mutation, polymorphism and comparative genome analysis.
These data have been successfully superimposed upon the firm foundations of
evolutionary biology already established by a synthesis of population genetics,
molecular biology and phylogenetics (Li, 1997). Within the next 5 years, the
Human Genome Project should yield the entire sequence of the human genome,
providing us with an unparalleled opportunity to understand the structure and
function of our genome as well as its evolution (Clark, 1999). To appreciate fully
the relationship between the structure and function of our genes and the proteins
they encode, we must be aware of their evolutionary past and their molecular
ontogeny. Conversely, to have any real understanding of the evolutionary pathways taken, we must also understand gene structure and function.

The aim in writing this book was to bring together the highly dispersed
literature on human gene structure, function and expression, to integrate this
with our emerging knowledge of chromosome and genome structure, and to draw

comparisons both within and between paralogous and orthologous gene sequences in order to establish the nature of the mutational mechanisms responsible for their evolutionary divergence. Such an approach, encompassing a large number of human genes and their extended families, is essential for any attempt to discern underlying evolutionary principles.

This volume is divided into three parts. The first, comprising two chapters, is intended to serve as an introduction to the structure, function and evolution of the human genome. The second, containing four chapters, focuses on the evolution of gene structure, organization and expression, the origins of human genes and gene inactivation. The third part, containing the remaining four chapters, discusses the wide range of mutational mechanisms that have created and fashioned extant human genes and fine-tuned their structure, function and expression.

A central theme of this volume is that mutations in human gene pathology and evolution represent two sides of the same coin in that those same mutational mechanisms that have frequently been implicated in human pathology have also been involved in potentiating evolutionary change. In order to illustrate this parallelism, a large number of examples of different types of mutational lesion which have occurred during molecular evolution have been collated. Regardless of whether they are advantageous, disadvantageous or neutral, these mutational changes and their putative underlying causal mechanisms are very similar. This book therefore constitutes a companion volume to *Human Gene Mutation* (Cooper and Krawczak, 1993), which investigated the causes and consequences of pathological gene lesions and demonstrated that the nonrandomness of mutation is determined largely by the local DNA sequence environment.

It is now clear that the gene has often been a dynamic entity over evolutionary time, not a static one. Many genes have undergone gross rearrangement as a result of the action of any one of a number of mutational processes such as insertion, inversion, duplication, repeat expansion, translocation or deletion. It turns out that even relatively conserved genes do not necessarily always change by a slow incremental process of single base-pair substitution; rather, such genes may acquire multiple nucleotide substitutions simultaneously by mechanisms such as gene conversion.

Ten years ago, in order to illustrate principles of human gene evolution, it would have been necessary to draw extensively on examples from organisms such as *Drosophila* and yeast. In recent years, however, the literature has expanded to such an extent that I have been able in most cases to use human genes as examples. Where this has not been possible, I have tried to quote examples from other mammals or, failing that, examples from other vertebrates. Wherever possible, data from nonhuman primates have been included in order to place human gene evolution in its proper context. For genes with greater antiquity, comparisons have been made with orthologues from other taxa or even other phyla. This notwithstanding, I have tried to confine my treatment of the subject area to the evolution of human gene sequences and have declined to stray too far from the gene itself, the primary target of mutation and source of hereditary variation. Some topics, such as the evolution of the mitochondrial genome, repetitive DNA sequences or human populations could easily have had whole volumes devoted specifically to them but have of necessity been treated somewhat summarily. Protein evolution also has a vast literature and I have deliberately not attempted to cover this topic in any detail here. There are numerous texts that discuss the evolution of protein folds, protein sequences and the relationship between protein structure and function; the interested reader is referred to these specialized texts for detailed information.

Every effort has been made to name the human genes cited according to their encoded products and also to specify their symbols (as currently recommended by the *Human Gene Nomenclature Committee*). In addition, their chromosomal location has been provided wherever possible, not only because of its potential functional significance but also to avoid ambiguity if and when gene symbols are altered. Altogether, more than 1600 different human genes are referred to specifically in the text and these have been listed in separate indices.

Despite the recent rapid advances in molecular evolutionary genetics, I have been conscious in places of being able only to describe or list, without being able to discern any underlying principle or illuminate a question by providing an explanatory mechanism. I am also acutely aware of the danger of supposing that we understand a given process when in fact all that we have done is to have collated the basic information necessary for us to derive explanations. As Morgan put it:

> When the biologist thinks of animals and plants, ... he runs the risk of thinking that he is explaining evolution when he is only describing it.
> T.H. Morgan *Evolution and Genetics* (1925)

It is of course almost impossible not to view evolution from a human perspective, that of the only organism able to contemplate its own origins and perhaps its own demise. Indeed, Haldane was aware of our tendency to anthropomorphize:

> I have been using such words as 'progress', 'advance', and 'degeneration', as I think one must in such a discussion, but I am well aware that such terminology represents rather a tendency of man to pat himself on the back than any clear scientific thinking. The change from monkey to man might well seem a change for the worse to a monkey ... we must remember that when we speak of progress in evolution we are already leaving the relatively firm ground of scientific objectivity for the shifting morass of human values.
> J.B.S. Haldane *The Causes of Evolution* (1932)

At the heart of the way in which we conceptualize the process of evolution is an apparent dichotomy: on the one hand, sequence conservation is usually held to imply functionality, on the other, the emergence of novel functions implies change. Thus, selection may act conservatively so as to retain features of structural or functional importance (negative or purifying selection), or act so as to favor changes that confer some advantageous characteristic (positive selection). Evolution can, however, also proceed in a stochastic neutralist fashion, some features being adopted not necessarily because of any selective advantage accruing to the organism, but rather owing to the vagaries of population size, structure or dynamics over an extended period of time. Further, it is now clear that some DNA sequences either possess or acquire a mutational momentum of their own and, in the absence of negative selection, will drive the process of evolutionary change without the necessary involvement of selection. In practice, these diverse mechanisms are acting in different combinations and permutations at many loci simultaneously. Evolution has no foresight, however, and the ultimate arbiter of evolutionary success is always the number of surviving, reproducing, offspring.

In the years ahead, one of the major challenges will be to determine how we differ from our closest living relatives, the chimpanzees and the other great apes; to explain in genetic terms the differences, not only in anatomy but also in the intellect and, in particular, language, that have come about over the last 5–7 million years since the divergence of the human and chimpanzee lineages. It appears unlikely that such major phenotypic differences can be explained simply in terms of incremental changes in the structure of specific genes leading to relatively subtle changes in protein structure and function. Instead, it is anticipated that we shall need to locate key changes in regulatory genes that have served to alter the tissue and developmental specificity of gene expression, or the expression of multiple downstream genes, gene pathways and even ultimately gene networks.

The story of our evolutionary past, written in the coded language of DNA, is told in our genome sequence. We have learned how to access it, we shall soon be able to read it, but the most demanding task of all still remains, that of interpreting it.

References

Clark, M.S. (1999) Comparative genomics: the key to understanding the Human Genome Project. *BioEssays* **21**: 121–130.

Cooper, D.N. and Krawczak, M. (1993) *Human Gene Mutation*. BIOS Scientific Publishers, Oxford.

Human Gene Nomenclature Committee: http://www.gene.ucl.ac.uk/nomenclature/

Li, W.-H. (1997) *Molecular Evolution*. Sinauer Associates, MASS.

McConkey, E.H. and Goodman, M. (1997) A Human Genome Evolution Project is needed. *Trends Genet.* **13**: 350–351.

Acknowledgements

I am most grateful to Beryl and Neil Cooper, Michael Krawczak, Annie Procter, David Rubinsztein, Tom Strachan, Nick Thomas, Peter Thompson, Meena Upadhyaya, Adam Wacey and Johannes Wienberg for their helpful comments, criticisms and suggestions on the draft manuscript and to Peter Harper for his enthusiasm and encouragement. I would also like to thank Margaret Winstanley for her invaluable assistance with indexing and in the preparation of figure legends, Julia White of the HUGO/GDB *Human Gene Nomenclature Committee* for helping to provide correct gene symbols and Jonathan Ray and Andrea Bosher of BIOS Scientific Publishers for their much appreciated practical support.

Abbreviations

Nucleotide symbols

A: adenine
C: cytosine
G: guanine
T: thymine
U: uracil
R: G/A (purine)
Y: C/T (pyrimidine)
K: G/T
M: A/C
S: G/C
W: A/T
B: G/T/C
D: G/A/T
H: A/C/T
V: G/C/A
N: A/G/C/T

Amino acid symbols

A: alanine
B: aspartate/asparagine
C: cysteine
D: aspartate
E: glutamate
F: phenylalanine
G: glycine
H: histidine
I: isoleucine
K: lysine
L: leucine
M: methionine
N: asparagine
P: proline
Q: glutamine
R: arginine
S: serine
T: threonine
V: valine
W: tryptophan
Y: tyrosine
Z: glutamate/glutamine

Websites cited in the text

Alternatively Spliced Genes Database (ASDB) **http://cbcg.nersc.gov/asdb**

C. elegans Genome Project **http://www.sanger.ac.uk/Projects/C_elegans/**

Codon Usage Tabulated from GenBank
http://www.dna.affrc.go.jp/~nakamura/CUTG.html

Eukaryotic Promoter Database **http://www.epd.isb-sib.ch**

Expressed Sequence Tags Database (dbEST)
http://www.ncbi.nlm.nih.gov/dbEST/index.html

FlyBase **http://flybase.bio.indiana.edu/**

GenBank **http://www.ncbi.nlm.nih.gov/**

Gene Map of the Human Genome **http://www.ncbi.nlm.nih.gov/genemap98/**

Genome Database (GDB) **http://gdbwww.gdb.org/**

Homology-derived Structures of Proteins **http://www.embl-ebi.ac.uk/dali/**

Human Gene Mutation Database **http://www.uwcm.ac.uk/uwcm/mg/hgmd0.html**

Human Gene Nomenclature Committee **http://www.gene.ucl.ac.uk/nomenclature/**

Human Genic Bi-Allelic Sequences (HGBASE) **http://hgbase.interactiva.de/intro.html**

Human Genome Project, Sanger Centre **http://www.sanger.ac.uk/HGP**

Human/Mouse Homology Relationships **http://www.ncbi.nlm.nih.gov/Homology/**

Joint Genome Institute **http://www.jgi.doe.gov/**

Mammalian Homology Database **http://www.informatics.jax.org/homology.html**

Munich Information Center for Protein Sequences **http://www.mips.biochem.mpg.de/**

Primate classification **http://nmnhwww.si.edu/cgi-bin/wdb/msw/children/query/11002**

Protein Domain Database **http://protein.toulouse.inra.fr/prodom.html**

Protein Folds and Families Database **http://www.embl-ebi.ac.uk/dali/**

Saccharomyces Genome Database **http://genome-www.stanford.edu/Saccharomyces/**

Single Nucleotide Polymorphism Database **http://www.ncbi.nlm.nih.gov/SNP/**

Transcription Regulatory Regions Database **http://www.bionet.nsc.ru/trrd**

Washington University Genome Sequencing Center

http://genome.wustl.edu/gsc/index.shtml

Whitehead Institute/MIT Genome Sequencing Project http://www-seq.wi.mit.edu/

UniGene http://www.ncbi.nlm.nih.gov/UniGene/index.html

PART 1
INTRODUCTION AND
OVERVIEW

Structure and function in the human genome

1.1 Introduction

> Nothing in biology makes sense except in the light of evolution.
> Theodosius Dobzhansky (1973)

The haploid human genome comprises 23 chromosomes containing between them some 3.2×10^9 bp of DNA. The bulk of the genome comprises DNA sequence of varying degrees of repetitivity whilst about 10% represents unique (single copy) sequence containing perhaps 70 000 genes. The repetitive portion contains a mixture of DNA sequence elements which may have a structural or regulatory role or may be merely 'junk DNA' without obvious function (Zuckerkandl, 1997). Human genes contain within them all the genetic information necessary to specify the encoded proteins but they also contain further information, in the form of a large number of different DNA sequence motifs that serve to control mRNA expression, splicing, transport, and stability. This chapter constitutes a short introduction to the structure, function and expression of the human genome together with a brief description of the types of mutational lesion in human genes that have been found to be responsible for inherited disease.

1.1.1 Chromosome structure and function

> DNA plays a role in life rather like that played by the telephone directory in the social life of London: you can't do anything much without it, but, having it, you need a lot of other things – telephones, wires, and so on – as well.
> C.H. Waddington (1968)

Chromatin structure. Human chromosomes contain DNA in a highly coiled and condensed form, organized and packaged by structures known as *nucleosomes*. Chains of nucleosomes comprise a '10 nm fibre' and this is coiled to form the '30 nm fibre' which is in turn further coiled to form *chromatin* (reviewed by Kornberg and Lorch, 1992; Paranjape *et al.*, 1994; Wolffe, 1992). The highest degree of condensation is found in transcriptionally inactive regions, those regions in which mRNA synthesis from specific genes is turned off. To allow transcriptional activation of a gene (i.e. to allow the enzyme RNA polymerase to initiate mRNA synthesis), chromatin must be uncoiled, a process which occurs in at least three stages: the unfolding of large chromosomal domains (25–100 kb), the remodeling of the chromatin structure of gene regulatory regions and the alteration of nucleosome structure in transcribed regions (Jackson, 1997). Unfolding reveals binding sites on the chromosomal DNA for activator proteins. Once bound, these proteins alter nucleosome positioning and reveal further binding sites for activator proteins. Decondensed chromatin thus provides an accessible template for the assembly of the transcriptional initiation complex.

Centromeres. The centromere is essential for normal disjunction of the chromosomes following cellular division at meiosis and mitosis. Centromeric DNA consists of arrays of tandemly repeated DNA sequences which have undergone homogenization by repeated genetic exchanges (Lee *et al.*, 1997; Warburton *et al.*, 1993). Alphoid satellite DNA is the only satellite DNA family known to be present in the centromeric regions of all human chromosomes. The basic 169 and 172 bp repeats of this primate-specific satellite DNA comprise the bulk of human centromeric DNA and contain a 17 bp binding site for the centromere-specific protein CENP-B. This binding site motif is present at the centromeres in the chromosomes of all the anthropoid apes but is absent from the genomes of Old and New World monkeys and prosimians (Haaf *et al.*, 1995). It is also absent from human Y chromosomal alphoid satellite DNA (Jørgensen, 1997).

Human alphoid satellite DNA may be divided into subsets which are largely chromosome-specific (Jørgensen, 1997). They may however be grouped into four supra-chromosomal families each characterized by a specific monomer type (Jørgensen, 1997). Homologies exist between alphoid satellite DNAs from different ape species but these are not associated with homologous chromosomes (Samonte *et al.*, 1997; Warburton *et al.*, 1996). Although alphoid satellite repeats may have evolved from a common ancestral repeat monomer (Haaf and Willard, 1998), they have also been subject to concerted evolution between homologous chromosomes within a given species (Haaf and Willard, 1997; Jørgensen *et al.*, 1992).

Using a combination of oligonucleotide primer extension and immunocytochemistry, Mitchell *et al.* (1992) showed that the alphoid repeats were closely associated with the *kinetochore* (the structural element on the chromosome that binds to the mitotic spindle). The presence of (AATGG)n (CCATT)n repeats in the centromeric region suggests that stem-loop structures might form which could serve as specific recognition sites for kinetochore function (Catasti *et al.*, 1994). Alphoid satellite DNA sequences are not however restricted to centromeric regions (Baldini *et al.*, 1993).

Telomeres. Telomeres allow the end of the chromosomal DNA to be replicated completely without the loss of bases at the termini (reviewed by Blackburn, 1994; Gilson *et al.*, 1993). They are the sites at which the pairing of homologous chromosomes is initiated and in humans contain long arrays (averaging about 10–15 kb) of minisatellite DNA comprising tandem hexanucleotide repeats, most frequently TTAGGG (Brown, 1989). Other telomeric hexanucleotide repeats (e.g. TTGGGG, TGAGGG) are also known (Allshire *et al.*, 1989; Brown, 1989). These sequences and their variants are tandemly repeated but are nonrandomly distributed and polymorphic in terms of their location. Chimpanzees also possess the TTAGGG telomeric repeat (Luke and Verma, 1993) but differ from humans in terms of their subterminal satellite sequences (Royle *et al.*, 1994).

The simple sequence of telomeres is synthesized by the ribonucleoprotein polymerase, *telomerase* (Blackburn, 1992), and is thought to protect the ends of chromosomes from degradation during DNA replication. Telomere length decreases with age and number of cell divisions. Some chromosomes (e.g. 9p, 12p, 14p, 17p, 21p and 9q) have telomeres shorter than the average whereas others (e.g. 4q, 5p, 18q and Xp) have telomeres which are longer (Martens *et al.*, 1998). Telomere length polymorphisms are apparent for some chromosomes, for example 11p (Martens *et al.*, 1998).

Sites of recombination. Chiasmata represent the cytological evidence for recombination. At meiosis, each pair of homologous chromosomes possesses at least one chiasma and the number of chiasma per pair is proportional to the size of the chromosome. Chiasma frequency is a function both of distance from the centromere (Laurie and Hulten, 1985) and chromosome band 'flavor': the G+C rich T bands exhibit a six-fold higher chiasma frequency than G bands (Holmquist, 1992). Since G+C content is in general positively correlated to chiasma frequency (Eyre-Walker, 1993), this could explain why gene density is higher in T bands (see section 1.1.1, *Gene distribution and density*).

Recombination is considerably higher in human females than in males, as evidenced by the average distance between two markers in the female genetic map being 85% longer than in males. Chiasma frequency is a function of distance from the centromere (Laurie and Hulten, 1985). Recombination therefore tends to increase towards the telomeres, the distal 15% of chromosomes containing 40% of the chiasmata. Finally, it may also be pertinent to consider that intrachromosomal homologous recombination may be enhanced by transcription in mammalian cells (Nickoloff, 1992).

Obligatory recombination occurs during male meiosis within the pseudoautosomal region, a 2.6 Mb stretch of homologous sequence at the tip of the short arms of the X and Y chromosomes (Petit *et al.*, 1988; Ellis and Goodfellow, 1989). Genes in this region escape X-inactivation and the boundaries of this region appear to have been conserved evolutionarily between Old World monkeys and human (Ellis *et al.*, 1990).

Regions of sex-specific hypo- and hyper-recombination have been reported in a study which compared genetic and physical maps of human chromosome 19 (Mohrenweiser *et al.*, 1998). Other recombination hotspots have been characterized in specific human genes including those encoding the T-cell receptor beta chain (*TCRB*; 7q35; Seboun *et al.*, 1993) and HLA-associated ATP transporter 2 (*TAP2*; 6p21.3; Cullen *et al.*, 1995) loci. Several different types of DNA sequence have been proposed to be recombinational hotspots in the genomes of mice and men. $(CAGA)_6$ and $(CAGG)_{7-9}$ represent hotspots of recombination in the murine MHC gene cluster (Steinmetz, 1987). Other sequences thought to promote recombinational instability are alphoid repeats (Heartline *et al.*, 1988), a *mariner* transposon-like element (Reiter *et al.*, 1996), Z-DNA ('left-handed DNA'; Wahls *et al.*,

1990a) and minisatellite sequences (Chandley and Mitchell, 1988; Wahls *et al.*, 1990b). Recombinational breakpoints have also been found to be associated with topoisomerase I cleavage sites in the rat genome (Bullock *et al.*, 1985). The majority of these cleavage sites contain the sequences CTT and GTT. It may therefore be that the process of nonhomologous recombination is mediated by topoisomerase I.

Gene distribution and density. Several thousand genes have now been mapped to within single chromosome bands. Some 80% map to the G-C rich R bands whilst 20% map to G bands (Bickmore and Sumner, 1989; Craig and Bickmore, 1993). A similar distribution is apparent for CpG islands (*see* Section 1.1.1, *CpG islands*): 86% are located in R bands (Craig and Bickmore, 1994; Larsen *et al.*, 1992). 'Housekeeping' genes are strictly confined to the R bands together with about half of the tissue-specific genes (Holmquist, 1992) whereas the remainder of the tissue-specific genes are present in the G bands. One of the four recognized types or 'flavors' of R band, known as T bands, are often found at telomeres, exhibit the highest G+C content and contain between 58% and 68% of R band genes as well as the majority of CpG islands (Collins *et al.*, 1996; Holmquist, 1992).

Chromosomes 13 and 18 appear to possess a relatively low gene density and chromosome 19 a relatively high density as evidenced by the chromosomal assignment of some 320 cDNAs derived from a human brain cDNA library (Polymeropoulos *et al.*, 1993). Interestingly, DNA excision repair may be preferentially directed toward regions of high gene density (Surrales *et al.*, 1997), a reflection perhaps of the preferential repair of actively transcribed gene sequences.

Isochores. The human genome is a mosaic of large (>300 kb) DNA segments or isochores that are compositionally homogeneous and which can be subdivided into a small number of families characterized by different degrees of GC-richness (30–60%) (Bernardi *et al.*, 1993a,b). Five families have been identified: L1 and L2 which are GC-poor and comprise 62% of the genome and H1, H2, and H3 which are GC-rich and represent 22%, 9%, and 3% of the genome respectively. Gene concentration varies between isochores: 34% of human genes are located in L1 and L2 isochores, 38% in the H1 and H2 isochores, and 28% in the H3 isochores (Mouchiroud *et al.*, 1991; Saccone *et al.*, 1996). The banding pattern of chromosomes reflects the isochore organization: thus, the G bands are formed by L1 and L2 isochores whilst the T bands are formed by the H2 and H3 families. A recent study of two primate globin pseudogenes which reside in different isochore compartments has provided evidence that isochores have arisen as a result of mutational bias rather than from the action of selection (Francino and Ochman, 1999).

Matrix attachment regions. Chromatin is attached to the nuclear matrix or scaffold at specific sites known as matrix or scaffold attachment regions (MARs/SARs). The organization of chromatin with respect to the nuclear scaffold is thought to determine chromosome architecture in terms of its functional domains; this in turn influences gene activity (reviewed by Dillon and Grosveld, 1994 and Walter *et al.*, 1998). Indeed, MARs may function so as to place genes at the nuclear scaffold in order to facilitate their transcription. MARs do not appear to share extensive sequence homology but often comprise 200 bp of AT-rich DNA, for example the sequence AATATTTTT in the murine immunoglobulin κ gene locus (Cockerill and Garrard, 1986). A number of MAR-binding proteins are known to bind to MARs including the attachment region-binding protein, ARBP and histone H1.

MARs appear to be preferentially associated with topoisomerase II cleavage sites (reviewed by Laemmli *et al.*, 1992) and share sequence homology with binding sites for homeobox proteins (Boulikas, 1992). Topoisomerase II plays a role in the segregation of daughter chromosomes after DNA replication and also in chromosome condensation; it binds preferentially to MARs (Adachi *et al.*, 1989). Vertebrate topoisomerase II cleavage sites also occur in association with MARs and manifest a consensus sequence, A/G N T/C N N C N N G T/C N G G/T T N T/C N T/C (Spitzner and Muller, 1988).

MARs also appear to be preferentially associated with enhancer-type elements. Indeed, MARs stimulate heterologous gene expression in reporter gene experiments. A *cis*-acting regulatory element 3′ to the human ᴬγ-globin (*HBG1*) gene, known to be associated with the nuclear matrix, has been shown to bind specifically to an AT-rich binding protein (SATB1) that binds to MARs (Cunningham *et al.*, 1994).

Origins of DNA replication. Chromosomal DNA replication initiates at specific points (*origins*) and proceeds outward bidirectionally from specific loci. Although a number of putative origins of replication have been identified in mammalian species (reviewed by Coverley and Laskey, 1994; De Pamphilis, 1993), data are still sparse. One of the best characterized is that found in human between the δ- (*HBD*) and β-globin (*HBB*) genes on chromosome 11 (Kitsberg *et al.*, 1993a). This replication origin is bidirectional and functional regardless of the transcriptional state of the β-globin gene. From the study of six putative origins of replication (including one in the human c-*myc* oncogene), Dobbs *et al.* (1994) claimed to have derived a consensus sequence, albeit a fairly redundant one: A/T A A/T T T A/G/T A/G/T A/T A/T A/T A/G/T A/C/T A/T G A/T A/C/T A/C A A/T T T. However, there are probably several different classes of replication origins which possess different sequence characteristics (Boulikas, 1996).

Replication origins are often associated with CpG-rich regions (Delgado *et al.*, 1998; Rein *et al.*, 1997; Tasheva and Roufa, 1995) and may sometimes be located in the vicinity of matrix attachment regions (Section 1.1.1, *Matrix attachment regions*) (Lagarkova *et al.*, 1998). In the human genome, the units of DNA replication range in size from 50 kb to 600 kb and are often clustered (Hand, 1978). For instance, there are at least six such *replicons* within the human dystrophin (*DMD*; Xp21) gene (Verbovaia and Razin, 1997). The gene-rich R bands replicate early in S phase whilst the G bands replicate late. Housekeeping genes invariably replicate early whilst tissue-specific genes can be early or late replicating (some replicate earlier when transcriptionally active) (Goldman *et al.*, 1984; Hatton *et al.*, 1988). Nontranscribed genes on the X chromosome also replicate late (Torchia *et al.*, 1994).

1.1.2 Gene organization and transcriptional regulation

Gene structure and regulation. The coding portion of the human genome, roughly 5% of the total DNA complement, probably contains some 70 000 different gene sequences (Fields *et al.*, 1994). Thus, the smallest human chromosome, 21, may contain as many as 2000 genes.

It is now known that most genes in higher organisms are not contiguous but rather are a complex mosaic of protein coding (*exon*) and intervening non-coding (*intron*) sequences (*Figure 1.1*). The exons represent that portion of the gene which encodes the amino acid sequence of the protein product plus 5' and 3' noncoding regions. Initially, both exons and introns are *transcribed* into mRNA but the intronic portion is ultimately removed during mRNA maturation by a process known as *splicing* (see section 1.1.2, *Sequence motifs involved in mRNA splicing and processing*). The mature mRNA is then *translated* into the amino acid sequence of a protein on the ribosome. Although the central dogma of molecular biology was therefore once summarized as 'DNA makes RNA makes protein,' the reverse flow of genetic information is also possible by *reverse transcription* of RNA into DNA (copy DNA or cDNA).

Each individual gene differs not only with respect to its DNA sequence specifying the amino acid sequence of the protein it encodes, but also with respect to its structure. A few human genes are devoid of introns (e.g. thrombomodulin (*THBD*) which spans 3.7 kb) whereas others may possess a considerable number, for example 79 in the 2.4 Mb dystrophin (*DMD*) gene and as many as 118 in the α1(VII) collagen (*COL7A1*) gene (Christiano *et al.*, 1994). Introns may be classified according to whether they interrupt the reading frame of the encoded protein. Thus phase 0 denotes that the intron lies between two codons, phase 1 between the first and second nucleotides of a codon, and phase 2 between the second and third nucleotides of a codon.

Some introns may be huge as in the case of the first intron of the human *COL5A1* gene (~600 kb; Takahara *et al.*, 1995). The average length of a vertebrate intron has been estimated to be ~620 bp (Hawkins, 1988) but introns separating exons preceding the coding ones are often rather longer with an average length of >1800 bp (Hawkins, 1988). This suggests that evolution may sometimes have had to trawl quite far upstream of a gene to recruit appropriate DNA sequence motifs to act as promoter/regulatory elements within the 5' untranslated region. The average length of an internal exon is ~140 bp (Hawkins, 1988) but this average figure conceals some very large exons, for example in the human factor VIII (*F8C*; Xq28) [3106 bp], apolipoprotein B (*APOB*; 2p23-p24) [7572 bp] and mucin 5B (*MUC5B*; 11p15.5) [10 690 bp] genes.

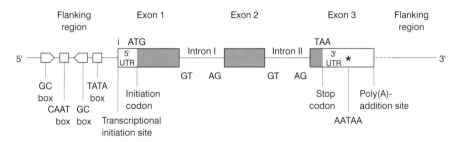

Figure 1.1 Schematic structure of an archetypal human protein-coding gene. UTR, untranslated region.

Some human genes encode very large proteins, for example the cardiac titin (*TTN*; 2q31) cDNA is 82 kb in length and predicts a 27 000 amino acid protein with a molecular weight of nearly 3000 kDa (Labeit and Kolmerer, 1995). Large mRNAs are also generated from the dystrophin (*DMD*; Xp21.2–21.3; 14 kb), apolipoprotein B (*APOB*; 2p23–24; 14 kb), and mucin (membrane-associated glycoprotein) genes *MUC2, MUC5AC, MUC5B, MUC6* (11p15.5; 15–18 kb), *MUC3* (7q22; 17 kb), and *MUC4* (3q24; 18–24 kb) (Debailleul *et al.*, 1998).

Only ~10% of human genes encode proteins of known function. The sequences of many of these genes can be found in GenBank (**http://www.ncbi.nlm.nih.gov/**). In addition, hundreds of thousands of 'expressed sequence tags' (ESTs; Gerhold and Caskey, 1996) have been characterized which together represent a nonredundant set of >45 000 human genes (UniGene; **http://www.ncbi.nlm.nih.gov/UniGene/index.html**).

Precisely what constitutes a gene is somewhat contentious (Epp, 1997) but a crude working definition might be: *a transcription unit plus associated regulatory sequences which together serve to specify both the sequence and the expression pattern of a protein product*. The term 'gene' cannot, however, be restricted to protein coding sequences since some genes (e.g. snRNA, rRNA, tRNA, *XIST, H19, IPW*) encode RNA molecules with a variety of biological functions and which are not translated into protein. A simple universally applicable definition of a gene is difficult to derive owing to the existence of exceptions to almost any rule that one might devise. Thus, some transcription units may encode multiple unrelated proteins with different functions as a result of *alternative splicing* (see section 1.1.2, *Sequence motifs involved in mRNA splicing and processing*; Figure 3.1 in Chapter 3). As a consequence, the notion of a gene becomes somewhat elastic. Some genes occur within the introns of other genes, for example *OMG, EVI2A, EVI2B* within the neurofibromatosis type 1 (*NF1*; 17q11.2; Viskochil *et al.*, 1991) gene, *F8A* within the factor VIII (*F8C*; Xq28) gene (Levinson *et al.*, 1990), and U21 within the L5 genes of chickens and mammals (Qu *et al.*, 1994). The genes of most known vertebrate small nucleolar mRNAs (snoRNAs) are located within the introns of other genes (Maxwell and Fournier 1995), two human examples being the *RNE1* and *RNE2* genes which are located within the mitotic regulator (*CHC1*; 1p36.1) and the 67 kDa laminin receptor (*LAMR1*; 3p21.3) genes, respectively. The realization that some genes reside within the introns of other genes makes the concept of the gene that much more diffuse.

Human genes can also overlap in a number of different ways. Thus, two genes encoding *erbA* homologues, *ear*-1 (*THRAL*) and *ear*-7 (*THRA*), located at the same locus on chromosome 17, possess overlapping exons but are transcribed from opposite DNA strands (Miyajima *et al.*, 1989). Similarly, the tenascin-X (*TNXA*; 6p21.3) gene overlaps with the last exon of the cytochrome P450 (*CYP21*; 6p21.3) gene on the opposite DNA strand (Speek *et al.*, 1996). Other examples of overlapping human genes transcribed from opposite DNA strands are provided by the *PMS2* (7p22) gene and a gene encoding a 34.5 kDa polypeptide (Nicolaides *et al.*, 1995), the CD3 ζ/η/θ (*CD3Z*) and Oct1 transcription factor (*POU2F1*) genes on 1q22-q23 (Lerner *et al.*, 1993), and the cytochrome *c* oxidase subunit X (*COX10*) gene and a partially characterized cDNA (C17ORF1) on chromosome 17p12-p11.2 (Kennerson *et al.*, 1997). An example of overlapping human genes transcribed from the same strand is provided by the prematurely termed growth hormone gene-derived transcriptional activator (*GHDTA*) gene and the growth hormone 1 (*GH1*; 17q22-q24) gene (Labarrière *et al.*, 1995). The *GHDTA* gene is transcribed from position -197 in the *GH1* gene promoter and contains an open reading frame that extends from the ATG at -151 to a Stop codon in exon 2 of the *GH1* gene. Another kind of overlap is exemplified by the transglutaminase 1 (*TGM1*; 14q11.2) gene some of whose *cis*-acting regulatory sequences reside within the transcribed portion of the functionally unrelated Rab geranylgeranyl transferase α subunit (*RABGGTA*; 14q11.2) gene located <2kb upstream of the *TGM1* transcriptional initiation site (van Bokhoven *et al.*, 1996). Clearly, the existence of overlapping genes can hamper any attempt to demarcate precisely and unambiguously where one gene ends and another begins.

Some genes such as the immunoglobulin and T-cell receptor genes (see Chapter 4, section 4.2.4) differ in structure between different cell types. Which is the gene; that which is present in the germline or that which is rearranged to perform a specific function in the soma?

Should distant regulatory sequences (such as LCRs, see section 1.1.2, *Locus control regions*) and protein binding sites that maintain chromosomal conformation be included within the boundaries of a gene? Indeed, if we are prepared to entertain radical redefinition of what constitutes a gene, should all parts of a gene necessarily be contiguous on the same DNA strand or even the same chromosome?

At the time of writing, the majority of human gene transcripts still remain to be characterized. However, Adams *et al.* (1995) attempted to ascribe functions to cDNAs by limited DNA sequence analysis and detection of homologies to known proteins. These data are summarized in *Table 1.1*. Of the human cDNAs studied by Adams *et al.* (1995), which probably represent no more than 10% of the total number, only eight of the corresponding genes were expressed in all 30 tissues examined. Some 227 genes were expressed in >20 tissues whilst some 4300 genes were found to be expressed in only one tissue. Clearly such data are extremely useful for any conceptual discussion of what we mean by tissue specificity of gene expression.

Although the primary control of gene expression is usually exerted at the level of transcription, the regulation of gene expression may also occur at several other different stages in the pathway including transcriptional activation, mRNA splicing, stability, export, translation (synthesis of the protein product), post-translational processing, and export of the mature protein (reviewed by Atwater *et al.*,

Table 1.1. Putative functions of a sample of human genes (after Adams *et al.*, 1995)

Putative function	Proportion of transcripts (%)
Cell signalling	12
RNA synthesis/processing	6
Protein synthesis/processing	15
Metabolism	16
Cell division/DNA synthesis	4
Cell structure/mobility	8
Cell/organism defence/homeostasis	12
Unclassified	24

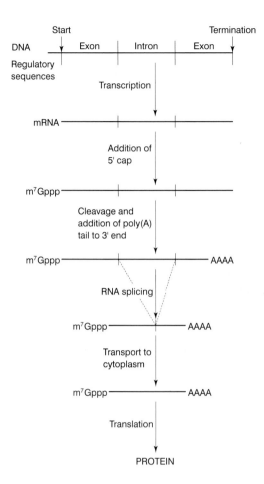

Figure 1.2. Expression pathway of a human protein-coding gene

1990; Hentze 1991; *Figure 1.2*). A variety of sequences within a typical gene region are required for normal and appropriate expression to occur. These are described briefly in subsequent sections.

Polymorphisms

> Variation, whatever may be its cause, and however it may be limited, is the essential phenomenon of evolution. Variation, in fact, *is* evolution. The readiest way, then, of solving the problem of evolution is to study the facts of variation.
> William Bateson (1894) *Materials for the Study of Variation*

The term *polymorphism* has been defined (Vogel and Motulsky, 1986) as a 'mendelian trait that exists in the population in at least two phenotypes, neither of which occurs at a frequency of less than 1%'. Some DNA polymorphisms are neutral single base-pair changes detected by virtue of the consequent introduction or removal of a restriction enzyme recognition site and are accordingly termed Restriction Fragment Length Polymorphisms (RFLPs). They are inherited as simple Mendelian traits since two alleles are generated as a consequence of the presence or absence of each restriction site. RFLPs are not rare, being distributed throughout the genome at a frequency of between 1/200 and 1/1000 bp (Collins *et al.*, 1997; Cooper *et al.*, 1985; Li and Sadler, 1991; Wang *et al.*, 1998). Some 25% of single nucleotide polymorphisms (SNPs) in higher primates occur in CpG dinucleotides (Savatier *et al.*, 1985; Yang *et al.*, 1996), consistent with a model of methylation-mediated deamination.

Not unexpectedly, the vast majority of polymorphisms occur in introns or inter-genic regions rather than within coding sequences and may thus be expected to be neutral with respect to fitness (Bowcock *et al.*, 1991). Those polymorphisms that occur either within coding regions (see Chapter 2, section 2.3.7) or in the promoter region (see Chapter 5, section 5.1.9) may however affect either the structure or function of the gene product or the expression of the gene and may therefore have the potential to be of phenotypic or even pathological significance (Cooper and Krawczak, 1993). Those coding sequence polymorphisms that alter the amino acid sequence of the encoded protein are found at a lower rate and with lower allele frequencies than silent substitutions (Cargill *et al.*, 1999). This probably reflects the action of negative selection on deleterious alleles during human evolution.

Some polymorphisms may be missense mutations, for example those underlying the Lewis *Le* alleles in the *FUT3* gene (19p13.3; Nishihara *et al.*, 1994) or the Arg/Gln 353 polymorphism in the factor VII (*F7*; 13q34) gene (see Chapter 2, section 2.3.7). Others are nonsense mutations that serve to inactivate the gene in question, for example the secretor *se* allele in the *FUT2* gene (19cen-qter) present in 20% of the population (Kelly *et al.*, 1995). Further types of gene-associated polymorphism in the human genome include triplet repeat copy number (see Chapter 8, section 8.9), gene deletions (see Chapter 8, section 8.1), gene duplications (see Chapter 8, section 8.5), intragenic duplications (see Chapter 8, section 8.6), micro-insertions (see Chapter 8, section 8.3), inversions (see Chapter 9, section 9.1), gene fusion (see Chapter 9, section 9.3), and gene copy number (see Chapter 8, sections 8.1 and 8.5). Various databases of human DNA polymorphisms are available for online consultation: The Genome Database (**http://gdbwww.gdb.org/**), the Database of Single Nucleotide Polymorphisms (**http://www.ncbi.nlm.nih.gov/SNP/**), and the database of Human Genic Bi-Allelic Sequences (**http://hgbase.interactiva.de/intro.html**).

The mechanisms by which polymorphisms are maintained in human populations are likely to be varied. The neutralist theory assumes no selection on the alleles of a polymorphic locus and the frequency of an allele may therefore increase simply by *genetic drift* (the change of allele frequency due to random sampling). Such 'transient polymorphisms' often remain at a low frequency in the population before being lost or may instead increase in frequency under the influence of either genetic drift or positive selection until one allele reaches fixation. Most known polymorphisms are probably of this type. However, if the alternative alleles are not neutral with respect to fitness (see Chapter 2, section 2.3.7), the DNA polymorphisms may be maintained by selection pressure, possibly overdominant selection ('balanced polymorphisms'). Finally, a 'hitchhiker effect' may operate if the polymorphisms are closely linked to a locus which is itself under strong selection. In some special cases, such as polymorphisms within CpG dinucleotides, recurrent mutation may serve to maintain the allele frequency in the population.

The alleles of closely linked polymorphisms often occur in specific combinations or *haplotypes*. In the β-globin (*HBB*; 11p15) gene, for example, some haplotypes have been identified in Europeans, Asians, Blacks, and Chinese, indicating that their origin must have predated racial divergence (Antonarakis *et al.*, 1985). By contrast, other haplotypes are population-specific or 'private' (e.g. Schneider *et al.*, 1998; Wainscoat *et al.*, 1986). In some cases, the number of different haplotypes can be quite high, the result of the shuffling of different single base-pair substitutions by such processes as recombination and gene conversion (Fullerton *et al.*, 1994).

In their analysis of 8 kb of intronic sequence of the Duchenne muscular dystrophy (*DMD*; Xp21) genes of individuals from 13 different human populations [European, Papua New Guinean, African (6), Asian (3) and Amerindian (2)], Zietkiewicz *et al.* (1997) identified 36 polymorphisms. Of these, 15 were shared among most of the populations screened, 13 were confined to Africans and four were confined to non-Africans. A detailed study of human DNA polymorphism has also been performed on a 9734 bp region of the human lipoprotein lipase (*LPL*; 8p22) gene, comprising 8736 bp intronic sequence and 998 bp coding sequence, performed on 142 chromosomes from 71 individuals from three distinct populations: Finns, European-Americans and African-Americans (Nickerson *et al.*, 1998). A total of 88 sites were found to vary (79 single base-pair substitutions and nine microdeletions or microinsertions) representing an average of one variable site every 500 bp. Only seven of the 88 variable sites were found in the coding region. At 34 of the 88 sites, the variation was found in only one of the three populations reflecting the differing population and mutational histories. A total of 88 unique haplotypes were identified in the 142 chromosomes sampled, probably a reflection of the complex historical interplay of recombination and mutation. Finally, a study of a total of 87 kb of human chromosome Xq22 revealed 102 polymorphisms, seven of which were shared by Europeans, Ashkenazim, and pygmies, two by pygmies and Europeans, and 19 by Ashkenazim and Europeans (Anagnostopoulos *et al.*, 1999).

Some polymorphisms within human gene coding regions also occur in the orthologous genes of the great apes. In some cases, this may imply an origin for the polymorphism that predated the adaptive radiation of the higher primates ('trans-species polymorphism'). Perhaps the best example of this phenomenon is provided by the ABO blood group locus (*ABO*; 9q34.1-q34.2). In humans, there are four amino acid substitutions (in positions 176, 235, 266 and 268) that distinguish the A- and B-transferases (respectively R, G, L, and G in A-transferase and G, S, M, and A in B-transferase) whilst the O allele is characterized by a single nucleotide (G_{261}) frameshift deletion that inactivates the protein (Yamamoto *et al.*, 1990). The gorilla and chimpanzee possess types B and A respectively whilst the baboon (*Papio cynocephalus*) exhibits both A and B (Yamamoto *et al.*, 1990; Kominato *et al.*, 1992; Martinko *et al.*, 1993). Certain alleles of the major histocompatibility complex (MHC) are also very ancient. Indeed, it would appear that some of the class I MHC allelic lineages are shared by human, chimpanzee and gorilla, implying that these polymorphisms were present in the common ancestor of the three species (see Chapter 4, section 4.2.1, *Genes of the major histocompatibility complex*). Another possible example of trans-species polymorphism is to be found in the complement C4 genes (*C4A*, *C4B*; 6p21.3) of humans, chimpanzees and orangutans (Kawaguchi *et al.*, 1990; Kawaguchi and Klein 1992; Paz-Artal *et al.*, 1994). Finally, trans-species polymorphism may also be present in the cone visual pigment genes, *RCP* and *GCP* (Xq28; Deeb *et al.*, 1994) but in this case, the evidence must be regarded as more equivocal. For polymorphisms to have survived the speciation process, they must have resisted fixation even during severe population bottlenecks. Neutral alleles would have been unlikely to survive the process of speciation and so the long-term persistence of polymorphisms must be explicable by selection. In the case of MHC and ABO alleles, allelic lineages have predated human speciation and pathogen-driven *overdominant selection* (heterozygote advantage) has probably been the factor that has served to maintain the presence of alternative alleles (see Chapter 4, section 4.2.1, *Genes of the major histocompatibility complex*). In other cases, it is conceivable that identical mutations could have occurred independently in the different lineages thereby mimicking true trans-species polymorphism.

Trans-specific polymorphism is likely to be the exception rather than the rule since, in the absence of positive selection, polymorphisms are most unlikely to survive speciation. Thus an average of $4N_e$ generations (N_e = effective population size) are required for a newly emerged allele to become fixed in a population as a consequence of random drift. Since until recently the effective population size of modern humans has been ~10 000 individuals, most neutral human polymorphisms may be expected to be less than 800 000 years old (assuming a 20 year generation time), rather less than the 5–6 Myrs that have elapsed since the divergence of the human and chimpanzee lineages. Hacia *et al.* (1999) studied the orthologous sequences from chimpanzees (both species) and gorilla for a total of 397 human single nucleotide polymorphisms (SNPs). They were able to determine the alleles in 214 of these SNP sites and of these, three segregated for the same nucleotides in both humans and pygmy chimpanzees whilst two segregated in humans and gorillas. However, 4/5 of the shared polymorphic sites occurred at hypermutable CpG dinucleotides suggesting that recurrent mutation rather than identity-by-descent was the cause of the shared polymorphism.

Functional organization of human genes. Well before the organization of the human genome is known in its entirety, some trends governing the distribution of genes have become apparent (McKusick, 1986; Schinzel *et al.*, 1993; Strachan, 1992) and these are discussed briefly below.

Genes which encode the same product (e.g. ribosomal RNA (*RNR*), histones (*H1F2, H1F3, H1F4, H2A, H2B, H3F2, H4F2*), HLA, homeobox proteins (*HOXA, HOXB, HOXC*, and *HOXD*), immmunoglobulins (*IGK, IGL, IGH*)) *are often clustered*. However, these clusters are usually distributed between several different chromosomes (e.g. *RNR*, chromosomes 13, 14, 15, 21, and 22; histones, chromosomes 1, 6, and 12; *HOX*, chromosomes 2, 7, 12, and 17). Thus, multigene families have evolved by duplication and divergence but the duplicated copies may no longer be *syntenic* (i.e. linkage on the same chromosome conserved) owing to the translocation of discrete chromosomal regions during evolution. The likelihood of synteny may well be related to the time that has elapsed since duplication.

Genes which encode tissue-specific protein isoforms or isoenzymes are sometimes clustered (e.g. pancreatic (*AMY2A, AMY2B*) and salivary (*AMY1A*) amylase genes on chromosome 1p21) but sometimes not (e.g. cardiac (*ACTC*), skeletal muscle (*ACTA1*), smooth muscle, aorta (*ACTA2*) and smooth muscle, enteric (*ACTG2*) α-actin genes which are located on chromosomes 15, 1, 10, and 2, respectively). Again, the time that has elapsed since the duplication event is likely to be an important factor in determining whether or not the duplicated genes have remained syntenic.

Genes encoding isozymes specific for different subcellular compartments are usually not syntenic [e.g. soluble/extracellular and mitochondrial forms of superoxide dismutase (*SOD1/SOD3*, *SOD2* on chromosomes 21/4 and 6, respectively), aconitase (*ACO1*, *ACO2* on chromosomes 9 and 22) and thymidine kinase (*TK1*, *TK2* on chromosomes 17 and 16)].

Genes encoding enzymes catalyzing successive steps in a particular metabolic pathway are usually not syntenic. Thus, the five enzymes of the urea cycle are encoded by genes (*ARG1*, *ASL*, *ASS*, *CPS1*, *OTC*) on chromosomes 6, 7, 9, 2, and X and four enzymes involved in galactose metabolism are encoded by genes (*GALE*, *GALK1*, *GALK2*, *GALT*) on chromosomes 1, 17, 15, and 9. However, there are exceptions to this rule: four genes encoding enzymes of the glycolytic pathway (*TPI1*, *GAPD*, *ENO2*, *LDHB*) are located on the short arm of chromosome 12 in the region p13-p12. Similarly, the *GDH* and *PGD* genes encoding enzymes of the phosphogluconate pathway are encoded by linked genes on chromosome 1. The reasons for this syntenic organization, when it occurs, are usually unclear but we may nevertheless surmise the evolutionary history of the genes involved. Indeed, as early as 1945, Horowitz proposed that genes encoding enzymes of metabolic pathways could have arisen by serial duplication. His idea was that the protein that would eventually become the terminal enzyme of a given metabolic pathway would possess a binding site for the substrate that it used. A novel protein derived from the duplicated gene could still use this binding site for interaction with the same substrate molecule but could evolve the capability of producing it as a product from another source. Thus, the substrate for the terminal enzyme would become an intermediate in the developing metabolic pathway. In this way, the various enzymes of a pathway would evolve from each other in the reverse order to that in which they appear in the modern pathway. The synteny observed in the cases cited above might be a consequence of conservation resulting from a requirement for coordinate regulation of the loci concerned. Some of these principles also appear to hold for the genes encoding the enzymes of the coagulation cascade (see Chapter 10).

Genes encoding different subunits of a heteromeric protein are often not syntenic (e.g. the genes encoding α-globin (*HBA1*, *HBA2*, chromosome 16) and β-globin (*HBB*, chromosome 11), lactate dehydrogenases A (*LDHA*, chromosome 11) and B (*LDHB*, chromosome 12), factor XIII subunits a (*F13A*, chromosome 6) and b (*F13B*, chromosome 1), the immunoglobulin light chains (*IGK*, *IGL*, chromosomes 2 and 22), and heavy chains (*IGH*, chromosome 14). However, several cases of synteny are known e.g. the genes encoding the three chains of fibrinogen (*FGA*, *FGB*, *FGG*, all closely linked on chromosome 4), the α- and β-chains of C4b-binding protein (*C4BPA* and *C4BPB*, closely linked on chromosome 1q32), the complement component 1Q α- and β-chains (*C1QA*, *C1QB*, linked on chromosome 1p) and the platelet membrane glycoproteins IIb and IIIa (*ITGA2B*, *ITGA3*, closely linked on chromosome 17q21-q22). The clustering of these genes may be important for their coordinate regulation by common control elements (see Chapter 5, section 5.1.14). The various subunits of the T-cell antigen receptor are intriguing: the α- and δ-subunits are encoded by genes (*TCRA, TCRD*) on chromosome 14, the β- and γ-subunit genes (*TCRB, TCRG*) are on chromosome 7 whereas the ε-subunit gene (*TCRE*) lies on chromosome 11. Whilst synteny probably implies a common evolutionary origin, lack of synteny does not necessarily argue against it.

Clustering of genes of similar function and common evolutionary origin is common, for example the genes encoding blood coagulation factors VII (*F7*) and X (*F10*) on chromosome 13q34, the γ-crystallin (*CRYG*) gene cluster on chromosome 2q33 and the six alcohol dehydrogenase (*ADH1, ADH2, ADH3, ADH4, ADH5, ADH7*) genes on chromosome 4q22. Many more examples are given in Chapter 4.

Genes do not usually exhibit chromosomal clustering with respect to the structure/function of particular organs or subcellular organelles (e.g. mitochondria). However, various genes encoding proteins expressed in the course of epidermal differentiation [involucrin (*IVL*), loricrin (*LOR*), filaggrin (*FLG*), the small proline-rich proteins (*SPRR1A, SPRR1B, SPRR2A*, and *SPRR3*), trichohyalin (*THH*)] are clustered together on chromosome 1q21 thereby betraying their common evolutionary origin (Volz *et al.*, 1993). A considerable number of the genes encoding various cytokines (including several hematopoietic growth factors) and their receptors are clustered on the long arm of chromosome 5: granulocyte-macrophage colony-stimulating factor (*GMCSF*), macrophage colony-stimulating factor (*CSF2*), the CSF1 receptor (*CSF1R*, colony-stimulating factor-1 receptor, also known as c-*fms*), interleukins 3, 4, 5, 9, 12B, and 13 (*IL3, IL4, IL5, IL9, IL12B, IL13*), platelet-derived growth factor receptor-β (*PDGFRB*), acidic fibroblast growth factor (*FGF1*) and fibroblast growth factor receptor 4 (*FGFR4*). The genes encoding the IL3 receptor α-chain (*IL3RA*) and the GMCSF receptor α-chain (*CSF2RA*) both map to the pseudoautosomal region of the sex chromosomes. Synteny betrays the common evolutionary origin of the genes as well as the probable mechanism—tandem duplication.

Genes encoding ligands and their associated receptors are sometimes syntenic, for example the genes encoding transferrin (*TF*) and its receptor (*TFRC*) are both located on chromosome 3q whilst the genes encoding apolipoprotein E (*APOE*) and the low density lipoprotein receptor (*LDLR*) are both located on chromosome 19. However, not surprisingly, this is far from always the case, for example insulin (*INS*, chromosome 11) and insulin receptor (*INSR*, chromosome 19); epidermal growth factor (*EGF*, chromosome 4), epidermal growth factor receptor (*EGFR*, chromosome 7); growth hormone (*GH1, GH2*, chromosome 17) and growth hormone receptor (*GHR*, chromosome 5); interferons α, β, γ and ω1 (*IFNA, IFNB1, IFNG, IFNW1*, chromosomes 9, 8 and 9, 12, 9), interferon receptors α/β/ω and γ (*IFNAR1, IFNGR1*, chromosome 6).

The linear order of members of a family of related genes can reflect the order in which they become activated during development, for example the *HBE1* (embryonic), *HBG2, HBG1* (fetal), *HBD, HBB* (postnatal) genes of the human β-globin cluster. The expression of these genes is controlled by an upstream locus control region (LCR; see section 1.1.2, *Locus control regions*) and correct gene order is required for the normal temporal pattern of developmental expression

(Hanscombe *et al.*, 1991). Another example of this phenomenon is provided by the *HOX* genes which are organized chromosomally according to their order of expression (van der Hoeven *et al.*, 1996). Intriguingly, the encoded Hox proteins vary in their affinity for their target DNA sequences and these affinities also correlate with the linear order of the genes (Pellerin *et al.*, 1994). By contrast, the order of the human *ADH* genes on chromosome 4q21, 5'-*ADH3-ADH2-ADH1*-3', is opposite to their order of transcriptional activation in hepatic development (Yasunami *et al.*, 1990) although the significance of this observation, if any, is unknown. The reflection of temporal order of expression in the physical order on the chromosome is not however a universal phenomenon as is evidenced by the human and murine myosin heavy chain (*MYH*) gene clusters on chromosomes 17 and 11, respectively (Weiss *et al.*, 1999). Finally, it may be significant that the order of four human mucin genes (*MUC2, MUC5AC, MUC5B, MUC6*) on chromosome 11p15 corresponds to their order in terms of the anterior-posterior axis of the epithelial areas where they are preferentially expressed (Pigny *et al.*, 1996).

Pseudogenes. Pseudogenes are DNA sequences which are closely related to functional genes but which are incapable of encoding a protein product on account of the presence of deletions, insertions and nonsense mutations which abolish the reading frame or otherwise prevent gene expression (reviewed by Wilde, 1985; see Chapter 6 for an in-depth treatment). Some human pseudogenes are transcribed (e.g. Bristow *et al.*, 1993; Nguyen *et al.*, 1991; Takahashi *et al.*, 1992) but these transcripts are not translated. There are two major types of pseudogene: the first arises through the duplication and subsequent inactivation of a gene (see Chapter 6, section 6.1.1). This type of pseudogene retains the exon/intron organization of the parental gene and are often closely linked to the parental gene. Examples include the pseudogenes in the α- and β-globin clusters (e.g. Cheng *et al.*, 1988). The second type of pseudogene contains only the exons of the parental gene, usually possess a poly(A) tail at the 3' end and are dispersed randomly in the genome (see Chapter 6, section 6.1.2). These *processed* genes are thought to have originated as mRNAs which have then become integrated into the genome by *retrotransposition* (i.e. the reverse transcription of the mRNA and the integration of the resulting cDNA).

Pseudogenes are relatively common in the human genome (McAlpine *et al.*, 1993; many hundreds are known) and may be especially prevalent in multigene families (e.g. β-globin, actin, HLA, interferons, snRNAs, keratins, T-cell receptors, immunoglobulin gene clusters; see Chapter 6). However, single copy genes may also have multiple pseudogenes (e.g. prohibitin, *PHB*, four pseudogenes; argininosuccinate synthetase, *ASS*, 14 pseudogenes).

Promoter elements. The archetypal gene contains promoter elements upstream (5') of the transcriptional initiation site (i.e. the beginning of the mRNA) which serve to specify the temporal and spatial pattern of expression of the downstream gene and define its potential for induction by external stimuli (*Figure 1.3*). Some genes contain multiple alternative promoters which are utilized in a tissue-specific fashion (e.g. *DMD*; Nishio *et al.*, 1994; *Figure 1.4*; reviewed by Ayoubi and van de Ven, 1996). Some promoters may be located within an intron (e.g. within the first

HSPA1A

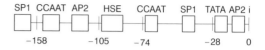

MT2A

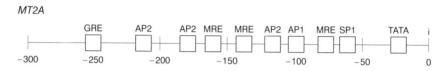

Figure 1.3. Transcriptional control elements upstream of the transcriptional initiation site in the human heat shock protein 1 (*HSPA1A*; 6p21.3) and metallothionein 2A (*MT2A*; 16q13) genes. The TATA, SP1, and CCAAT boxes bind transcription factors that are involved in constitutive transcription whilst the glucocorticoid response element (GRE), metal response element (MRE), heat shock element (HSE), and the AP1 and AP2 sites bind factors involved in the induction of gene expression in response to specific stimuli.

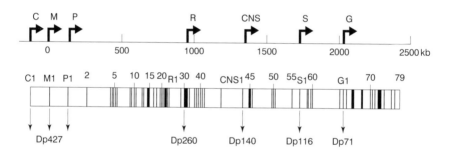

Figure 1.4. Alternative promoter usage in the human dystrophin (DMD) gene (redrawn from Strachan and Read 1996). The alternative promoters are: C, cortical; M, muscle; P, Purkinje; R, retinal; CNS, central nervous system; S, Schwann cell; G, Glial cell. The 79 exons of the DMD gene are denoted by bars. The first exon used to encode each isoform is given the suffix 1, that is S1 denotes the first exon incorporated as a result of Schwann cell promoter usage. Dystrophin isoforms are denoted by Dp acronyms.

intron of the human *ADA* gene; Aronow *et al.*, 1989). Some promoters may even be found buried within the introns of other genes (e.g. an element of the human *CYP21* gene promoter lies within intron 35 of the *C4A* gene; Tee *et al.*, 1995).

The transcriptional initiation site is usually preceded by *constitutive promoter elements* of defined sequence, for example the TATA box (TATAAA; 25–30 bp 5′ to the cap site), initiator element (Py Py A$_{+1}$ N T/A Py Py) which serves as a functional analogue of the TATA box in TATA-less promoters (Lo and Smale 1996) and the CCAAT motif (~90 bp 5′ to the cap site) which potentiate a basal level of gene expression. Further upstream regulatory motifs bind tissue-specific transcription factors, some binding different factors in different tissues. *Response elements*, as the name suggests, are able to confer transcriptional responsiveness to various external trigger stimuli such as temperature, hormones and growth factors. In this way, gene expression is rendered responsive to both the internal (cellular) and external (to the organism) environment. Together, these upstream DNA sequences are involved in controlling gene regulation and induction and conferring tissue-specificity of expression (reviewed by Mitchell and Tjian, 1989).

A large number of DNA sequence motifs have now been identified within gene promoters which represent binding sites for DNA-binding proteins (Johnson and McKnight, 1989; Kel *et al.*, 1995; see section 1.1.2, *Trans-acting protein factors*). These protein-DNA interactions are required to confer appropriate regulation upon the genes bearing them (reviewed by Clark and Doherty, 1993; Faisst and Meyer, 1992; Freemont *et al.*, 1991; Latchman, 1990). Removal of these DNA sequence motifs abolishes the gene's specific pattern of spatial and/or temporal expression.

Enhancers. Enhancers are DNA sequences that are present 5′ or 3′ to a gene (or within an exon or an intron) and which are capable of activating the transcription of the gene in a tissue-specific fashion, independently of their orientation and distance from the correct initiation site (Müller *et al.*, 1988; Wasylyk, 1988). Collins *et al.* (1998) reported the unusual case of the mouse thrombospondin 3 (*Thbs3*) enhancer that is located far upstream of the *Thbs3* gene, within intron 6 of the divergently transcribed metaxin (*Mtx*) gene.

Enhancers function by acting as templates for the assembly of multiprotein complexes on the gene promoter. Thompson and McKnight (1992) likened enhancer-protein interactions to a three-dimensional jigsaw puzzle:

> 'the arrangement of regulatory motifs forms the puzzle template. Specific regulatory proteins, by forming contours of appropriate fit for both DNA template and neighbouring proteins, constitute the puzzle pieces.'

Mechanistic models involving the looping out of DNA between the enhancer and the transcriptional initiation complex have been proposed in order to explain how enhancers manage to influence the activity of their target promoters at considerable distances (Ptashne, 1988).

Negative regulatory elements. The negative regulation of transcription by *repressors* or *silencers* is now so well documented (reviewed by Clark and Doherty, 1993; Herschbach and Johnson, 1993; Jackson, 1991; Levine and Manley, 1989) that it is likely that many if not most genes are subject to their inhibitory influence. Indeed, every gene promoter region is likely to possess its own unique combination of positive and negative regulatory elements which serves to determine its temporal and spatial pattern of expression. These elements permit the binding of a specific set of DNA-binding proteins which are thereby brought into sufficiently close proximity so as to allow their interaction both with each other and with the RNA polymerase in order to influence transcription either positively or negatively. Negative regulation therefore serves to prevent the expression of a gene in an inappropriate tissue or at an inappropriate time or at an inappropriate level. They also potentiate the down-regulation of the expression of a gene following its transient induction. Negative regulatory elements have been found not merely in the promoter regions of human genes as in the α1-antitrypsin (*PI*; 14q32.1) gene (De Simone and Cortese, 1989) but also in the first exon [osteocalcin (*BGLAP*; 1q25-q31; Li *et al.*, 1995) gene], the first intron [(*CD4*; 12pter-p12; Donda *et al.*, 1996) gene], or even the third intron [type IV collagen (*COL4A2*; 13q34; Haniel *et al.*, 1995) gene] of a gene.

Locus control regions. Regulatory elements that exert their effects on the expression of downstream genes over great distances have been described in the human α-globin (*HBA1*; 16p13.3), β-globin (*HBB*; 11p15), growth hormone (*GH1*; 17q22-q24) and red/green cone pigment (*RCP*, *GCP*; Xq28) genes among others (Hanscombe *et al.*, 1991; Jarman *et al.*, 1991; Jones *et al.*, 1995; Wang *et al.*, 1992; Figure 6.2 in Chapter 6). These are known as *locus control regions* (LCRs). The LCR located 40 kb upstream of the *HBB* gene is essential for the high level, tissue-specific expression of the *HBB* gene. It is believed to contain an enhancer capable of controlling the replication timing of the β-globin gene cluster, organizing it into an active chromatin domain and directing the expression of downstream sequences in an erythroid-specific and developmental stage-specific fashion (Higgs, 1998; Orkin, 1995). Although conserved between mammals, little homology is apparent between mammalian and avian β-globin LCR regions (Hardison, 1998).

Trans-acting protein factors. As we have seen, the transcriptional activation of eukaryotic genes is made possible by the interaction of *trans*-acting protein factors with *cis*-acting DNA sequence motifs including enhancers. These transcription factors typically contain a sequence-specific DNA-binding domain, a multimerization domain which allows the formation of either homomultimers or heteromultimers, and a transcriptional activation domain. These domains can be combined in modular fashion to generate an array of different transcription factors (Latchman, 1998; Tjian and Maniatis, 1994).

It is the *cis*-acting DNA sequences within a gene promoter that allow transcription factors to be brought into close proximity so that they may either interact with each other in the transcriptional initiation complex or combine together in an enhancer complex. No one enhancer-binding protein can act on its own, rather it must act in concert with other enhancer-binding proteins. For example, one factor may induce a bend in the DNA thereby promoting the interaction of two already bound proteins with each other. Once the enhancer complex is assembled, it must be able to interact either directly or indirectly with the basal transcription apparatus via its activation domain (Ptashne, 1988; Ptashne and Gann, 1990; Pugh and Tjian, 1990).

Transcription factors can be grouped into families of related proteins whose relatedness extends to homology in their DNA-binding domains and therefore an ability to bind to related DNA sequences (reviewed by Pabo and Sauer, 1992). Specific transcription factors can thus bind to more than one DNA sequence. Conversely, a single DNA sequence motif may sometimes be bound by more than one transcription factor. DNA-binding domains fall into one of four main groups defined by homologous amino acid sequences that give rise to a particular structure capable of binding DNA: homeodomain, zinc finger, leucine zipper and helix-loop-helix (Pabo and Sauer, 1992). These domains usually bind the negatively charged DNA molecule through basic amino acid residues. Activation domains of transcription factors also come in four types: rich in either glutamine, proline, serine/threonine, or acidic amino acids. Examples of these types are SP1, CTF/NF-1, Pit-1, and GATA-1, respectively (reviewed by Ptashne, 1988; Ptashne and Gann, 1990).

Transcriptional repressor proteins may bind directly to DNA or exert their repressive effects indirectly by interacting with basal transcription factors, transcriptional activators and co-repressor proteins to inhibit transcription by RNA polymerases (Hanna-Rose and Hansen, 1996).

The study of DNA-protein interactions has been enormously facilitated by the use of two techniques: gel retardation analysis (also termed band or mobility shift assays; Dent and Latchman, 1993) and DNase I footprinting (Lakin, 1993). The former technique is extremely useful for searching for DNA-binding proteins in crude nuclear extracts whereas the latter method provides information as to the precise location of the binding site on the DNA sequence under study.

5′ and 3′ untranslated regions. The presence of the 5′ untranslated region (5′ UTR; the sequence lying between the transcriptional initiation site and the translational start codon, ATG; *Figure 1.1*) is often essential for the normal expression of a gene. Sequences in the 5′ UTRs of various genes are thought to play a role in controlling the translation of the encoded mRNA (reviewed by Curtis *et al.*, 1995; Melefors and Hentze, 1993; Pesole *et al.*, 1997; Sachs, 1993). Perhaps the best characterized posttranscriptional control mechanism involving the 5′ UTR is that of the iron response element (IRE). The IRE is found in the 5′ UTRs of several human genes [e.g. ferritin (*FTH1*), transferrin (*TF*), erythroid 5-aminolevulinate acid synthase (*ALAS1*); Cox and Adrian 1993; Bhasker *et al.*, 1993] and is capable of adopting a stem-loop structure that interacts with a cytosolic RNA-binding protein thereby inhibiting mRNA translation. The post-transcriptional regulation of several other human genes [e.g. transforming growth factor-β1 (*TGFB1*; Kim *et al.*, 1992b) and basic fibroblast growth factor (*FGF2*; Prats *et al.*, 1992)] also appears to involve regulatory elements that modulate the efficiency of translation.

Zhang (1998) examined the nature of translational start codons in human genes. The relative frequencies of start codons occurring as the first, second, third or fourth ATG codons in the reading frame were 474, 51, 5, and 0, respectively. There are, however, quite a few examples of more than one ATG being used in the same gene. Thus, the alternative use of two ATG codons 84 bp apart in the human peroxisome proliferator-activated receptor (*PPARG*; 3p25) gene serves to generate two distinct protein isoforms that differ in length by 28 amino acids at the amino terminal end of the protein (Elbrecht *et al.*, 1996). The alternative use of different ATG initiation codons has also been reported for the human von Hippel Lindau (*VHL*; 3p25-p26) gene which leads to the production of two distinct protein products that differ in length by 53 amino acids (Blankenship *et al.*, 1999).

Zhang (1998) reported that the ratio of the frequencies of stop codons TAA, TAG and TGA in human genes was about 1:1:2. The 3′ untranslated region is that region of a gene which lies downstream of the stop codon. The AATAAA sequence located downstream of the stop codon (TAA in *Figure 1.1*) of ~90% of genes (i.e. in the 3' untranslated region) controls cleavage/polyadenylation (addition of a poly A tail 10–30 bp downstream) which is vital for mRNA stability. The AAUAAA motif is thought to interact with at least two RNA-binding proteins, cleavage-polyadenylation specificity factor (CPSF) and cleavage-stimulatory factor (CstF), which are thought to promote assembly of the polyadenylation complex (Manley and Proudfoot 1994). Some genes possess multiple *polyadenylation sites* which can be used alternatively (Edwalds-Gilbert *et al.*, 1997).

Other sequences have been implicated in determining mRNA stability, for example a GU-rich motif (consensus sequence, YGUGUUYY), present 20–40 bases downstream of the cleavage site in ~70% of mammalian pre-mRNAs, that plays a crucial role in 3' end formation (Decker and Parker, 1994; Manley and Proudfoot, 1994; Ross, 1996; Sachs, 1993). A further U-rich element, upstream of the AAUAAA poyladenylation motif is also involved. Some short-lived mRNAs possess AU-rich regions which specify instability (Chen and Shyu, 1995). Still other sequences may be involved in mRNA export (Izaurralde and Moottaj, 1995), nucleocytoplasmic transfer (Colgan and Manley, 1997), intracellular localization (St Johnston, 1995) and in promoting translational efficiency (Curtis *et al.*, 1995). A wide variety of RNA-binding proteins are now known which play an important role in the post-transcriptional regulation of gene expression (reviewed by Burd and Dreyfuss, 1994; Siomi and Dreyfuss, 1997).

Boundary elements. The position and orientation independence of enhancers and LCRs clearly raises the potential problem of the inappropriate activation of promoters from neighboring genes. Sequences which constrain the activity of enhancers were originally reported in mice and *Drosophila* (Kellum and Schedl, 1991; reviewed by Eissenberg and Elgin, 1991). Termed 'boundary elements' or 'insulators,' these sequences serve to insulate a gene from the effects of either enhancer or suppressor elements emanating from the surrounding chromatin. This insulator function appears to work either by blocking interactions with other sequences past the boundary element and/or by limiting the influence of an enhancer to the locality of its target gene(s) (Geyer, 1997). Boundary elements therefore serve as functional barriers when inserted between an enhancer and its downstream reporter gene. However, bracketing an enhancer–gene combination with boundary elements serves to confine the enhancer activity to that gene's promoter such that its expression is maintained or even increased. MARs (see section 1.1.1, *Matrix attachment regions*) may act as boundary elements as evidenced by their ability to establish transcriptional domains around transgenes and act as buffers to shield these transgenes from position effects (McKnight *et al.*, 1992).

A boundary element has been described 5' to the LCR of the chicken β-globin gene cluster (hypersensitive site 4); it serves to insulate genes 5' to it from the influence of the LCR, a function which it also manifests in both human and *Drosophila* cells (Chung *et al.*, 1993). DNA around the DNaseI hypersensitive site 5 of the human β-globin gene appears to function in a similar fashion (Li and Stamatoyannopoulos, 1994).

Another possible example of a boundary element in humans has been noted upstream of the coagulation factor X (*F10*;13q34) gene (Miao *et al.*, 1992). This gene lies 2.8 kb downstream of the gene (*F7*) encoding the homologous clotting protein, factor VII. Three positive regulatory elements (*FXP3*, *FXP2*, and *FXP1*) have been characterized upstream of the *F10* gene. *FXP1* and *FXP2* act in an orientation- and position-independent fashion whilst *FXP1* and *FXP3* are responsible for directing liver-specific gene expression. The putative boundary element just upstream of the *FXP3* sequence is thought to prevent transcriptional activation of the *F7* gene by the *F10* gene enhancers.

Sequence motifs involved in mRNA splicing and processing. One of the characteristics of eukaryotic genes that distinguishes them from their prokaryotic counterparts is the production of large pre-mRNAs which contain intervening non-coding sequences (introns) that are removed by a highly accurate cleavage/ligation reaction known as splicing before the mRNA is transported to the cytoplasm for translation (reviewed by Green, 1986; Padgett *et al.*, 1986). Splicing not only permits the removal of introns from the primary transcript but also allows the generation of different mRNAs from the same gene by *alternative splicing*, an important mechanism for tissue-specific or developmental regulation of gene expression and a very economical means of generating biological diversity (Nadal-Ginard *et al.*, 1987; Norton, 1994). Alternative splicing may be regulated by variation in the intracellular levels of antagonistic splicing factors (Caceres *et al.*, 1994).

The splicing of a eukaryotic mRNA appears to occur as a two-stage process. In the case of a simple two exon gene, the pre-mRNA is first cleaved at the 5′ (donor) splice site to generate two splicing intermediates, an exon-containing RNA species and a lariat RNA species containing the second exon plus intervening intron. Cleavage at the 3′ (acceptor) splice site and ligation of the exons then occurs resulting in the excision of the intervening intron in the form of a lariat. Splicing efficiency is critically dependent upon the accuracy of cleavage and rejoining. This accuracy appears to be determined, at least in part, by the virtually invariant GT and AG dinucleotides present at the 5′ and 3′ exon/intron junctions respectively. However, more extensive consensus sequences spanning the 5′ and 3′ splice junctions are evident (Mount, 1982; Padgett *et al.*, 1986) and the coding sequence flanking intron junctions exhibits some degree of conservation (Long *et al.*, 1998). More recently, Zhang (1998) has proposed AG I GTRAGT as a consensus sequence for donor splice sites and $(Y)_n$NCAG I G (where n has a mean of nine) as a consensus for acceptor splice sites. Stephens and Schneider (1992) noted the similarity between the donor and acceptor sites and suggested that these junctions may have been derived from a common 'proto-splice site' ancestor (*Figure 1.5*). During evolution, the emphasis of the sequence information at each site has shifted to the intronic side of the junction (*Figure 1.5*).

A few human nuclear genes possess introns with noncanonical terminal dinucleotide sequences (e.g. 'AT-AC introns'; Tarn and Steitz, 1997); these include the transcription factor *E2F1* (20q11.2) gene, the cartilage matrix protein (*MATN1*; 1p25) gene, the Hermansky–Pudlak syndrome (*HPS*; 10q23) gene and the paralogous sodium channel α-subunit genes (*SCN4A*, 17q23.1-q25.3; *SCN5A*, 3p24-p21; *SCN8A*, 12q13). The removal of AT-AC introns requires two low-abundance snRNAs, U4atac and U6atac, not found in the major spliceosome (Tarn and Steitz, 1996). There is now good evidence for the existence of two distinct splicing systems

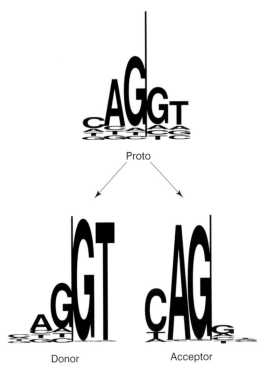

Proto

Donor   Acceptor

Figure 1.5. Putative schema for the evolution of donor and acceptor splice sites from a single 'proto-splice site' ancestor (from Stephens and Schneider, 1992). Sequence logos for proto-, donor and acceptor splice sites are shown with the heights of the individual letters being proportional to their frequencies at each position. On the donor side, the proto-splice site 'cag' has lost emphasis to become a smaller 'cag' whilst the 'gt' has become a more strongly conserved 'GT'. On the acceptor side, the proto-splice site 'cag' has gained emphasis to become a 'CAG' whilst the 'gt' has become even smaller.

in eukaryotes which, having evolved in separate lineages, came together in a eukaryotic progenitor (Dietrich *et al.*, 1997; Wu and Krainer, 1997; Burge *et al.*, 1998). Some variant splice sites are evolutionarily conserved, for example the highly unusual GA donor splice site present in exon 10 of the *Xenopus* and human fibroblast growth factor receptor 1 (*FGFR1*; 8p11) genes as well as in the paralogous *FGFR2* (10q26) and *FGFR3* (4p16) genes of mouse and human (Twigg *et al.*, 1998).

A further conserved sequence element, the 'branch-point', has been identified in the introns of eukaryotic genes and, when it occurs, is usually located some 18–40 bp upstream of the 3′ splice site (Green, 1986). Although this sequence appears to play a role in forming a branch with the 5′ terminus of the intron, it exhibits a rather weak consensus sequence (Y_{81} N Y_{100} T_{87} R_{81} A_{100} Y_{94}; Krainer and Maniatis, 1988). The sequence UACUAAC appears to be the most efficient branch site for mammalian mRNA splicing both *in vitro* and *in vivo* (Zhuang *et al.*, 1989). The branch-point sequences in human can best be described by the consensus Y U V A Y for loci with low G+C content and C U G/C A Y for loci with high G+C content (Zhang, 1998). Whereas both the length and location of the pyrimidine tract may be important determinants of branch-point and acceptor splice site utilization, the 3′ acceptor splice site itself appears to possess little specificity and may serve merely as the first AG dinucleotide downstream of the branch-point/pyrimidine tract.

A further category of sequences required for alternative splicing are the *splicing enhancers* (Lopez, 1998). Several types of splicing enhancer have been recognized. These sequences occur in either exons or introns and serve to promote the use of neighboring weak 5′ or 3′ splice sites. They do this through the binding of serine-arginine (SR) proteins, a family of modular splicing factors which contain one or more RNA-binding domains.

Splicing occurs within the *spliceosome*, a complex assembly of small ribonucleoprotein particles (snRNPs) composed of a variety of snRNAs and associated proteins (reviewed by Lührmann *et al.*, 1990). The pre-mRNA is folded in such a way that splice sites are optimally aligned for cleavage and ligation. In this process, the snRNAs play a vital role. Our current understanding of mRNA splicing suggests that the formation of the 5' splice site complex is contingent upon the prior formation of the 3' splice site complex (Smith *et al.*, 1993). Spliceosomal structure and function has been conserved in eukaryotes from yeast to humans (Wentz-Hunter and Potashkin, 1995).

Polycistronic and polyprotein genes. In most cases, human genes adhere to the principle of 'one gene, one polypeptide.' However, some multi-chain protein genes encode more than one polypeptide. In such cases, synthesis of the mature protein involves the post-translational cleavage of a polypeptide precursor followed by association of the constituent chains. Thus, the human insulin (*INS*; 11p15) gene encodes an A chain, a B chain, and a connecting peptide which is thought to be important in maintaining the conformation of the two chains. More often than not, however, multichain proteins are encoded by different genes, for example the α- and β-globins. Nevertheless, even although the microtubule-associated proteins 1A and 1B (*MAP1A*, 15q13-qter; *MAP1B*, 5q13) are encoded by different genes, both genes encode heavy and light chains that must be proteolytically processed from a precursor polypeptide (Hammarback *et al.*, 1991; Fink *et al.*, 1996).

Several human genes are *polycistronic* in that they encode different peptides with different functions, for example calcitonin/calcitonin gene-related peptide α (*CALCA*, 11p15; Allison *et al.*, 1981), pancreatic polypeptide/pancreatic icosapeptide (*PPY*; 17p11-qter; Boel *et al.*, 1984) and vasoactive intestinal polypeptide/PHM27 (*VIP*; 6q26-q27; Bodner *et al.*, 1985; Tsukada *et al.*, 1985). A further putative polycistronic gene has been characterized in both mouse and human: the growth/differentiation factor-1 (*GDF1*) gene is cotranscribed with a gene of unknown function (*Uog1*) separated from each other by a 269 bp intercistronic region (Lee, 1991). Finally, Reiss *et al.* (1998) described the polycistronic human molybdopterin synthase (*MOCOD*; 6p21.3) gene which contains two overlapping open reading frames (ORF) encoding distinct polypeptides required for the synthesis of molybdenum cofactor. The two polypeptides are encoded by the same 3.2 kb mRNA, the first by exons 1–9, the second by exon 10. Stallmeyer *et al.* (1999) have demonstrated that each ORF is translated independently.

More dramatic is the case of the human glycinamide ribonucleotide synthetase (*GART*; 21q22.1) gene which generates an mRNA encoding a trifunctional protein with three distinct enzymatic activities (glycinamide ribonucleotide synthetase, glycinamide ribonucleotide formyltransferase and aminoimidazole ribonucleotide synthetase which encode respectively the second, third and fifth enzymatic steps in the *de novo* synthesis of purines) (Brodsky *et al.*, 1997). These activities are separate in bacteria and partly separate in yeast implying that ancient recombination events may have fused the coding regions together

(Davidson and Peterson, 1997). Fusion could have occurred as a means to allow coordinate regulation of the three enzymatic activities.

One way in which polycistronic mRNAs might have arisen during evolution is by inefficient RNA polymerase termination of transcription of one mRNA species allowing the cotranscription of a second mRNA species from a downstream gene. Such a phenomenon has been shown to occur in normal human cells and involves the *fusion splicing* of two adjacent (9p13) genes, those encoding galactose-1-phosphate uridylyl-transferase (*GALT*) and interleukin-11 receptor α-chain (*IL11RA*) (Magrangeas *et al.*, 1998).

Some human genes encode *polyproteins* and comprise multiple tandem repeats of the same coding region on the same transcript, for example those encoding the ubiquitins (*UBA52*, 19p13; *UBB*, 17p11-p12; *UBC*, 12q24; Wiborg *et al.*, 1985) and filaggrin (*FLG*; 1q21; Gan *et al.*, 1990; McKinley-Grant *et al.*, 1989).

1.1.3 DNA methylation

5-methylcytosine (5mC) is the most common form of DNA modification in eukaryotic genomes. Soon after DNA synthesis is complete, target cytosines are modified by a DNA methyltransferase using S-adenosylmethionine as methyl donor. In humans, between 70% and 90% of 5mC occurs in CpG dinucleotides, the majority of which appear to be methylated (Cooper, 1983).

Whereas the vertebrate genome is heavily methylated, methylation is virtually undetectable in insects and other arthropods (Cooper, 1983). An intermediate level of methylation is exhibited by the echinoderms, coelenterates and molluscs whose genomes are characterized by the presence of long methylated and unmethylated tracts (Cooper, 1983). The transition from a fractional to a global methylation pattern appears to have occurred close to the origin of the vertebrates since the cephalochordate *Amphioxus* exhibits a typically invertebrate pattern of genome methylation whereas the jawless vertebrates hagfish and lamprey possess a vertebrate pattern (Tweedie *et al.*, 1997). The increase in size of the methylated compartment characteristic of vertebrate genomes correlates with the sharp increase in gene number during the invertebrate-vertebrate transition (Bird, 1995). DNA methylation may thus have been recruited as a transcriptional regulatory mechanism (Colot and Rossignol, 1999).

DNA methylation is essential for normal mammalian development (Razin and Shemer, 1995). It is thought to play a role in both gene regulation (Kass *et al.*, 1997) and imprinting (Jaenisch, 1997; see section 1.1.3, *Imprinting and imprinted genes*), may serve as a cue for strand specificity in DNA replication and repair (Hare and Taylor 1988) and could conceivably serve as a self-defence mechanism to silence transposable elements and proviral DNAs integrated into the genome during evolution (Yoder *et al.*, 1997; Simmen *et al.*, 1999). This post-synthetic modification occurs almost exclusively in CpG dinucleotides of which between 60% and 90% are methylated in mammalian tissues (Jost and Bruhat, 1997).

Although it is as yet unclear how tissue-specific methylation patterns are established (Bestor and Tycko 1996; Turker and Bestor, 1997), they are nevertheless heritable and reproducible after transmission through the germline (Pfeifer *et al.*, 1990; Silva and White, 1988). The establishment of cell type-specific methylation patterns in both the soma and the germline begins with global methylation of non CpG island sequences in the embryo whilst the final methylation patterns are determined by a specific and highly regulated process of demethylation (Razin and Shemer, 1995).

In humans, differences in site-specific methylation patterns between different ethnic groups have not so far been apparent in studies of a variety of different types of DNA sequence (Behn-Krappa *et al.*, 1991; Bottema *et al.*, 1990; Kochanek *et al.*, 1990). In keeping with these reports, Millar *et al.* (1998) found no evidence in sperm DNA for variable methylation status in the factor VIII (*F8C*; Xq28) gene between European Caucasians and Asians. However, the methylation status of specific CpG sites in the *F8C* gene did exhibit significant inter-individual variation (Millar *et al.*, 1998).

CpG islands. Spatially, the distribution of CpG appears to be nonrandom in the human genome; about 1% of the genome consists of stretches very rich in CpG which together account for roughly 15% of all CpG dinucleotides (Bird, 1986). In contrast to most of the scattered CpG dinucleotides, these *CpG islands* represent unmethylated domains and comprise ~50% of all unmethylated CpGs in the genome (Bird *et al.*, 1985). CpG islands occur, on average, every 100 kilobases in the murine genome (Brown and Bird, 1986) and, in human, are often located immediately 5' to gene coding regions (Gardiner *et al.*, 1990; Larsen *et al.*, 1992; *Figure 1.6*). The methylation of CpG islands may modulate gene expression through its influence on chromatin organization (Kundu and Rao, 1999).

Not all vertebrate genes, however, possess CpG islands (Bird, 1986; Gardiner-Garden and Frommer, 1987; *Figure 1.6*) and many are partially or even heavily methylated. In general, gene promoters containing CpG islands are unmethylated regardless of expression whereas promoters lacking CpG

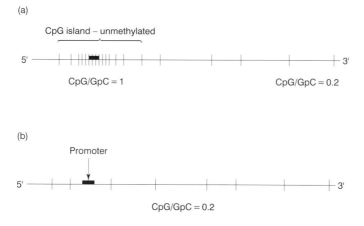

Figure 1.6. The distribution of CpG within genes. (a) A typical CpG island containing gene as found in many 'housekeeping' genes. (b) A typical tissue-specific gene lacking a CpG island.

islands tend to lose their methylation upon transcription. CpG islands some-times also occur within the coding regions of genes as in the case of the human apolipoprotein E (*APOE*; 19q13) gene in which exon 4 constitutes a 940 bp CpG island. Other examples of human genes with CpG islands within their coding regions are the *CDKN2A* (9p21), *MYOD1* (11p15.4), and *PAX6* (11p13) genes.

CpG islands are predominantly found in early replicating (R band) as opposed to late replicating (G band) regions of the genome (Craig and Bickmore, 1994). They may be considered to be the last remnants of the unmethylated domains that once dominated the vertebrate genome. The evolution of the heavily methy-lated vertebrate genome has been accompanied by a progressive loss of CpG din-ucleotides as a direct consequence of their methylation in the germline. Although CpG islands are usually unmethylated and therefore relatively immune to muta-tional decay (Luoh *et al.*, 1995; see section 1.1.3, *CpG suppression in the vertebrate genome and its origin*), there is nevertheless some evidence for their gradual erosion over evolutionary time (Matsuo *et al.*, 1993).

CpG suppression in the vertebrate genome and its origin. Methylation of cytosine results in a high level of mutation due to the propensity of 5mC to undergo deamination to form thymidine. Deamination of 5mC probably does not occur during the enzymatic replication of the methylation pattern which appears to be a high fidelity process. Indeed, 5mC deamination probably occurs with the same frequency as the deamination of cytosine to uracil. However, whereas uracil DNA glycosylase activity in eukaryotic cells is able to recognize and excise uracil, thymine being a 'normal' DNA base is thought to be both less readily detectable and removable by cellular DNA repair mechanisms.

One consequence of the hypermutability of 5mC is the paucity of CpG in the genomes of many eukaryotes (Setlow, 1976), the heavily methylated vertebrate genomes exhibiting the most extreme 'CpG suppression' (Bird, 1980; Jabbari *et al.*, 1997; Schorderet and Gartler, 1992). A first estimate of the *in vivo* rate at which 5mC is deaminated and fixed as thymidine was arrived at by extrapolation from *in vitro* data (Cooper and Krawczak, 1989). To this end, the deamination rate of 5mC as measured under laboratory conditions in single stranded DNA, was modified so as to be consistent with the observed spectrum of point mutations found to have caused human genetic disease. The rate estimate of $1.66 \times 10^{-16} sec^{-1}$ was consistent with the rate calculated from the evolutionary pattern of CpG substitution exhibited by β-globin gene and pseudogene sequences in human, chimpanzee and macaque (Cooper and Krawczak, 1989). Mathematical modelling allowed a detailed study of the dynamics underlying the CpG sup-pression currently found in the bulk DNA of vertebrate genomes (Cooper and Krawczak, 1989). It was inferred that the time span required in order to create the currently observed CpG frequency (0.01) was 50 to 100 Myrs. On the other hand, the process of CpG loss must have lasted for approximately 450 Myrs in order for the mononucleotide frequencies to have attained their present levels. This time span corresponds closely to the estimated time since the emergence and adaptive radiation of the vertebrates and thus coincides with the probable advent of heavily methylated genomes. These data are therefore consistent with patterns of vertebrate gene methylation having been comparatively stable over relatively long periods of evolutionary time.

The CpG dinucleotide as a mutation hotspot. CpG has been found to be a hotspot for mutation in a wide range of different human genes (reviewed by Cooper and Krawczak, 1993). Of the single base-pair substitutions so far reported to cause human genetic disease (Krawczak *et al.*, 1998), ~30% involve a CpG dinucleotide, and ~79% (~23% of the total) of these are either C→T or G→A transitions. These data imply that the rate at which CpG mutates to either TpG or CpA is some five-fold higher than the basal mutation rate per nucleotide. Methylation-mediated deamination of 5mC therefore represents a major cause of gene mutability leading to human genetic disease.

The proportion of human gene mutations compatible with a model of methylation-mediated deamination (CG→TG, CA) is however very much an average figure and provides no information on individual genes. When these values are recalculated on a gene-specific basis, considerable variation becomes apparent. The frequency of CG→TG and CG→CA mutations may be smaller than 10% (*HBB* (4%), *HPRT1* (5%), *TTR* (9%)) or larger than 50% (*MYH7* (67%), *ADA* (63%), *RB1* (57%)). In the assumed absence of a detection bias (Cooper and Krawczak, 1993), we may surmise that this variation is due either to differences in (a) germline DNA methylation and/or (b) relative intragenic CpG frequency (itself dependent upon (a)). Imbalances in the propensity to transition at CpG dinucleotides exist even between the two DNA strands. For the human hypoxanthine phosphoribosyltransferase (*HPRT1*) gene, Skandalis *et al.* (1994) have noted a significant strand bias in mutations recovered at CpG sites, thereby confirming our own findings (Cooper and Krawczak, 1993; Krawczak *et al.*, 1998).

CpG hypermutability in inherited disease implies that the CpG sites in question are methylated in the germline thereby rendering them prone to 5mC deamination. However, on its own, CpG hypermutability still represents very indirect evidence for CpG methylation. That 5mC deamination is itself directly responsible for these mutational events is evidenced by the fact that several cytosine residues known to have undergone a germline mutation in the low density lipoprotein receptor (*LDLR*; hypercholesterolemia), p53 (*TP53*; various types of tumor), factor VIII (*F8C*; hemophilia A), and neurofibromatosis type 1 (*NF1*) genes are indeed methylated in sperm (Rideout *et al.*, 1990; Andrews *et al.*, 1996; Millar *et al.*, 1998; El-Maarri *et al.*, 1998). As yet, however, these are the only studies to have attempted to correlate CpG hypermutability with DNA methylation directly for specific CpG dinucleotides.

Imprinting and imprinted genes. DNA methylation is thought to be involved in *imprinting* which was defined by Monk (1988) as the 'differential modification of the maternal and paternal contributions to the zygote, resulting in the differential expression of parental alleles during development and in the adult.' This differential modification appears to be essential for normal mammalian development since parthenogenetic embryos (whether diploid paternal or diploid maternal) do not survive to term: in diploid maternal embryos, fetal development is normal but development of the extraembryonic membranes is abnormal. In diploid paternal embryos, it is the other way around (reviewed by Monk, 1988). Clearly, maternal and paternal chromosomes must differ epigenetically and in such a way that different developmental programmes are followed.

Only some 100–200 genes in mammalian genomes are imprinted and these are often clustered (Barlow, 1995). Human examples include the *INS*, *H19* (D11S878E) and *IGF2* (11p15.5), *SNRPN* (15q11-q12), *WT1* (11p13), *IGF2R* (6q25-q27), and *XIST* (Xq13.2) genes. The *XIST* gene is essential for X-inactivation (a special case of imprinting) and is expressed exclusively from the inactive X chromosome (Jamieson *et al.*, 1996; Kay 1998). It gives rise to a non-coding mRNA product that controls the production of an inactivation signal which spreads along the chromosome, silencing all but a handful of genes (Heard *et al.*, 1997).

A common function of imprinted genes is in the control of embryonic growth with paternally expressed genes (e.g. *IGF2*) tending to enhance growth rates and maternally expressed genes (e.g. *H19*, *IGF2R*) reducing them. This dichotomy has led to the proposal of the *genetic conflict hypothesis* (Haig and Trivers 1995) which attempts to explain the evolution of imprinting in terms of a conflict of interest between the maternal and paternal genes of an individual. This theory might predict antagonistic coevolution between maternally and paternally derived genes for the control of fetal growth. However, contrary to the expectations of the conflict hypothesis, the rate of evolution of imprinted genes is not significantly different from that exhibited by non-imprinted genes encoding receptors (McVean and Hurst, 1997).

It has been suggested that the imprinted expression of genes is usually conserved between human and mouse (Barlow, 1995) but whilst the maternal-specific expression of the *Igf2r* gene is seen in mouse (Barlow *et al.*, 1991), imprinting of the *IGF2R* gene is only apparent in a minority of the human population (Xu *et al.*, 1993; Ogawa *et al.*, 1993). Such an 'imprinting polymorphism' has also been noted for the *WT1* gene in human populations (Nishiwaki *et al.*, 1997).

Imprinted genes may have fewer and smaller introns than nonimprinted genes (Hurst *et al.*, 1996; Haig, 1996). Whether this is an adaptation to allow these genes to be transcribed rapidly or whether it is merely a property of the chromosomal region in question is however unknown.

It is at present unclear if imprinting is confined to mammals. Parent-of-origin-specific effects have been claimed in the zebrafish, *Danio rerio* (McGowan and Martin, 1997). If confirmed, this would be reconcilable with the observed survival of androgenetic zebrafish only if the expression of the imprinted genes were not required for development.

1.1.4 Repetitive sequence elements

> There is repetition everywhere, and nothing is found only once in the world.
> J. G. v. Goethe

Repetitive DNA comprises the bulk (>90%) of the human genome. A large number of different types of repetitive sequence element are found in the human genome (reviewed by Jelinek and Schmid, 1982; Vogt, 1990) and their analysis may go some way toward explaining the patterns of chromosome bands noted after the staining of metaphase chromosomes. Three main categories are now recognized: (i) a highly repetitive class including sequence families with $>10^5$ copies per haploid genome, (ii) a middle repetitive sequence class (10^2–10^5 copies), and (iii) a low repetitive class whose members possess between two and 100 copies per haploid genome. The different classes of highly repetitive satellite DNA (tandem repeats) are briefly reviewed here together with *Alu* repeats and LINE (L1) elements, the two most abundant and best characterized of the interspersed middle repetitive sequence families in the human genome.

Tandem repeats. Tandemly repetitive DNA comprises satellite DNA, minisatellite DNA and microsatellite DNA. Satellite DNA comprises the majority of heterochromatin and is clustered in tandem arrays of up to several megabases (Mb) in length. A number of different families [e.g. simple sequence (5–25 bp repeats), alphoid (169 bp and 172 bp repeat), *Sau*3A (~68 bp repeat)] have been identified (reviewed by Vogt, 1990) which are largely confined to the centromeres. One function of satellite DNA could be to maintain regions of late replication thereby ensuring that the centromere is the last region to replicate on a chromosome (Csink and Henikoff, 1998).

The hypervariable minisatellite sequences (about 10^4 copies/genome) share a core consensus sequence [GGTGGGCAGARG] which is reminiscent of the *Escherichia coli* Chi element known to be a signal for generalized recombination (Jeffreys, 1987). These minisatellites exhibit substantial copy number variability in terms of the number of constituent repeat units and are often telomeric in location (see section 1.1.1, *Telomeres*).

Microsatellite DNA families are simple sequence repeats, the most common being $(A)_n/(T)_n$, $(CA)_n/(TG)_n$ and $(CT)_n/(AG)_n$ types (Beckmann and Weber, 1992; Vogt *et al.*, 1990). Minisatellites and microsatellites account for between 0.2% and 0.5% of the human genome, respectively and are widely scattered on many chromosomes. Their high copy number variability and association with a considerable number of different genes has meant that they provide a very valuable source of highly informative markers for disease analysis and diagnosis (Bruford and Wayne, 1993). Since their presence at specific homologous chromosomal locations is often evolutionarily conserved in primates (Coote and Bruford, 1996; Crouau-Roy *et al.*, 1996; Morin *et al.*, 1998), they may also be useful in both population genetic and evolutionary studies.

More recently, a further type of tandemly reiterated DNA sequence has been described (Gondo *et al.*, 1998). The 4746 bp RS447 repeat sequence, quaintly named 'megasatellite' DNA, is repeated between 50 and 70 times on chromosome 4p15. It is evolutionarily conserved among mammals, polymorphic in humans and since it contains an open reading frame, it may encode a protein product.

Alu sequences and other SINEs. The *Alu* family of short interspersed repeated elements (SINEs) is present in all primates. Up to 900 000 copies are thought to exist in the human genome (some 5% of the total DNA complement) with an average spacing of 4 kb (Hwu *et al.*, 1986). Most occur in noncoding DNA but some are known to be located in untranslated regions (Makalowski *et al.*, 1994) or even coding regions (Margalit *et al.*, 1994).

Alu repeats share a recognizable consensus sequence but the extent of homology to this consensus varies from 72% to 99% (Kariya *et al.*, 1987; Batzer *et al.*, 1990). Human *Alu* sequences are ~300 bp in length, are polyadenylated and consist of two related sequences each between 120 and 150 bp long separated by an A-rich region (*Figure 1.7*). *Alu* sequences appear to be degenerate forms of 7SL RNA (*RN7SL*) that have been reverse transcribed and integrated into the genome (Ullu and Tschudi, 1984). As to a possible function for *Alu* sequences, the question still remains open (Schmid, 1998).

Several reports of transcription of *Alu* sequences either by RNA polymerase II or III have appeared (Maraia *et al.*, 1993) but their transcription is often silenced by DNA methylation (Liu *et al.*, 1994) and/or nucleosome positioning (Englander *et al.*, 1993). *Alu* sequences contain an internal RNA polymerase III promoter (Jelinek and Schmid 1982), a functional retinoic acid response element (Vansant and Reynolds, 1995) and a regulatory element that can confer positive or negative regulation of transcription upon a variety of promoters *in vitro* (Brini *et al.*, 1993). The presence of the polIII promoter is important since it ensures high expression in the germline, a prerequisite for efficient transposition.

There are at least four different types of *Alu* sequence which belong to two distinct subfamilies (Jurka and Smith, 1988; Britten *et al.*, 1988; Deininger and Batzer, 1993). Some types of *Alu* sequence are human-specific (Batzer *et al.*, 1990; Batzer and Deininger, 1991) and these appear to be derived from a number of different but closely related master copies or 'source genes' (Matera *et al.*, 1990a). The vast majority of human *Alu* sequences appear to be transcriptionally inert. Some of the human-specific subfamilies are transpositionally competent and these may also be transcriptionally active (Matera *et al.*, 1990b; Sinnett *et al.*, 1992).

Alu repeats are concentrated in R bands in metaphase chromosomes (Korenberg and Rykowski, 1988) whereas these repeats are under-represented in other regions (e.g. centric heterochromatin) (Moyzis *et al.*, 1989). R bands are G+C rich, replicate their DNA early in S phase and condense late in mitotic prophase. R bands also contain the bulk of active gene sequences. One consequence of the nonrandom

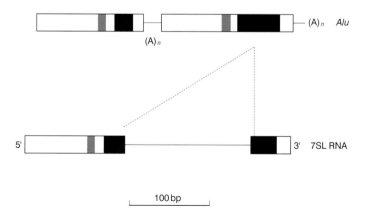

Figure 1.7. Structure of the human *Alu* repeat element compared with the related 7SL RNA. A 155 bp portion of the 7SL RNA is absent from the *Alu* sequence as indicated. Poly(A) stretches are denoted by (A)$_n$.

distribution pattern of *Alu* sequences in the genome is that procedures which screen for these repeats in genomic DNA clones will preferentially locate gene sequences.

SINEs have a long evolutionary history, being present in all vertebrate classes as well as being found in molluscs (Gilbert and Labuda, 1999). SINEs in vertebrate genomes include various types of DNA transposable element such as *Tiggers* and *mariners* which are characterized by the possession of terminal inverted repeats and target site duplications evident as flanking direct repeats (Smit and Riggs, 1996; Robertson and Martos, 1997). The *Tigger* element closely resembles *pogo*, a DNA transposon in *Drosophila*. Most copies of the *mariner* family of transposable elements, of which there are probably more than 1000 in the human genome, probably originated between 80 and 50 Myrs ago (Robertson and Zumpano, 1997; Robertson and Martos, 1997). In addition, the MIR transposable element which has a simple tRNA-like internal polymerase III promoter, was amplified to several hundred thousand copies before the adaptive radiation of the mammals (Murnane and Morales, 1995; Smit and Riggs, 1995).

LINE elements. Some 10^5 copies of long interspersed repeat elements (LINES) are present in the human genome (Hwu *et al.*, 1986). LINES are derived from polII transcripts and account for perhaps 2–3% of the total DNA complement (reviewed by Skowronski and Singer, 1986; Singer *et al.*, 1993). LINE elements have been found in all mammalian species so far examined and may be traced back to an ancestral LINE element that originated before the adaptive radiation of mammals (Furano and Usdin, 1995). Human LINE elements vary in size from as little as 60 bp up to 6–7 kb; about 95% are truncated at their 5′ ends but they mostly appear to contain the same 3′ sequences as well as a poly(A) tail of variable length (*Figure 1.8*). Each individual LINE element differs from the consensus sequence (Scott *et al.*, 1987) by ~13% although many exhibit internal deletions and rearrangements. The majority of human LINE elements appear to have been generated within the last 30 Myrs (Scott *et al.*, 1987).

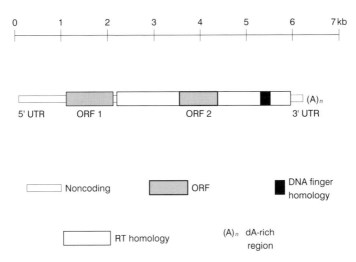

Figure 1.8 Structure of a full-length human LINE (L1) element. UTR: Untranslated region. ORF: Open reading frame. RT: Reverse transcriptase.

LINE elements are confined to the G/Q (Giemsa/Quinacrine) bands of the euchromatin (Korenberg and Rykowski, 1988). G/Q bands are A + T rich, replicate their DNA late during the DNA synthetic period, condense early during mitosis and are relatively poor in expressed genes.

LINES probably represent processed pseudogene-like copies of reverse transcripts which have been re-integrated into the genome (Hattori *et al.*, 1986). A full-length LINE element possesses two open-reading frames, ORF1 (1 kb) and ORF2 (4 kb); the latter possesses reverse transcriptase activity (Mathias *et al.*, 1991) which may serve to mediate the retrotransposition not only of the LINE elements themselves (Sassaman *et al.*, 1997; DeBerardinis *et al.*, 1998) but also of other retroposons such as *Alu* sequences (Jurka 1997; Gilbert and Labuda, 1999).

LINE element transcripts have been found in undifferentiated teratocarcinoma cells (Skowronski and Singer, 1985; Skowronski *et al.*, 1988) suggesting that they may be expressed early on in mammalian development. A promoter at the 5′ end appears to be responsible for the specific expression of LINE elements in teratocarcinoma cells (Swergold, 1990). However, only a small subset of all LINE elements is capable of being transcribed.

Endogenous retroviral sequences and transposons. Human transposable elements are essentially of two kinds, those that undergo transposition through a DNA intermediate (*transposons*) and those that undergo transposition through an RNA intermediate (*retroelements*). In excess of 10% of the human genome comprises integrated copies of RNA molecules including retroviruses, retroviral-like DNAs, retroposons and retrotranscripts (reviewed by Cohen and Larsson, 1988; Amariglio and Rechavi, 1993; McDonald, 1993; Leib-Mösch and Seifarth, 1996). This represents more than 500,000 separate integration events. Nonviral retroposons include the *Alu* sequences and LINE elements discussed above. A number of endogenous retroviral or retroviral-like sequence families have been identified and characterized. These include HERV-K (Ono *et al.*, 1987; Goodchild *et al.*, 1995; Mayer *et al.*, 1997a; 1997b, 1999), RTVL-1 (Maeda and Kim, 1990), RTVL-H (Wilkinson *et al.*, 1993; Goodchild *et al.*, 1993), MaLR (Smit, 1993), LTR13 (Liao *et al.*, 1998), and the immunoglobulin gene-related human transposon (THE1) (Deka *et al.*, 1988; Fields *et al.*, 1992; Hakim *et al.*, 1994). The long terminal repeats of the HERV-K family are capable of binding human host cell nuclear proteins (Akopov *et al.*, 1998).

Retroviral sequences have sometimes become integrated into human genes (e.g. the endogenous retrovirus HRES-1 lies within the coding sequence of the transaldolase gene (*TALDO*; 11p15; Banki *et al.*, 1994; see Chapter 9, section 9.4) and once transposed, they may even be recruited to play a role in the transcriptional regulation of a gene [e.g. as postulated for the human salivary amylase (*AMY1C*) gene (Ting *et al.*, 1992); see Chapter 5, section 5.1.13].

1.1.5 Genes, mutations and disease

A central message of this volume is that the same mutational mechanisms that are responsible for disrupting the structure and function of human genes causing inherited disease are also responsible for having both created and fashioned these same genes over millions of years of evolutionary time. Since Chapters 7–9 are devoted to discussing mutational mechanisms in evolution, it is pertinent to devote at least some space here to summarizing briefly what is known of mutational mechanisms underlying human genetic disease.

In general, there are three types of mutation that give rise to an inherited disease: (i) mutations which lead to a loss of function, (ii) mutations which lead to a gain of function that is deleterious, and (iii) dominant negative mutations that adversely affect protein subunit activity or assembly. Characterized gene mutations causing genetic disease have been found to occur within coding sequences, untranslated sequences, promoter and locus control regions, in splice junctions, within introns and in polyadenylation sites [reviewed in detail by Cooper and Krawczak, 1993; see also the *Human Gene Mutation Database* at **http://www.uwcm.ac.uk/uwcm/mg/hgmd0.html,** an information resource which currently contains details of >18 500 different mutations in >900 different genes; Cooper *et al.*, 1998]. Indeed, they may interfere with any stage in the pathway of expression from gene to protein product.

Table 1.2 presents a basic classificatory system of mutation types by reference to the nature and position of the gene lesion and the stage in the expression pathway

Table 1.2. A classification of types of mutation found to cause human single gene defects through either reduced synthesis of a normal protein or normal synthesis of an abnormal protein

(a) **Reduced synthesis of a normal gene product**
Defect in

Promoter function	Binding of positive regulatory protein reduced or abolished
	Binding of negative regulatory protein increased
Gene structure	Deletions (frameshift)
	Insertions, duplications, inversions (frameshift).
RNA processing	Mutations in transcriptional initiation site causing failure to
stability	initiate transcription
	Splice junction mutations resulting in exon skipping and/or
	cryptic splice site utilization
	Activation of cryptic splice sites
	Polyadenylation/cleavage signal mutations
	Mutations in 3′ untranslated region
Translation	Initiation and termination codon mutations
	Mutations in 5′ untranslated region
	Nonsense mutations

(b) **Synthesis of structurally/functionally abnormal gene product**
Gene structure defect resulting in

Shortened gene product	Deletions (in-frame), nonsense mutations
Fusion genes	Deletions involving two linked genes
Elongated gene product	Insertions, duplications (in-frame)
	Termination codon mutations
Defective post-translational modification or processing, instability of protein product, impaired assembly/secretion, altered substrate/cofactor/receptor affinity.	Missense mutations

(*Figure 1.2*) with which it interferes. Mutations are firstly divided on the basis of whether they result in the reduced synthesis of a gene product (A) or the synthesis of a structurally/functionally abnormal gene product (B). Mutations are then secondarily divided into four categories; promoter function, gene structure, RNA processing and translation. Some gene lesions may be placed into more than one category. For example, missense mutations with drastic effects on protein structure and stability or which serve to activate an exonic cryptic splice site can fall into both categories. Similarly, a missense mutation close to an intron/exon splice junction could affect mRNA splicing efficiency as well as protein structure. The effects of specific amino acid substitutions on protein structure are the subject of several reviews (Pakula and Sauer, 1989; Alber, 1989; Wacey *et al.*, 1994).

In the context of human pathology, by far the most frequent genetic lesions in the genome are point mutations and deletions. The remainder comprise a mixed assortment of insertions, duplications, inversions, sequence amplifications and complex rearrangements. A brief review of the different types of known pathological lesion will be given.

Single base-pair substitutions within the coding region. Some 23% of point mutations are CG→TG or CG→CA transitions, representing a five-fold higher frequency for this dinucleotide than that predicted from random expectation (Krawczak *et al.*, 1998). This is thought to be due to the hypermutability of the methylated dinucleotide CpG; spontaneous deamination of 5-methylcytosine (5mC) to thymidine in this doublet gives rise to C→T or G→A transitions depending upon the strand in which the 5mC is mutated. CpG hypermutability in inherited disease implies that the CpG sites in question are methylated in the germline and thereby rendered prone to 5mC deamination.

The spectrum of point mutations occurring outwith CpG dinucleotides is also nonrandom (Cooper and Krawczak, 1993; Krawczak *et al.*, 1998). In principle, the nonrandomness of the initial mutation event, the nonrandomness of the DNA sequences under study, differences in the relative efficiency with which certain mutations are repaired, differences in phenotypic effect (and hence selection), or a bias in the clinical detection of such variants, may all play a role in determining the observed mutational spectrum.

The majority of single base-pair substitutions causing human genetic disease alter the amino acid encoded (missense mutations) but a sizeable proportion result in the introduction of a termination codon (nonsense mutations). The likelihood of clinical detection is estimated to be about three times as high for nonsense mutations as for missense mutations (Krawczak *et al.*, 1998). Using a multidomain molecular model of the human factor IX protein, Wacey *et al.* (1994) have shown that the likelihood that a factor IX gene (*F9*; Xq28) mutation (causing hemophilia B) will come to clinical attention is a complex function of the sequence characteristics of the *F9* gene, the nature of the amino acid substitution, its precise location and immediate environment within the protein molecule, and its resulting effects on the structure and function of the protein.

Single base-pair substitutions within splice sites. Splicing defects have been estimated to make up between 8% and 15% of all single base-pair substitutions causing human genetic disease (Krawczak *et al.*, 1992). Mutations appear to occur disproportionately at the most evolutionarily conserved positions within the splice site (Krawczak *et al.*, 1992). This is due to a detection bias resulting from the relative phenotypic severity of these lesions. Phenotypic consequences of splice site mutations include exon skipping and cryptic splice site utilization.

Single base-pair substitutions within promoter regions. A wide variety of mutations are now known which occur within gene promoter regions (Cooper and Krawczak, 1993). These lesions disrupt the normal processes of gene activation and transcriptional initiation and serve to decrease (or rather less often, increase) the level of mRNA/protein product synthesized. What these lesions have in common is their ability to alter or abolish the binding capacity of *cis*-acting DNA sequence motifs for the *trans*-acting protein factors that normally interact with them.

Gross gene deletions. Gross gene deletions may arise through a number of different recombinational mechanisms including homologous unequal recombination (occurring either between related gene sequences or repetitive elements). *Alu* sequences have been noted to flank deletion breakpoints in a considerable number of human genetic conditions and may represent hotspots for gene deletions (Cooper and Krawczak, 1993).

Microdeletions. Deletion breakpoint junctions flanking short (<20 bp) human gene deletions are non-random both at the nucleotide and dinucleotide levels, an observation consistent with a sequence-directed mechanism of mutagenesis (Cooper and Krawczak, 1993). Direct repeats flanking the deleted sequence are a common finding, consistent with a model of slipped mispairing at the replication fork. Two specific types of sequence have been found at high frequency in the vicinity of short gene deletions: polypyrimidine runs of at least 5 bp (YYYYY) and a 'deletion hotspot consensus sequence' (TGRRKM) (Cooper and Krawczak, 1993).

Insertions. That insertional mutagenesis might be as intrinsically non-random as point mutations and gene deletions was strongly suggested by the findings of Fearon *et al.* (1990), who reported 10 independent examples of DNA insertion within the same 170 bp intronic region of the *DCC* (18q21) gene (a locus which has been proposed to play an important role in human colorectal neoplasia). Insertional mutations involving the introduction of <10 bp DNA sequence into a gene coding region are (i) nonrandom and appear to be highly dependent upon the local DNA sequence context and (ii) may be explained by those mechanisms held to be responsible for gene deletions (Cooper and Krawczak, 1993).

Inversions. Inversions are a highly unusual mutational mechanism causing human genetic disease. The best known example is that found recently in the factor VIII (*F8C*) gene causing hemophilia A: this rearrangement occurs in about 40% of severely affected patients and recurs at high frequency (Lakich *et al.*, 1993; Naylor *et al.*, 1993). The mechanism responsible is thought to be homologous intrachromosomal recombination between a gene (*F8A*) located in intron 22 of the *F8C* gene and two additional homologues of the *F8A* gene situated 500 kb upstream of the *F8C* gene.

Expansion of unstable repeat sequences. A recently recognized mutational mechanism involves the instability of certain specific trinucleotide repeat sequences (reviewed by Timchenko and Caskey, 1996). This mechanism was first reported as a cause of the fragile X mental retardation syndrome. The brain-expressed *FMR1* gene responsible was found to contain an ususual (CGG)n repeat which exhibited copy number variation of between 6 and 54 in normal healthy controls, between 52 and >200 in phenotypically normal transmitting males (the 'premutation') and between 300 and >1000 in affected males (the 'full mutation'). Expansion of premutations to full mutations occurs only during female meiotic transmission whilst the probability of repeat expansion correlates with repeat copy number, consistent with a mechanism of slipped mispairing during replication. Expansion of a sequence can thus itself lead to further expansion, a process termed 'dynamic mutation' by Richards *et al.* (1992). The discovery of this novel mutational mechanism soon led to the recognition that the expansion of unstable repeats is responsible for a number of other human inherited diseases, almost all neuromuscular (see Chapter 8, section 8.9.1).

The triplet repeats (CGG)n and (CAG)n are very abundant in the human genome (Stallings, 1994; Han *et al.*, 1994). A considerable number of human genes, both ubiquitously expressed and tissue-specific, have now been identified as containing such triplet repeats (Riggins *et al.*, 1992; Karlin and Burge, 1996). Many of these repeats are highly polymorphic and may thus represent examples of more subtle triplet repeat expansions. Whether specific polymorphic alleles are associated with any particular phenotype, however, remains to be seen.

This chapter has attempted to provide the reader with a brief introduction to what is known of structure–function relationships in the human genome. Specific DNA sequences have clearly evolved for different cellular functions. Some sequences represent gene coding regions that contain the genetic information necessary to direct the synthesis of proteins. Other sequences in gene promoter or untranslated regions serve to direct appropriate gene expression or are involved in mRNA processing, stability and nucleocytoplasmic transport. Some highly repetitive sequences may play an important role in chromosome architecture whereas others may almost be commensally parasitic in that they cohabit in the genome having increased in copy number under their own mutational momentum. Some sequences are modified by DNA methylation which itself may influence function. Other sequences, by their very nature, are more mutable than others and may be capable of rapid change. Whatever their cellular role, many DNA sequences are able to interact with proteins or mRNA molecules in order to effect their function. It follows that the imminent analysis of the human genome sequence should reveal the existence of new sequence codes (and perhaps codes within codes) to add to that first elucidated in the early 1960s.

References

Adachi Y., Käs E., Laemmli U.K. (1989) Preferential, cooperative binding of DNA topoisomerase II to scaffold-associated regions. *EMBO J.* **8**: 3997–4006.

Adams M.D. *et al.* (1995) Initial assessment of human gene diversity and expression patterns based upon 83 million nucleotides of cDNA sequence. *Nature* **377** Suppl.: 3–174.

Akopov S.B. Nikolaev L.G. Khil P.P., Lebedev Y.B., Sverdlov E.D. (1998) Long terminal repeats of human endogenous retrovirus K family (HERV-K) specifically bind host cell nuclear proteins. *FEBS Letts.* **421**: 229–233.

Alber T. (1989) Mutational effects on protein stability. *Ann. Rev. Biochem.* **58**: 765–798.

Allison J., Hall L., MacIntyre I., Craig R.K. (1981) The construction and partial characterization of plasmids containing complementary DNA sequences to calcitonin precursor polyprotein. *Biochem. J.* **199**: 725–731.

Allshire R.C., Dempster M., Hastie N.D. (1989) Human telomeres contain at least three types of G rich repeat distributed non-randomly. *Nucleic Acids Res.* **17**: 4611–4627.

Amariglio N., Rechavi G. (1993) Insertional mutagenesis by transposable elements in the mammalian genome. *Environ. Molec. Mutagen.* **21**: 212–218.

Anagnostopoulos T., Green P.M., Rowley G., Lewis C.M., Giannelli F. (1999) DNA variation in a 5-Mb region of the X chromosome and estimates of sex-specific/type-specific mutation rates. *Am. J. Hum. Genet.* **64**: 508–517.

Andrews J.D., Mancini D.N., Singh S.M., Rodenhiser D.I. (1996) Site and sequence specific DNA methylation in the neurofibromatosis (*NF1*) gene includes C5839T: the site of the recurrent substitution mutation in exon 31. *Hum. Mol. Genet.* **5**: 503–507.

Antonarakis S.E., Kazazaian H.H., Orkin S.H. (1985) DNA polymorphism and molecular pathology of the human globin gene cluster. *Hum. Genet.* **69**: 1–14.

Aronow B., Lattier D., Silbiger R., Dusing M., Hutton J., Jones G. Stock J., McNeish J., Potter S., Witte D., Wiginton D. (1989) Evidence for a complex regulatory array in the first intron of the human adenosine deaminase gene. *Genes Devel.* **3**: 1384–1400.

Atwater J.A., Wisdom R., Verma I.M. (1990) Regulated mRNA stability. *Ann. Rev. Genet.* **24**: 519–541.

Ayoubi T.A., van de Ven W.J. (1996) Regulation of gene expression by alternative promoters. *FASEB J.* **10**: 453–460.

Baldini A., Ried T., Shridhar V., Ogura K., D'Aiuto L., Rocchi M., Ward D.C. (1993) An alphoid DNA sequence conserved in all human and great ape chromosomes: evidence for ancient centromeric sequences at human chromosomal regions 2q21 and 9q13. *Hum. Genet.* **90**: 577–583.

Banki K., Halladay D., Perl A. (1994) Cloning and expression of the human gene for transaldolase. A novel highly repetitive element constitutes an integral part of the coding sequence. *J. Biol. Chem.* **269**: 2847–2851.

Barlow D.P. (1995) Gametic imprinting in mammals. *Science* **270**: 1610–1613.

Barlow D.P., Stoger R., Herrmann B.G., Saito K., Schweifer N. (1991) The mouse insulin-like growth factor type-2 receptor is imprinted and closely linked to the *Tme* locus. *Nature* **349**: 84–87.

Batzer M.A., Kilroy G.E., Richard P.E., Shaikh T.H., Desselle T.D., Hoppens C.L., Deininger P.L. (1990) Structure and variability of recently inserted *Alu* family members. *Nucleic Acids Res.* **18**: 6793–6798.

Batzer M.A., Deininger P.L. (1991) A human-specific subfamily of *Alu* sequences. *Genomics* **9**: 481–487.

Beckmann J.S. Weber, J.L. (1992) Survey of human and rat microsatellites. *Genomics* **12**: 627–631.

Behn-Krappa A., Hölker I., Sandaradura de Silva U., Doerfler W. (1991) Patterns of DNA methylation are indistinguishable in different individuals over a wide range of human DNA sequences. *Genomics* **11**: 1–7.

Bernardi G. (1993a) The vertebrate genome: isochores and evolution. *Mol. Biol. Evol.* **10**: 186–204.

Bernardi G. (1993b) The isochore organization of the human genome and its evolutionary history – a review. *Gene* **135**: 57–66.

Bestor T.H., Tycko B. (1996) Creation of genomic methylation patterns. *Nature Genet.* **12**: 363–367.

Bhasker C.R., Burgiel G., Neupert B., Emery-Goodman A., Kuhn L.C., May B.K. (1993) The putative iron-responsive element in the human erythroid 5-aminolevulinate synthase mRNA mediates translational control. *J. Biol. Chem.* **268**: 12 699–12 705.

Bickmore W.A., Sumner A.T. (1989) Mammalian chromosome banding—an expression of genome organization. *Trends Genet.* **5**: 144–148.

Bird A.P. (1980) DNA methylation and the frequency of CpG in animal DNA. *Nucleic Acids Res.* **8**: 1499–1504.

Bird A.P. (1986) CpG-rich islands and the function of DNA methylation. *Nature* **321**: 209–213.

Bird A.P. (1992) The essentials of DNA methylation. *Cell* **70**: 5–8.

Bird A.P. (1995) Gene number, noise reduction and biological complexity. *Trends Genet.* **11**: 94–100.

Bird A.P., Taggart M., Frommer M., Miller O.J., Macleod D. (1985) A fraction of the mouse genome that is derived from islands of nonmethylated CpG-rich DNA. *Cell* **40**: 91–99.

Birnstiel M.L., Busslinger M., Strub K. (1985) Transcription termination and 3′ processing: the end is in sight. *Cell* **41**: 349–359.

Blackburn E.H. (1992) Telomerases. *Annu. Rev. Biochem.* **61**: 113–129.

Blackburn E.H. (1994) Telomeres: no end in sight. *Cell* **77**: 621–623.

Blankenship C., Naglich J.G., Whaley J.M., Seizinger B., Kley N. (1999) Alternate choice of initiation codon produces a biologically active product of the von Hippel Lindau gene with tumor suppressor activity. *Oncogene* **18**: 1529–1535.

Bodner M., Fridkin M., Gozes I. (1985) Coding sequences for vasoactive intestinal peptide and PHM-27 peptide are located on two adjacent exons in the human genome. *Proc. Natl Acad. Sci. USA* **82**: 3548–3551.

Boel E., Schwartz T.W., Norris K.E., Fiil N.P. (1984) A cDNA encoding a small common precursor for human pancreatic polypeptide and pancreatic icosapeptide. *EMBO J.* **3**: 909–912.

Bottema C.D.K., Ketterling R.P., Yoon H.-S., Sommer S.S. (1990) The pattern of factor IX germ-line mutation in Asians is similar to that of Caucasians. *Am. J. Hum. Genet.* **47**: 835–841.

Boulikas T. (1992) Homeotic protein binding sites, origins of replication, and nuclear matrix anchorage sites share the ATTA and ATTTA motifs. *J. Cell. Biochem.* **50**: 111–116.

Boulikas T. (1996) Common structural features of replication origins in all life forms. *J. Cell. Biochem.* **60**: 297–316.

Bowcock A.M., Kidd J.R., Mountain J.L., Hebert J.M., Carotenuto L., Kidd K.K., Cavalli-Sforza L.L. (1991) Drift, admixture, and selection in human evolution: a study with DNA polymorphisms. *Proc. Natl Acad. Sci. USA* **88**: 839–843.

Brini A.T., Lee G.M., Kinet J.-P. (1993) Involvement of *Alu* sequences in the cell-specific regulation of transcription of the γ chain of Fc and T cell receptors. *J. Biol. Chem.* **268**: 1355–1361.

Bristow J., Gitelman S.E., Tee M.K., Staels B., Miller W.L. (1993) Abundant adrenal-specific transcription of the human P450c21A 'pseudogene'. *J. Biol. Chem.* **268**: 12 919–12 924.

Britten R.J., Baron W.F., Stout D.B., Davidson E.H. (1988) Sources and evolution of human *Alu* repeated sequences. *Proc. Natl Acad. Sci. USA* **85**: 4770–4774.

Brodsky G., Barnes T., Bleskan J., Becker L., Cox M., Patterson D. (1997) The human GARS-AIRS-GART gene encodes two proteins which are differentially expressed during human brain development and temporally overexpressed in cerebellum of individuals with Down syndrome. *Hum. Molec. Genet.* **6**: 2043–2050.

Brown W.R.A. (1989) Molecular cloning of human telomeres in yeast. *Nature* **338**: 774–776.

Brown W.R.A., Bird A.P. (1986) Long-range restriction site mapping of mammalian genomic DNA. *Nature* **322**: 477–481.

Bruford M.W., Wayne R.K. (1993) Microsatellites and their application to population genetic studies. *Curr. Opin. Genet. Devel.* **3**: 939–943.

Bullock P., Champoux J.J., Botchan M. (1985) Association of crossover points with topoisomerase I cleavage sites: a model for nonhomologous recombination. *Science* **230**: 954–957.

Burd C.G., Dreyfuss G. (1994) Conserved structures and diversity of functions of RNA-binding proteins. *Science* **265**: 615–621.

Burge C.B., Padgett R.A., Sharp P.A. (1998) Evolutionary fates and origins of U12-type introns. *Mol. Cell* **2**: 773–785.

Caceres J.F., Stamm S., Helfman D.M., Krainer A.R. (1994) Regulation of alternative splicing *in vivo* by overexpression of antagonistic splicing factors. *Science* **265**: 1706–1710.

Cargill M., Altshuber D., Ireland J., *et al.* (1999) Characterization of single-nucleotide polymorphisms in coding regions of human genes. *Nature Genet.* **22**:231–238.

Catasti P., Gupta G., Garcia A.E., Ratliff R., Hong L., Yau P., Moyzis R.K., Bradbury E.M. (1994) Unusual structures of the tandem repetitive DNA sequences located at human centomeres. *Biochemistry* **33**: 3819–3830.

Chandley A.C., Mitchell A.R. (1988) Hypervariable minisatellite regions are sites for crossing over at meiosis in man. *Cytogenet. Cell Genet.* **48**: 152–155.

Chen C.-Y.A., Shyu A.-B. (1995) AU-rich elements: characterization and importance in mRNA degradation. *Trends Biochem. Sci.* **20**: 465–470.

Cheng J.F., Krane D.E., Hardison R.C. (1988) Nucleotide sequence and expression of rabbit globin genes ζ1, ζ2, and ζ3. Pseudogenes generated by block duplications are transcriptionally competent. *J. Biol. Chem.* **263**: 9981–9993.

Christiano A.M., Hoffman G.G., Chung-Honet L.C., Lee S., Cheng W., Uitto J., Greenspan D.S. (1994) Structural organization of the human type VII collagen gene (*COL7A1*), composed of more exons than any previously characterized gene. *Genomics* **21**: 169–179.

Chung J.H., Whiteley M., Felsenfeld G. (1993) A 5′ element of the chicken β-globin domain serves as an insulator in human erythroid cells and protects against position effect in *Drosophila*. *Cell* **74**: 505–514.

Clark A.R., Docherty K. (1993) Negative regulation of transcription in eukaryotes. *Biochem. J.* **296**: 521–541.

Cockerill P.N., Garrard W.T. (1986) Chromosomal loop anchorage of the kappa immunoglobulin gene occurs next to the enhancer in a region containing topoisomerase II sites. *Cell* **44**: 273–282.

Cohen M., Larsson E. (1988) Human endogenous retroviruses. *BioEssays* **9**: 191–196.

Colgan D.F., Manley J.L. (1997) Mechanism and regulation of mRNA polyadenylation. *Genes Dev.* **11**: 2755–2766.

Collins A., Frezal J., Teague J., Morton N.E. (1996) A metric map of humans: 23 500 loci in 850 bands. *Proc. Natl Acad. Sci. USA* **93**: 14 771–14 775.

Collins F.S., Guyer M.S., Chakravarti A. (1997) Variations on a theme: cataloguing human DNA sequence variation. *Science* **278**: 1580–1581.

Collins M., Rojnuckarin P., Zhu Y.-H., Bornstein P. (1998) A far upstream, cell type-specific enhancer of the mouse thrombospondin 3 gene is located within intron 6 of the adjacent metaxin gene. *J. Biol. Chem.* **273**: 21816–21824.

Colot V., Rossignol J.-L. (1999) Eukaryotic DNA methylation as an evolutionary device. *Bioessays*, **21**: 402–411.

Cooper D.N. (1983) Eukaryotic DNA methylation. *Hum. Genet.* **64**: 315–333.

Cooper D.N., Krawczak M. (1989) Cytosine methylation and the fate of CpG dinucleotides in vertebrate genomes. *Hum. Genet.* **83**: 181–189.

Cooper D.N., Krawczak M. (1993) *Human Gene Mutation*. BIOS Scientific Publishers, Oxford.

Cooper D.N., Ball E.V., Krawczak M. (1998) The human gene mutation database. *Nucleic Acids Res.* **26**: 285–287.

Cooper D.N., Smith B.A., Cooke H.J., Niemann S., Schmidtke J. (1985) A estimate of unique DNA sequence heterozygosity in the human genome. *Hum. Genet.* **69**: 201–205.

Coote T., Bruford M.W. (1996) Human microsatellites applicable for analysis of genetic variation in apes and Old World monkeys. *J. Hered.* **87**: 406–410.

Coverley D., Laskey R.A. (1994) Regulation of eukaryotic DNA replication. *Annu. Rev. Biochem.* **63**: 745–776.

Cox L.A., Adrian G.S. (1993) Posttranscriptional regulation of chimeric human transferrin genes by iron. *Biochemistry* **32**: 4738–4745.

Craig J.M., Bickmore W.A. (1993) Chromosome bands—flavours to savour. *BioEssays* **15**: 349–354.

Craig J.M., Bickmore W.A. (1994) The distribution of CpG islands in mammalian chromosomes. *Nature Genet.* **7**: 376–379.

Crouau-Roy B., Service S., Slatkin M., Freimer N. (1996) A fine-scale comparison of the human and chimpanzee genomes: linkage, linkage disequilibrium and sequence analysis. *Hum. Mol. Genet.* **5**: 1131–1137.

Csink A.K., Henikoff S. (1998) Something from nothing: the evolution and utility of satellite repeats. *Trends Genet.* **14**: 200–204.

Cullen M., Erlich H., Klitz W., Carrington M. (1995) Molecular mapping of a recombination hotspot located in the second intron of the human TAP2 locus. *Am. J. Hum. Genet.* **56**: 1350–1358.

Cunningham J.M., Purucker M.E., Jane S.M., Safer B., Vanin E.F., Ney P.A., Lowrey C.H., Nienhuis A.W. (1994) The regulatory element 3′ to the ᴬγ-globin gene binds to the nuclear matrix and interacts with special A-T-rich binding protein 1 (SATB1), an SAR/MAR-associating region DNA binding protein. *Blood* **84**: 1298–1308.

Curtis D., Lehmann R., Zamore P.D. (1995) Translational regulation in development. *Cell* **81**: 171–178.

Davidson J.N., Peterson M.L. (1997) Origin of genes encoding multi-enzymatic proteins in eukaryotes. *Trends Genet.* **13**: 281–285.

De Pamphilis M.L. (1993) Eukaryotic DNA replication – anatomy of an origin. *Annu. Rev. Biochem.* **62**: 29–63.

De Simone V., Cortese R. (1989) A negative regulatory element in the promoter of the human α1-antitrypsin gene. *Nucleic Acids Res.* **17**: 9407–9414.

Debailleul V., Laine A., Huet G., Mathon P., d'Hooghe M.C., Aubert J.P., Porchet N. (1998) Human mucin genes *MUC2, MUC3, MUC4, MUC5AC, MUC5B* and *MUC6* express stable and extremely large mRNAs and exhibit a variable length polymorphism. *J. Biol. Chem.* **273**: 881–890.

DeBerardinis R.J., Goodier J.L., Ostertag E.M., Kazazian H.H. (1998) Rapid amplification of a retrotransposon subfamily is evolving the mouse genome. *Nature Genet.* **20**: 288–290.

Decker C.J., Parker R. (1994) Mechanisms of mRNA degradation in eukaryotes. *Trends Biochem. Sci.* **19**: 336–340.

Deeb S.S., Jørgensen A.L., Battisti L., Iwasaki L., Motulsky A.G. (1994) Sequence divergence of the red and green visual pigments in great apes and humans. *Proc. Natl Acad. Sci. USA* **91**: 7262–7266.

Deininger P.L., Batzer M.A. (1993) Evolution of retroposons. Chap. 5 In: *Evolutionary Biology*, vol. 27 (ed. MK Hecht). Plenum Press, New York. pp. 157–196.

Deka N., Willard C.R., Wong E., Schmid C.W. (1988) Human transposon-like elements insert at a preferred target site: evidence for a retrovirally mediated process. *Nucleic Acids Res.* **16**: 1143–1151.

Delgado S., Gómez M., Bird A., Antequera F. (1998) Initiation of DNA replication at CpG islands in mammalian chromosomes. *EMBO J.* **17**: 2426–2435.

Dietrich R.C., Incorvaia R., Padgett R.A. (1997) Terminal intron dinucleotide sequences do not distinguish between U2- and U12-dependent introns. *Mol. Cell* **1**: 151–160.

Dillon N., Grosveld F. (1994) Chromatin domains as potential units of eukaryotic function. *Curr. Opin. Genet. Devel.* **4**: 260–264.

Dobbs D.L., Shaiu W.-L., Benbow R.M. (1994) Modular sequence elements associated with origin regions in eukaryotic chromosomal DNA. *Nucleic Acids Res.* **22**: 2479–2489.

Donda A., Schulz M., Burki K., De Libero G., Uematsu Y. (1996) Identification and characterization of a human *CD4* silencer. *Eur. J. Immunol.* **26**: 493–500.

Edwalds-Gilbert G., Veraldi K.L., Milcarek C. (1997) Alternative poly(A) site selection in complex transcription units: means to an end? *Nucleic Acids Res.* **25**: 2547–2561.

Eissenberg J.C., Elgin S.C.R. (1991) Boundary functions in the control of gene expression. *Trends Genet.* **7**: 335–340.

El-Maarri O., Olek A., Balaban B., Montag M., van der Ven H., Urman B., Olek K., Caglayan S.H., Walter J., Oldenburg J. (1998) Methylation levels at selected CpG sites in the factor VIII and *FGFR3* genes in mature female and male germ cells: implications for male-driven evolution. *Am. J. Hum. Genet.* **63**: 1001–1008.

Elbrecht A., Chen Y., Cullinan C.A., Hayes N., Leibowitz M.D., Moller D.E., Berger J. (1996) Molecular cloning, expression and characterization of human peroxisome proliferator activated receptors γ1 and γ2. *Biochem. Biophys. Res. Commun.* **224**: 431–437.

Ellis N., Goodfellow P.N. (1989) The mammalian pseudoautosomal region. *Trends Genet.* **5**: 406–409.

Ellis N., Yen P., Neiswanger K., Shapiro L.J., Goodfellow P.N. (1990) Evolution of the pseudoautosomal boundary in Old World monkeys and great apes. *Cell* **63**: 977–986.

Englander E.W., Wolffe A.P., Howard B.H. (1993) Nucleosome interactions with a human *Alu* element. *J. Biol. Chem.* **268**: 19 565–19 573.

Epp C.D. (1997) Definition of a gene. *Nature* **389**: 537.

Eyre-Walker A. (1993) Recombination and mammalian genome evolution. *Proc. R. Soc. Lond. B* **252**: 237–243.

Faisst S., Meyer S. (1992) Compilation of vertebrate-encoded transcription factors. *Nucleic Acids Res.* **20**: 3–26.

Fearon F.R., Cho K.R., Nigro J.M., Kern S.E., Simons J.W., Ruppert J.M., Hamilton S.R., Presinger A.C., Thomas G., Kinzler K.W., Vogelstein B. (1990) Identification of a chromosome 18q gene that is altered in colorectal cancers. *Science* **247**: 49–56.

Fields C.A., Grady D.L., Moyzis R.K. (1992) The human THE-LTR(O) and *Mst*II interspersed repeats are subfamilies of a single widely distributed highly variable repeat family. *Genomics* **13**: 431–436.

Fields C., Adams M.D., White O., Venter J.C. (1994) How many genes in the human genome? *Nature Genet.* **7**: 345–346.

Fink J.K., Jones S.M., Esposito C., Wilkowski J. (1996) Human microtubule-associated protein 1α (*MAP1A*) gene: genomic organization, cDNA sequence, and developmental- and tissue-specific expression. *Genomics* **35**: 577–585.

Francino M.P., Ochman H. (1999) Isochores result from mutation not selection. *Nature* **400**:30–31.

Freemont P.S., Lane A.N., Sanderson M.R. (1991) Structural aspects of protein-DNA recognition. *Biochem. J.* **278**: 1–23.

Fullerton S.M., Harding R.M., Boyce A.J., Clegg J.B. (1994) Molecular and population genetic analysis of allelic sequence diversity at the human β-globin locus. *Proc. Natl Acad. Sci. USA* **91**: 1805–1809.

Furano A.V., Usdin K. (1995) DNA 'fossils' and phylogenetic analysis. *J. Biol. Chem.* **270**: 25 301–25 304.

Gan S.Q., McBride O.W., Idler W.W., Markova N., Steinert P.M. (1990) Organization, structure and polymorphisms of the human profilaggrin gene. *Biochemistry* **29**: 9432–9440.

Gardiner K., Horisberger M., Kraus J., Tantravahi U., Korenberg J., Rao V., Reddy S., Patterson D. (1990) Analysis of human chromosome 21: correlation of physical and cytogenetic maps; gene and CpG island distributions. *EMBO J.* **9**: 25–34.

Gardiner-Garden M., Frommer M. (1987) CpG islands in vertebrate genomes. *J. Mol. Biol.* **196**: 261–282.

Gerhold D., Caskey C.T. (1996) It's the genes! EST access to human genome content. *BioEssays* **18**: 973–981.

Geyer P.K. (1997) The role of insulator elements in defining domains of gene expression. *Curr. Opin. Genet. Devel.* **7**: 242–248.

Gilbert N., Labuda D. (1999) CORE-SINEs: eukaryotic short interspersed retroposing elements with common sequence motifs. *Proc. Natl Acad. Sci. USA* **96**: 2869–2874.

Gilson E., Laroche T., Gasser S.M. (1993) Telomeres and the functional architecture of the nucleus. *Trends Cell Biol.* **3**: 128–134.

Goldman M.A., Holmquist G.P., Gray M.C., Caston L.A., Nag A. (1984) Replication timing of genes and middle repetitive sequences. *Science* **224**: 686–692.

Gondo Y., Okada T., Matsuyama N., Saitoh Y., Yanagisawa Y., Ikeda J.-E. (1998) Human megasatellite DNA RS447: copy-number polymorphisms and interspecies conservation. *Genomics* **54**: 39–49.

Goodchild N.L., Wilkinson D.A., Mager D.L. (1993) Recent evolutionary expansion of a subfamily of RTV$_L$-H human endogenous retrovirus-like elements. *Virology* **196**: 778–788.

Goodchild N.L., Freeman J.D., Mager D.L. (1995) Spliced HERV-H endogenous retroviral sequences in human genomic DNA: evidence for amplification via retrotransposition. *Virology* **206**: 164–173.

Green M.R. (1986) Pre-mRNA splicing. *Ann. Rev. Genet.* **20**: 671–708.

Haaf T., Willard H.F. (1997) Chromosome-specific alpha-satellite DNA from the centromere of chimpanzee chromosome 4. *Chromosoma* **106**: 226–232.

Haaf T., Willard H.F. (1998) Orangutan alpha-satellite monomers are closely related to the human consensus sequence. *Mammalian Genome* **9**: 440–447.

Haaf T., Mater A.G., Wienberg J., Ward D.C. (1995) Presence and absence of CENP-B box sequences in great ape subsets of primate-specific alpha-satellite DNA. *J. Mol. Evol.* **41**: 487–491.

Hacia J.G., Fan J.-B., Ryder O. *et al.* (1999) Determination of ancestral alleles for human single-nucleotide polymorphisms using high-density oligonucleotide arrays. *Nature Genet.* **22**:164–167.

Haig D. (1996) Do imprinted genes have few and small introns? *BioEssays* **18**: 351–353.

Haig D., Trivers R. (1995) The evolution of parental imprinting: a review of hypotheses. Chap. 2 In: *Genomic Imprinting: Causes and Consequences* (eds R. Ohlsson, K. Hall and M. Ritzen), Cambridge University Press.

Hakim I., Amariglio N., Grossman Z., Simoni-Brok F., Ohno S., Rechavi G. (1994) The genome of the THE I human transposable repetitive elements is composed of a basic motif homologous to an ancestral immunoglobulin gene sequence. *Proc. Natl Acad. Sci. USA* **91**: 7967–7969.

Hammarback J.A., Obar R.A., Hughes S.M., Vallee R.B. (1991) MAP1B is encoded as a polyprotein that is processed to form a complex N-terminal microtubule-binding domain. *Neuron* **7**: 129–139.

Han J., Hsu C., Zhu Z., Longshore J.W., Finley W.H. (1994) Over-representation of the disease associated (CAG) and (CGG) repeats in the human genome. *Nucleic Acids Res.* **22**: 1735–1740.

Hand R. (1978) Eukaryotic DNA: organization of the genome for replication. *Cell* **15**: 317–325.

Haniel A., Welge-Lussen U., Kuhn K., Poschl E. (1995) Identification and characterization of a novel transcriptional silencer in the human collagen type IV gene *COL4A2*. *J. Biol. Chem.* **270**: 11209–11215.

Hanna-Rose W., Hansen U. (1996) Active repression mechanisms of eukaryotic transcription repressors. *Trends Genet.* **12**: 229–234.

Hanscombe O., Whyatt D., Fraser P., Yannoutsos N., Greaves D., Dillon N., Grosveld F. (1991) Importance of globin gene order for correct developmental expression. *Genes Devel.* **5**: 1387–1394.

Hardison R. (1998) Hemoglobins from bacteria to man: evolution of different patterns of gene expression. *J. Exp. Biol.* **201**: 1099–1117.

Hare J.T., Taylor J.H. (1988) Hemi-methylation dictates strand selection in repair of G/T and A/C mismatches in SV40. *Gene* **74**: 159–161.

Hatton K.S., Dhar V., Brown E.H., Iqbal M.A., Stuart S., Didano V.T., Schildkraut C.L. (1988) Replication program of active and inactive multigene families in mammalian cells. *Mol. Cell. Biol.* **8**: 2149–2158.

Hattori M., Kuhara S., Takenaka O., Sakaki Y. (1986) L1 family of repetitive DNA sequences in primates may be derived from a sequence encoding a reverse-transcriptase-related protein. *Nature* **321**: 625–628.

Hawkins J.D. (1988) A survey on intron and exon lengths. *Nucleic Acids Res.* **16**: 9893–9908.

Heard E., Clerc P., Avner P. (1997) X-chromosome inactivation in mammals. *Ann. Rev. Genet.* **31**: 571–610.

Heartlein M.W., Knoll J.H.M., Latt S.A. (1988) Chromosome instability associated with human alphoid DNA transfected into the Chinese hamster genome. *Molec. Cell. Biol.* **8**: 3611–3618.

Hentze M.W. (1991) Determinants and regulation of cytoplasmic mRNA stability in eukaryotic cells. *Biochim. Biophys. Acta* **1090**: 281–292.

Herschbach B.M., Johnson A.D. (1993) Transcriptional repression in eukaryotes. *Annu. Rev. Cell Biol.* **9**: 479–509.

Higgs D.R. (1998) Do LCRs open chromatin domains? *Cell* **95**: 299–302.

Hofferbert S., Schanen N.C., Chehab F., Francke U. (1997) Trinucleotide repeats in the human genome: size distributions for all possible triplets and detection of expanded disease alleles in a group of Huntington disease individuals by the Repeat Expansion Detection method. *Hum. Molec. Genet.* **6**: 77–83.

Holmquist G.P. (1992) Chromosome bands, their chromatin flavors and their functional features. *Am. J. Hum. Genet.* **51**: 17–37.

Horowitz N.H. (1945) The evolution of biochemical systems. *Proc. Natl Acad. Sci. USA* **31**: 153–157.

Hurst L.D., McVean G.T., Moore T. (1996) Imprinted genes have few and small introns. *Nature Genet.* **12**: 234–237.

Hwu H.R., Roberts J.W., Davidson E.H., Britten R.J. (1986) Insertion and/or deletion of many repeated DNA sequences in human and higher ape evolution. *Proc. Natl Acad. Sci. USA* **83**: 3875–3879.

Izaurralde E., Mattaj I.W. (1995) RNA export. *Cell* **81**: 153–159.

Jabbari K., Caccio S., Pais de Barros J.P., Desgres J., Bernardi G. (1997) Evolutionary changes in CpG and methylation levels in the genome of vertebrates. *Gene* **205**: 109–118.

Jackson M.E. (1991) Negative regulation of eukaryotic transcription. *J. Cell. Sci.* **100**: 1–7.

Jackson D.A. (1997) Chromatin domains and nuclear compartments: establishing sites of gene expression in eukaryotic nuclei. *Mol. Biol. Rep.* **24**: 209–220.

Jaenisch R. (1997) DNA methylation and imprinting: why bother? *Trends Genet.* **13**: 323–329.

Jamieson R.V., Tam P.P.L., Gardiner-Garden M. (1996) X-chromosome activity: impact of imprinting and chromatin structure. *Int. J. Dev. Biol.* **40**: 1065–1080.

Jarman A.P., Wood W.G., Sharpe J.A., Gourdon G., Ayyub H., Higgs D.R. (1991) Characterization of the major regulatory element upstream of the human α-globin gene cluster. *Molec. Cell. Biol.* **11**: 4679–4689.

Jeffreys A. (1987) Highly variable minisatellites and DNA fingerprints. *Biochem. Soc. Transact.* **15**: 309–317.

Jelinek W.R., Schmid C.W. (1982) Repetitive sequences in eukaryotic DNA and their expression. *Ann. Rev. Biochem.* **51**: 813–844.

Johnson P.F., McKnight S.L. (1989) Eukaryotic transcriptional regulatory proteins. *Ann. Rev. Biochem.* **58**: 799–839.

Jones B.K., Monks B.R., Liebhaber S.A., Cooke N.E. (1995) The human growth hormone gene is regulated by a multicomponent locus control region. *Molec. Cell. Biol.* **15**: 7010–7021.

Jørgensen A.L. (1997) Alphoid repetitive DNA in human chromosomes. *Dan. Med. Bull.* **44**: 522–534.

Jørgensen A.L., Laursen H.B., Jones C., Bak A.L. (1992) Evolutionarily different alphoid repeat DNA on homologous chromosomes in human and chimpanzee. *Proc. Natl Acad. Sci. USA* **89**: 3310–3314.

Jost J.-P., Bruhat A. (1997) The formation of DNA methylation patterns and the silencing of genes. *Prog. Nucleic Acid Res. Mol. Biol.* **57**: 217–248.

Jurka J. (1997) Sequence patterns indicate an enzymatic involvement in integration of mammalian retroposons. *Proc. Natl Acad. Sci. USA* **94**: 1872–1877.

Jurka J., Smith T. (1988) A fundamental division in the *Alu* family of repeated sequences. *Proc. Natl Acad. Sci. USA* **85**: 4775–4778.

Kariya Y., Kato K., Hayashizaki Y., Himeno S., Tarui S., Matsubara K. (1987) Revision of consensus sequence of human *Alu* repeats—a review. *Gene* **53**: 1–10.

Karlin S., Burge C. (1996) Trinucleotide repeats and long homopeptides in genes and proteins associated with nervous system disease and development. *Proc. Natl Acad. Sci. USA* **93**: 1560–1565.

Kass S.U., Pruss D., Wolffe A.P. (1997) How does DNA methylation repress transcription? *Trends Genet.* **13**: 444–449.

Kawaguchi H., Klein J. (1992) Organization of the *C4* and *CYP21* loci in gorilla and orangutan. *Hum. Immunol.* **33**: 153–162.

Kawaguchi H., Golubic M., Figueroa F., Klein J. (1990) Organization of the chimpanzee *C4-CYP21* region: implications for the evolution of human genes. *Eur. J. Immunol.* **20**: 739–749.

Kay G.F. (1998) Xist and X chromosome inactivation. *Mol. Cell. Endocrinol.* **140**: 71–76.

Kel O.V., Romaschenko A.G., Kel A.E., Wingender E., Kolchanov N.A. (1995) A compilation of composite regulatory elements affecting gene transcription in vertebrates. *Nucleic Acids Res.* **23**: 4097–4103.

Kellum R., Schedl P. (1991) A position-effect assay for boundaries of higher order chromosomal domains. *Cell* **64**: 941–950.

Kelly R.J., Rouquier S., Giorgi D., Lennon G.G., Lowe J.B. (1995) Sequence and expression of a candidate for the human secretor blood group α1,2-fucosyltransferase gene (*FUT2*). *J. Biol. Chem.* **270**: 4640–4649.

Kennerson M.L., Nassif N.T., Dawkins J.L., DeKroon R.M., Yang J.G., Nicholson G.A. (1997) The Charcot-Marie-Tooth binary repeat contains a gene transcribed from the opposite strand of a partially duplicated region of the *COX10* gene. *Genomics* **46**: 61–69.

Kim S.J., Park K., Koeller D., Kim K.Y., Wakefield L.M., Sporn M.B., Roberts A.B. (1992) Post-transcriptional regulation of the human transforming growth factor-beta 1 gene. *J. Biol. Chem.* **267**: 13702–13707.

Kitsberg D., Selig S., Keshet I., Cedar H. (1993) Replication structure of the human β-globin gene domain. *Nature* **366**: 588–590.

Kochanek S., Toth M., Dehmel A., Renz D., Doerfler W. (1990) Inter-individual concordance of methylation profiles in human genes for tumor necrosis factors α and β. *Proc. Natl Acad. Sci. USA* **87**: 8830–8834.

Kominato Y., McNeill P.D., Yamamoto M., Russell M., Hakomori S., Yamamoto F. (1992) Animal histo-blood group ABO genes. *Biochem. Biophys. Res. Commun.* **189**: 154–164.

Korenberg J.R., Rykowski M.C. (1988) Human genome organization: *Alu*, LINES, and the molecular structure of metaphase chromosome bands. *Cell* **53**: 391–400.

Kornberg R.D., Lorch Y. (1992) Chromatin structure and transcription. *Annu. Rev. Cell Biol.* **8**: 563–587.

Krainer A.R., Maniatis T. (1988) RNA splicing. Chap. 4 In: *Transcription and Splicing* (eds B.D. Hames and D.M. Glover). IRL Press, Oxford. pp. 131–206.

Krawczak M., Reiss J., Cooper D.N. (1992) The mutational spectrum of single base-pair substitutions in mRNA splice junctions of human genes: causes and consequences. *Hum. Genet.* **90**: 41–54.

Kundu T.K., Rao M.R.S. (1999) CpG islands in chromatin organization and gene expression. *J. Biochem.* **125**:217–222.

Labarrière N., Selvais P.L., Lemaigre F.P., Michel A., Maiter D.M., Rousseau G.G. (1995) A novel transcriptional activator originating from an upstream promoter in the human growth hormone gene. *J. Biol. Chem.* **270**: 19 205–19 208.

Labeit S., Kolmerer B. (1995) Titins: giant proteins in charge of muscle ultrastructure and elasticity. *Science* **270**: 293–297.

Laemmli U.K., Ks E., Poljak L., Adachi Y. (1992) Scaffold-associated regions: *cis*-acting determinants of chromatin structural loops and functional domains. *Curr. Op. Genet. Devel.* **2**: 275–285.

Lagarkova M.A., Svetlova E., Giacca M., Falaschi A., Razin S.V. (1998) DNA loop anchorage region colocalizes with the replication origin located downstream to the human gene encoding lamin B2. *J. Cell. Biochem.* **69**: 13–18.

Lakich D., Kazazian H.H., Antonarakis S.E., Gitschier J. (1993) Inversions disrupting the factor VIII gene are a common cause of severe haemophilia A. *Nature Genet.* **5**: 236–241.

Lakin N.D. (1993) Determination of DNA sequences that bind transcription factors by DNA footprinting. Chap. 2 In: *Transcription Factors; A Practical Approach* (ed. D.S. Latchman). IRL Press Oxford, pp. 27–47.

Larsen F., Gundersen G., Lopez R., Prydz H. (1992) CpG islands as gene markers in the human genome. *Genomics* **13**: 1095–1107.

Latchman D.S. (1990) Eukaryotic transcription factors. *Biochem. J.* **270**: 281–289.

Latchman D.S. (1998) *Eukaryotic Transcription Factors.* 3rd edn. Academic Press, San Diego.

Laurie D.A., Hulten M.A. (1985) Further studies on chiasma distribution and interference in the human male. *Ann. Hum. Genet.* **49**: 203–214.

Lee S.-J. (1991) Expression of growth/differentiation factor 1 in the nervous system: conservation of a bicistronic structure. *Proc. Natl Acad. Sci. USA* **88**: 4250–4254.

Lee C., Wevrick R., Fisher R.B., Ferguson-Smith M.A., Lin C.C. (1997) Human centromeric DNAs. *Hum. Genet.* **100**: 291–304.

Leib-M'sch C., Seifarth W. (1996) Evolution and biological significance of human retroelements. *Virus Genes* **11**: 133–145.

Lerner A., D'Adamio L., Diener A.C., Clayton L.K., Reinherz E.L. (1993) CD3 zeta/eta/theta locus is colinear with and transcribed antisense to the gene encoding the transcription factor Oct-1. *J. Immunol.* **151**: 3152–3162.

Levine M., Manley J.L. (1989) Transcriptional repression of eukaryotic promoters. *Cell* **59**: 405–408.

Li W.-H., Sadler L.A. (1991) Low nucleotide diversity in man. *Genetics* **129**: 513–523.

Li Q., Stamatoyannopoulos G. (1994) Hypersensitive site 5 of the human β locus control region functions as a chromatin insulator. *Blood* **84**: 1399–1401.

Li Y.-P., Chen W., Stashenko P. (1995) Characterization of a silencer element in the first exon of the human osteocalcin gene. *Nucleic Acids Res.* **23**: 5064–5072.

Liao D., Pavelitz T., Weiner A.M. (1998) Characterization of a novel class of interspersed LTR elements in primate genomes: structure, genomic distribution, and evolution. *J. Mol. Evol.* **46**: 649–660.

Liu W.-M., Maraia R.J., Rubin C.M., Scmid C.W. (1994) *Alu* transcripts: cytoplasmic localisation and regulation by DNA methylation. *Nucleic Acids Res.* **22**: 1087–1095.

Lo K., Smale S.T. (1996) Generality of a functional initiator consensus sequence. *Gene* **182**: 13–19.

Long M., de Souza S.J., Rosenberg C., Gilbert W. (1998) Relationship between "proto-splice sites" and intron phases: evidence from dicodon analysis. *Proc. Natl Acad. Sci. USA* **95**: 219–223.

Lopez A.J. (1998) Alternative splicing of pre-mRNA: developmental consequences and mechanisms of regulation. *Ann. Rev. Genet.* **32**: 279–305.

Lührmann R., Kastner B., Bach M. (1990) Structure of spliceosomal snRNPs and their role in pre-mRNA splicing. *Biochim. Biophys. Acta* **1087**: 265–292.

Luke S., Verma R.S. (1993) Telomeric repeat [TTAGGG]n sequences of human chromosomes are conserved in chimpanzee (*Pan troglodytes*). *Mol. Gen. Genet.* **237**: 460–462.

Luoh S.W., Jegalian K., Lee A., Chen E.Y., Ridley A., Page D.C. (1995) CpG islands in human *ZFX* and *ZFY* and mouse *Zfx* genes; sequence similarities and methylation differences. *Genomics* **29**: 353–363.

McAlpine P.J., Shows T.B., Povey S., Carritt B., Pericak-Vance M.A., Boucheix C., Anderson W.A., White J.A. (1993) The 1993 catalog of approved genes and report of the nomenclature committee. In: *Human Gene Mapping; A Compendium* (eds A.J. Cuticchia and P.L. Pearson), Johns Hopkins University Press, pp. 6–124.

McDonald J.F. (1993) Evolution and consequences of transposable elements. *Curr. Opin. Genet. Devel.* **3**: 855–864.

McGowan R.A., Martin C.C. (1997) DNA methylation and genome imprinting in the zebrafish, Danio rerio: some evolutionary ramifications. *Biochem. Cell Biol.* **75**: 499–506.

McKinley-Grant L.J., Idler W.W., Bernstein I.A., Parry D.A.D., Cannizzaro L., Croce C.M., Huebner K., Lessin S.R., Steinert P.M. (1989) Characterization of a cDNA clone encoding human filaggrin and localization of the gene to chromosome region 1q21. *Proc. Natl Acad. Sci. USA* **86**: 4848–4852.

McKnight R.A., Shamay A., Sankaran L., Wall R.J., Hennighausen L. (1992) Matrix-attachment regions can impart position-independent regulation of a tissue-specific gene in transgenic mice. *Proc. Natl Acad. Sci. USA* **89**: 6943–6947.

McKusick V.A. (1986) The morbid anatomy of the human genome. *Medicine* **65**: 1–33.

McLaughlan J., Gaffney D., Whitton J.L., Clements J.B. (1985) The consensus sequence YGTGTTYY located downstream from the AATAAA signal is required for efficient formation of mRNA 3' termini. *Nucleic Acids Res.* **13**: 1347–1368.

McVean G.T., Hurst L.D. (1997) Molecular evolution of imprinted genes: no evidence for antagonistic coevolution. *Proc. R. Soc. Lond. Series B* **264**: 739–746.

Maeda N., Kim H.S. (1990) Three independent insertions of retrovirus-like sequences in the haptoglobin gene cluster of primates. *Genomics* **8**: 671–683.

Magrangeas F., Pitiot G., Dubois S., Bragado-Nilsson E., Chérel M., Jobert S., Lebeau B., Boisteau O., Lethé B., Mallet J., Jacques Y., Minivielle S. (1998) Cotranscription and intergenic splicing of human galactose-1-phosphate uridylyltransferase and interleukin-11 receptor α-chain genes generate a fusion mRNA in normal cells. *J. Biol. Chem.* **273**: 16005–16010.

Makalowski W., Mitchell G.A., Labuda D. (1994) *Alu* sequences in the coding regions of mRNA: a source of protein variability. *Trends Genet.* **10**: 188–193.

Manley J.L., Proudfoot N.J. (1994) RNA 3' ends: formation and function – meeting review. *Genes Devel.* **8**: 259–264.

Maraia R.J., Driscoll C.T., Bilyeu T., Hsu K., Darlington G.J. (1993) Multiple dispersed loci produce small cytoplasmic *Alu* RNA. *Molec. Cell Biol.* **13**: 4233–4241.

Margalit H., Nadir E., Ben-Sasson S.A. (1994) A complete *Alu* element within the coding sequence of a central gene. *Cell* **78**: 173–174.

Martens U.M., Zijlmans J.M.J.M., Poon S.S.S., Dragowska W., Yui J., Chavez E.A., Ward R.K., Lansdorp P.M. (1998) Short telomeres on human chromosome 17p. *Nature Genet.* **18**: 76–80.

Martinko J.M., Vincek V., Klein D., Klein J. (1993) Primate ABO glycosyltransferases: evidence for trans-species evolution. *Immunogenetics* **37**: 274–278.

Matera A.G., Hellmann U., Hintz M.F., Schmid C.W. (1990a) Recently transposed *Alu* repeats result from multiple source genes. *Nucleic Acids Res.* **18**: 6019–6023.

Matera A.G., Hellmann U., Schmid C.W. (1990b) A transpositionally and transcriptionally competent *Alu* subfamily. *Mol. Cell. Biol.* **10**: 5424–5432.

Mathias S.L., Scott A.F., Kazazian. H.H., Boeke J.D., Gabriel A. (1991) Reverse transcriptase encoded by a human transposable element. *Science* **254**: 1808–1810.

Matsuo K., Clay O., Takahashi T., Silke J., Schaffner W. (1993) Evidence for erosion of mouse CpG islands during mammalian evolution. *Somat. Cell. Molec. Genet.* **19**: 543–555.

Maxwell E.S., Fournier M.J. (1995) The small nucleolar RNAs. *Ann. Rev. Biochem.* **64**: 897–934.

Mayer J., Meese E., Mueller-Lantzsch N. (1997a) Multiple human endogenous retrovirus (HERV-K) loci with *gag* open reading frames in the human genome. *Cytogenet. Cell Genet.* **78**: 1–5.

Mayer J., Meese E., Mueller-Lantzsch N. (1997b) Chromosomal assignment of human endogenous retrovirus K (HERV-K) *env* open reading frames. *Cytogenet. Cell Genet.* **78**: 157–161.

Mayer J., Sauter M., Racz A., Scherer D., Mueller-Lantzsch N., Meese E. (1999) An almost intact human endogenous retrovirus K on human chromosome 7. *Nature Genet.* 21: 257–258.

Meehan R.R., Lewis J.D., McKay S., Kleiner E.L., Bird A.P. (1989) Identification of a mammalian protein that binds specifically to DNA containing methylated CpGs. *Cell* 58: 499–507.

Melefors Ö., Hentze M.W. (1993) Translational regulation by mRNA/protein interactions in eukaryotic cells: ferritin and beyond. *BioEssays* 15: 85–90.

Miao C.H., Leytus S.P., Chung D.W., Davie E.W. (1992) Liver-specific expression of the gene coding for human factor X, a blood coagulation factor. *J. Biol. Chem.* 267: 7395–7401.

Millar D.S., Krawczak M., Cooper D.N. (1998) Variation of site-specific methylation patterns in the factor VIII (*F8C*) gene in human sperm DNA. *Hum. Genet.* 103: 228–233.

Mitchell P.J., Tjian R. (1989) Transcriptional regulation in mammalian cells by sequence-specific DNA binding proteins. *Science* 245: 371–378.

Mitchell A., Jeppesen P., Hanratty D., Gosden J. (1992) The organization of repetitive DNA sequences on human chromosomes with respect to the kinetochore analysed using a combination of oligonucleotide primers and CREST anticentromere serum. *Chromosoma* 101: 333–341.

Miyajima N., Horiuchi R., Shibuya Y., Fukushige S., Matsubara K., Toyoshima K., Yamamoto T. (1989) Two *erbA* homologs encoding proteins with different T_3 binding capacities are transcribed from opposite DNA strands of the same genetic locus. *Cell* 57: 31–39.

Mohrenweiser H.W., Tsujimoto S., Gordon L., Olsen A.S. (1998) Regions of sex-specific hypo- and hyper-recombination identified through integration of 180 genetic markers into the metric map of human chromosome 19. *Genomics* 47: 153–162.

Monk M. (1988) Genomic imprinting. *Genes Dev.* 2: 921–925.

Morin P.A., Mahboubi P., Wedel S., Rogers J. (1998) Rapid screening and comparison of human microsatellite markers in baboons: allele size is conserved, but allele number is not. *Genomics* 53: 12–20.

Mouchiroud D., D'Onofrio G., Aissani B., Macaya C., Gautier C., Bernardi G. (1991) The distribution of genes in the human genome. *Gene* 100: 181–187.

Mount S.M. (1982) A catalogue of splice junction sequences. *Nucleic Acids Res.* 10: 459–472.

Moyzis R.K., Torney D.C., Meyne J., Buckingham J.M., Wu J.-R., Burks C., Sirotkin K.M., Goad W.B. (1989) The distribution of interspersed repetitive DNA sequences in the human genome. *Genomics* 4: 273–289.

Muiznieks I., Doerfler W. (1994) The impact of 5'-CG-3' methylation on the activity of different eukaryotic promoters: a comparative study. *FEBS Letts.* 344: 251–254.

Müller M.M., Gerster T., Schaffner W. (1988) Enhancer sequences and the regulation of gene transcription. *Eur. J. Biochem.* 176: 485–495.

Murakami T., Nishiyori A., Takiguchi M., Mori M. (1990) Promoter and 11kb upstream enhancer elements responsible for hepatoma cell-specific expression of the rat ornithine transcarbamylase gene. *Mol. Cell. Biol.* 10: 1180–1191.

Murnane J.P., Morales J.F. (1995) Use of a mammalian interspersed repetitive (MIR) element in the coding and processing sequences of mammalian genes. *Nucleic Acids Res.* 23: 2837–2839.

Nadal-Ginard B., Gallego M.E., Andreadis A. (1987) Alternative splicing: mechanistic and biological implications of generating multiple proteins from a single gene. In: *Genetic Engineering. Principles and Methods* vol. 9 (ed. J.K. Setlow). Plenum Press, New York.

Naylor J., Brinke A., Hassock S., Green P.M., Giannelli F. (1993) Characteristic mRNA abnormality found in half the patients with severe haemophilia A is due to large DNA inversions. *Hum. Molec. Genet.* 2: 1773–1778.

Nguyen T., Sunahara R., Marchese A., Van Tol H.H., Seeman P., O'Dowd B.F. (1991) Transcription of a human dopamine D5 pseudogene. *Biochem. Biophys. Res. Commun.* 181: 16–21.

Nickerson D.A., Taylor S.L., Weiss K.M., Clark A.G., Hutchinson R.G., Stengrd J., Salomaa V., Vartiainen E., Boerwinkle E., Sing C.F. (1998) DNA sequence diversity in a 9.7-kb region of the human lipoprotein lipase gene. *Nature Genet.* 19: 233–240.

Nickoloff J.A. (1992) Transcription enhances intrachromosomal homologous recombination in mammalian cells. *Mol. Cell. Biol.* 12: 5311–5318.

Nicolaides N.C., Kinzler K.W., Vogelstein B. (1995) Analysis of the 5' region of *PMS2* reveals heterogeneous transcripts and a novel overlapping gene. *Genomics* 29: 329–334.

Nishihara S., Narimatsu H., Iwasaki H., Yazawa S., Akamatsu S., Ando T., Seno T., Narimatsu I. (1994) Molecular genetic analysis of the human Lewis histo-blood group system. *J. Biol. Chem.* **269**: 29 271–29 278.

Nishio H., Takeshima Y., Narita N., Yanagawa H., Suzuki Y., Ishikawa Y., Ishikawa Y., Minami R., Nakamura H., Matsuo M. (1994) Identification of a novel first exon in the human dystrophin gene and of a new promoter located more than 500 kb upstream of the nearest known promoter. *J. Clin. Invest.* **94**: 1037–1042.

Nishiwaki K., Niikawa N., Ishikawa M. (1997) Polymorphic and tissue-specific imprinting of the human Wilms tumor gene, *WT1. Jpn. J. Hum. Genet.* **42**: 205–211.

Norton P.A. (1994) Alternative pre-mRNA splicing: factors involved in splice site selection. *J. Cell Sci.* **107**: 1–7.

Ogawa O., McNoe L.A., Eccles R., Morrison I.M., Reeve A.E. (1993) Human insulin-like growth factor type I and type II receptors are not imprinted. *Hum. Molec. Genet.* **2**: 2163–2165.

Ono M., Kawakami M., Takezawa T. (1987) A novel human nonviral retroposon derived from an endogenous retrovirus. *Nucleic Acids Res.* **15**: 8725–8737.

Orkin S.H. (1995) Regulation of globin gene expression in erythroid in erythroid cells. *Eur. J. Biochem.* **231**: 271–281.

Pabo C.O., Sauer R.T. (1992) Transcription factors: structural families and principles of DNA recognition. *Annu. Rev. Biochem.* **61**: 1053–1095.

Padgett R.A., Grabowski P.J., Konarska M.M., Seiler S., Sharp P.A. (1986) Splicing of messenger RNA precursors. *Ann. Rev. Biochem.* **55**: 1119–1150.

Pakula A.A., Sauer R.T. (1989) Genetic analysis of protein stability and function. *Ann. Rev. Genet.* **23**: 289–310.

Paranjape S.M., Kamakaka R.T., Kadonaga J.T. (1994) Role of chromatin structure in the regulation of transcription by RNA polymerase II. *Annu. Rev. Biochem.* **63**: 265–297.

Paz-Artal E., Corell A., Alvarez M., Varela P., Allende L., Madrono A., Rosal M., Arnaiz-Villena A. (1994) C4 gene polymorphism in primates: evolution, generation, and Chido and Rodgers antigenicity. *Immunogenetics* **40**: 381–396.

Pellerin I., Schnabel C., Catron K.M., Abate C. (1994) Hox proteins have different affinities for a consensus DNA site that correlate with the positions of their genes on the *hox* cluster. *Mol. Cell. Biol.* **14**: 4532–4545.

Pesole G., Liuni S., Grillo G., Saccone C. (1997) Structural and compositional features of untranslated regions of eukaryotic mRNAs. *Gene* **205**: 95–102.

Petit C., Levilliers J., Weissenbach J. (1988) Physical mapping of the human pseudoautosomal region; comparison with genetic linkage map. *EMBO J.* **7**: 2369–2376.

Pfeifer G.P., Steigerwald S.D., Hansen R.S., Gartler S.M., Riggs A.D. (1990) Polymerase chain reaction-aided genomic sequencing of an X chromosome-linked CpG island: methylation patterns suggest clonal inheritance, CpG site autonomy and an explanation of activity state stability. *Proc. Natl Acad. Sci. USA* **87**: 8252–8256.

Pigny P., Guyonnet-Duperat V., Hill A.S., Pratt W.S., Galiegue-Zouitina S., d'Hooge M.C., Laine A., van Seuningen I., Degand P., Gum J.R., Kim Y.S., Swallow D.M., Aubert J.-P., Porchet N. (1996) Human mucin genes assigned to 11p15.5: identification and organization of a cluster of genes. *Genomics* **38**: 340–352.

Polymeropoulos M.H., Xiao H., Sikela J.M., Adams M., Venter J.C., Merril C.R. (1993) Chromosomal distribution of 320 genes from a brain cDNA library. *Nature Genet.* **4**: 381–386.

Prats A.C., Vagner S., Prats H., Amalric F. (1992) *Cis*-acting elements involved in the alternative translation initiation process of human basic fibroblast growth factor mRNA. *Molec. Cell. Biol.* **12**: 4796–4805.

Ptashne M. (1988) How eucaryotic transcriptional activators work. *Nature* **335**: 683–689.

Ptashne M., Gann A.A.F. (1990) Activators and targets. *Nature* **346**: 329–331.

Pugh B.F., Tjian R. (1990) Mechanism of activation by Sp1: evidence for coactivators. *Cell* **61**: 1187–1197.

Qu L.-H., Nicoloso M., Michot B., Azum M.-C., Caizergues-Ferrer M., Renalier M.-H., Bachellerie J.-P. (1994) U21, a novel small nucleolar RNA with a 13 nt complementarity to 28S rRNA, is encoded in an intron of ribosomal protein L5 gene in chicken and mammals. *Nucleic Acids Res.* **22**: 4073–4081.

Razin A., Shemer R. (1995) DNA methylation in early development. *Hum. Mol. Genet.* **4**: 1751–1755.

Rein T., Zorbas H., DePamphilis M.L. (1997) Active mammalian replication origins are associated with a high-density cluster of mCpG dinucleotides. *Mol. Cell. Biol.* **17**: 416–426.

Reiss J., Cohen N., Dorche C., Mandel H., Mendel R.R., Stallmeyer B., Zabot M.-T., Dierks T. (1998) Mutations in a polycistronic nuclear gene associated with molybdenum cofactor deficiency. *Nature Genet.* **20**: 51–53.

Reiter L.T., Murakami T., Koeuth T., Pentao L., Muzny D.M., Gibbs R.A., Lupski J.R. (1996) A recombination hotspot responsible for two inherited peripheral neuropathies is located near a *mariner* transposon-like element. *Nature Genet.* **12**: 288–297.

Richards R.I., Sutherland G.R. (1992) Dynamic mutations: a new class of mutations causing human disease. *Cell* **70**: 709–712.

Rideout W.M., Coetzee G.A., Olumi A.F., Jones P.A. (1990) 5-Methylcytosine as an endogenous mutagen in the human LDL receptor and p53 genes. *Science* **249**: 1288–1290.

Riggins G.J., Lokey L.K., Chastain J.L., Leiner H.A., Sherman S.L., Wilkinson K.D., Warren S.T. (1992) Human genes containing polymorphic trinucleotide repeats. *Nature Genet.* **2**: 186–191.

Robertson H.M., Martos R. (1997) Molecular evolution of the second ancient human *mariner* transposon, Hsmar2, illustrates patterns of neutral evolution in the human genome lineage. *Gene* **205**: 219–228.

Robertson H.M., Zumpano K.L. (1997) Molecular evolution of an ancient *mariner* transposon, *Hsmar1*, in the human genome. *Gene* **205**: 203–217.

Ross J. (1996) Control of messenger RNA stability in higher eukaryotes. *Trends Genet.* **12**: 171–175.

Royle N.J., Baird D.M., Jeffreys A.J. (1994) A subterminal satellite located adjacent to telomeres in chimpanzees is absent from the human genome. *Nature Genet.* **6**: 52–56.

St Johnston D. (1995) The intracellular localization of messenger RNAs. *Cell* **81**: 161–170.

Saccone S., Caccio S., Kusuda J., Andreozzi L., Bernardi G. (1996) Identification of the gene-richest bands in human chromosomes. *Gene* **174**: 85–94.

Sachs A.B. (1993) Messenger RNA degradation in eukaryotes. *Cell* **74**: 413–421.

Samonte R.V., Ramesh K.H., Verma R.S. (1997) Comparative mapping of human alphoid satellite DNA repeat sequences in the great apes. *Genetica* **101**: 97–104.

Sassaman D.M., Dombroski B.A., Moran J.V., Kimberland M.L., Naas T.P., DeBerardinis R.J., Gabriel A., Swergold G.D., Kazazian H.H. (1997) Many human L1 elements are capable of retrotransposition. *Nature Genet.* **16**: 37–42.

Savatier P., Trabuchet G., Fauré C., Chebloune Y., Gouy M., Verdier G., Nigon V.M. (1985) Evolution of the primate beta-globin gene region. High rate of variation in CpG dinucleotides and in short repeated sequences between man and chimpanzee. *J. Mol. Biol.* **182**: 21–29.

Schinzel A., McKusick V.A., Francomano C., Pearson P.L. (1993) Report of the committee for clinical disorders and chromosomal aberrations. In: *Human Gene Mapping.* Johns Hopkins University Press, Baltimore, pp. 735–772.

Schmid CW. (1998) Does SINE evolution preclude *Alu* function? *Nucleic Acids Res.* **26**: 4541–4550.

Schneider A., Forman L., Westwood B., Yim C., Lin J., Singh S., Beutler E. (1998) The relationship of the -5, -8 and -24 variant alleles in African Americans to triosephosphate isomerase (TPI) enzyme activity and to TPI deficiency. *Blood* **92**: 2959–2962.

Schorderet D.F., Gartler S.M. (1992) Analysis of CpG suppression in methylated and nonmethylated species. *Proc. Natl Acad. Sci. USA* **89**: 957–961.

Scott A.F., Schmeckpeper B.J., Abdelrazik M., Comey C.T., O'Hara B., Rossiter J.P., Cooley T., Heath P., Smith K.D., Margolet L. (1987) Origin of the human L1 elements: proposed progenitor genes deduced from a consensus DNA sequence. *Genomics* **1**: 113–125.

Seboun E., Houghton L., Hatem C.J., Lincoln R., Hauser S.L. (1993) Unusual organization of the human T-cell receptor beta-chain gene complex is linked to recombination hotspots. *Proc. Natl Acad. Sci. USA* **90**: 5026–5029.

Setlow P. (1976) Nearest neighbour frequencies in deoxyribonucleic acids. In: *CRC Handbook of Biochemistry and Molecular Biology*, vol 2: *Nucleic Acids*, 3rd edn (ed. G.D. Fasman). CRC Press, Cleveland, Ohio, pp. 312–318.

Silva A.J., White R. (1988) Inheritance of allelic blueprints for methylation patterns. *Cell* **54**: 145–152.

Simmen M.W., Leitgeb S., Charlton J., Jones S.J.M., Harris B.R., Clark V.H., Bird A. (1999) Nonmethylated elements and methylated genes in a chordate genome. *Science* **283**: 1164–1169.

Singer M.F., Krek V., McMillan J.P., Swergold G.D., Thayer R.E. (1993) LINE-1: a human transposable element. *Gene* **135**: 183–188.

Sinnett D., Richer C., Deragon J.-M., Labuda D. (1992) *Alu* RNA transcripts in human embryonal carcinoma cells. *J. Mol. Biol.* **226**: 689–706.

Siomi H., Dreyfuss G. (1997) RNA-binding proteins as regulators of gene expression. *Curr. Opin. Genet. Devel.* **7**: 345–353.

Sippel A.E., Saueressig H., Winter D., Grewal T., Faust N., Hecht A., Bonifer C. (1992) The regulatory domain organization of eukaryotic genomes: implications for stable gene transfer. Chap. 1 In: *Transgenic Animals* (eds F. Grosveld and G. Kollias). Academic Press, London. pp. 1–26.

Skowronski J., Singer M.F. (1986) The abundant LINE-1 family of repeated DNA sequences in mammals: genes and pseudogenes. *Cold Spring Harb. Symp. Quant. Biol.* **51**: 457–464.

Skowronski J., Fanning T.G., Singer M.F. (1988) Unit-length LINE-1 transcripts in human teratocarcinoma cells. *Mol. Cell. Biol.* **8**: 1385–1397.

Smit A.F. (1993) Identification of a new, abundant superfamily of mammalian LTR-transposons. *Nucleic Acids Res.* **21**: 1863–1872.

Smit A.F., Riggs A.D. (1995) MIRs are classic, tRNA-derived SINEs that amplified before the mammalian radiation. *Nucleic Acids Res.* **23**: 98–102.

Smit A.F., Riggs A.D. (1996) Tiggers and other DNA transposon fossils in the human genome. *Proc. Natl Acad. Sci. USA* **93**: 1443–1448.

Speek M., Barry F., Miller W.L. (1996) Alternate promoters and alternate splicing of tenascin-X, a gene with 5′ and 3′ ends buried in other genes. *Hum. Molec. Genet.* **5**: 1749–1758.

Spitzner J.R., Muller M.T. (1988) A consensus sequence for cleavage by vertebrate DNA topoisomerase II. *Nucleic Acids Res.* **16**: 5533–5538.

Stallings R.L. (1994) Distribution of trinucleotide microsatellites in different categories of mammalian genomic sequence: implications for human genetic diseases. *Genomics* **21**: 116–121.

Stallmeyer B., Drugeon G., Reiss J., Haenni A.L., Mendel R.R. (1999) Human molybdopterin synthase gene: identification of a bicistronic transcript with overlapping reading frames. *Am. J. Hum. Genet.* **64**: 698–705.

Steinmetz M., Uematsu Y., Lindahl K.F. (1987) Hotspots of homologous recombination in mammalian genomes. *Trends Genet.* **3**: 7–10.

Stephens R.M., Schneider T.D. (1992) Features of spliceosome evolution and function inferred from an analysis of the information at human splice sites. *J. Mol. Biol.* **228**: 1124–1136.

Strachan T. (1992) *The Human Genome.* BIOS Scientific Publishers, Oxford.

Strachan T., Read A.P. (1996) *Human Molecular Genetics.* 1st edn., BIOS Scientific Publishers, Oxford.

Surralles J., Sebastian S., Natarajan A.T. (1997) Chromosomes with high gene density are preferentially repaired in human cells. *Mutagenesis* **12**: 437–442.

Swergold G.D. (1990) Identification, characterization and cell specificity of a human LINE-1 promoter. *Mol. Cell. Biol.* **10**: 6718–6729.

Takahara K., Hoffman G.G., Greenspan D.S. (1995) Complete structural organization of the human α1(V) collagen gene (*COL5A1*): divergence from the conserved organization of the other characterized fibrillar collagen genes. *Genomics* **29**: 588–597.

Takahashi N., Nagai Y., Ueno S., Saeki Y., Yanagihara T. (1992) Human peripheral blood lymphocytes express D5 dopamine receptor gene and transcribe the two pseudogenes. *FEBS Letts.* **314**: 23–25.

Tarn W.-Y., Steitz J.A. (1996) Highly diverged U4 and U6 small nuclear RNAs required for splicing rare AT-AC introns. *Science* **273**: 1824–1832.

Tarn W.-Y., Steitz J.A. (1997) Pre-mRNA splicing: the discovery of a new spliceosome doubles the challenge. *Trends Biochem. Sci.* **22**: 132–137.

Tasheva E.S., Roufa D.J. (1995) A densely methylated DNA island is associated with a chromosomal replication origin in the human RPS14 locus. *Somat. Cell Molec. Genet.* **21**: 369–383.

Tee M.K., Babalola G.O., Aza-Blanc P., Speek M., Gitelman S.E., Miller W.L. (1995) A promoter within intron 35 of the human C4A gene initiates abundant adrenal-specific transcription of a 1 kb RNA: location of a cryptic *CYP21* promoter element? *Hum. Molec. Genet.* **4**: 2109–2116.

Thompson C.C., McKnight S.L. (1992) Anatomy of an enhancer. *Trends Genet.* **8**: 232–236.

Timchenko L.T., Caskey C.T. (1996) Trinucleotide repeat disorders in humans: discussions of mechanisms and medical issues. *FASEB J.* **10**: 1589–1597.

Ting C.N., Rosenberg M.P., Snow C.M., Samuelson L.C., Meisler M.H. (1992) Endogenous retroviral sequences are required for tissue-specific expression of a human salivary amylase gene. *Genes Devel.* **6**: 1457–1465.

Tjian R., Maniatis T. (1994) Transcriptional activation: a complex puzzle with few easy pieces. *Cell* **77**: 5–8.

Torchia B.S., Call L.M., Migeon B.R. (1994) DNA replication analysis of FMR1, XIST and Factor 8C loci by FISH shows nontranscribed X-linked genes replicate late. *Am. J. Hum. Genet.* **55**: 96–104.

Tsukada T., Horovitch S.J., Montminy M.R., Mandel G., Goodman R.H. (1985) Structure of the human vasoactive intestinal polypeptide gene. *DNA* **4**: 293–300.

Turker M.S., Bestor T.H. (1997) Formation of methylation patterns in the mammalian genome. *Mutation Res.* **386**: 119–130.

Tweedie S., Charlton J., Clark V., Bird A. (1997) Methylation of genomes and genes at the invertebrate-vertebrate boundary. *Mol. Cell. Biol.* **17**: 1469–1475.

Twigg S.R.F., Burns H.D., Oldridge M., Heath J.K., Wilkie A.O.M. (1998) Conserved use of a non-canonical 5' splice site (/GA) in alternative splicing by fibroblast growth factor receptors 1, 2 and 3. *Hum. Molec. Genet.* **7**: 685–691.

Ullu E., Tschudi C. (1984) *Alu* sequences are processed 7SL RNA genes. *Nature* **312**: 171–172.

van Bokhoven H., Rawson R.B., Merkx G.F., Cremers F.P., Seabra M.C. (1996) cDNA cloning and chromosomal localization of the genes encoding the α- and β-subunits of human geranylgeranyl transferase: the 3' end of the α-subunit gene overlaps with the transglutaminase 1 gene promoter. *Genomics* **38**: 133–140.

van der Hoeven F., Zakany J., Duboule D. (1996) Gene transpositions in the *HoxD* complex reveal a hierarchy of regulatory controls. *Cell* **85**: 1025–1035.

Vansant G., Reynolds W.F. (1995) The consensus sequence of a major *Alu* subfamily contains a functional retinoic acid response element. *Proc. Natl Acad. Sci. USA* **92**: 8229–8233.

Verbovaia L.V., Razin S.V. (1997) Mapping of replication origins and termination sites in the Duchenne muscular dystrophy gene. *Genomics* **45**: 24–30.

Viskochil D., Cawthon R., O'Connell P., Xu G., Stevens J., Culver M., Carey J., White R. (1991) The gene encoding the oligodendrocyte-myelin glycoprotein is embedded within the neurofibromatosis type 1 gene. *Molec. Cell. Biol.* **11**: 906–912.

Vogel F., Motulsky A.G. (1986) *Human Genetics—Problems and Approaches* 2nd edn. Springer, Berlin.

Vogt P. (1990) Potential genetic functions of tandem repeated DNA sequence blocks in the human genome are based on a highly conserved "chromatin folding code". *Hum. Genet.* **84**: 301–336.

Volz A., Korge B.P., Compton J.G., Ziegler A., Steinert P.M., Mischke D. (1993) Physical mapping of a functional cluster of epidermal differentiation genes on chromosome 1q21. *Genomics* **18**: 92–99.

Wacey A.I., Krawczak M., Kakkar V.V., Cooper D.N. (1994) Determinants of the factor IX mutational spectrum in haemophilia B: an analysis of missense mutations using a multi-domain molecular model of the activated protein. *Hum. Genet.* **94**: 594–608.

Wada-Kiyama Y., Kiyama R. (1994) Periodicity of DNA bend sites in human ε-globin gene region. *J. Biol. Chem.* **269**: 22 238–22 244.

Wahle E. (1992) The end of the message: 3'-end processing leading to polyadenylated messenger RNA. *BioEssays* **14**: 113–118.

Wahle E., Keller W. (1992) The biochemistry of 3'-end cleavage and polyadenylation of messenger RNA precursors. *Annu. Rev. Biochem.* **61**: 419–440.

Wahls W.P., Wallace L.J., Moore P.D. (1990a) The Z-DNA motif d(TG)30 promotes reception of information during gene conversion events while stimulating homologous recombination in human cells in culture. *Molec. Cell. Biol.* **10**: 785–793.

Wahls W.P., Wallace L.J., Moore P.D. (1990b) Hypervariable minisatellite DNA is a hotspot for homologous recombination in human cells. *Cell* **60**: 95–103.

Wainscoat J.S., Hill A.V.S., Boyce A.L., Flint J., Hernandez M., Thein S.L., Old J.M., Lynch J.R., Falusi A.G., Weatherall D.J., Clegg J.B. (1986) Evolutionary relationships of human populations from an analysis of nuclear DNA polymorphisms. *Nature* 319: 491–493.

Walter W.R., Singh G.B., Krawetz S.A. (1998) MARs mission update. *Biochem. Biophys. Res. Commun.* 242: 419–422.

Wang Y., Macke J.P., Merbs S.L., Zack D.J., Klaunberg B., Bennett J., Gearhart J., Nathans J. (1992) A locus control region adjacent to the human red and green visual pigment genes. *Neuron* 9: 429–440.

Wang D.G., Fan J.-B., Siao C.-J., et al. (1998) Large-scale identification, mapping, and genotyping of single-nucleotide polymorphisms in the human genome. *Science* 280: 1077–1082.

Warburton P.E., Waye J.S., Willard H.F. (1993) Nonrandom localization of recombination events in human alpha satellite repeat unit variants: implications for higher order structural characteristics within centromeric heterochromatin. *Mol. Cell. Biol.* 13: 6520–6529.

Warburton P.E., Haaf T., Gosden J., Lawson D., Willard H.F. (1996) Characterization of a chromosome-specific chimpanzee alpha satellite subset: evolutionary relationship to subsets on human chromosomes. *Genomics* 33: 220–228.

Wasylyk B. (1988) Enhancers and transcription factors in the control of gene expression. *Biochim. Biophys. Acta* 951: 17–35.

Weiss A., McDonough D., Wertman B., Acakpo-Satchivi L., Montgomery K., Kucherlapati R., Leinwand L., Krauter K. (1999) Organization of human and mouse skeletal myosin heavy chain gene clusters is highly conserved. *Proc. Natl Acad. Sci. USA* 96: 2958–2963.

Wentz-Hunter K., Potashkin J. (1995) The evolutionary conservation of the splicing apparatus between fission yeast and man. *Nucleic Acids Symposium Series* 33: 226–228.

Wiborg O., Pedersen M.S., Wind A., Berglund L.E., Marcker K.A., Vuust J. (1985) The human ubiquitin multigene family: some genes contain multiple directly repeated ubiquitin coding sequences. *EMBO J.* 4: 755–759.

Wilde C.D. (1985) Pseudogenes. *CRC Crit. Rev. Biochem.* 19: 323–352.

Wilkinson D.A., Goodchild N.L., Saxton T.M., Wood S., Mager DL. (1993) Evidence for a functional subclass of the RTVL-H family of human endogenous retrovirus-like sequences. *J. Virol.* 67: 2981–2989.

Wolffe A. (1992) *Chromatin. Structure and Function.* Academic Press, London.

Wu Q., Krainer A.R. (1997) Splicing of a divergent subclass of AT-AC introns requires the major spliceosomal snRNAs. *RNA* 3: 586–601.

Xu W., Goodyer C.G., Deal C., Polychronakos C. (1993) Functional polymorphism in the parental imprinting of the human *IGF2R* gene. *Biochem. Biophys. Res. Commun.* 197: 747–754.

Yamamoto F., Clausen H., White T., Marken J., Hakomori S. (1990) Molecular genetic basis of the human histo-blood group ABO system. *Nature* 345: 229–233.

Yang A.S., Gonzalgo M.L. Zingg J.M., Millar R .P., Buckley J.D., Jones P.A. (1996) The rate of CpG mutation in *Alu* repetitive elements within the p53 tumor suppressor gene in the primate germline. *J. Mol. Biol.* 258: 240–250.

Yasunami M., Kikuchi I., Sarapata D.E., Yoshida A. (1990) The human class I alcohol dehydrogenase gene cluster: three genes are tandemly organized in an 80-kb-long segment of the genome. *Genomics* 7: 152–158.

Yoder J.A., Walsh C.P., Bestor T.H. (1997) Cytosine methylation and the ecology of intragenomic parasites. *Trends Genet.* 13: 335–340.

Zhang M.Q. (1998) Statistical features of human exons and their flanking sequences. *Hum. Molec. Genet.* 7: 919–932.

Zhuang Y., Goldstein A.M., Weiner A.M. (1989) UACUAAC is the preferred branch site for mammalian mRNA splicing. *Proc. Natl Acad. Sci. USA* 86: 2752–2756.

Zietkiewicz E., Yotova V., Jarnik M., Korab-Laskowska M., Kidd K.K., Modiano D., Scozzari R., Stoneking M., Tishkoff S., Batzer M., Labuda D. (1997) Nuclear DNA diversity in worldwide distributed human populations. *Gene* 205: 161–171.

Zuckerkandl E. (1997) Junk DNA and sectorial gene repression. *Gene* 205: 323–343.

2

Evolution of the human genome

2.1 Ancient genome duplications at the dawn of vertebrate evolution

2.1.1 Evidence for an ancient genome duplication

In evolution, novel genes have arisen either by whole genome duplication or by regionally localized duplication events (see Chapter 9, section 9.5). Both mechanisms give rise to *paralogous* genes, genes that occur within the same species and which have a common ancestor. Thus, members of multigene families and superfamilies are paralogous and are distinct from *orthologous* genes which are found in different species and have diverged from their common ancestor over evolutionary time. In practice, the consequences of whole genome duplication and regionally localized gene duplication are often very hard to distinguish. Skrabanek and Wolfe (1998) elegantly explained the problem by analogy:

> 'Take four, or maybe eight, decks of 52 playing cards. Shuffle them all together and then throw some cards away. Pick 20 cards at random and drop the rest on the floor. Give the 20 cards to some evolutionary biologists and ask them to figure out what you've done. For encouragement, tell them they can have the cards on the floor in 2005 (the estimated date of completion of the human genome sequence)'.

Ohno (1970) first proposed that the vertebrate genome had evolved to its present size through an ancient tetraploidization event. Evidence supporting this hypothesis now comes from a variety of different sources. Comparative nuclear DNA measurements in higher organisms are compatible with the idea of successive genome doublings (Ohno, 1970; Sparrow and Nauman, 1976). So are cytogenetic data, since the human karyotype can be divided into similar pairs of chromosomes on the basis of structure and banding patterns viz. chromosomes 1 and 2, 4 and 5, 7 and 8, 11 and 12, 14 and 15, 16 and 17, 19 and 20, 21 and 22 (Comings 1972; McKusick, 1980). The postulated ancient tetraploidization event also appears to be reflected in the genetic information content of these chromosome pairs. This is exemplified by the distribution of the insulin and *RAS* oncogene

Human Gene Evolution, David N. Cooper.
© 1999 BIOS Scientific Publishers Ltd, Oxford.

family members in the human genome: the insulin (*INS*), insulin-like growth factor 2 (*IGF2*) and Harvey RAS (*HRAS*) genes are located on chromosome 11p whilst the insulin-like growth factor 1 (*IGF1*) and Kirsten RAS (*KRAS2*) genes are located on chromosome 12 (Hoppener *et al.*, 1985).

That the genome duplications occurred prior to the adaptive radiation of the vertebrates is evidenced by the similar gene number exhibited by both fish and mammals (Elgar *et al.*, 1996), the survival of syntenic linkage groups over quite long periods of evolutionary time (Lundin 1993; Trower *et al.*, 1996) and the tetralogy exhibited by many vertebrate loci. This tetralogy is evident from the observation that the basic set of ~15 000 genes found in all primitive metazoans varies only moderately from *Caenorhabditis elegans* (Ahringer, 1997) to *Drosophila* (Miklos and Rubin, 1996) to *Ciona intestinalis*, a tunicate (Simmen *et al.*, 1998) but this gene complement is approximately four-fold smaller than the total number of genes in vertebrate genomes. At the level of the individual gene, there are numerous instances where an invertebrate (*Drosophila*) gene has up to four related genes (paralogues) in vertebrates, implying two rounds of genome duplication (Spring, 1997). These paralogues represent quadruplicated loci on different chromosomes that are more similar to each other than they are to members of other tetralogous groups. In humans, these so-called 'tetralogues' include the *HOX* genes (*HOXA*, 7p14-p15; *HOXB*, 17q21-q22; *HOXC*, 12q12-q13; *HOXD*, 2q31), the epidermal growth factor receptor genes (*EGFR*, 7p12; *ERBB2*, 17q11.2-q12; *ERBB3*, 12q13; *ERBB4*, 2q34), the Jak family of tyrosine kinases (*JAK1*, 1p31.3–32.3; *JAK2*, 9p24; *JAK3*, 19p12-p13; *TYK2*, 19p13.2), the MADS box enhancing factors (*MEF2A*, 15q26; *MEF2B*, 19p12; *MEF2C*, 5q14; *MEF2D*, 1q12-q23), the Src family of nonreceptor tyrosine kinases (*SRC*, 20q11.2; *YES1*, 18p11.22-p11.31; *FGR*, 1p36.1-p36.2; *FYN*, 6q21) and the syndecans (*SDC1*, 2p; *SDC2*, 8q22-q23; *SDC3*, 1p32; *SDC4*, 20q12-q13) (Spring 1997; Pebusque *et al.*, 1998). Many further examples of triplicated loci occur in the human genome (e.g. Katsanis *et al.*, 1996; Plummer and Meisler, 1999). In these cases, it may be that one paralogue has been lost or alternatively, still remains to be characterized (Spring, 1997).

The results of studies of the genes encoding the homeobox proteins, insulin-like growth factor genes and high mobility group proteins (Chan *et al.*, 1990; Garcia-Fernandez and Holland, 1994; Holland, 1991; Pendleton *et al.*, 1993; Sharman *et al.*, 1996; see Chapter 4, sections 4.2.1, *Homeobox genes* and 4.2.3, *Insulin and insulin-like growth factor genes*) in the primitive chordates *Amphioxus* and *Ciona*, and a jawless vertebrate *Lampetra fluviatilis* (lamprey), are consistent with the occurrence of one genome duplication in the common ancestor of all jawed and jawless vertebrates after the lineage leading to *Amphioxus* had diverged, and a second genome duplication occurring in the common ancestor of jawed vertebrates after their divergence from the jawless vertebrates (Sidow, 1996). Studies of *HOX* gene number suggest that the duplication events are likely to have occurred before the radiation of the teleosts (Amores *et al.*, 1998).

Gene number does not however automatically distinguish between tandem duplications and polyploidization events. Postlethwait *et al.* (1998) mapped 144 zebrafish (*Danio rerio*) genes; comparison of the resulting map with their mammalian counterparts led to the identification of orthologous chromosome segments for at least three chromosome paralogy groups in zebrafish and mammals. This finding is consistent with the hypothesis that these segments were duplicated prior to the divergence of zebrafish and mammals. The presence of more than two copies of each paralogous chromosomal segment, is suggestive of at least two rounds of duplication which would have occurred after the divergence of the cephalochordates and cranial chordates, but before the divergence of the ray-finned and lobe-finned fishes, which is thought to have occurred about 420 Myrs ago.

Several extensive regions of paralogy have been identified in the human genome which have been claimed to result from ancient tetraploidization events. The 13 groups of paralogous genes found on chromosomes 4 and 5 (*Table 2.1*) provide one example (Lundin, 1993). Lundin (1993) identified several other possible examples of paralogous pairs or groups of genes on different human chromosomes: (i) parts of chromosomes 2, 7, 12, 14, and 17, (ii) parts of chromosomes 8, 10, and 16, and (iii) parts of chromosomes 1, 11, 12, 15, and 19 (*Table 2.2*). Although the extensive paralogy noted between chromosomes 11 and 12 is explicable by a model of chromosome duplication resulting from tetraploidization, there are some discrepancies in the locations of genes on these chromosomes. These can however be accounted for by the occurrence of a pericentric inversion on chromosome 12.

Paralogy may be explained by mechanisms other than tetraploidization. Indeed, some paralogous gene loci are explicable by regional duplication (see Chapter 9, section 9.5); *Table 2.3* lists those identified on human chromosome 1. The relative importance of regional duplication/translocation as compared to tetraploidization is unclear and ambiguity even extends to individual cases. Thus, the relative locations of the tyrosine hydroxylase (*TH*; 11p15.5), tryptophan hydroxylase (*TPH*; 11p14.3-p15.1) and phenylalanine hydroxylase (*PAH*; 12q22-q24) genes have been explained in terms of both mechanisms (Craig *et al.*, 1986; Ledley *et al.*, 1987; Lundin, 1993).

The two highly related regions on the proximal and distal long arms of human chromosome 21 (21q22.1 and 21q11.2) appear to have arisen as a result of an intra-chromosomal duplication of >200 kb (Dutriaux *et al.*, 1994). This duplication is thought to have arisen between 15 and 30 Myrs ago after the separation of the orangutan from the other great apes (Orti *et al.*, 1998). By contrast, the origin of the paralogous 2–20 Mb segments on human chromosomes 1, 6, and 9 (Banyer *et al.*, 1998; Endo *et al.*, 1998; Katsanis *et al.*, 1996) is unclear. Regardless of the mechanism, at least two intra-chromosomal duplications must have occurred resulting in the triplication of a series of genes, for example the retinoid X receptor genes *RXRG* (1q22-q23), *RXRB* (6p21.3) and *RXRA* (9q34.31), the pre-B cell leukemia transcription factor genes *PBX1* (1q23), *PBX2* (6p21.3), and *PBX3* (9q34), and the tenascin genes *TNR* (1q25-q31), *TNXA* (6p21.3), and *HXB* (9q32-q34) (Katsanis *et al.*, 1996). Interestingly, *Alu*- and LINE-dense clusters flank the boundaries of the 6p21.3 segment, a finding which may be significant in view of the recombinogenic potential of these sequence elements. In this context, it may be significant that a sequence related to the pseudoautosomal boundary of the human sex chromosomes (see Section 2.3.4) has also been noted at the centromeric boundary of the 6p21.3 segment (Fukagawa *et al.*, 1996).

2.1.2 Consequences of genome duplications for gene evolution

In principle, the genetic redundancy created by a genome duplication would have

Table 2.1. Possible paralogies between parts of human chromosomes 4 and 5 (after Lundin *et al.*, 1993)

4	5
FGFR3	FGFR4
	HTR1A
	ADRB2
ADRA2C	ADRA1B
DRD5	DRD1
QDPR	DHFR
GABRA2	GABRA1
GABRB1	
STATH	SPARC
KIT	PDGFRB
PDGFRA	CSF1R
AREG	C7
EGF	C9
AGA	HEXB
FGF5	
FGF2	FGF1
IF	F12
F11	GZMA
KLK3	
	CSF2
IL2	IL3
	IL4
	IL5
	IL9
	CSF1
MLR	GRL
ANX3	
ANX5	ANX6

allowed evolutionary experimentation, in that while one gene copy continued to function as before, the other was freed to acquire mutations, irrespective of whether they were adaptive or inactivating (Ohta, 1989; 1991). If the newly duplicated gene acquired mutations that modified either the expression pattern of the encoded gene or the function of the encoded protein in an advantageous way, the novel allele could have become fixed in the population. Ohno (1970) expressed this idea rather elegantly:

> An escape from the ruthless pressure of natural selection is provided by the mechanism of gene duplication. By duplication, a redundant copy of a locus is created. Natural selection often ignores such a redundant copy, and, while being ignored, it accumulates formerly forbidden mutations and is reborn as a new gene locus with a hitherto non-existent function. Thus, gene duplication emerges as the major force of evolution.

In this context, evidence for positive selection (Chapter 7, section 7.1.3) has come from the observation of accelerated evolution in some genes subsequent to gene

Table 2.2. Possible paralogies between parts of human chromosomes 1, 11, 12, 15 and 19 (after Lundin *et al.*, 1993)

1	11	12	15	19
RYR2		CACNA1C		RYR1
		BCAT1		BCAT2
EN01		EN02		
GDH			SORD	
EBVS1	EBVM1			
				CEAL1
				CEA
PTPRF	NCAM1			MAG
				NCA
				PSG1–13
TNFR2		TNFR1		
SLC2A1		SLC2A3		
SLC2A5				
GOT2L1		GOT2L3		
GOT2L2				
GNAI3		GNA12L		
GNAT2				
GNB1		GNB3		
	LMO1	LMO3		
	LMO2			
MYCL1		MYF5		LYL1
MYOG	MYOD1	MYF6		TCF3
CHRM3	CHRM1,4		CHRM5	
	FGF3	FGF6		
	FGF4			
		A2M		
		PZP		
	ESA4	ELA1		C3
	F2	C1S		KLK1
		C1R		KLK2
			LIPC	LIPE
		TPI1	MPI	GPI
LDHAL2	LDHA, LDHC	LDHB		
NRAS	HRAS	KRAS2		
RAP1A		RAP1B		RRAS
RAB3B				RAB3A
RAB4				
	CALCA	IAPP		
	CALCB			
	PTH	PTHLH		
FGR				
LCK	SEA		FES	
ABL2				
JUN				JUNB
				JUND

Table 2.2 *continued*

1	11	12	15	19
INSRR		ERBB3	IGF1R	INSR
NTRK1			LTK	TYK2
C8A, C8B		LRP1	THBS1	LDLR
GSTM1	GSTP1	MGST1		
C1QA, C1QB				
COL11A1		COL2A1		
COL8A2				
H1F2		H1F4		
	PGR	HMR		
		RARG	CRABP1	
		VDR		
HKR3	WT1	GLI		HKR1
				HKR2
PEPC		PEPB	ANPEP	PEPD
KCNC4		KCNA1,2,5		KCNA7
		KCNC2		KCNC3
ACADM		ACADS	IVD	
	TH, TPH	PAH		
NGFB	INS			
	IGF2	IGF1		
	INSL2			
	TRV2	TRV3		
	FRV1	FRV3		
PLA2G2A		PLA2G1B		
PLA2G2C				
TSHB	FSHB			LHB
				CGB
	ACP2		ACP5	
PKLR			PKM2	
ATP1A1				ATP1A3
ATP1A2				
ATP1B1				
ATP1AL2		ATP2A2		
ATP2B2		ATP2B1		
TRAP2		TRA1	TRAP1	
CD48	THY1			
	CD3D			
	CD3E			
CD3Z	CD3G			A1BG
CD1A-E	FCER1B			
FCER1A				
FCER1G				FCER2
FCGR2A, 2B,				
2C, 3A, 3B		CD4		
PIGR				
	APOA4			APOC1
APOA2	APOA1			APOC2
	APOC3			APOE

Table 2.2 *continued*

1	11	12	15	19
	FUT4			FUT1
	SIAT4C			FUT2
			MANA1	MANB
			CKMT1	CKM
	GANAB		GANC	
CAPN2	CAPN1			
			CAPN3	CAPN4
MUC1	MUC2, MUC5B MUC5AC			
AT3	C1NH			
AGT				
PFKM		PFKX		
FDPSL1			CHR39B	
ACTA1			ACTC	
POU2F1				POU2F2
TNNI1				TNNT1
SNRPE			SNRPN	SNRPA
				SNRP70
TGFB2				AMH
				TGFB1
CTSE	CTSD		CTSH	
REN	PGA3–5			

duplication (Burger *et al.*, 1994; Hill and Hastie 1987; Kurihara *et al.*, 1997; Ohta 1994; Wallis 1993). In such studies, positive selection is implicated in the process of evolutionary change by the observation of a higher frequency of non-synonymous over synonymous substitutions (Hughes and Nei, 1988). Changes in the expression patterns of the duplicated genes are sometimes also apparent as in the neuronal and muscle expressed genes of the nicotinic acetylcholine receptor family (Le Novre and Chaneux, 1995). In principle, gene conversion (see Chapter 9, section 9.5) may either promote diversification of proteins encoded by duplicated genes (Ohta and Basten, 1992) or promote homogenization (Sidow and Thomas, 1994) in which case the molecular evolutionary record is automatically erased. In practice, however, at least for the HLA and immunoglobulin genes, gene conversion is notable more by its absence: instead new genes tend to be created by a 'birth-and-death' process of duplication and deletion (Nei *et al.*, 1997).

A more likely scenario, however, is that the duplicated gene rapidly acquires inactivating mutations and becomes a pseudogene. Indeed, assuming that a gene duplication is not selectively disadvantageous, the duplicated genes can survive in the genome for quite long periods (Clark, 1994; Loomis and Gilpin, 1986; Nowak *et al.*, 1997). Arguably the best available model system to assess whether duplicated genes will be retained or inactivated over evolutionary time is yeast (*Saccharomyces cerevisiae*), the organism with the best characterized genome

Table 2.3. Locations of human gene loci indicating large regional duplications of chromosome 1 (after Lundin *et al.*, 1993)

TRN	tRNA, asparagine	1p36
TRNL	tRNA, asparagine-like	1q12-q22
TRE	tRNA, glutamic acid	1p36
TREL1	tRNA, glutamic acid-like 1	1q21-q22
RNU1	Small nuclear U1 RNA	1p36
RNU1P1-4	Small nuclear U1 RNA pseudogenes 1-4	1q12-q22
FGR	Feline sarcoma oncogene	1p36
LCK	Lymphocyte protein tyrosine kinase	1p32-p35
ABLL	Abelson murine leukemia oncogene-like	1q24-q25
C8A, C8B	Complement component, 8, α/β chains	1p22-p36
C1QA, C1QB	Complement component 1q, α/β chains	1p
C4BPA, C4BPB	Complement component 4 binding protein	1q32
CR1, CR2	Complement component receptors 1/2	1q32
AK2	Adenylate kinase 2	1p34
GUK1	Guanylate kinase 1	1q32-q42
GOT2L1	Glutamic-oxaloacetic transaminase 2-like 1	1p32-p33
GOT2L2	Glutamic-oxaloacetic transaminase 2-like 2	1q25-q31
RAB3B	Member RAS oncogene family	1p31-p32
NRAS	Neuroblastoma RAS oncogene	1p13
RAP1A	Member RAS oncogene family	1p12-p13
RAB4	Member RAS oncogene family	1q42-q43
FTHL1	Ferritin, heavy polypeptide-like 1	1p22-p31
FTHL2	Ferritin, heavy polypeptide-like 2	1q32-q42
CD58	Lymphocyte function-associated antigen	1p13
DAF	Decay accelerating factor for complement	1q32
ATP1A1	ATPase, Na$^+$K$^+$, α1 polypeptide	1p13
ATP1A2	ATPase, Na$^+$K$^+$, α2 polypeptide	1q21-q23
ATP1B1	ATPase, Na$^+$K$^+$, β polypeptide	1q22-q25

duplication. Sequencing data from the 12 Mb yeast genome (Mewes *et al.*, 1997) are consistent with the occurrence of a whole genome duplication that occurred ~100 Myrs ago, after the divergence of *S. cerevisiae* from *Kluyveromyces* (Wolfe and Shields 1997). Most duplicated genes were subsequently deleted; only 13% of yeast proteins are now represented as homologous pairs encoded by homologous genes. Of these homologous gene pairs, only a few possess functions that have clearly diverged under the influence of selection viz. the mitochondrial and peroxisomal isozymes of citrate synthase (*CIT1* and *CIT2*), *RAS1* and *RAS2*, the transcription factors *ACE2* and *SWI5*, the phosphatidylinositol kinases *TOR1* and *TOR2*, and the myosins *MYO3* and *MYO5*. Conclusions drawn from yeast may not however be applicable to vertebrates and in this phylum, many more genes may have survived the aftermath of duplication to acquire new functions. Thus, Nadeau and Sankoff (1997) analyzed the frequency distribution of family size for gene families present in humans and mice and which arose putatively by genome duplication early in vertebrate evolution. They concluded that duplicated genes were as likely to have survived and acquired a novel function as to have been lost through the acquisition of inactivating mutations. In agreement with these findings, studies of fish (Bailey *et al.*, 1978) and *Xenopus* (Hughes and Hughes, 1983) have both suggested that about 50% of newly duplicated genes are retained after tetraploidization.

2.2 Mammalian genome evolution

The best estimates of divergence times for the various mammalian orders and the other major vertebrate lineages have come from the use of extant gene sequences to calibrate a 'molecular clock' of vertebrate evolution (Kumar and Hedges, 1998). Gene-specific evolutionary rates often vary quite dramatically (see Chapter 7, section 7.1.3) but the use of multiple genes to derive mean divergence times should yield more accurate and reliable estimates. Kumar and Hedges (1998) therefore employed 658 genes from 207 vertebrate species to derive a molecular timescale for vertebrate evolution (*Figure 2.1*). The divergence times corresponded well to previous estimates based upon the fossil record. Thus the calculated divergence time for the jawless fish (Agnatha) was 564 Myrs ago in the Precambrian era. Interestingly, the molecular data indicated that at least five major lineages of placental mammals [Edentata (armadillos, anteaters and sloths), Hystricognathi (porcupines and guinea pigs), Sciurognathi (squirrels), Paenungulata (hyraxes) and Ferungulata (carnivores)] could have arisen in the early to middle Cretacious between 130 and 90 Myrs ago. This represents an important revision of previous estimates of the timing of the adaptive radiation of the mammals [estimated by Novacek (1992) to have occurred between 100 Myrs and 65 Myrs ago]. Since this now appears to have predated the Cretaceous/Tertiary extinction of the dinosaurs 65 Myrs ago, the adaptive radiation of the mammals could not have been simply a consequence of the filling of niches vacated by the departing super-lizards. Other factors such as climatic change and the continental breakup must also have played a role (Hedges *et al.*, 1996). This question notwithstanding, the adaptive radiation of the mammals has been very successful, resulting in the emergence of >4600 living species that occupy a very diverse range of habitats and environments.

It has been suggested that the rate of mammalian speciation may have been influenced by the rate of karyotypic change (Bush *et al.*, 1977; Qumsiyeh, 1994; Wilson *et al.*, 1977). However, mammalian genomes still contain significant regions of genetic linkage that have been conserved through evolutionary time. Indeed, appreciation of the phenomenon of linkage group conservation between different mammalian species has led to the identification and chromosomal localization of novel genes in the human genome. Such studies also promise to aid greatly our understanding of mammalian genome evolution by, for example, revealing chromosomal inversions or translocations, duplications of genes or gene regions, or more subtle lesions that may have led to the functional divergence and specialization of proteins.

Genomic mapping projects are underway for a variety of mammals including the mouse, rat, cow, sheep, and pig but by far the most data are available for the mouse (Edwards *et al.*, 1994; Eppig, 1996; Nadeau *et al.*, 1995; Nadeau and Sankoff, 1998; O'Brien *et al.*, 1993). Comparative mapping data for a range of mammals including a number of primates are available at **http://www.informatics.jax.org/homology.html**. The comparative analysis of the human and murine

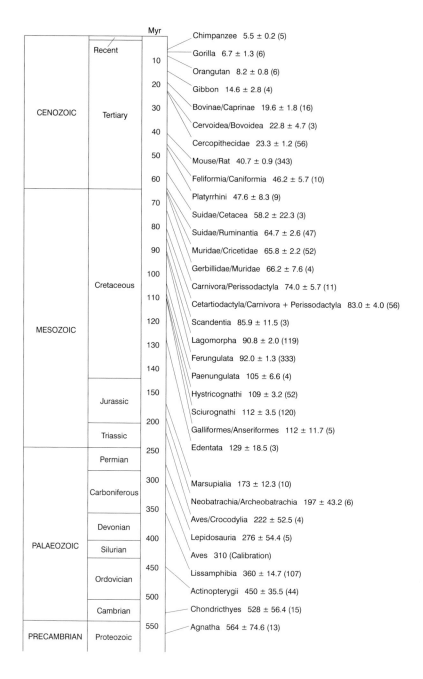

Figure 2.1. A molecular timescale for vertebrate evolution (after Kumar and Hedges 1998). All times indicate Myrs separating humans (or the largest sister group containing humans) and the group shown, except when the comparative groups are separated by a slash (/). Time estimates are shown with ±s.e.m. and the number of genes used is given in parentheses. Three groups of mammalian orders are Archonta (Primates, Scandentia, Dermoptera, Chiroptera and Lagomorpha), Ferungulata (Carnivora, Cetartiodactyla and Perissodactyla) and Paenungulata (Hyracoidea, Proboscidea, Sirenia).

genomes is especially important because the mouse represents a genetic system with a large number of spontaneous mutants (some of which are relevant to the study of human genetic disease), numerous inbred strains and enormous potential for the application of transgenic technology (Rubin and Barsh, 1996).

The latest comparative genetic map for human and mouse contains nearly 1800 genes mapped to over 200 different syntenic chromosomal groups [DeBry and Seldin, 1996; **http://www.ncbi.nlm.nih.gov/Homology/**]. Despite a timespan of between 100 (Kumar and Hedges, 1998) and 80 Myrs (Collins and Jukes, 1994) since the divergence of the two species, some syntenic groups have maintained gene content, order and spacing over considerable genetic and physical distances, for example the 1q21-q23 region of human and mouse chromosomes 1 (Oakey *et al.*, 1992) and a portion of mouse chromosome 2 with the entire human chromosome 20 (DeBry and Seldin, 1996). Other syntenic groups, by contrast, have accumulated substantial differences in gene order between the two species, for example human chromosome 19q13 and a segment of the homologous murine chromosome 7 which exhibit nine separate conserved linkage groups (DeBry and Seldin, 1996). The human and mouse X chromosomes contain a minimum of eight conserved syntenic groups (Blair *et al.*, 1994). The possible events which may have led to the current arrangement of homologous segments on the human and murine X chromosomes are shown in *Figure 2.2* and discussed further in Section 2.3.4.

Syntenic regions need not, however, necessarily be comparable in size. Thus, the human T-cell receptor β (*TCRB*; 7q35) region (800 kb in length) is considerably larger than its murine counterpart (500 kb) as a result of repeated duplications of specific Vβ segments in the primate lineage (Hood *et al.*, 1993). Similarly, the human HLA class II region (6p21.3) at ~900 kb is approximately three times the length of its mouse equivalent as a result of the duplication of specific HLA-DP, -DQ and -DR members of this multigene family in the primate lineage (Amadou *et al.*, 1995; Hanson and Trowsdale, 1991). By contrast, the class III HLA regions of human and mice are remarkably similar in structure (Peelman *et al.*, 1996).

Many syntenic blocks extend across human centromeres (Moseley and Seldin, 1989). Thus, gene order in the pericentric regions of human chromosomes 1, 2, 4, 6, 11, and 19 is conserved in the homologous regions of murine chromosomes 3, 6, 5, 1, 2, and 8, respectively (DeBry and Seldin, 1996). The majority of rearrangements appear to be due to inversions within chromosomes rather than rearrangements between chromosomes but more detailed mapping data will be required to obtain the map resolution necessary for a definitive assessment. One *caveat* which should be borne in mind is that the linkage map may have been broken up to a larger extent in rodents than in primates as compared to the ancestral mammalian genome (Lundin, 1993). One way to study this type of rearrangement is by the comparative analysis of the chromosomal locations of the individual gene loci involved. For example, Maresco *et al.* (1998) determined the locations of the high-affinity immunoglobulin receptor genes (*FCGR1A*, 1q21; *FCGR1B*, 1p12; *FCGR1C*, 1q21) in the rhesus monkey (*Macaca mulatta*), baboon (*Papio papio*) and chimpanzee (*Pan troglodytes*) thereby providing evidence for the occurrence of two pericentric inversions during the evolution of human chromosome 1.

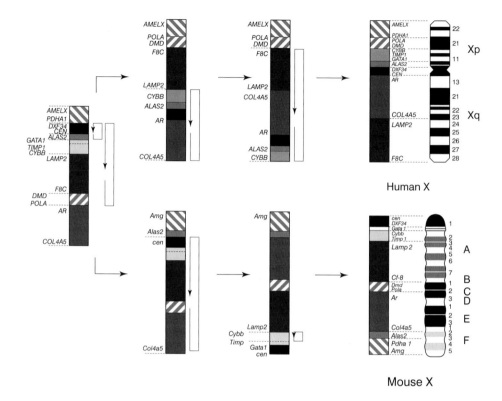

Figure 2.2. Possible events that may have led to the current arrangement of homologous blocks on the human and mouse X chromosomes (from Blair *et al.*, 1994). On the left, a postulated ancestral X chromosome is shown. Three inversion events are indicated. Each homologous block is shaded in the same manner throughout and the proximal and distal loci that define the blocks are given. Dotted lines between loci indicate the positions of evolutionary breakpoints. The comparative maps of the human and mouse X chromosomes are shown alongside their cytogenetic maps.

Conservation of synteny may extent beyond the mammals. For example, 11/18 genes from the chicken Z chromosome have orthologues on human chromosome 9pter-q22, albeit in a different order (Nanda *et al.*, 1999). What are the reasons for this degree of conservation? One reason may have been functional, for example in order to allow coordinate regulation of the genes involved (see Chapter 8, section 8.5) or in order to avoid possible meiotic disturbance consequent to a major chromosomal rearrangement. In the case of the HLA system, synteny may have indirectly promoted the generation of diversity by optimizing the potential for gene conversion. Alternatively, it is possible that insufficient time has passed for ancestral linkages to have been broken up completely. Synteny may be conserved in one phylogenetic group but not in another. Thus, the surfeit genes are tightly clustered in human (*SURF1, SURF2, SURF3, SURF4, SURF5*; 9q34.1) and chicken but not in invertebrates where the *Surf* genes are unlinked (Duhig *et al.*, 1998).

Assuming a human–rodent divergence time of 80 Myrs, Collins and Jukes (1994) estimated the rate of silent substitution to be 2.9×10^{-9} site^{-1} year^{-1}.

However, this is very much an average figure arrived at by comparison of the 4-fold degenerate sites in the protein-coding sequences of 337 human/rodent gene comparisons. Sequence conservation does vary quite considerably between proteins: indeed a study of 1196 orthologous mouse and human protein sequences revealed sequence conservation of between 36% and 100% with an average of 85% (Makalowski *et al.*, 1996).

Large scale genomic DNA sequence comparisons between human and mouse are still difficult owing to the paucity of orthologous pairs of sequences >20kb in length. However, Koop (1995) noted three distinct patterns of sequence divergence in the noncoding DNA sequences of human and rodent genomes:

(i)   A high level of sequence similarity in gene regions contrasting with divergent non-coding regions, for example β-globin (*HBB*; 11p15.5) and γ-crystallin (*CRYGA*; 2q33-q35) genes.

(ii) A conserved pattern of sequence similarity in noncoding regions, for example T-cell receptor-α (*TCRA*; 14q11.2) and -δ (*TCRD*; 14q11.2) genes and α- and β-myosin heavy chain (*MYH6, MYH7*; 14q11.2-q13) genes (Koop and Hood, 1994). At least some of this conservation may be attributed to regions which bind T-cell nuclear proteins which may play a role in the control of gene transcription (Koop *et al.*, 1994).

(iii) A mixed pattern of sequence similarity, for example the immunoglobulin heavy chain J-Cμ-Cδ gene region (14q32.33): the J-Cμ portion exhibits ~64% sequence homology between human and mouse whilst the Cδ region shows little if any sequence conservation.

This 'mosaic model' of genome evolution may reflect differing rates of mutation or differential repair efficiencies between different regions of the genome. As the sequences of further syntenic regions become available [e.g. the Bruton's tyrosine kinase (*BTK*; Xq21.33-q22) gene region; Oeltjen *et al.*, 1997], this question can be addressed.

2.3 Primate evolution

> I believe that our Heavenly Father invented man because he was disappointed in the monkey.
> Mark Twain

2.3.1 Adaptation and adaptive radiation

The order of primates originated in the Palaeocene some 50–60 Myrs ago and contains the monkeys, apes and humans (Martin 1993; *Figure 2.3*). Originally adapted for arboreal life, extant primates include low canopy runners (guenons), high canopy acrobats (spider monkeys) and the brachiating great apes whilst some have become exclusively terrestrial (baboons, mandrills, and humans). These habits are responsible for the adaptation of the skeleto-muscular system to allow jumping, swinging and grasping. Primates exhibit a wide variety of characteristics which have equipped them to exploit their various niches optimally.

The Anthropoidea, with their relatively large brain size, possess binocular vision and high visual acuity but a relatively poor olfactory sense. Their high intelligence allows rapid and measured reaction to external stimuli. The extension of the time periods for gestation, developmental growth and parental care are associated with increased powers of learning whilst communication by use of elaborate vocal signals has allowed the development of complex patterns of social behavior. The menstrual cycle with ovulation has been accompanied by the development of sexual behavior and signaling in the female. Finally, the omnivorous diet of primates may have been responsible for the development of their characteristic manual dexterity which has allowed them both to explore and manipulate their environment.

The order Primates contains about 200 living species which are divided into two suborders: the Prosimii and the Anthropoidea (*Table 2.4*). The Prosimii, which originated in the Palaeocene (*Figure 2.3*), have retained the insectivoran characteristics of long face, lateral eyes and small brain. The Anthropoidea comprise the Old World monkeys, the New World monkeys and the great apes. The New World or platyrrhine (flat nosed) monkeys are thought to have been isolated since the Eocene (*Figure 2.3*). The Old World or catarrhine monkeys share a common ancestor in the late Eocene (*Figure 2.3*) and do not differ markedly in either habits or organization from the New World monkeys. The great apes include the gibbon and orangutan from East Asia and the chimpanzee and gorilla from Africa.

2.3.2 Primate phylogeny

> The fact that we are able to classify organisms at all in accordance with the structural characteristics which they present is due to the fact of their being related by descent.
> D.W. Thompson (1917)

The phylogeny of the hominoid primates was initially investigated by means of DNA-DNA hybridization (Sibley and Ahlquist, 1984, 1987). These studies employed single copy nuclear DNA to calculate the temperature ($T_{50}H$) in degrees Celsius at which 50% of all single copy DNA sequences were in the hybrid form and 50% had dissociated (*Figure 2.4*). The delta $T_{50}H$ between chimpanzee and human is 1.6. Assuming a relationship of delta $T_{50}H = 1\%$ base mismatches, this translates into $\sim 3.2 \times 10^7$ mismatches between the chimpanzee and human genomes. Sibley and Ahlquist (1987) estimated times of divergence for higher primates as: Old World monkeys, 25–34 Myrs ago; gibbons, 16.4–23 Myrs ago; orangutan, 12.2–17 Myrs ago; gorilla, 7.7–11 Myrs ago, chimpanzees-humans, 5.5–7.7 Myrs ago. It is now recognized that the chimpanzee is actually represented by two distinct species, the common chimpanzee (*Pan troglodytes*) and the pigmy chimpanzee (*Pan paniscus*) which diverged from each other ~2.3 Myrs ago.

The studies of Sibley and Ahlquist (1984, 1987) received support from Caccone and Powell (1989). However, the validity of conclusions drawn from DNA–DNA hybridization data has been challenged and the interpretation of these studies is still somewhat contentious (Marks *et al.*, 1988; Sarich *et al.*, 1989; Sibley *et al.*, 1990). Sibley and Ahlquist's scheme is nevertheless in broad agreement with the fossil record, comparative morphology, immunological studies (Gingerich, 1984)

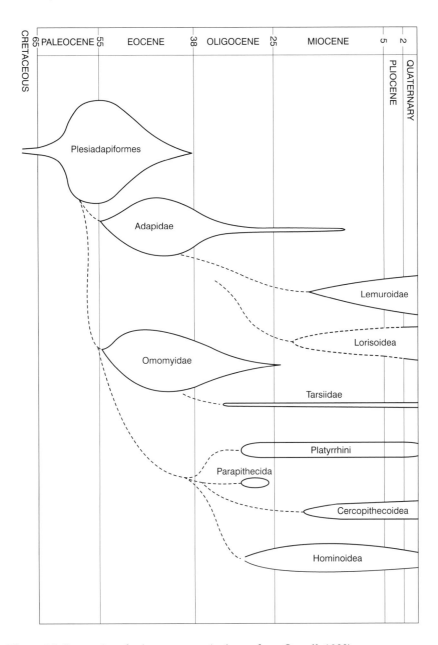

Figure 2.3. Longevity of primate groups (redrawn from Carroll, 1988)

and chromosome phylogeny (Yunis and Prakash, 1982; see Section 2.3.3) as well as being compatible with studies of individual DNA sequences. Thus, a similar picture has emerged from maximum parsimony analysis of DNA sequences derived from primate β-globin gene regions (Goodman *et al.*, 1989, 1990; Hasegawa *et al.*, 1987; Koop *et al.*, 1989; Maeda *et al.*, 1988; Miyamoto *et al.*, 1987, 1988), the c-*myc* oncogene (Mohammad-Ali *et al.*, 1995), the ribosomal DNA genes (Gonzalez *et al.*, 1990), the α-1,3-galactosyltransferase gene (Galili and

Table 2.4. A simplified classification of the primates (derived from Young, 1973 and Fleagle, 1988)

Order: Primates
  Suborder: Prosimii

    Infraorder: Lemuriformes
      Superfamily: Lemuroidea
        Family: Omomyidae
        Family: Adapidae (extinct)
        Family: Lemuridae (lemurs)
        Family: Indriidae (indris)
        Family: Daubentoniidae [aye-aye (*Daubentonia*)]
        Family: Lepilemuridae (*Lepilemur*)
      Superfamily: Lorisoidea
        Family: Cheirogaleidae
        Family: Galagidae [bushbaby (*Galago*)]
        Family: Lorisidae (*Loris*, potto)
    Infraorder: Tarsiiformes
      Family: Tarsiidae (tarsier)

  Suborder: Anthropoidea
    Infraorder: Platyrrhini (New World monkeys)
      Superfamily: Ceboidea
        Family: Callitrichidae (marmosets)
        Family: Cebidae [capuchins (*Cebus*), owl monkeys (*Aotus*)]
        Family: Atelidae [spider monkeys (*Ateles*), howler monkeys (*Alouatta*)]
    Infraorder: Catarrhini (Old World monkeys)
      Superfamily: Cercopithecoidea
        Family: Parapithecidae (extinct)
        Family: Cercopithecidae
          Subfamily: Cercopithecinae [macaques (*Macaca*), baboon (*Papio*), guenon
              (*Cercopithecus*), mandrill, langurs (*Presbytis*)]
          Subfamily: Colobinae [colobus monkeys (*Colobus*)]
      Superfamily: Hominoidea
        Family: Hylobatidae [gibbon (*Hylobates*)]
        Family: Pongidae [Apes; orangutan (*Pongo pygmaeus*), gorilla (*Gorilla gorilla*), chimpanzee (*Pan troglodytes*), pigmy chimpanzee (*Pan paniscus*)]
        Family: Hominidae [Human (*Homo*)]

Readers requiring a much more detailed classification of the primates are referred to Groves (1997) or to the Smithsonian Institution Website at **http://nmnhwww.si.edu/cgi-bin/wdb/msw/children/query/11002**.

Swanson, 1991), the cytochrome P450 *CYP21* gene (Kawaguchi *et al.*, 1992) and mitochondrial DNA (Bailey *et al.*, 1991; Brown *et al.*, 1982; Hixson and Brown, 1986; Horai *et al.*, 1992; Perrin-Pecontal *et al.*, 1992; Ruvolo *et al.*, 1991; Saitou 1991; Saitou and Nei, 1986) as well as from transversion rates in nuclear and mitochondrial DNA (Holmquist *et al.*, 1988).

However, divergence data from primate protamine P1 (Retief *et al.*, 1993) and α-fetoprotein (Nishio *et al.*, 1995) gene sequences place gorilla closer to human than to chimpanzee. These gene sequences are known to have evolved extremely rapidly and the observed differences may well have been stochastic changes with no biological consequences for the species concerned or implications for their phylogeny. Similar explanations probably also apply to other examples of apparent gorilla–human closeness, for example studies of the mitochondrial genome

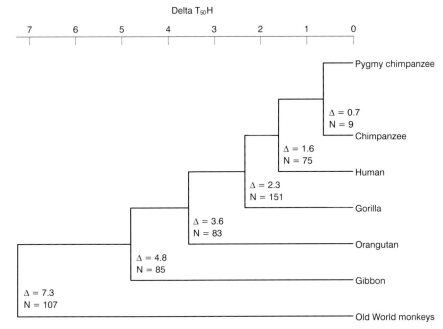

Figure 2.4. Phylogeny of the hominoid primates as determined by average linkage clustering of delta $T_{50}H$ values derived from DNA–DNA hybridization (redrawn from Sibley and Ahlquist, 1987).

(Ruvolo *et al.*, 1991; Horai *et al.*, 1992) and a polymorphism at the *HLA-DQA* locus (Gyllensten and Erlich 1989). Studies of immunoglobulin ε pseudogenes have been equivocal (Ueda *et al.*, 1985, 1989). Taken together, it is clear is that the split between humans, chimps and gorillas was very close and that it is not unreasonable to expect that different sequences will have diverged to variable extents in the different lines. Thus the importance of the recent data of Kumar and Hedges (1998) discussed in Section 2.2. These data yielded slightly revised divergence times for higher primates with tighter error margins: gibbons, 14.6 ± 2.8 Myrs ago; orangutan, 8.2 ± 0.8 Myrs ago; gorilla, 6.7 ± 1.3 Myrs ago, chimpanzees-humans, 5.5 ± 0.2 Myrs ago.

The rates of accumulation of mutations appear to vary by as much as seven-fold between different primate lineages (Koop *et al.*, 1989). The line of descent from the primate node to humans shows a slowdown in evolutionary rates from 7.7×10^{-9} fixed changes $\text{site}^{-1}\ \text{year}^{-1}$ for the first 15 Myrs (55–40 Myrs ago) to 1.3×10^{-9} for the next 15 Myrs (40–25 Myrs ago) to 1.0×10^{-9} for the last 25 Myrs (Koop *et al.*, 1989). The average evolutionary rate for the hominoids (1.1×10^{-9}) is lower than the rates for macaque, a catarrhine (1.9×10^{-9}) and for spider monkey, a platyrrhine (1.8×10^{-9}). By comparison, the line of descent from primate node to *Tarsius* shows an evolutionary rate of 3.4×10^{-9} fixed changes $\text{site}^{-1}\ \text{year}^{-1}$ which is approximately half the stem-simian rate (Koop *et al.*, 1989). The *hominoid slowdown* is at its greatest in human (Li and Tanimura, 1987) although anatomically, humans are quite divergent. Clearly, changes in certain key genes must have assumed a critical importance. As if perhaps to emphasize this point, some gene sequences appear to buck the trend; thus, the evolutionary rate of the noncoding region of the immunoglobulin-α gene is greater in hominoids than in Old World monkeys (Kawamura *et al.*, 1991).

2.3.3 Chromosome evolution in primates

Old World primates

> In dim outline, evolution is evident enough but that particular and essential bit of the theory of evolution which is concerned with the origin and nature of species remains utterly mysterious.
> W. Bateson *William Bateson, Naturalist* (1928)

The evolution and probable phylogeny of primate chromosomes has been extensively reviewed by both Rumpler and Dutrillaux (1990) and Clemente *et al.* (1990). The interested reader is referred to these reviews for detailed accounts. Some chromosomes appear to have been relatively protected from change during primate evolution, for example human chromosomes 19 and X. By contrast, other chromosomes have been prone to significant reorganization, for example human chromosomes 1, 3, and 7 (*Figure 2.5*).

The most frequent types of chromosomal change detected in primate evolution are inversions (especially pericentric, see Haaf and Bray-Ward, 1996), changes in the amount and localization of heterochromatin, fusions and fissions, and changes in the location of centromeres due to activation/inactivation. Reciprocal translocations, deletions and insertions are much less frequent. Human chromosome 18 differs from the homologous chromosomes in the great apes by a pericentric inversion and it is thought that one inversion breakpoint may have been located at or within the centromere (McConkey, 1997). Pericentric inversion breakpoints have also been identified on the chimpanzee equivalents of human chromosomes 4 (4p14, 4q21), 9 (9q22), and 12 (12p12 and 12q15) and these appear to coincide with the locations of either fragile sites or tumor-associated breakpoints (Nickerson and Nelson, 1998). Pericentric inversions may have played an important role in establishing reproductive isolation and speciation during the evolution of the higher primates.

There are at least a dozen blocks of X-Y sequence homology outwith the pseudoautosomal region of humans but these blocks occur in a very different order and orientation on these chromosomes (Vogt *et al.*, 1997). This may be accounted for in terms of the occurrence of a number of different inversions, transpositions and other rearrangements during primate evolution (Bickmore and Cooke, 1987; Lambson *et al.*, 1992; Mumm *et al.*, 1997; Page *et al.*, 1984; Yen *et al.*, 1988). One example of a human-specific inversion is that involving the short arm of the Y chromosome (Schwartz *et al.*, 1998). Since humans from different racial groups (Caucasian, African, and Asian) all possess this Yp inversion, the rearrangement must have occurred prior to the divergence of the human racial groups.

Studies of chromosome banding patterns and hybridization homologies between ape and human chromosomes have provided evidence for human chromosome 2 having arisen from the fusion of two ancestral simian chromosomes. IJdo *et al.* (1991) showed that this probably occurred by telomere–telomere fusion

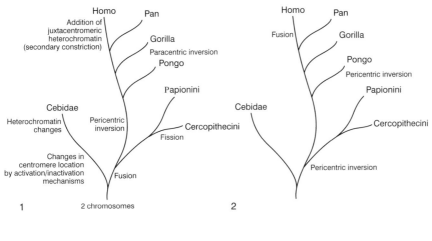

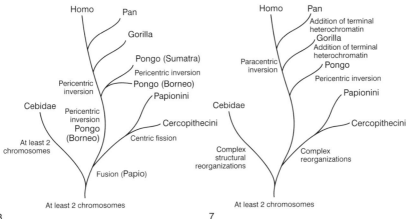

Figure 2.5. Chromosome evolution in primates. Evolution of chromosomes 1, 2, 3, and 7. The main reorganizations that took place are indicated in the phylogenetic trees (redrawn from Clemente *et al.*, 1990).

at 2q13 rather than by translocation after chromosome breakage. This fusion has subsequently been confirmed by chromosome painting (Arnold *et al.*, 1995; Wienberg *et al.*, 1994) and since it accounts for the reduction in chromosome number from 24 pairs in the great apes (chimpanzee, orangutan, and gorilla) to 23 pairs in humans, it must have been a relatively recent event. Fusion must have been accompanied or followed by inactivation or removal of one of the ancestral centromeres. Consistent with this postulate, IJdo *et al.* (1991) found evidence by hybridization for the residual presence of an ancestral centromere at 2q21. Clearly, this reduction in chromosome number would have been a critical event during the speciation process; if it was not in itself responsible for bringing about reproductive isolation, it would certainly have helped to maintain it.

Another major chromosomal rearrangement to have occurred in the great apes is to be found in the gorilla. Using human chromosome-specific libraries as probes for *in situ* hybridization, Stanyon *et al.* (1992) described a reciprocal translocation in the gorilla lineage which is not present in the chimpanzee. Thus, chromosomes 4 and 19 of the gorilla were derived from a reciprocal translocation between the ancestral chromosomes homologous to human chromosomes 5 and 17 (*Figure 2.6*). Wienberg *et al.* (1990) used chromosomal *in situ* suppression hybridization to demonstrate that the centromere of human chromosome 17, the long arm of the chromosome and a small part of the short arm all contribute to gorilla chromosome 4 whilst most of the short arm of chromosome 17 contributes to gorilla chromosome 19.

Probably the best understood human chromosome in terms of its evolution is chromosome 21 (Richard and Dutrillaux, 1998; *Figure 2.7*). The equivalent of human chromosome 21 (HSA21) formed a large and unique chromosome together with chromosome 3 (HSA3) in the eutherian ancestor. This chromosome was conserved without significant alterations only in lemurs, the civet and the pig. It underwent inversions in the tree shrew and the cow. Various translocations involving the portion corresponding to HSA3 occurred in the brown lemur, cat, rabbit and mouse. In the primates, two independent fissions occurred. In New World monkeys, a small segment of HSA3 remained attached to HSA21 and this chromosome then underwent further rearrangements: an inversion in the marmoset, the addition of heterochromatin in the capuchin monkey and a translocation in the saki monkey. HSA21 was formed in the common ancestor of Old World monkeys and underwent translocations with various equivalents of human chromosomes in all the Cercopithecidae. HSA21 was conserved without visible alteration in the black gibbon and the great apes.

One technique which is proving extremely useful in primate cytological studies is cross-species chromosome painting (CSCP; also known as comparative painting or ZOO-FISH). CSCP involves the hybridization of a chromosome-specific paint from one species (usually human) onto metaphase spreads of another

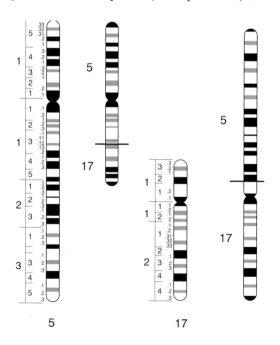

Figure 2.6. Idiograms of (from left to right) human chromosome 5, gorilla chromosome 19, human chromosome 17, and gorilla chromosome 4 (redrawn from Stanyon *et al.*, 1992). The gorilla chromosomes are numbered to show their origin by comparison with the human chromosomes.

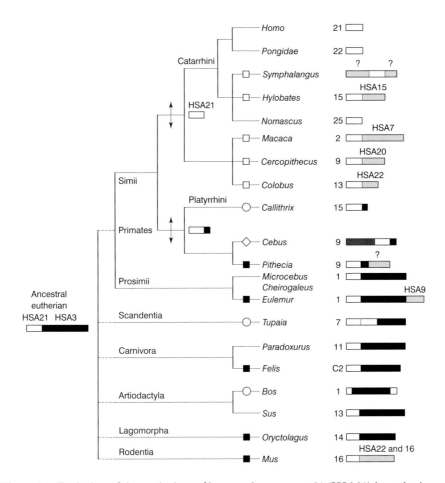

Figure 2.7. Evolution of the equivalent of human chromosome 21 (HSA21) in eutherian mammals (after Richard and Dutrillaux, 1998). HSA21 (open rectangle) formed a large and unique chromosome with HSA3 (dark rectangle) in the eutherian ancestor. This chromosome was conserved without significant alterations only in *Microcebus murinus* (lesser mouse lemur) and *Cheirogaleus major* (greater dwarf lemur), *Paradoxurus hermaphroditus* (common palm civet) and *Sus scrofa* (pig). It underwent inversions (open circles) in *Tupaia glis* (tree shrew) and *Bos taurus* (cow) or various translocations involving the portion corresponding to HSA3 (dark squares) in *Eulemur fulvus* (brown lemur), *Felis catus* (cat), *Oryctolagus cuniculus* (rabbit), and *Mus musculus* (mouse). In the primates, two independent fissions (arrowed) occurred. In the Platyrrhini (New World monkeys), a small segment of HSA3 remained attached to HSA21 and this chromosome then underwent further rearrangements: an inversion in *Callithrix jacchus* (marmoset), the addition of heterochromatin in *Cebus capuchinus* (capuchin monkey) and translocation in *Pithecia pithecia* (saki). HSA21 was formed in the common ancestor of all Catarrhini (Old World monkeys) and underwent translocations (open squares) with various equivalents of human chromosomes (grey rectangles) in all Cercopithecidae viz. *Macaca sylvana* (Barbary ape), *Cercopithecus mona* (Mona monkey), *Colobus abyssinicus* (northern black and white colobus), *Symphalangus syndactylus* (simiang) and *Hylobates lar* (white-handed gibbon). HSA21 was conserved without visible alteration in *Nomascus concolor* (black gibbon) and in the great apes (*Gorilla gorilla*, *Pan troglodytes* and *Pongo pygmeus*). The number of the carrier chromosome of each species is indicated.

species. This approach allows the rapid construction of chromosome maps from the primate species in question which can reveal cytogenetic homologies with the human karyotype (Wienberg and Stanyon, 1997). By these means, a high degree of synteny has been found between human and baboon chromosomes (Rogers *et al.*, 1995). By contrast, numerous translocations are apparent in the gibbon (*Hylobates lar*, 2n = 44) genome as compared to human and the great apes (Jauch *et al.*, 1992; Arnold *et al.*, 1996); the 22 human autosomes have to be divided into 51 elements in order to recombine them into the 21 gibbon autosomes. Similarly, in the Concolor gibbon (*Hylobates concolor*, 2n = 52), the 22 human autosomes have to be divided into 63–67 segments in order to recombine them into the 25 gibbon autosomes (Koehler *et al.*, 1995).

Despite the degree of genetic similarity between the great apes, the chimpanzee genome is approximately 10% larger than that of the human or gorilla (Pellicciari *et al.*, 1982). Nevertheless, the structure of the human and chimpanzee genomes is highly conserved at both chromosomal and sub-chromosomal levels (Jauch *et al.*, 1992; Ried *et al.*, 1993). Even at lower resolution, the extent of evolutionary conservation is readily apparent. Human microsatellite DNA sequences are sufficiently well conserved in chimpanzees that human PCR primers can be used to amplify $(CA)_n$ repeats in the chimpanzee (Blanquer-Maumont and Crouau-Roy, 1995; Deka *et al.*, 1994; Garza *et al.*, 1995). Differences in microsatellite allele length between humans and other primates have however been noted (Ellegren *et al.*, 1995; Garza and Freiner, 1996; Rubinztein *et al.*, 1995). Crouau-Roy *et al.* (1996) studied microsatellites within a 30 cM region of human chromosome 4p and found that all informative loci which are linked in human were also linked in the chimpanzee, indicating that evolutionary conservation extends to the locus level. In general, heterozygosity was found to be greater in chimpanzees, a reflection perhaps of the greater genetic diversity in chimpanzee populations. Some loci, however, appeared to be less heterozygous than in human, a phenomenon that appears to be caused by interruptions of the repeat elements at these loci.

Human chromosomes are not always identical. Indeed many exhibit heteromorphism, especially in the centromeric and satellite regions of the acrocentric chromosomes. Chromosomes 1, 9, 13, 14, 15, 16, 19, 21, 22, and Y are the most heteromorphic whilst chromosomes 2–8 and X are the least heteromorphic (Park *et al.*, 1998; Samonte *et al.*, 1996; Trask *et al.*, 1989a). Inter-chromosomal variation can be substantial; two homologues of chromosome 21 having been noted to vary in healthy individuals by as much as 21 Mb or 40% of the length of the chromosome (Trask *et al.*, 1989a). Family studies have shown that such heteromorphisms are not artefactual and can be inherited in mendelian fashion (Trask *et al.*, 1989b). It would appear that this variation can be largely ascribed to variation in the size of repeat sequence arrays and probably results from unequal crossing over between different classes of repetitive element. Bivariate flow karyotyping has been used to study the relative DNA content of homologous chromosome pairs in individuals from different racial groups (Mefford *et al.*, 1997). Significant variation in DNA content, ranging from 10 to 40%, was found for chromosomes 1, 13, 14, 15, 16, 19, 21, 22, and Y. However, the spectrum of variation observed in the different racial groups was very similar.

New World primates. Like gibbons, many New World primates possess highly rearranged genomes. All New World primates have a translocation between chromosomes 8 and 18. The Cebidae possess translocations between chromosomes 10/16 and 2/16 (Richard *et al.*, 1996) whilst the Atelidae are characterized by translocations between 3/15 and 4/15. Significant synteny is nevertheless apparent between human and the capuchin monkey, *Cebus* (Richard *et al.*, 1996), between human and the Colobus monkey (Bigoni *et al.*, 1997), between human and the howler monkey (Consigliere *et al.*, 1996), between human and the black-handed spider monkey, *Ateles geoffroyi* (Morescalchi *et al.*, 1997) and between human and the marmoset, *Callithrix* (Sherlock *et al.*, 1996).

2.3.4 Evolution of the human sex chromosomes and the pseudoautosomal regions

The human sex chromosomes are heteromorphic; the X chromosome contains ~160 Mb DNA and perhaps 3000 genes whereas the Y chromosome contains only 60 Mb DNA and probably only a handful of genes. The Y chromosome is largely composed of constitutive heterochromatin harboring different families of repetitive DNA (Wolf *et al.*, 1992). Despite the size difference between the sex chromosomes, they are nevertheless able to pair successfully during meiosis. Recombination, however, is largely confined to the two *pseudoautosomal regions* (PARs); a major PAR (2.6 Mb in size) at the tips of the short arms of the X and Y chromosomes which is the site of an obligate crossover during male meiosis (Simmler *et al.*, 1995), and a minor PAR (320 kb) at the tips of the long arms of the X and Y chromosomes (Kvaloy *et al.*, 1994). Regular X-Y recombinational exchanges have served to maintain homology between the chromosomes in the PAR regions (Graves *et al.*, 1998a).

Gene mapping studies have shown that part of the eutherian (placental) mammalian X chromosome ('conserved region'; XCR) is shared by the X chromosomes of marsupials and monotremes (reviewed in Wilcox *et al.*, 1996). Since a series of genes on the short arm of the human X are clustered in two autosomal groups in marsupials and monotremes, these loci define a region (XRA) that has been recently added to the X chromosome in eutherian mammals. Thus, the X chromosome of the common mammalian ancestor was smaller than that found in extant eutherians and at least two autosomal regions have since been added to it. The location of this X-autosome fusion corresponds to the border between the XCR and the XRA and lies at Xp11.23 (Wilcox *et al.*, 1996; *Figure 2.8*).

The human X and Y chromosomes also exhibit substantial homology outwith the pseudoautosomal region (Bardoni *et al.*, 1992; Lambson *et al.*, 1992; Vogt *et al.*, 1997). Although this homology is consistent with these chromosomes having once constituted a homomorphic pair (Graves *et al.*, 1998b), the observed homologies also reflect intra-chromosomal duplication events followed by inter-chromosomal translocations. For example, the minor PAR is thought to have originated during human evolution by the translocation of 320 kb of X chromosomal sequence to the Y chromosome, an event which may have been mediated by recombination between LINE elements (Kvaloy *et al.*, 1994).

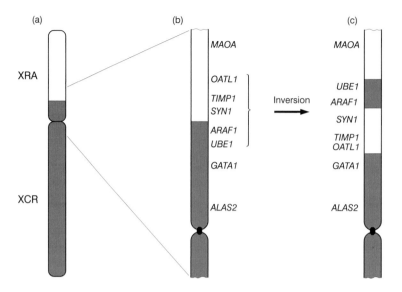

Figure 2.8. (a) Evolutionary origins of the human X chromosome. XCR, conserved region of the X. XRA, recently added region. (b) Details of original gene order at the fusion point. (c) Possible inversion event that placed the XCR genes, *UBE1* and *ARAF1*, between the XRA genes, *SYN1* and *MAOA* (redrawn from Wilcox *et al.*, 1996).

The sex chromosomes however vary dramatically in terms of their evolutionary conservation. The Y chromosome has undergone significant changes during mammalian evolution (Glaser *et al.*, 1998). By contrast, the X chromosome exhibits conservation of synteny between human and mouse (Ohno's Law; Ohno, 1967). This is considered to be a consequence of X inactivation because X-autosome translocations would have tended to be disadvantageous by virtue of their interference with the dosage compensation mechanism. Despite this, gene order on the X chromosome has changed significantly between mouse and human with numerous inversions altering the relative position of genes (Bardoni *et al.*, 1991).

The human Y chromosome contains a number of active genes (Lahn and Page 1997; Vogt *et al.*, 1997). Most notably it hosts the rapidly evolving sex determining gene, *SRY* (Yp11.3; Pamilo and O'Neil 1997; Whitfield *et al.*, 1993) which occurs in the sex chromosome-specific region, about 5 kb from its boundary with the major PAR. In human, a few X-linked genes have functional homologues on the Y chromosome (e.g. *CSF2RA* (Xp22.32, Yp11.3), *MIC2* (Xp22.33, Yp11.3), *RPS4X* (Xq13.1) and *RPS4Y* (Yp11.3), *ZFX* (Xp21.3-p22.1) and *ZFY* (Yp11.3), *AMELX* (Xp22.1-p22.31) and *AMELY* (Yp11.2)). Some Y chromosome homologues are however nonfunctional, for example the *KAL1* pseudogene on Yq11.2 (Incerti *et al.*, 1992) or the *XG* pseudogene on Yq11.21 (Weller *et al.*, 1995). One consequence of the sequence identity of the PARs between the sex chromosomes is that the X chromosomal homologues escape X inactivation in females thereby ensuring that gene dosage is maintained at the level found in males.

Genes identified in the nonrecombining portion of the Y chromosome (NRY) appear to fall into two distinct classes (Lahn and Page, 1997):

(i) Those that are specifically or predominantly expressed in the testis [sex determining gene (*SRY*; Yp11.3), deleted in azoospermia (*DAZ*; Yq11), RNA-binding motif protein 1 (*RBM1*; Yq11), testis-specific protein (*TSPY*), chromodomain Y (*CDY*), basic proteins Y1 and Y2 (*BPY1*, *BPY2*), XK related Y (*XKRY*), tyrosine phosphatase PTP-BL related Y (*PRY*) and testis transcripts Y1 and Y2 (*TTY1*, *TTY2*)]. That these genes have not only been retained on the Y chromosome, but in two cases have also been amplified, may have been part of an evolutionary strategy to optimize male reproductive fitness.

(ii) Those that are widely or ubiquitously expressed and which have closely related counterparts on the X chromosome [dead box Y (*DBY*), thymosin β4 (*TB4Y*), translation initiation factor 1A (*EIF1AY*), ubiquitous TPR motif Y (*UTY*) and *Drosophila* fat facets-related (*DFFRY*; Yq11.2), *AMELY* (Yp11.2), *RPS4Y* (Yp11.3), zinc finger protein Y (*ZFY*; Yp11.3) and *SMCY*; Lahn and Page, 1997]. Conservation of specific X–Y gene pairs may have been associated with a requirement to maintain comparable expression levels for certain housekeeping genes between males and females. Consistent with the predictions of this postulate, the X chromosome homologues of these Y-borne genes escape X inactivation.

The mammalian sex chromosomes are thought to be descended from a homologous pair of autosomes (reviewed by Ellis 1996; Graves *et al.*, 1998a, 1998b; Wolf *et al.*, 1992). This process could have been initiated with the evolutionary appearance of the testis-determining *SRY* gene on the nascent Y chromosome, probably by duplication, translocation and subsequent divergence of the X-linked *SOX3* (Xq26-q27) gene. Suppression of recombination with its homologous chromosome (the nascent X) then led to the gradual degeneration of the Y chromosome owing to its inability to segregate genes carrying deleterious alleles ('Müller's ratchet'; Charlesworth, 1978). Evidence for this degenerative process comes from several sources. The rate of nucleotide substitution in Y chromosome genes appears to be ~2-fold higher than the rate exhibited by X chromosomal genes (Pamilo and Bianchi, 1993; Shimmin *et al.*, 1993) although the frequency of DNA sequence polymorphism may be lower in the sex-specific region of the Y chromosome than in the PARs (Allen and Oster, 1994; Whitfield *et al.*, 1995). The Y chromosome also exhibits a high frequency of retroviral insertion in humans, chimpanzees and orangutans (Kjellman *et al.*, 1995). Finally, there is emerging evidence for gene loss from the Y chromosome during mammalian evolution. In mouse and human, the ubiquitin-activating enzyme (*UBE1*) gene is located on the X chromosome (Xp11.23-p11.3). A copy of the *Ube1* gene is also located on the Y chromosome in the mouse, ring-tailed lemur, squirrel monkey (*Saimiri sciureus*) and marmoset (*Callithrix jacchus*) but not in the Old World monkeys, chimpanzee or human, indicating loss of the Y-linked gene >35 Myrs ago during the evolution of the primates (Mitchell *et al.*, 1998). Similarly, the Y-linked copy of the *EIF2S3* gene (Xp22.1-p22.2) encoding the eukaryotic translation initiation factor EIF-2γ was lost 35–60 Myrs ago in a common ancestor of the simian primates (Ehrmann *et al.*, 1998). The gradual degeneration of the Y chromosome implies that the retention of functional gene copies on this chromosome for a significant period of evolutionary time could have conferred some selective advantage.

It should however be noted that the Y chromosome can also acquire genetic material from other chromosomes. For example, the multicopy *DAZL1* gene (Yq11.23; deleted in azoospermia) was transposed to the Y chromosome from an autosome during primate evolution (Glaser *et al.*, 1998; Saxena *et al.*, 1996; Shan *et al.*, 1996) as was the multicopy RNA-binding motif (*RBM1*; Yq11) gene (Chai *et al.*, 1998; Delbridge *et al.*, 1997). It may be that transfer to a male-specific location provided protection against inactivation or loss. Another example of the duplicational transposition of a gene to the Y chromosome is that of *AMELX* (Xp22.1-p22.31) and its Y-chromosome counterpart *AMELY* (Yp11.2); the latter gene, which appears to be fully functional, is present on the Y chromosomes of bovids and primates but not rodents thereby dating the transpositional event to at least 40 Myrs ago (Toyosawa *et al.*, 1998).

In humans, the *XG* blood group gene (Xp22.32) spans the major PAR on the X chromosome—the first three exons are pseudoautosomal whereas the remaining seven are X chromosome-specific (Weller *et al.*, 1995). In humans and the great apes, an *Alu* sequence is located at the boundary between the major PAR and the Y chromosome-specific DNA (Ellis *et al.*, 1990) but this sequence is not present in Old World monkeys. The *Alu* sequence was therefore inserted into the pre-existing boundary after the divergence of the great apes from the Old World monkeys. Although it did not create the boundary, the *Alu* sequence does serve to demarcate it.

Ellis *et al.* (1994) proposed a model for the formation of the boundary of the major PAR. They hypothesized a pericentric inversion of the Y chromosome with one breakpoint in the ancestral *XG* gene and the other breakpoint 5 kb distal to the ancestral *SRY* gene. In a refinement of this postulate, Fukagawa *et al.* (1996) suggested that the inversion occurred by illegitimate recombination between two PAR boundary sequences, one in the ancestral *XG* gene and the other near the ancestral *SRY* gene.

The PAR has undergone quite rapid change during mammalian evolution involving both gene duplication and translocation events in the region (e.g. *STS*, *MIC2*, *XG*, *CSF2RA*, *IL3RA*, *ARSD*, *ARSE*; Meroni *et al.*, 1996; Ried *et al.*, 1998) and resulting in the movement of the PAR boundary to create X-unique regions (Perry *et al.*, 1998). The evolution of the PAR and the divergence of the mammalian X and Y chromosomes may be viewed in terms of the 'addition-attrition' hypothesis (Graves, 1995; Graves *et al.*, 1998a). This states that the incorporation of autosomal sequences into the PAR of either the X or Y chromosome initially served to generate homologous regions which could pair at meiosis. Recombination with an homologous partner could then result in PAR enlargement. Alternatively, the steadily accumulating mutations on the Y chromosome would have served to decrease the level of homology to the X chromosome thereby reducing PAR size. Fukagawa *et al.* (1996) proposed a further twist to this argument in that once divergence had reached a certain level, recombination frequency would have decreased thereby further increasing the rate of divergence.

Evidence in favor of the addition-attrition theory comes from the dynamic nature of the major PAR region during mammalian evolution. Thus, the *STS* gene which is X-linked in humans and the great apes (Xp22.32) is autosomal in prosimians as is the *ANT3* gene which is pseudoautosomal (Yp11.3) in humans

(Toder *et al.*, 1995). Similarly, the human pseudoautosomal genes *CSF2RA* (Xp22.32/Yp11.3), *SHOX* (Xp22/Yp11.3) and *IL3RA* (Xp22.3/Yp11.3), are autosomal in the mouse (Ellis, 1996). Further evidence for the process of attrition may come from the finding that whereas the *Fxy* gene spans the PAR on the murine X chromosome, its human counterpart (*FXY*; Xp22.3) lies proximal to the human PAR (Perry *et al.*, 1998).

The 'X-driven' hypothesis of Graves (1995; 1998) essentially proposes that the rapid evolutionary spreading of X inactivation preceded the decay of Y chromosomal genes and even drove its initial steps. This hypothesis predicts that inactivated X-linked genes with functionally comparable Y-linked homologues should exist as evolutionary intermediates, but as yet no such gene has been found in any mammalian species. Indeed, a considerable number of human X-linked genes escape X-inactivation but have no detectable Y-borne counterpart.

An alternative 'Y-driven' pathway of X–Y gene evolution has been proposed by Jegalian and Page (1998). Briefly, these authors suggested that many extant genes represent intermediates on a general pathway by which X–Y genes or gene clusters evolved from autosomal genes (*Figure 2.9*). Autosomal genes would have entered the pathway either by virtue of their presence on the emergent sex chromosomes or via translocation of an autosomal gene. This would have been followed by suppression of X–Y recombination. These steps occurred either at the chromosomal or sub-chromosomal level and gave rise to functionally equivalent X-linked (but not inactivated) genes and Y-linked genes. Subsequently, three different processes (Y gene decay, upregulation of X-linked gene expression, and X-inactivation) interacted resulting in an inactivated X-linked gene accompanied by the loss of the Y gene. Expression of the X-linked gene then increased as an adaptation to the reduced or restricted expression of its Y-linked counterpart. This compensated for the loss of Y gene function and restored optimal expression levels in males. X-inactivation, on the other hand, may be viewed as a counter-response which restored optimal expression levels in females. This explanation could in principle account for most X-linked and X-Y homologous genes in extant mammals, many of which exist at intermediate steps in the pathway. Jegalian and Page (1998) pointed out that only one gene cannot be accommodated in their pathway schema: the human pseudoautosomal gene, *SYBL1*, which is X-inactivated and transcriptionally silenced on the Y chromosome. A model for the evolution of the mammalian sex chromosomes summarizing the above processes is presented in *Figure 2.10*.

2.3.5 Evolution of the mitochondrial genome

The 16 569 bp of the human mitochondrial genome encodes 13 polypeptides, all subunits of the enzyme complexes of the pathway of oxidative phosphorylation, and a total of 22 tRNAs. The mitochondrial genome is characterized by its high proportion of coding DNA, the paucity of repetitive DNA sequence, the absence of introns within its genes and its own genetic code distinct from that of the nuclear genome (Kurland, 1992). Mitochondrial genes experience a mutation rate that has been estimated to be up to 17 times higher than the corresponding rate for nuclear genes (Wallace *et al.*, 1987). This is thought to be due to the fact that

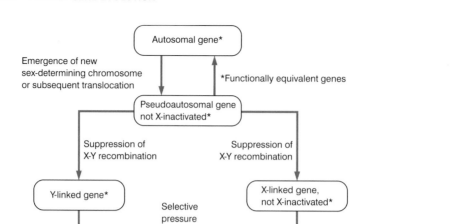

Figure 2.9. A proposed Y-driven pathway for X-Y gene evolution in mammals (after Jegalian and Page, 1998).

the mitochondrial genome undergoes many more rounds of replication but may also be a consequence of the relative error proneness of the mitochondrial DNA polymerase γ and increased exposure to the potentially mutagenic products of oxidative metabolism. Whatever the explanation, one consequence of the higher mutation rate has been that the mitochondrially encoded subunits of the oxidative phosphorylation enzyme complexes have evolved at a much higher rate than their nuclear encoded counterparts (Shoffner and Wallace, 1990).

The origin of mitochondria is thought to have been through endocytosis by an anaerobic eukaryote of an aerobic eubacterium possessing an oxidative phosphorylation system (Gray, 1992). During the evolutionary transformation of endosymbiont to organelle, there has been significant transfer of DNA sequences from the mitochondrial to the nuclear genome (Hu and Thilly, 1994; Sadlock *et al.*, 1993; Sorenson and Fleischer, 1996). Interestingly, the mitochondrial genome of the slime mould *Dictyostelium* has retained a gene encoding a NADH: ubiquinone oxidoreductase subunit which was transferred to the nuclear genome in the common ancestor of other eukaryotes (Cole *et al.*, 1995). Some mitochondrial DNA sequences present in the human genome as pseudogenes have been incorporated only relatively recently in primate evolution (Collura and Stewart 1995; Wallace *et al.*, 1997; Zischler *et al.*, 1995). It has been suggested that coevolution may have occurred between nuclear

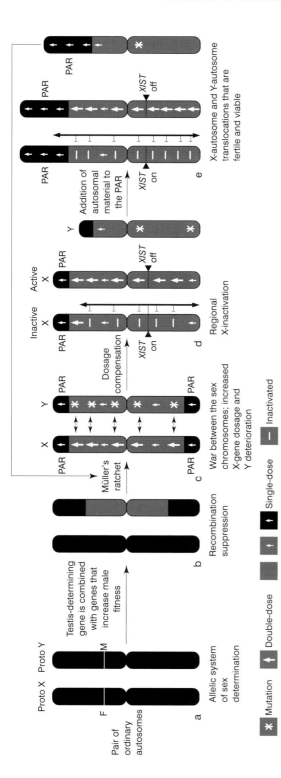

Figure 2.10. Model for the evolution of the mammalian sex chromosomes (redrawn from Ellis, 1998). (a) Mammalian sex chromosomes evolved from a pair of ordinary autosomes. Initially, sex was determined by a simple diallelic system, F and M, with the male being the heterogametic sex. (b) Sex chromosome differentiation began when the proto-Y chromosome acquired at least one additional gene that together with the M allele conferred a selective advantage upon males. Combinations of syntenic genes were then selected for a meiotic system in which recombination was suppressed. (c) The consequence of recombination suppression was the acquisition by the Y chromosome of recessive mutations in the many genes unrelated to sexual development by the process known as Müller's ratchet. Recessive mutations in the recombination-suppressed segment are not subject to selection. The loss of fitness of males resulting from hemizygosity of an increasing number of X chromosomal genes was counterbalanced by increased expression of the X chromosomal genes, thereby raising their level of expression in females. Crossing over is confined to the pseudoautosomal regions (PAR) and is required for the segregation of the sex chromosomes at meiosis. (d) The inactivation of one X chromosome evolved in females. Initiation of the inactivation process was controlled by the *XIST* locus. (e) PARs can be sites to which autosomal material can be translocated. The newly translocated material becomes subject to Müller's ratchet.

and mitochondrial genes, for example cytochrome c oxidase subunit IV (*COX4*; 16q22-qter and mt genome; Lomax *et al.*, 1992).

Mitochondrial DNA variation has been very informative for the study of the evolution of human populations (see Section 2.3.6). The interested reader is referred to Stoneking (1996) for a review of the topic.

2.3.6 The evolution of human populations

> Darwinian Man, though well behaved, at best is only a monkey shaved!
> W.S.Gilbert *Princess Ida* (1884)

Discussion of the fossil record of early hominids is outwith the remit of this volume and the interested reader is referred to Wood (1992, 1996) for readable préces. Similarly, the origins of modern humans and the movement of human populations have been well reviewed by a number of authors including Cavalli-Sforza *et al.* (1994), Cavalli-Sforza (1998), von Haeseler *et al.* (1995), Jones, Martin, and Pilbeam (1992) and Lewin (1998).

There are currently two different views of the origins of modern humans. The first, which is not inconsistent with the fossil record, proposes that different populations ('races') of *Homo sapiens* evolved independently from their ancestor *Homo erectus* in different parts of the Old World ('multiregional model'). Migration of *H. erectus* from Africa to the rest of the Old World may have occurred ~1 Myrs ago. The alternative hypothesis is that *H. sapiens* arose once in Africa ('Out of Africa model') and that the species may have been subjected at some stage to a severe population bottleneck. With both models, the geographical separation of populations has led to the emergence of morphological differences although continued gene flow between populations has served to ensure that human gene diversity remains graded rather than discrete.

One of the most dramatic demonstrations so far of the utility of molecular genetics to the study of human gene evolution has been the analysis of mitochondrial DNA (mtDNA) from a fossil Neanderthal specimen (Krings *et al.*, 1997). Neanderthals and humans are considered to have shared a common ancestor between 550 000 and 690 000 years ago. Since the Neanderthal mtDNA sequence was found to be quite distinct from those of modern humans, it would appear as if Neanderthals became extinct without contributing mtDNA to human populations, a finding consistent with the Out of Africa hypothesis.

The question of the age of the human gene pool has been approached by studying variation associated either with the mitochondrial genome or with Y chromosome-derived DNA sequences (mtDNA exhibits a high rate of evolutionary change and, along with most of the Y chromosome, is transmitted without recombination). Studies of mtDNA have yielded dates of the order of 200 000 years ago for the origin of modern humans (Hey, 1997; Horai *et al.*, 1995; Ruvolo *et al.*, 1993; Vigilant *et al.*, 1991), broadly consistent with estimates derived from Y-chromosome-derived DNA sequence data (Agulnik *et al.*, 1998; Brookfield, 1995; Dorit *et al.*, 1995; Fu and Li, 1996; Hammer, 1995; Jobling, 1996; Mitchell and Hammer, 1996; Weiss and von Haeseler, 1996). Use of intronic variation in the *ZFX* (Xp21.3-p22.2) gene yielded a figure of 306 000 years (Huang *et al.*, 1998) that is almost certainly discrepant since the 95% confidence interval was extremely broad. Using 30 microsatellite loci to construct a phylogenetic tree for 14 different human populations, Goldstein et al. (1995) estimated that the time since divergence of African and nonAfrican populations was 156 000 years (95% CI, 75 000–287 000 years). The above estimates are broadly compatible with those derived from polymorphisms associated with the *CD4* (12p12-pter) gene (Tishkoff et al., 1996), microsatellite data (Nei and Takezaki, 1996) and protein polymorphism data (Nei 1995; Nei and Takezaki, 1996). It must be remembered, however, that these results simply reflect the time elapsed since the most recent common ancestor for the sample population rather than the most recent common ancestor of all humans. Small populations and/or population bottlenecks (Ambrose, 1998) will have served to obscure the actual timing of the 'origin' of modern humans.

As to the place of origin of modern humans, Wainscoat et al. (1986) claimed that the relative frequencies of haplotypes of the β-globin (*HBB*; 11p15.5) gene in African and nonAfrican populations provided evidence for a migration out of Africa by a fairly small population. Further evidence for a recent African origin for modern humans comes from the observation that African populations have a ~20% greater microsatellite sequence diversity as compared with Asian and European poulations (Armour et al., 1996; Bowcock et al., 1994; Jorde et al., 1997; 1998; Nei 1995; Perez-Lezaun et al., 1997; Tishkoff et al., 1996). Studies of mitochondrial DNA have also shown that there is greater genetic diversity between African populations than among Asian or European populations (Comas et al., 1997; Merriwether et al., 1991). Indeed, these authors showed that genetic variation among humans on all continents are subsets of the variation present in Africans. However, some polymorphism lineages do not show deep branches for African populations which has made the Out of Africa hypothesis somewhat contentious (Jorde et al., 1995; Harding et al., 1997). The higher level of genetic diversity manifested by African populations may simply be a reflection of their greater population size over the last million years (Relethford and Harpending, 1995).

Y chromosome variants appear to be more highly clustered geographically than those of mtDNA (Cavalli-Sforza and Minch 1997; Ruiz-Linares et al., 1996; Underhill et al., 1997). One explanation for this difference could be that male migration has been more limited than that of women (Seielstad et al., 1998).

About 84% of human genetic diversity exists as differences between individuals within populations but the remaining 16% can be used to distinguish between populations (Barbujani et al., 1997). By comparison with apes, the extent of the genetic variation exhibited by modern humans is relatively low (Ferris et al., 1981; Jorde et al., 1998; Li and Sadler 1991). This lack of genetic diversity is likely to be a reflection of long-term small population size [before the introduction of agriculture 10 000 years ago, the entire human population probably did not exceed 100 000 and is thought to have been around 10 000 for most of its history (Erlich et al., 1996; Harpending et al., 1998; Takahata, 1993; Zietkiewicz et al., 1998)], the effects of past population bottlenecks and the explosive population growth particularly during the last 10 000 years (Ambrose, 1998; Cavalli-Sforza et al., 1993; Harding et al., 1997; Knight et al., 1996; Reich and Goldstein, 1998; Xiong et al., 1991).

In human history, demographic expansions have often occurred as a result of technological developments affecting food availability and transportation fuelled by the pursuit of military or economic objectives. In terms of establishing genetic differences between populations, linguistic barriers have also been important (Cavalli-Sforza 1997).

2.3.7 The action of natural selection in human populations

> From the time when the ancestral man first walked erect, with hands freed from any active part in locomotion, and when his brain-power became sufficient to cause him to use his hands in making weapons and tools, houses and clothing, to use fire for cooking, and to plant seeds or roots to supply himself with stores of food, the power of natural selection would cease to act in producing modifications of his body, but would continuously advance his mind through the development of its organ, the brain.
> Alfred Russel Wallace *Darwinism* (1889)

Evidence for the recent effects of natural selection on human populations comes indirectly from observed variation in allele frequencies (Cavalli-Sforza *et al.*, 1994) with infectious disease often serving as the selecting agent (Hill 1996a; 1996b; Levin *et al.*, 1999). The classical example of pathogen-driven selection is the heterozygote advantage accruing to carriers of the Glu6→Val sickle cell mutation in the β-globin (*HBB*) gene which confers resistance to *Falciparum* malaria (Vogel and Motulsky, 1997). It has been suggested that the high frequency of certain diseases in specific populations may be explained in similar ways (Motulsky 1995; Zlotogora 1994).

Heterozygote advantage has been invoked to account for the spread of the common cystic fibrosis mutation (δF508) in the *CFTR* (7q31.3) gene in Caucasian populations. The basis for heterozygote advantage was proposed to be increased fitness of heterozygous carriers during cholera epidemics (Gabriel *et al.*, 1994). However, it is difficult to see how such overdominant selection could have brought about the extremely high prevalence of one specific *CFTR* lesion (δF508) relative to the large number of alternative *CFTR* mutations known. It would also be difficult to explain the gradient in δF508 frequency across Europe unless there was also a gradient of selective pressure. Thus, the most parsimonious explanation is probably genetic drift.

A large body of data has accumulated on HLA polymorphism and its relationship to disease susceptibility, resistance and progression (Bodmer, 1996; Hall and Bowness, 1996). Polymorphic alleles at both HLA class I and II loci have been shown to be under selection (Begovich *et al.*, 1992; Hill *et al.*, 1991; Hughes 1988, 1989). One recent example is selection for specific HLA class II (*HLA-DR* and *HLA-DQB*; 6p21.3) alleles as a result of hepatitis B virus infection (Thursz *et al.*, 1997). More than 90% of the 135 known *HLA-DRB1* alleles appear to have been generated since the divergence of human and chimpanzee (Bergström *et al.*, 1998); such changes appear to have arisen both by point mutation and by gene conversion and are consistent with the existence of substantial selective pressure.

Pathogen-driven selection may also have been responsible for increasing the frequency of genetic variants at other loci, for example the *CCR2* (3p21; Smith *et al.*,

1997) and *CCR5* (3p21; Dean *et al.*, 1996; Libert *et al.*, 1998) chemokine receptor genes and the stromal cell-derived (*SDF1*; 10q11.2; Winkler *et al.*, 1998) gene. With the advent of the HIV epidemic, these variants may become advantageous in that they appear able to restrict HIV-1 infection and decrease the progression of HIV-1 infection to AIDS. Genetic susceptibility to parasitic infections may be under oligogenic control (e.g. *Leishmania*, Shaw *et al.*, 1995; *Schistosoma mansoni*, Marquet *et al.*, 1996; *Mycobacterium tuberculosis*, Bellamy *et al.*, 1998; 1999; Shaw *et al.*, 1997; *Plasmodium falciparum*; Rihet *et al.*, 1998) implying that a number of different variants at different loci may be subject to selection. Infectious and parasitic disease is currently estimated to kill up to 20 million people in the world every year with acute respiratory infection, tuberculosis, diarrhoea, malaria, measles, hepatitis B, whooping cough and tetanus responsible for 3/4 of this toll. Clearly, it is likely that selection will continue to operate on existing genetic variation at a wide range of genetic loci that serve to determine both the host's susceptibility and resistance to the disease in question.

Polymorphism of the human ABO blood group system (*ABO*; 9q34) may owe its origins to balancing (overdominant) selection mediated by infectious agents (Eder and Spitalnik, 1997). Intriguingly, the same substitutions that differentiate the A and B alleles in humans are also present in the great apes and Old World monkeys (see Chapter 1, section 1.2.2) leading to speculation that these polymorphic antigens may have arisen early in primate evolution (Kominato *et al.*, 1992; Matinko *et al.*, 1993). However, intronic sequence data point instead to an origin for the human alleles about 3 Myrs ago (O'Huigen *et al.*, 1997) which argue for a model of convergent evolution of the blood group antigens in the higher primates.

Whilst pathogen-driven selection may be an important factor in increasing the frequency of certain genetic variants at specific human loci, susceptibility to infectious disease is unlikely to be the sole means by which natural selection influences allele frequencies. One common genetic variant not thought to be associated with infectious disease has been found in factor VII. Plasma levels of factor VII vary significantly in the general population (Howard *et al.*, 1994) and are known to be influenced by a number of different environmental factors including sex, age, cholesterol, and triglyceride levels (Scarabin *et al.*, 1996). An Arg/Gln polymorphism at residue 353 of factor VII (*F7*; 13q34) which occurs with a frequency of about 10% in various populations, is associated with a 20–25% reduction in the level of plasma factor VII activity as a result of the impaired secretion of the Gln variant (Cooper *et al.*, 1997). This high frequency is suggestive of a balanced polymorphism and could indicate that the Gln variant confers some benefit, for example protection against thrombosis, myocardial infarction or arterial disease (Escoffre *et al.*, 1995). In support of this postulate, Silveira *et al.* (1994) have shown that the Gln allele is associated with a reduction in the amount of activated factor VII (FVIIa) generated in response to fat intake; individuals with the Arg/Gln genotype were found to possess FVIIa levels 48% of that exhibited by individuals homozygous for the Arg allele. Interestingly, a decanucleotide insertion polymorphism at -323 in the *F7* gene promoter has been shown to be associated with a 33% reduction in promoter activity *in vitro* and a lower level of plasma factor VII activity and antigen *in vivo* (Pollak *et al.*, 1996; Humphries *et al.*, 1996). The occurrence of two functionally significant polymorphisms in the same gene is consistent with the idea of a selective advantage accruing to individuals with reduced factor VII activity.

A similar hemostatic system polymorphism is factor V Leiden. This variant, which underlies the phenomenon of activated protein C resistance, results from the substitution of Arg506 by Gln in coagulation factor V (*F5*; 1q23). Factor Va serves as a cofactor in the activation of prothrombin by factor Xa and the factor V Leiden variant is relatively resistant to activated protein C-mediated inactivation. Between 1% and 7% of the Caucasian population possess the factor V Leiden mutation (Cooper and Krawczak, 1997) which may therefore be regarded as a fairly frequent polymorphism with phenotypic effect. Since the factor V Leiden mutation is also associated with a relative risk of ~6.0 for venous thrombosis, this is also a polymorphism with clinical effect. Why is this factor V variant so common? Its high frequency in the general population suggests that it confers, or has conferred, some selective advantage on its bearers. Dahlbäck (1994) speculated that a slight hypercoagulable state associated with possession of the factor V Leiden variant might have been advantageous in certain situations such as traumatic injury and childbirth. Consistent with this postulate, Lindqvist *et al.* (1998) have shown that carriers of the factor V Leiden variant have a significantly reduced risk of bleeding during childbirth. It may be that other common polymorphic variants in hemostatic factor genes e.g. the G20210A transition in the 3' untranslated region of the prothrombin (*F2*; 11p11-q12; Zivelin *et al.*, 1998) gene, are explicable by similar models.

Other examples of polymorphic variants which may have conferred a selective advantage on carriers are the 'insertion' (I) allele of the angiotensin-converting enzyme (*DCP1*; 17q23) gene which appears to be associated with improved human endurance (Montgomery *et al.*, 1998) and the Glu487/Lys polymorphism in the aldehyde dehydrogenase 2 (*ALDH2*; 12q24) gene which is associated with alcohol sensitivity and alcohol avoidance (Goedde *et al.*, 1992). Selection is likely to have also operated on a variety of other human characteristics including cognitive ability (McClearn *et al.*, 1997), body form, skin pigmentation (Smith, 1993) and pharmacogenetic variation (Kalow, 1997).

2.4 Sequencing the genomes of model organisms and humans

The characterization of the genomes of a number of different and disparate species should aid significantly our understanding of the human genome, its structure, function and evolution. Such species ('model organisms') include a bacterium (*Escherichia coli*), a yeast (*Saccharomyces cerevisiae*), a nematode (*Caenorhabditis elegans*), the fruitfly (*Drosophila melanogaster*), the pufferfish (*Fugu rubripes*), the mouse, and the rat. Sequencing of the genomes of these model organisms is essential for the discovery, description and characterization of all genes within those genomes and the proteins that these genes encode. It will provide information not only on the chromosomal organization of genes and gene families but also on the elements that control their expression.

The 4.64 Mb genome of *E. coli* has been sequenced and encodes 4288 protein-coding genes (Blattner *et al.*, 1997). By comparison, the 12.1 Mb genome of *S. cerevisiae*, which is organized into 16 chromosomes, contains 5885 protein-coding genes, ~140 ribosomal RNA genes, 40 snRNA genes and 275 tRNA genes (Goffeau *et al.*, 1996). The Saccharomyces Genome Database is available online at **http://genome-www.stanford.edu/Saccharomyces/**.

The entire 97 Mb genome of *C. elegans* has also been sequenced and represents the first fully characterized genome of a multicellular eukaryote (*C. elegans* Sequencing Consortium, 1998). This sequence predicts a total of 19 099 protein-coding genes and at least several hundred further genes specifying noncoding RNAs. At least 36% of *C. elegans* proteins exhibit a match in humans whilst 74% of characterized human proteins exhibit a match with a *C. elegans* protein. Each *C. elegans* gene has an average of five introns and the exons constitute some 27% of the nematode genome. Sequence data are available through the *C. elegans* Genome Project Website at **http://www.sanger.ac.uk/Projects/C_elegans/**.

Comparison of the complete gene/protein sets of yeast and nematode has revealed that for a substantial proportion of the two organisms' genes, one-to-one orthologous relationships are identifiable (Chervitz *et al.*, 1998). This suggests that the functions of many gene products were already established in the common ancestor of fungi and the metazoa. By contrast, most of the *C. elegans* signaling and regulatory genes that are known or expected to be involved in multicellularity have no yeast orthologue even though they may contain domain sequences present that are in yeast (Chervitz *et al.*, 1998).

The possession of the complete genome sequences of various model organisms is proving of enormous benefit in identifying the human homologues of genes that are shared between these organisms and humans. Thus, the expressed sequence tags (EST) database (dbEST; **http://www.ncbi.nlm.nih.gov/dbEST/index.html**) can be screened using model organism genes as 'probes'. An example of this approach (termed 'cyberscreening' or '*in silico* cloning') is provided by the cloning of five human orthologues of yeast genes encoding proteins of the mitochondrial respiratory chain complex (Petruzzella *et al.*, 1998).

Within the next 5 years, the 3200 Mb human genome sequence should also become available (Rowen *et al.*, 1997). This will permit integration of cytogenetic, genetic, physical and transcriptional maps of the genome, information on inter-individual polymorphic variation, and the genotype-phenotype relationship particularly in the context of complex traits. Progress made in mapping and sequencing the human genome may be followed at the following websites: **http://www.ncbi.nlm.nih.gov/genemap98/** (National Center for Biotechnology Information), **http://www.sanger.ac.uk/HGP** (Sanger Centre), **http://genome.wustl.edu/gsc/index.shtml** (Washington University Genome Sequencing Center), **http://www-seq.wi.mit.edu/** (Whitehead Institute/MIT Genome Sequencing Project) and **http://www.jgi.doe.gov/** (Joint Genome Institute). The availability of the human genome sequence will lead to the identification of novel genes encoding new proteins and the characterization of disease genes which should provide new insights into mechanisms of disease. Comparative genome mapping will also provide important insights into the evolution of the mammalian genome, its chromosomal architecture and its genes and gene families.

References

Agulnik A.I., Zharkikh A., Boettger-Tong H., Bourgeron T., McElreavey K., Bishop C.E. (1998) Evolution of the *DAZ* gene family suggests that Y-linked *DAZ* plays little, or a limited, role in spermatogenesis but underlines a recent African origin for human populations. *Hum. Molec. Genet.* 7: 1371–1377.

Ahringer J. (1997) Turn to the worm! *Curr. Opin. Genet. Devel.* 7: 410–415.

Allen B.S., Ostrer H. (1994) Conservation of human Y chromosome sequences among male great apes: implications for the evolution of Y chromosomes. *J. Mol. Evol.* 39: 13–21.

Amadou C., Ribouchon M.T., Mattei M.G., Jenkins N.A., Gilbert D.J., Copeland N.G., Avoustin P., Pontarotti P. (1995) Localization of new genes and markers to the distal part of the human major histocompatibility complex (MHC) region and comparison with the mouse: new insights into the evolution of mammalian genomes. *Genomics* 26: 9–20.

Ambrose S.H. (1998) Late Pleistocene human population bottlenecks, volcanic winter, and differentiation of modern humans. *J. Hum. Evol.* 34: 623–651.

Amores A., Force A., Yan Y.-L., Joly L., Amemiya C., Fritz A., Ho R.K., Langeland J., Prince V., Wang Y.-L., Westerfield M., Ekker M., Postlethwait J.H. (1998) Zebrafish *hox* clusters and vertebrate genome evolution. *Science* 282: 1711–1714.

Armour J.A.L., Anttinen T., May C.A., Vega E.E., Sajantila A., Kidd J.R., Kidd K.K., Bertranpetit J., Pbo S., Jeffreys A.J. (1996) Minisatellite diversity supports a recent African origin for modern humans. *Nature Genet.* 13: 154–159.

Arnold N., Wienberg J., Ermert K., Zachau H.G. (1995) Comparative chromosome mapping by *in situ* hybridization of DNA probes derived from the V kappa immunoglobulin genes. *Genomics* 26: 147–150.

Arnold N., Stanyon R., Jauch A., O'Brien P., Wienberg J. (1996) Identification of complex chromosome rearrangements in the gibbon by fluorescent *in situ* hybridization (FISH) of a human chromosome 2q specific microlibrary, yeast artificial chromosomes, and reciprocal chromosome painting. *Cytogenet. Cell Genet.* 74: 80–85.

Bailey G.S., Poulter R.T., Stockwell P.A. (1978) Gene duplication in tetraploid fish: model for gene silencing at unlinked duplicated loci. *Proc. Natl Acad. Sci. USA* 75: 5575–5579.

Bailey W.J., Fitch D.H.A., Tagle D.A., Czelusniak J., Slightom J.L., Goodman M. (1991) Molecular evolution of the χη gene locus: gibbon phylogeny and the hominid slowdown. *Mol. Biol. Evol.* 8: 155–184.

Banyer J.L., Goldwurm S., Cullen L., van der Griend B., Zournazi A., Smit D.J., Powell L.W., Jazwinska E.C. (1998) The spinal muscular atrophy gene region at 5q13.1 has a paralogous chromosomal region at 6p21.3. *Mamm. Genome* 9: 235–239.

Barbujani G., Magagni A., Minch E., Cavalli-Sforza L.L. (1997) An aportionment of human DNA diversity. *Proc. Natl Acad. Sci. USA* 94: 4516–4515.

Bardoni B., Zuffardi O., Guioli S., Ballabio A., Simi P., Cavalli P., Grimoldi M.G., Fraccaro M., Camerino G. (1991) A deletion map of the human Yp11 region: implications for the evolution of the Y chromosome and tentative mapping of a locus involved in spermatogenesis. *Genomics* 11: 443–451.

Begovich A.B., McClure G.R., Suraj V.C., Helmuth R.C., Fildes N., Bugawan T.L., Erlich H.A., Klitz W. (1992) Polymorphism, recombination, and linkage disequilibrium within the HLA class II region. *J. Immunol.* 148: 249–258.

Bellamy R., Ruwende C., Corrah T., McAdam K.P.W.J., Whittle H.C., Hill A.V.S. (1998) Variations in the *NRAMP1* gene and susceptibility to tuberculosis in West Africans. *New Engl. J. Med.* 338: 640–644.

Bellamy R., Ruwende C., Corrah T., McAdam K.P.W.J., Thursz M., Whittle H.C., Hill A.V.S. (1999) Tuberculosis and chronic hepatitis B virus infection in Africans and variation in the vitamin D receptor gene. *J. Infect. Dis.* 179: 721–724.

Bergström T.F., Josefsson A., Erlich H.A., Gyllensten U. (1998) Recent origin of *HLA-DRB1* alleles and implications for human evolution. *Nature Genet.* 18: 237–242.

Bickmore W.A., Cooke H.J. (1987) Evolution of homologous sequences on the human X and Y chromosomes, outside of the meiotic pairing segment. *Nucleic Acids Res.* 15: 6261–6271.

Bigoni F., Koehler U., Stanyon R., Wienberg J. (1997) Mapping homology between human and black and white Colobine monkey chromosomes by fluorescence *in situ* hybridization. *Am. J. Primatol.* **42**: 289–298.

Blair H.J., Reed V., Laval S.H., Boyd Y. (1994) New insights into the man-mouse comparative map of the X chromosome. *Genomics* **19**: 215–220.

Blanquer-Maumont A., Crouau-Roy B. (1995) Polymorphism, monomorphism and sequences in conserved microsatellites in pimate species. *J. Mol. Evol.* **41**: 491–497.

Blattner F.R., Plunkett G., Bloch C.A., Perna N.T., Burland V., Riley M., Collado-Vides J., Glasner J.D., Rode C.K., Mayhew G.F., Gregor J., Davis N.W., Kirkpatrick H.A., Goeden M.A., Rose D.J., Mau B., Shao Y. (1997) The complete genome sequence of *Escherichia coli* K-12. *Science* **277**: 1453–1461.

Bodmer J. (1996) World distribution of HLA alleles and implications for disease. In: *Variation in the Human Genome*. Ciba Foundation Symposium 197. John Wiley, Chichester. pp 233–258.

Bowcock A.M., Ruiz-Linares A., Tomfohrde J., Minch E., Kidd J.R., Cavalli-Sforza L.L. (1994) High resolution of human evolutionary trees with polymorphic microsatellites. *Nature* **368**: 455–458.

Brookfield J.F.Y. (1995) Y-chromosome clues to human ancestry. *Current Biol.* **5**: 1114–1115.

Brown W.M., Prager E.M., Wang A., Wilson A.C. (1982) Mitochondrial DNA sequences of primates: tempo and mode of evolution. *J. Mol. Evol.* **18**: 225–239.

Burger H., Wagemaker G., Leunissen J.A.M., Dorssers L.C.J. (1994) Molecular evolution of interleukin-3. *J. Mol. Evol.* **39**: 255–267.

Bush G.L., Case S.M., Wilson A.C., Patton J.L. (1977) Rapid speciation and chromosomal evolution in mammals. *Proc. Natl Acad. Sci. USA* **74**: 3942–3946.

C. elegans Sequencing Consortium. (1998) Genome sequence of the nematode *C. elegans*: a platform for investigating biology. *Science* **282**: 2012–2018.

Caccone A., DeSalle R., Powell J.R. (1988) Calibration of the change in thermal stability of DNA duplexes and degree of base pair mismatch. *J. Mol. Evol.* **27**: 21–216.

Carroll RL. (1988) *Vertebrate Paleontology and Evolution*. WH Freeman, New York.

Cavalli-Sforza L.L., Minch E. (1997) Paleolithic and Neolithic lineages in the European mitochondrial gene pool. *Am. J. Hum. Genet.* **61**: 247–250.

Cavalli-Sforza L.L. (1998) The DNA revolution in population genetics. *Trends Genet.* **14**: 60–65.

Cavalli-Sforza L.L., Menozzi P., Piazza A. (1993) Demic expansions and human evolution. *Science* **259**: 639–646.

Cavalli-Sforza L.L., Menozzi P., Piazza A. (1994) Genetic history of world populations. In: *The History and Geography of Human Genes*. Princeton University Press, New Jersey. Chap. 2, pp 60–157.

Cavalli-Sforza L.L. (1997) Genes, peoples, and languages. *Proc. Natl Acad. Sci. USA* **94**: 7719–7724.

Chai N.N., Zhou H., Hernandez J., Najmabadi H., Bhasin S., Yen P.H. (1998) Structure and organization of the *RBMY* genes on the human Y chromosome: transposition and amplification of an ancestral autosomal hnRNPG gene. *Genomics* **49**: 283–289.

Chan S.J., Cao Q.P., Steiner D.F. (1990) Evolution of the insulin superfamily: cloning of a hybrid insulin/insulin-like growth factor cDNA from amphioxus. *Proc. Natl Acad. Sci. USA* **87**: 9319–9323.

Charlesworth B. (1978) Model for evolution of Y chromosomes and dosage compensation. *Proc. Natl Acad. Sci. USA* **75**: 5618–5622.

Chervitz S.A., Aravind L., Sherlock G., Ball C.A., Koonin E.V., Dwight S.S., Harris M.A., Dolinski K., Mohr S., Smith T., Weng S., Cherry J.M., Botstein D. (1998) Comparison of the complete protein sets of worm and yeast: orthology and divergence. *Science* **282**: 2022–2027.

Clark A.G. (1994) Invasion and maintenance of a gene duplication. *Proc. Natl Acad. Sci. USA* **91**: 2950–2954.

Clemente I.C., Ponsa M., Garcia M., Egozcue J. (1990) Evolution of the Simiiformes and the phylogeny of human chromosomes. *Hum. Genet.* **84**: 493–506.

Cole R.A., Slade M.B., Williams K.L. (1995) *Dictyosteliym discoideum* mitochondrial DNA encodes a NADH: ubiquinone oxidoreductase subunit which is nuclear encoded in other eukaryotes. *J. Mol. Evol.* **40**: 616–621.

Collins D.W., Jukes T.H. (1994) Rates of transition and transversion in coding sequences since the human-rodent divergence. *Genomics* **20**: 386–396.

Collura R.V., Stewart C.B. (1995) Insertions and duplications of mtDNA in the nuclear genomes of Old World monkeys and hominoids. *Nature* **378**: 485–489.

Comas D., Calafell F., Mateu E., Perez-Lezaun A., Bosch E., Bertranpetit J. (1997) Mitochondrial DNA variation and the origin of the Europeans. *Hum. Genet.* **99**: 443–449.

Comings D.E. (1972) Evidence for ancient tetraploidy and conservation of linkage groups in mammalian chromosomes. *Nature* **238**: 455–457.

Consigliere S., Stanyon R., Koehler U., Agoramoorthy G., Wienberg J. (1996) Chromosome painting defines genomic rearrangements between red howler monkey subspecies. *Chromosome Res.* **4**: 264–270.

Cooper D.N., Krawczak M. (1997) *Venous Thrombosis; From Genes to Clinical Medicine.* BIOS Scientific Publishers, Oxford.

Cooper D.N., Millar D.S., Wacey A., Banner D.W., Tuddenham E.G.D. (1997) Inherited factor VII deficiency: molecular genetics and pathophysiology. *Thromb. Haemost.* **78**: 151–160.

Craig S.P., Buckle V.J., Lamouroux A., Mallet J., Craig I. (1986) Localization of the human tyrosine hydroxylase gene to 11p15: gene duplication and evolution of metabolic pathways. *Cytogenet. Cell Genet.* **42**: 29–32.

Crouau-Roy B., Service S., Slatkin M., Freimer N. (1996) A fine-scale comparison of the human and chimpanzee genomes: linkage, linkage disequilibrium and sequence analysis. *Hum. Molec. Genet.* **5**: 1131–1137.

Dahlbäck B. (1994) Physiological anticoagulation. Resistance to activated protein C and venous thromboembolism. *J. Clin. Invest.* **94**: 923–927.

DeBry R.W., Seldin F. (1996) Human/mouse homology relationships. *Genomics* **33**: 337–351.

Dean M., Carrington M., Winkler C., Huttley G.A., Smith M.W., Allikmets R., Goedert J.J., Buchbinder S.P., Vittinghoff E., Gomperts E., Donfield S., Vlahov D., Kaslow R., Saah A., Rinaldo C., Detels R., O'Brien S.J. (1996) Genetic restriction of HIV-1 infection and progression to AIDS by a deletion allele of the CKR5 structural gene. *Science* **273**: 1856–1861.

Deka R., Shriver M.D., Yu L.M., Jin L., Aston C.E., Chakraborty R., Ferrell R.E. (1994) Conservation of human chromosome 13 polymorphic microsatellite $(CA)_n$ repeats in chimpanzees. *Genomics* **22**: 226–230.

Delbridge M.L., Harry J.L., Toder R., Waugh O'Neill R.J., Ma K., Chandley A.C., Graves J.A.M. (1997) A human candidate spermatogenesis gene, *RBM1*, is conserved and amplified on the marsupial Y chromosome. *Nature Genet.* **15**: 131–136.

Dorit R.L., Akashi H., Gilbert W. (1995) Absence of polymorphism at the ZFY locus on the human Y chromosome. *Science* **268**: 1183–1185.

Duhig T., Ruhrberg C., Mor O., Fried M. (1998) The human surfeit locus. *Genomics* **52**: 72–78.

Dutriaux A., Rossier J., Van Hul W., Nizetic D., Theophilli D., Delabar J.M., Van Broeckhoven C., Potier M.-C. (1994) Cloning and characterization of a 135- to 500-kb region of homology on the long arm of human chromosome 21. *Genomics* **22**: 472–477.

Eder A.F., Spitalnik S.L. (1997) Blood group antigens as receptors for pathogens. In: *Molecular Biology and Evolution of Blood Group and MHC Antigens in Primates* (eds A. Blancher, J. Klein and W.W. Socha. Springer-Verlag, Berlin, Chap 8, pp 268–304.

Edwards J.H. (1994) Comparative genome mapping in mammals. *Curr. Opin. Genet. Devel.* **4**: 861–867.

Ehrmann I.E., Ellis P.S., Mazeyrat S., Duthie S., Brockdorff N., Mattei M.G., Gavin M.A., Affara N.A., Brown G.M., Simpson E., Mitchell M.J., Scott D.M. (1998) Characterization of genes encoding translation initiation factor eIF-2γ in mouse and human: sex chromosome localization, escape from X-inactivation and evolution. *Hum. Molec. Genet.* **7**: 1725–1737.

Elgar G., Sandford R., Aparicio S., Macrae A., Venkatesh B., Brenner S. (1996) Small is beautiful: comparative genomics with the pufferfish (*Fugu rubripes*). *Trends Genet.* **12**: 145–150.

Ellegren H., Primmer C.K., Sheldon B.C. (1995) Microsatellite evolution – directionality or bias. *Nature Genet.* **11**: 360–362.

Ellis NA. (1996) Human sex chromosome evolution. In: *Human Genome Evolution.* BIOS Scientific Publishers, Oxford, Chap 10, pp 229–261.

Ellis N.A. (1998) The war of the sex chromosomes. *Nature Genet.* **20**: 9–10.

Ellis N.A., Yen P., Neiswanger K., Shapiro L.J., Goodfellow P.J. (1990) Evolution of the pseudoautosomal boundary in Old World monkeys and great apes. *Cell* **63**: 977–986.

Endo T., Imanishi T., Gojobori T., Inoko H. (1998) Evolutionary significance of intra-genome duplications on human chromosomes. *Gene* **205**: 19–27.

Eppig J.T. (1996) Comparative maps: adding pieces to the mamalian jigsaw puzzle. *Curr. Opin. Genet. Devel.* **6**: 723–730.

Erlich H.A., Bergstrom T.F., Stoneking M., Gyllensten U. (1996) HLA sequence polymorphism and the origin of humans. *Science* **274**: 1552–1554.

Escoffre M., Zini J.M., Schliamser L., Mazoyer E., Soria C., Tobelem G., Dupuy E. (1995) Severe arterial thrombosis in a congenitally factor VII deficient patient. *Br. J. Haematol.* **91**: 739–741.

Ferris S.D., Brown W.M., Davidson W.S., Wilson A.C. (1981) Extensive polymorphism in the mitochondrial DNA of apes. *Proc. Natl Acad. Sci. USA* **78**: 6319–6323.

Fleagle J.G. (1988) *Primate Adaptation and Evolution.* Academic Press, San Diego.

Fu Y.X., Li W.-H. (1996) Estimating the age of the common ancestor of men from the *ZFY* intron. *Science* **272**: 1356–1357.

Fukagawa T., Sugaya K., Matsumoto K.-I., Okumura K., Ando A., Inoko H., Ikemura T. (1995) A boundary of long range G+C% mosaic domains in the human MHC locus: pseudoautosomal boundary-like sequence exists near the boundary. *Genomics* **25**: 184–191.

Fukagawa T., Nakamura Y., Katsuzumi O., Nogami M., Ando A., Inoko H., Saitou N., Ikemura T. (1996) Human pseudoautosomal boundary-like sequences: expression and involvement in evolutionary formation of the present-day pseudoautosomal boundary of human sex chromosomes. *Hum. Molec. Genet.* **5**: 23–32.

Gabriel S.E., Brigman K.N., Koller B.H., Boucher R.C., Stutts M.J. (1994) Cystic fibrosis heterozygote resistance to cholera toxin in the cystic fibrosis mouse model. *Science* **266**: 107–109.

Galili U., Swanson K. (1991) Gene sequences suggest inactivation of α-1,3-galactosyltransferase in catarrhines after the divergence of apes from monkeys. *Proc. Natl Acad. Sci. USA* **88**: 7401–7404.

Garcia-Fernandez J.G., Holland P.W.H. (1994) Archetypal organization of the amphioxus *Hox* gene cluster. *Nature* **370**: 563–566.

Garza J.C., Freimer N.B. (1996) Homoplasmy for size at microsatellite loci in humans and chimpanzees. *Genome Res.* **6**: 211–217.

Garza J., Slatkin M., Freimer N. (1995) Microsatellite allele frequencies in humans and chimpanzees, with implications for constraints in allele size. *Mol. Biol. Evol.* **12**: 594–603.

Gingerich P.D. (1984) Primate evolution: evidence from the fossil record, comparative morphology, and molecular biology. *Yearbook Phys. Anthropol.* **27**: 57–72.

Glaser B., Grutzner F., Willmann U., Stanyon R., Arnold N., Taylor K., Rietschel W., Zeitler S., Toder R., Schempp W. (1998) Simian Y chromosomes: species-specific rearrangements of *DAZ*, *RBM*, and *TSPY* versus contiguity of PAR and *SRY*. *Mamm. Genome* **9**: 223–231.

Goedde H.W., Agarwal D.P., Fritze G., Meier-Tackmann D., Singh S., Beckmann G., Bhatia K., Chen L.Z., Fang B., Lisker R., Paik Y.K., Rothhammer F., Saha N., Segal B., Srivastava L.M., Czeizel A. (1992) Distribution of ADH$_2$ and ALDH2 genotypes in different populations. *Hum. Genet.* **88**: 344–346.

Goffeau A., Barrell B.G., Bussey H., Davis R.W., Dujon B., Feldman H., Galibert F., Hoheisel J.D., Jacq C., Johnston M., Louis E.J Mewes H.W., Murakami Y., Philippsen. P., Tettelin H., Oliver S.G. (1997) Life with 6000 genes. *Science* **274**: 563–567.

Goldstein D.B., Ruiz Linares A., Cavalli-Sforza L.L., Feldman M.W. (1995) Genetic absolute dating based on microsatellites and the origin of modern humans. *Proc. Natl Acad. Sci. USA* **92**: 6723–6727.

Gonzalez I.L., Sylvester J.E., Smith T.F., Stambolian D., Schmickel R.D. (1990) Ribosomal RNA gene sequences and hominoid phylogeny. *Mol. Biol. Evol.* **7**: 203–219.

Goodman M., Koop B.F., Czelusniak J., Fitch D.H.A., Tagle D.A., Slightom J.L. (1989) Molecular phylogeny of the family of apes and humans. *Genome* **31**: 316–335.

Goodman M., Tagle D.A., Fitch D.H.A., Bailey W., Czelusniak J., Koop B.F., Benson P., Slightom J.L. (1990) Primate evolution at the DNA level and a classification of hominoids. *J. Mol. Evol.* **30**: 260–266.

Graves J.A.M. (1995) The origin and function of the mammalian Y chromosome and Y-borne genes—an evolving understanding. *Bioessays* **17**: 311–321.

Graves J.A.M., Disteche C.M., Toder R. (1998a) Gene dosage in the evolution and function of mammalian sex chromosomes. *Cytogenet. Cell Genet.* **80**: 94–103.

Graves J.A.M., Wakefield M.J., Toder R. (1998b) The origin and evolution of the pseudoautosomal regions of human sex chromosomes. *Hum. Molec. Genet.* **7**: 1991–1996.

Gray M.W. (1993) Origin and evolution of organelle genomes. *Curr. Opin. Genet. Devel.* **3**: 884–890.

Groves C.P. (1997) Taxonomy and phylogeny of primates. In: *Molecular Biology and Evolution of Blood Group and MHC Antigens in Primates* (eds. A. Blancher, J. Klein and W.W. Socha). Springer-Verlag, Berlin, Chap 1, pp 3–23.

Gyllensten U.B., Erlich H.A. (1989) Ancient roots for polymorphism at the HLA-DQ alpha locus in primates. *Proc. Natl Acad. Sci. USA* **86**: 9986–9990.

Haaf T., Bray-Ward P. (1996) Region-specific YAC banding and painting probes for comparative genome mapping: implications for the evolution of human chromosome 2. *Chromosoma* **104**: 537–544.

Hall F.C., Bowness P. (1996) HLA and disease: from molecular function to disease association? In: *HLA and MHC: Genes, Molecules and Function* (eds M.J. Browning and A.J. McMichael). BIOS Scientific Publishers, Oxford, pp. 353–381.

Hammer M.F. (1995) A recent common ancestry for human Y chromosomes. *Nature* **378**: 376–378.

Hanson I.M., Trowsdale J. (1991) Colinearity of novel genes in the class II regions of the MHC in mouse and human. *Immunogenetics* **34**: 5–11.

Harding R.M., Fullerton S.M., Griffiths R.C., Bond J., Cox M.J., Schneider J.A., Moulin D.S., Clegg J.B. (1997) Archaic African *and* Asian lineages in the genetic ancestry of modern humans. *Am. J. Hum. Genet.* **60**: 772–789.

Harpending H.C., Batzer M.A., Gurven M., Jorde L.B., Rogers A.R., Sherry ST. (1998) Genetic traces of ancient demography. *Proc. Natl Acad. Sci. USA* **95**: 1961–1967.

Hasegawa M., Kishino H., Yano T. (1987) Man's place in hominoidea as inferred from molecular clocks of DNA. *J. Mol. Evol.* **26**: 132–147.

Hedges S.B., Parker P.H., Sibley C.G., Kumar S. (1996) Continental breakup and the ordinal diversification of birds and mammals. *Nature* **381**: 226–229.

Hey J. (1997) Mitochondrial and nuclear genes present conflicting portraits of human origins. *Mol. Biol. Evol.* **14**: 166–172.

Hill A.V. (1996a) Genetics of infectious disease resistance. *Curr. Op. Genet. Devel.* **6**: 348–353.

Hill A.V. (1996b) Genetic susceptibility to malaria and other infectious diseases: from the MHC to the whole genome. *Parasitology* **112** Suppl.: S75–S84.

Hill A.V.S., Allsopp C.E.M., Kwiatkowski D. (1991) Common West African HLA antigens are associated with protection from severe malaria. *Nature* **352**: 595–600.

Hill R.E., Hastie N.D. (1987) Accelerated evolution in the reactive centre regions of serine protease inhibitors. *Nature* **326**: 96–99.

Hixson J.E., Brown W.M. (1986) A comparison of the small ribosomal genes from the mitochondrial DNA of the great apes and humans: sequence, structure, evolution and phylogenetic implications. *Mol. Biol. Evol.* **3**: 1–18.

Holland P.W.H. (1991) Cloning and evolutionary analysis of *msh*-like homeobox genes from mouse, zebrafish and ascidian. *Gene* **98**: 253–257.

Holmquist R., Miyamoto M.M., Goodman M. (1988) Analysis of higher-primate phylogeny from transversion differences in nuclear and mitochondrial DNA by Lake's methods of evolutionary parsimony and operator metrics. *Mol. Biol. Evol.* **5**: 217–236.

Hoppener J.W.M., de Pagter-Holthuizen P., Geurts van Kessel A.H.M., Jansen M., Kittur S.D., Antonarakis S.E., Lips C.J.M., Sussenbach J.S. (1985) The human gene encoding insulin-like growth factor I is located on chromosome 12. *Hum. Genet.* **69**: 157–160.

Horai S., Satta Y., Hayasaka K., Kondo R., Inoue T., Ishida T., Hayashi S., Takahata N. (1992) Man's place in hominoidea revealed by mitochondrial DNA genealogy. *J. Mol. Evol.* **34**: 32–43.

Horai S., Hayasaka K., Kondo R., Tsugane K., Takahata N. (1995) Recent African origin of modern humans revealed by complete sequences of hominoid mitochondrial DNAs. *Proc. Natl Acad. Sci. USA* **92**: 532–536.

Howard P.R., Bovill E.G., Pike J., Church W.R., Tracy R.P. (1994) Factor VII antigen levels in a healthy blood donor population. *Thromb. Haemost.* **72**: 21–27.

Hu G., Thilly W.G. (1994) Evolutionary trail of the mitochondrial genome as based on human 16S rDNA pseudogenes. *Gene* **147**: 197–204.

Huang W., Fu Y.X., Chang B.H., Gu X., Jorde L.B., Li W.-H. (1998) Sequence variation in *ZFX* introns in human populations. *Mol. Biol. Evol.* **15**: 138–142.

Hughes M.K., Hughes A.L. (1993) Evolution of duplicate genes in a tetraploid animal, *Xenopus laevis*. *Mol. Biol. Evol.* **10**: 1360–1369.

Hughes A.L., Nei M. (1988) Pattern of nucleotide substitution at major histocompatibility complex loci reveals overdominant selection. *Nature* **335**: 167–170.

Hughes A.L., Nei M. (1989) Nucleotide substitution at major histocompatibility complex class II loci: evidence for overdominant selection. *Proc. Natl Acad. Sci. USA* **86**: 958–962.

Humphries S., Temple A., Lane A., Green F., Cooper J., Miller G. (1996) Low plasma levels of factor VIIC and antigen are more strongly associated with the 10 base pair promoter (-323) insertion than the glutamine 353 variant. *Thromb. Haemost.* **75**: 567–572.

IJdo J.W., Baldini A., Ward D.C., Reeders S.T., Wells R.A. (1991) Origin of human chromosome 2: an ancestral telomere-telomere fusion. *Proc. Natl Acad. Sci. USA* **88**: 9051–9055.

Incerti B., Guioli S., Pragliola A., Zanaria E., Borsani G., Tonlorenzi R., Bardoni B., Franco B., Wheeler D., Ballabio A., Camerino G. (1992) Kallmann syndrome gene on the X and Y chromosomes: implications for evolutionary divergence of human sex chromosomes. *Nature Genet.* **2**: 311–314.

Jauch A., Wienberg J., Stanyon R., Arnold N., Tofanelli S., Ishida R., Cremer T. (1992) Reconstruction of genomic rearrangements in great apes and gibbons by chromosome painting. *Proc. Natl Acad. Sci. USA* **89**: 8611–8615.

Jegalian K., Page D.C. (1998) A proposed path by which genes common to mammalian X and Y chromosomes evolve to become X inactivated. *Nature* **394**: 776–780.

Jobling M. (1996) Y chromosome sequences as a tool for studying human evolution. In: *Human Genome Evolution*, (eds M. Jackson, T. Strachan and G. Dover). BIOS Scientific Publishers, Oxford, Chap 12, pp 283–301.

Jones S., Martin R., Pilbeam D. (eds). (1992) *The Cambridge Encyclopedia of Human Evolution*. Cambridge University Press, Cambridge.

Jorde L.B., Bamshad M.J., Watkins W.S., Zenger R., Fraley A.E., Krakowiak P.A., Carpenter K.D., Soodyall H., Jenkins T., Rogers A.R. (1995) Origins and affinities of modern humans: a comparison of mitochondrial and nuclear genetic data. *Am. J. Hum. Genet.* **57**: 523–538.

Jorde L.B., Rogers A.R., Bamshad M., Watkins W.S., Krakwiak P., Sung S., Kere J., Harpending H.C. (1997) Microsatellite diversity and the demographic history of modern humans. *Proc. Natl Acad. Sci. USA* **94**: 3100–3103.

Jorde L.B., Bamshad M., Rogers A.R. (1998) Using mitochondrial and nuclear DNA markers to reconstruct human evolution. *BioEssays* **20**: 126–136.

Kalow W. (1997) Pharmacogenetics in biological perspective. *Pharmacol. Rev.* **49**: 369–379.

Katsanis N., Fitzgibbon J., Fisher E.M.C. (1996) Paralogy mapping: identification of a region in the human MHC triplicated onto chromosomes 1 and 9 allows the prediction and isolation of novel *PBX* and *NOTCH* loci. *Genomics* **35**: 101–108.

Kawaguchi H., O'Huigin C., Klein J. (1992) Evolutionary origin of mutations in the primate cytochrome P450c21 gene. *Am. J. Hum. Genet.* **50**: 766–780.

Kawamura A., Tanabe H., Watanabe Y., Kurosaki K., Saitou N., Ueda S. (1991) Evolutionary rate of immunoglobulin alpha non-coding region is greater in hominoids than in Old World monkeys. *Mol. Biol. Evol.* **8**: 743–752.

Kjellman C Sjogren H-O Widegren B. (1995) The Y chromosome: a graveyard for endogenous retroviruses. *Gene* **161**: 163–170.

Knight A., Batzer M.A., Stoneking M., Tiwari H.K., Scheer W.D., Herrera R.J., Deininger P.L. (1996) DNA sequences of *Alu* elements indicate a recent replacement of the human autosomal genetic complement. *Proc. Natl Acad. Sci. USA* **93**: 4360–4364.

Koehler U., Bigoni F., Wienberg J., Stanyon R. (1995) Genomic reorganization in the Concolor gibbon (*Hylobates concolor*) revealed by chromosome painting. *Genomics* **30**: 287–292.

Kominato Y., McNeill P.D., Yamamoto M., Russel M., Hakomori S.-I., Yamamoto F. (1992) Animal histo-blood group ABO genes. *Biochem. Biophys. Res. Commun.* **189**: 154–165.

Koop B.F. (1995) Human and rodent DNA sequence comparisons: a mosaic model of genomic evolution. *Trends Genet.* **11**: 367–371.

Koop B.F., Hood L. (1994) Striking sequence similarity over almost 100 kilobases of human and mouse T-cell receptor DNA. *Nature Genet.* **7**: 48–53.

Koop B.F., Tagle D.A., Goodman M., Slightom J.L. (1989) A molecular view of primate phylogeny and important systematic and evolutionary questions. *Mol. Biol. Evol.* **6**: 580–612.

Koop B.F., Rowen L., Wang K., Kuo C.L., Seto D., Lenstra J.A., Howard S., Shan W., Deshpande P., Hood L. (1994) The human T-cell receptor *TCRAC/TCRDC* (Cα/Cδ) region: organization, sequence and evolution of 97.6 kb of DNA. *Genomics* **19**: 478–493.

Krings M., Stone A., Schmitz R.W., Krainitzki H., Stoneking M., Pääbo S. (1997) Neandertal DNA sequences and the origin of modern humans. *Cell* **90**: 1–20.

Kumar S., Hedges S.B. (1998) A molecular timescale for vertebrate evolution. *Nature* **392**: 917–919.

Kurihara T., Sakuma M., Gojobori T. (1997) Molecular evolution of myelin proteolipid protein. *Biochem. Biophys. Res. Commun.* **237**: 559–561.

Kurland C.G. (1992) Evolution of mitochondrial genomes and the genetic code. *Bioessays* **14**: 709–714.

Kvaloy K., Galvagni F., Brown W.R.A. (1994) The sequence organization of the long arm pseudoautosomal region of the human sex chromosomes. *Hum. Molec. Genet.* **3**: 771–778.

Lahn B.T., Page D.C. (1997) Functional coherence of the human Y chromosome. *Science* **278**: 675–679.

Lambson B., Affara N.A., Mitchell M., Ferguson-Smith M.A. (1992) Evolution of DNA sequence homologies between the sex chromosomes in primate species. *Genomics* **14**: 1032–1040.

Le Novre N., Changeux J.-P. (1995) Molecular evolution of the nicotinic acetylcholine receptor: an example of multigene family in excitable cells. *J. Mol. Evol.* **40**: 155–172.

Ledley F.D., Grenett H.E., Bartos D.P., van Tuinen P., Ledbetter D.H., Woo S.L.C. (1987) Assignment of human tryptophan hydroxylase locus to chromosome 11: gene duplication and translocation in evolution of aromatic amino acid hydroxylases. *Somat. Cell Molec. Genet.* **13**: 575–580.

Levin B.R., Lipsitch M., Bonhoeffer S. (1999) Population biology, evolution, and infectious disease: convergence and synthesis. *Science* **283**: 806–809.

Lewin R. (1998) *Principles of Human Evolution*. Blackwell, Malden.

Li W.-H., Tanimura M. (1987) The molecular clock runs more slowly in man than in apes and monkeys. *Nature* **326**: 93–96.

Li W.-H., Sadler L.A. (1991) Low nucleotide diversity in man. *Genetics* **129**: 513–523.

Libert F., Cochaux P., Beckman G., Samson M., Aksenova M., Cao A., Czeizel A., Claustres M., de la Ra C., Ferrari M., Ferrec C., Glover G., Grinde B., Güran S., Kucinkas V., Lavinha J., Mercier B., Ogur G., Peltonen L., Rosatelli C., Schwatz M., Spitsyn V., Timar L., Beckman L., Parmentier M., Vassart G. (1998) The Δccr5 mutation conferring protection against HIV-1 in Caucasian populations has a single and recent origin in Northeastern Europe. *Hum. Molec. Genet.* **7**: 399–406.

Lindqvist P.G., Svensson P.J., Dahlbäck B., Marsal K. (1998) Factor V Q506 mutation (activated protein C resistance) associated with reduced intrapartum blood loss – a possible evolutionary selection mechanism. *Thromb. Haemost.* **79**: 69–73.

Lomax M.I., Hewett-Emmett D., Yang T.L., Grossman L.I. (1992) Rapid evolution of the human gene for cytochrome *c* oxidase subunit IV. *Proc. Natl Acad. Sci. USA* **89**: 5266–5270.

Loomis W.F., Gilpin M.E. (1986) Multigene families and vestigial sequences. *Proc. Natl Acad. Sci. USA* **83**: 2143–2147.

Lundin L.G. (1993) Evolution of the vertebrate genome as reflected in paralogous chromosomal regions in man and the house mouse. *Genomics* **16**: 1–19.

McClearn G.E., Johansson. B., Berg S., Pedersen N.L., Ahern F., Petrill S.A., Plomin R. (1997) Substantial genetic influence on cognitive abilities in twins 80 or more years old. *Science* **276**: 1560–1563.

McConkey E.H. (1997) The origin of human chromosome 18 from a human/ape ancestor. *Cytogenet. Cell Genet.* **76**: 189–191.

McKusick V. (1980) The anatomy of the human genome. *J. Hered.* **71**: 370–391.

Maeda N., Wu C.-I., Bliska J., Reneke J. (1988) Molecular evolution of intergenic DNA in higher primates: pattern of DNA changes, molecular clock, and evolution of repetitive sequences. *Mol. Biol. Evol.* **5**: 1–20.

Makalowski W., Zhang J., Boguski M.S. (1996) Comparative analysis of 1196 orthologous mouse and human full-length mRNA and protein sequences. *Genome Res.* **6**: 846–857.

Maresco D.L., Blue L.E., Culley L.L., Kimberly R.P., Anderson C.L., Theil K.S. (1998) Localization of FCGR1 encoding Fcγ receptor class I in primates: molecular evidence for two pericentric inversions during the evolution of human chromosome 1. *Cytogenet. Cell Genet.* **82**: 71–74.

Marks J., Schmid C.W., Sarich V.M. (1988) DNA hybridization as a guide to phylogeny: relations of the hominoidea. *J. Hum. Evol.* **17**: 769–786.

Marquet S., Abel L., Hillaire D., Dessein H., Kalil J., Feingold J., Weissenbach J., Dessein A.J. (1996) Genetic localization of a locus controlling the intensity of infection by *Schistosoma mansoni* on chromosome 5q31-q33. *Nature Genet.* **14**: 181–184.

Martin R.D. (1993) Primate origins: plugging the gaps. *Nature* **363**: 223–233.

Martinko J.M., Vincek V., Klein D., Klein J. (1993) Primate ABO glycosyltransferases: evidence for trans-specific evolution. *Immunogenetics* **37**: 274–278.

Mefford H., Van den Engh G., Friedman C., Trask B.J. (1997) Analysis of the variation in chromosome size among diverse human populations by bivariate flow karyotyping. *Hum. Genet.* **100**: 138–144.

Meroni G., Franco B., Archidiacono N., Messali S., Andolfi G., Rocchi M., Ballabio A. (1996) Characterization of a cluster of sulfatase genes on Xp22.3 suggests gene duplications in an ancestral pseudoautosomal region. *Hum. Molec. Genet.* **5**: 423–431.

Merriwether D.A., Clark A.G., Ballinger S.W., Schurr T.G., Soodyall H., Jenkins T., Sherry S.T., Wallace D.C. (1991) The structure of human mitochondrial DNA variation. *J. Mol. Evol.* **33**: 543–555.

Mewes H.W., Albermann K., Bahr M., Frishman D., Gleissner A., Hani J., Heumann K., Kleine K., Maieri A., Oliver S.G., Pfeiffer F., Zollner A. (1997) Overview of the yeast genome. *Nature* **387**: 7–8.

Miklos G.L.G., Rubin G.M. (1996) The role of the genome project in determining gene function: insight from model organisms. *Cell* **86**: 521–529.

Mitchell R.J., Hammer M.F. (1996) Human evolution and the Y chromosome. *Curr. Opin. Genet. Devel.* **6**: 737–742.

Mitchell M.J., Wilcox S.A., Watson J.M., Lerner J.L., Woods D.R., Scheffler J., Hearn J.P., Bishop C.E., Graves J.A.M. (1998) The origin and loss of the ubiquitin activating enzyme gene on the mammalian Y chromosome. *Hum. Molec. Genet.* **7**: 429–434.

Miyamoto M.M., Slightom J.L., Goodman M. (1987) Phylogenetic relations of humans and African apes from DNA sequences in the χη-globin region. *Science* **238**: 369–372.

Miyamoto M.M., Koop B.F., Slightom J.L., Goodman M., Tennant M.R. (1988) Molecular systematics of higher primates: genealogical relations and classification. *Proc. Natl Acad. Sci. USA* **85**: 7627–7631.

Mohammad-Ali K., Eladari M.-E., Galibert F. (1995) Gorilla and orangutan c-*myc* nucleotide sequences: inference on hominoid phylogeny. *J. Mol. Evol.* **41**: 262–276.

Montgomery H.E., Marshall R., Hemingway *et al.* (1998) Human gene for physical endurance. *Nature* **393**: 221–222.

Morescalchi M.A., Schempp W., Consigliere S., Bignoni. F., Wienberg J., Stanyon R. (1997) Mapping chromosomal homology between humans and the black-handed spider monkey by fluorescence *in situ* hybridization. *Chrom. Res.* **5**: 527–536.

Motulsky A.G. (1995) Jewish diseases and origins. *Nature Genet.* **9**: 99–101.

Mumm S., Molini B., Terrell J., Srivastava A., Schlessinger D. (1997) Evolutionary features of the 4-Mb Xq21.3 XY homology region revealed by a map at 60-kb resolution. *Genome Res.* **7**: 307–314.

Nadeau J.H., Sankoff D. (1997) Comparable rates of gene loss and functional divergence after genome duplications early in vertebrate evolution. *Genetics* **147**: 1259–1266.

Nadeau J.H., Sankoff D. (1998) Counting on comparative maps. *Trends Genet.* **14**: 495–501.

Nadeau J.H., Grant P.L., Mankala S., Reiner A.H., Richardson J.E., Eppig J.T. (1995) A rosetta stone of mammalian genetics. *Nature* **373**: 363–365.

Nanda I., Shan Z., Schartl M., *et al.* (1999) 300 million years of conserved synteny between chicken Z and human chromosome 9. *Nature Genet.* **21**: 258–259.

Nei M. (1995) Genetic support for the out-of-Africa theory of human evolution. *Proc. Natl Acad. Sci. USA* **92**: 6720–6722.

Nei M., Takezaki N. (1996) The root of the phylogenetic tree of human populations. *Mol. Biol. Evol.* **13**: 170–177.

Nei M., Gu X., Sitnikova T. (1997) Evolution by the birth-and-death process in multigene families of the vertebrate immune system. *Proc. Natl Acad. Sci. USA* **94**: 7799–7806.

Nickerson E., Nelson D.L. (1998) Molecular definition of pericentric inversion breakpoints occurring during the evolution of humans and chimpanzees. *Genomics* **50**: 368–372.

Nishio H., Gibbs P.E.M., Minghetti P.P., Zielinski R., Dugaiczyk A. (1995) The chimpanzee α-fetoprotein-encoding gene shows structural similarity to that of gorilla but distinct differences from that of human. *Gene* **162**: 213–220.

Novacek M.J. (1992) Mammalian phylogeny: shaking the tree. *Nature* **356**: 121–125.

Nowak M.A., Boerlijst M.C., Cooke J., Smith J.M. (1997) Evolution of genetic redundancy. *Nature* **388**: 167–170.

O'Brien S.J., Womack J.E., Lyons L.A., Moore K.J., Jenkins N.A., Copeland NG. (1993) Anchored reference loci for comparative genome mapping in mammals. *Nature Genet.* **3**: 103–112.

O'Huigin C., Sato A., Klein J. (1997) Evidence for convergent evolution of A and B blood group antigens in primates. *Hum. Genet.* **101**: 141–148.

Oakey R., Watson M.L., Seldin M.F. (1992) Construction of a physical map on mouse and human chromosome 1: comparison of 13 Mb of mouse and 11 Mb of human DNA. *Hum. Molec. Genet.* **1**: 613–620.

Oeltjen J.C., Malley T.M., Muzny D.M., Miller R.A., Gibbs R.A., Belmont J.W. (1997) Large-scale comparative sequence analysis of the human and murine Bruton's tyrosine kinase loci reveals conserved regulatory domains. *Genome Res.* **7**: 315–329.

Ohno S. (1967) *Sex Chromosomes and Sex-linked Genes*. Springer-Verlag, Berlin.

Ohno S. (1970) *Evolution by Gene Duplication*. Springer-Verlag, New York.

Ohta T. (1989) Role of gene duplication in evolution. *Genome* **31**: 304–310.

Ohta T. (1991) Multigene families and the evolution of complexity. *J. Mol. Evol.* **33**: 34–41.

Ohta T. (1994) Further examples of evolution by gene duplication revealed through DNA sequence comparisons. *Genetics* **138**: 1331–1337.

Ohta T., Basten C.J. (1992) Gene conversion generates hypervariability at the variable regions of kallikreins and their inhibitors. *Mol. Phylogenet. Evol.* **1**: 87–90.

Orti R., Potier M.C., Maunoury C., Prieur M., Créau N., Delabar J.M. (1998) Conservation of pericentromeric duplications of a 200-kb part of the human 21q22.1 region. *Cytogenet. Cell Genet.* **83**: 262–265.

Page D.C., Harper M.E., Love J., Botstein D. (1984) Occurrence of a transposition from the X-chromosome long arm to the Y-chromosome short arm during human evolution. *Nature* **311**: 119–123.

Pamilo P., Bianchi N.O. (1993) Evolution of the *Zfx* and *Zfy* genes: rates and interdependence between the genes. *Mol. Biol. Evol.* **10**: 271–281.

Pamilo P., Waugh, O'Neill R.J. (1997) Evolution of the *Sry* genes. *Mol. Biol. Evol.* **14**: 49–55.

Park J.P., Wojiski S.A., Spellman R.A., Rhodes C.H., Mohandas T.K. (1998) Human chromosome 9 pericentric homologies: implications for chromosome 9 heteromorphisms. *Cytogenet. Cell Genet.* **82**: 192–194.

Pebusque M.J., Coulier F., Birnbaum D., Pontarotti P. (1998) Ancient large-scale genome duplications: phylogenetic and linkage analyses shed light on chordate genome evolution. *Mol. Biol. Evol.* **15**: 1145–1159.

Peelman L.J., Chardon P., Vaiman M., Matthews M., Van Zeveren A., Van de Weghe A., Bouquet Y., Campbell R.D. (1996) A detailed physical map of the porcine major histocompatibility complex (MHC) class III region: comparison with human and mouse MHC class III regions. *Mamm. Genome* **7**: 363–367.

Pellicciari C., Formenti D., Redi C.A., Manfredi Romanini M.G. (1982) DNA content variability in primates. *J. Hum. Evol.* **11**: 131–141.

Pendleton J.W., Nagai B.K., Murtha M.T., Ruddle F.H. (1993) Expansion of the *Hox* gene family and the evolution of chordates. *Proc. Natl Acad. Sci. USA* **90**: 6300–6304.

Perez-Lezaun A., Calafell F., Mateu E., Comas D., Ruiz-Pacheco R., Bertranpetit J. (1997) Microsatellite variation and the differentiation of modern humans. *Hum. Genet.* **99**: 1–7.

Perrin-Pecontal P., Gouy M., Nigon V.-M., Trabuchet G. (1992) Evolution of the primate β-globin gene region: nucleotide sequence of the δβ-globin intergenic region of gorilla and phylogenetic relationship between African apes and man. *J. Mol. Evol.* **34**: 17–30.

Perry O., Feather S., Smith A., Palmer S., Ashworth A. (1998) The human *FXY* gene is located within Xp22.3: implications for evolution of the mammalian X chromosome. *Hum. Molec. Genet.* **7**: 299–305.

Petruzzella V., Tiranti V., Fernandez P., Ianna P., Carrozzo R., Zeviani M. (1998) Identification and characterization of human cDNAs specific to BCS1, PET112, SCO1, COX15, and COX11, five genes involved in the formation and function of the mitochondrial respiratory chain. *Genomics* **54**: 494–504.

Plummer N.W., Meisler M.H. (1999) Evolution and diversity of mammalian sodium channel genes. *Genomics* **57**: 323–331.

Pollak E.S., Hung H.-L., Godin W., Overton G.C., High K.A. (1996) Functional characterization of the human factor VII 5′-flanking region. *J. Biol. Chem.* **271**: 1738–1747.

Postlethwait J.H., Yan Y.-L., Gates M.A., et al. (1998) Vertebrate genome evolution and the zebrafish gene map. *Nature Genet.* **18**: 345–349.

Qumsiyeh M.B. (1994) Evolution of number and morphology of mammalian chromosomes. *J. Hered.* **85**: 455–465.

Reich D.E., Goldstein D.B. (1998) Genetic evidence for a Paleolithic human population expansion in Africa. *Proc. Natl Acad. Sci. USA* **95**: 8119–8123.

Relethford J.H., Harpending H.C. (1995) Ancient differences in population size can mimic a recent African origin of modern humans. *Curr. Anthropol.* **36**: 667–674.

Retief J.D., Winkfein R.J., Dixon G.H., Adroer R., Queralt R., Ballabriga J., Oliva R. (1993) Evolution of protamine P1 genes in primates. *J. Mol. Evol.* **37**: 426–434.

Richard F., Dutrillaux B. (1998) Origin of human chromosome 21 and its consequences: a 50-million-year-old story. *Chromosome Res.* **6**: 263–268.

Richard F., Lombard M., Dutrillaux B. (1996) ZOO-FISH suggests a complete homology between human and capuchin monkey (Platyrrhini) euchromatin. *Genomics* **36**: 417–423.

Ried T., Arnold C., Ward D.C., Wienberg J. (1993) Comparative high resolution mapping of human and primate chromosomes by fluorescence *in situ* hybridization. *Genomics* **18**: 381–386.

Ried K., Rao E., Schiebel K., Rappold G.A. (1998) Gene duplications as a recurrent theme in the evolution of the human pseudoautosomal region 1: isolation of the gene *ASMTL*. *Hum. Molec. Genet.* **7**: 1771–1778.

Rihet P., Traore Y., Abel L., Aucan C., Traore-Leroux T., Fumoux F. (1998) Malaria in humans: *Plasmodium falciparum* blood infection levels are linked to chromosome 5q31-q33. *Am. J. Hum. Genet.* **63**: 498–505.

Rogers J., Witte S.M., Kammerer C.M., Hixson J.E., MacCluer J.W. (1995) Linkage mapping in *Papio* baboons: conservation of a syntenic group of six markers on chromosome 1. *Genomics* **28**: 251–254.

Rowen L., Mahairas G., Hood L. (1997) Sequencing the human genome. *Science* **278**: 605–607.

Rubin E.M., Barsh G.S. (1996) Biological insights through genomics: mouse to man. *J. Clin. Invest.* **97**: 275–280.

Rubinztein D.C., Amos W., Leggo J., Goodurn S., Jain S., Li S., Margolis R.L., Ross C.A., Ferguson-Smith M.A. (1995) Microsatellite evolution—evidence for directionality and variation in rate between species. *Nature Genet.* **10**: 337–343.

Ruiz-Linares A., Nayar K., Goldstein D.B., Hebert J.M., Seielstad M.T., Underhill P.A., Lin A.A., Feldman M.W., Cavalli Sforza L.L. (1996) Geographic clustering of human Y-chromosome haplotypes. *Ann. Hum. Genet.* **60**: 401–408.

Rumpler Y., Dutrillaux B. (1990) *Chromosomal Evolution and Speciation in Primates. Revisiones Sobre Biologia Celular* vol. **23**. Springer International/University of the Basque Country.

Ruvolo M., Disotell T.R., Allard M.W., Brown W.M., Honeycutt R.L. (1991) Resolution of the African hominoid trichotomy by use of a mitochondrial gene sequence. *Proc. Natl Acad. Sci. USA* **88**: 1570–1574.

Ruvolo M., Zehr S., von Dornum M., Pan D., Chang B., Lin J. (1993) Mitochondrial COII sequences and modern human origins. *Mol. Biol. Evol.* **10**: 1115–1135.

Sadlock J.E., Lightowlers R.N., Capaldi R.A., Schon EA. (1993) Isolation of a cDNA specifying subunit VIIb of human cytochrome *c* oxidase. *Biochim. Biophys. Acta* 1172: 223–225.

Saitou N. (1991) Reconstruction of molecular phylogeny of extant hominoids from DNA sequence data. *Am. J. Phys. Anthropol.* **84**: 75–86.

Saitou N., Nei M. (1986) The number of nucleotides required to determine the branching order of three species with special reference to the human-chimpanzee-gorilla divergence. *J. Mol. Evol.* **24**: 189–204.

Samonte R.V., Conte R.A., Ramesh K.H., Verma R.S. (1996) Molecular cytogenetic characterization of breakpoints involving pericentric inversions of human chromosome 9. *Hum. Genet.* **98**: 576–580.

Sarich V.M., Schmid C.W., Marks J. (1989) DNA hybridization as a guide to phylogeny: a critical analysis. *Cladistics* 5: 3–32.

Saxena R., Brown L.G., Hawkins T., Alagappan R.K., Skaletsky H., Reeve M.P., Reijo R., Rozen R., Dinulos M.B., Disteche C.M., Page D.C. (1996) The DAZ gene cluster on the human Y chromosome arose from an autosomal gene that was transposed, repeatedly amplified and pruned. *Nature Genet.* **14**: 292–299.

Scarabin P.-Y., Vissac A.-M., Kirzin J.-M., Bourgeat P., Amiral J., Agher R., Guize L. (1996) Population correlates of coagulation factor VII. *Arterioscler. Thromb. Vasc. Biol.* **16**: 1170–1176.

Schwartz A., Chan D.C., Brown L.G., Alagappan R., Pettay D., Disteche C., McGillivray B., de la Chapelle A., Page D.C. (1998) Reconstructing hominid evolution: X-homologous block, created by X-Y transposition, was disrupted by Yp inversion through LINE-LINE recombination. *Hum. Molec. Genet.* **7**: 1–11.

Seielstad M.T., Minch E., Cavalli-Sforza L. (1998) Genetic evidence for a higher female migration rate in humans. *Nature Genet.* **20**: 278–280.

Shan Z., Hirschmann P., Seebacher T., Edelmann A., Jauch A., Morell J., Urbitsch P., Vogt P.H. (1996) A *SPGY* copy homologous to the mouse gene *dazla* and the *Drosophila* gene *boule* is autosomal and expressed only in the human male gonad. *Hum. Molec. Genet.* **5**: 2005–2011.

Sharman A.C., Hay-Schmidt A., Holland P.W.H. (1996) Cloning and analysis of an *HMG* gene from the lamprey *Lampetra fluviatilis*: gene duplication in vertebrate evolution. *Gene* 184: 99–105.

Shaw M.A., Davies C.R., Llanos-Cuentas E.A., Collins A. (1995) Human genetic susceptibility and infection with *Leishmania peruviana*. *Am. J. Hum. Genet.* **57**: 1159–1168.

Shaw M.A., Collins A., Peacock C.S., Miller E.N., Black G.F., Sibthorpe D., Lins-Lainson Z., Shaw J.J., Ramos F., Silveira F., Blackwell J.M. (1997) Evidence that genetic susceptibility to *Mycobacterium tuberculosis* is under oligogenic control: linkage study of the candidate genes NRAMP1 and TNFA. *Tubercle Lung Dis.* **78**: 35–45.

Sherlock J.K., Griffin D.K., Delhanty J.D.A., Parrington J.M. (1996) Homologies between human and marmoset (*Callithrix jacchus*) chromosomes revealed by comparative chromosome painting. *Genomics* 33: 214–219.

Shimmin L.C., Chang B.H., Li W.H. (1993) Male-driven evolution of DNA sequences. *Nature* **362**: 745–747.

Shoffner JM Wallace DC. (1990) Oxidative phosphorylation diseases. Disorders of two genomes. *Adv. Hum. Genet.* **19**: 267–330.

Sibley C.G., Ahlquist J.E. (1984) The phylogeny of the hominoid primates as indicated by DNA-DNA hybridization. *J. Mol. Evol.* **20**: 2–15.

Sibley C.G., Ahlquist J.E. (1987) DNA hybridization evidence of hominoid phylogeny: results from an expanded data set. *J. Mol. Evol.* **26**: 99–121.

Sibley C.G., Comstock J.A., Ahlquist J.E. (1990) DNA hybridization evidence of hominoid phylogeny: a reanalysis of the data. *J. Mol. Evol.* **30**: 202–236.

Sidow A. (1996) Gen(om)e duplications in the evolution of early vertebrates. *Curr. Opin. Genet. Devel.* **6**: 715–722.

Sidow A., Thomas W.K. (1994) An evolutionary framework for eukaryotic model organisms. *Curr. Biol.* **4**: 596–603.

Silveria A., Green F., Karpe F., Blombck M., Humphries S., Hamsten A. (1994) Elevated levels of factor VII activity in the postprandial state: effect of the factor VII Arg-Gln polymorphism. *Thromb. Haemost.* **72**: 734–739.

Simmen M.W., Leitgeb S., Clark V.H., Jones S.J., Bird A. (1998) Gene number in an invertebrate chordate, *Ciona intestinalis. Proc. Natl Acad. Sci. USA* **95**: 4437–4440.

Simmler M.-C., Rouyer F., Vergnaud G., Nystrom-Lahti M., Ngo K.Y., de la Chapelle A., Weissenbach J. (1985) Pseudoautosomal DNA sequences in the pairing region of the human sex chromosomes. *Nature* **317**: 692–697.

Skrabanek L., Wolfe K.H. (1998) Eukaryotic genome duplication—where's the evidence? *Curr. Opin. Genet. Devel.* **8**: 694–700.

Smith M.T. (1993) Genetic adaptation. In: *Human Adaptation* (ed. G.A. Harrison). Oxford University Press, Oxford, Chap 1, pp 1–54.

Smith M.W., Dean M., Carrington M., *et al.* (1997) Contrasting genetic influence of CCR2 and CCR5 variants on HIV-1 infection and disease progression. *Science* **277**: 959–965.

Sorenson M.D., Fleischer R.C. (1996) Multiple independent transpositions of mitochondrial DNA control region sequences to the nucleus. *Proc. Natl Acad. Sci. USA* **93**: 15 239–15 243.

Sparrow A.H., Nauman A.F. (1976) Evolution of genome size by DNA doublings. *Science* **192**: 524–529.

Spring J. (1997) Vertebrate evolution by interspecific hybridisation—are we polyploid? *FEBS Letts.* **400**: 2–8.

Stanyon R., Wienberg J., Romagno D., Bigoni F., Jauch A., Cremer T. (1992) Molecular and classical cytogenetic analyses demonstrate an apomorphic reciprocal chromosomal translocation in *Gorilla gorilla. Am. J. Phys. Anthropol.* **88**: 245–250.

Stoneking M. (1996) Mitochodrial DNA variation and human evolution. In: *Human Genome Evolution* (eds M. Jackson, T. Strachan and G. Dover). BIOS Scientific Publishers, Oxford, Chap 11, pp 263–281.

Sudhof T.C., Russell D.W., Goldstein J.L., Brown M.S., Sanchez-Pescador R., Bell G.I. (1985) Cassette of eight exons shared by genes for LDL receptor and EGF precursor. *Science* **228**: 893–895.

Takahata N. (1993) Allelic genealogy and human evolution. *Mol. Biol. Evol.* **10**: 2–22.

Thursz M.R., Thomas H.C., Greenwood B.M., Hill A.V.S. (1997) Heterozygote advantage for HLA class-II type in hepatitis B virus infection. *Nature Genet.* **17**: 11–12.

Tishkoff S.A., Dietzsch E., Speed W., Pakstis A.J., Kidd J.R., Cheung K., Bonné-Tamir B., Santachiara-Benerecetti A.S., Moral P., Krings M. (1996) Global patterns of linkage disequilibrium at the CD4 locus and modern human origins. *Science* **271**: 1380–1387.

Toder R., Rappold G., Schiebel K., Schempp W. (1995) *ANT3* and *STS* are autosomal in prosimian lemurs: implications for the evolution of the pseudoautosomal region. *Hum. Genet.* **95**: 22–28.

Toyosawa S., O'Huigin C., Figueroa F., Tichy H., Klein J. (1998) Identification and characterization of amelogenin genes in monotremes, reptiles, and amphibians. *Proc. Natl Acad. Sci. USA* **95**: 13 056–13 061.

Trask B., van den Engh G., Mayall B., Gray J.W. (1989a) Chromosome heteromorphism quantified by high-resolution bivariate flow karyotyping. *Am. J. Hum. Genet.* **45**: 739–752.

Trask B., van den Engh G., Gray J.W. (1989b) Inheritance of chromosome heteromorphisms analyzed by high-resolution bivariate flow karyotyping. *Am. J. Hum. Genet.* **45**: 753–760.

Trower M.K., Orton S.M., Purvis I.J., Sanseau P., Riley J., Christodoulou C., Burt D., See C.G., Elgar G., Sherrington R., Rogaev E.I., St George-Hyslop P., Brenner S., Dykes C.W. (1996) Conservation of synteny between the genome of the pufferfish (*Fugu rubripes*) and the region on human chromosome 14 (14q24.3) associated with familial Alzheimer disease (*AD3* locus). *Proc. Natl Acad. Sci. USA* **93**: 1366–1369.

Ueda S., Takenaka O., Honjo T. (1985) A truncated immunoglobulin ε pseudogene is found in gorilla and man but not in chimpanzee. *Proc. Natl Acad. Sci. USA* **82**: 3712–3715.

Ueda S., Watanabe Y., Saitou N., Omoto K., Hayashida H., Miyata T., Hisajima H., Honjo T. (1989) Nucleotide sequences of immunoglobulin-ε pseudogenes in man and apes and their phylogenetic relationships. *J. Mol. Biol.* **205**: 85–90.

Underhill P.A., Jin L., Lin A.A., Mehdi S.Q., Jenkins T., Vollrath D., Davis R.W., Cavalli-Sforza L.L., Oefner P.J. (1997) Detection of numerous Y chromosome biallelic polymorphisms by denaturing high-performance liquid chromatography. *Genome Res.* **7**: 996–1005.

Vigilant L., Stoneking M., Harpending H., Hawkes K., Wilson A.C. (1991) African popula… and the evolution of mitochondrial DNA. *Science* **253**: 1503–1507.

Vogel F., Motulsky A.G. (1997) *Human Genetics: Problems and Approaches*. 3rd edn, Springer, Berlin.

Vogt P.H., Affara N., Davey P., Hammer M., Jobling M.A., Lau Y.F.-C., Mitchell M., Schempp W., Tyler-Smith C., Williams G., Yen P., Rappold G.A. (1997) Report of the third international workshop on Y chromosome mapping 1997. *Cytogenet. Cell Genet.* **79**: 1–20.

von Haeseler A., Sajantila A., Pääbo S. (1995) The genetical archaeology of the human genome. *Nature Genet.* **14**: 135–140.

Wainscoat J.S., Hill A.V., Boyce A.L., Flint J., Hernandez M., Thein S.L., Old J.M., Lynch J.R., Falusi A.G., Weatherall D.J. (1986) Evolutionary relationships of human populations from an analysis of nuclear DNA polymorphisms. *Nature* **319**: 491–493.

Wallace D.C., Stugard C., Murdock D., Schurr T., Brown M.D. (1997) Ancient mtDNA sequences in the human nuclear genome: a potential source of errors in identifying pathogenic mutations. *Proc. Natl Acad. Sci. USA* **94**: 14 900–14 905.

Wallace D.C., Ye J.H., Neckelmann S.N., Singh G., Webster K.A., Greenberg B.D. (1987) Sequence analysis of cDNAs for the human and bovine ATP synthase beta subunit: mitochondrial DNA genes sustain seventeen times more mutations. *Curr. Genet.* **12**: 81–90.

Wallis M. (1993) Remarkably high rate of molecular evolution of ruminant placental lactogens. *J. Mol. Evol.* **37**: 86–88.

Weiss G., von Haeseler A. (1996) How many Ys? *Science* **272**: 1358–1360.

Weller P.A., Critcher R., Goodfellow P.N., German J., Ellis N.A. (1995) The human Y chromosome homologue of *XG*: transcription of a naturally truncated gene. *Hum. Molec. Genet.* **4**: 859–868.

Whitfield L.S., Lovell-Badge R., Goodfellow P.N. (1993) Rapid sequence evolution of the mammalian sex-determining gene *SRY*. *Nature* **364**: 713–715.

Whitfield L.S., Sulston J.E., Goodfellow P.N. (1995) Sequence variation of the human Y chromosome. *Nature* **378**: 379–382.

Wienberg J., Stanyon R. (1997) Comparative painting of mammalian chromosomes. *Curr. Opin. Genet. Devel.* **7**: 784–791.

Wienberg J., Jauch A., Stanyon R., Cremer T. (1990) Molecular cytotaxonomy of primates by chromosomal *in situ* suppression hybridization. *Genomics* **8**: 347–350.

Wienberg J., Jauch A., Lüdecke H.-J., Senger G., Hosthemke B., Claussen U., Cremer T., Arnold N., Lengauer C. (1994) The origin of human chromosome 2 analyzed by comparative chromosome mapping with a DNA microlibrary. *Chrom. Res.* **2**: 405–410.

Wilson A.C., Carlson S.S., White T.J. (1977) Biochemical evolution. *Annu. Rev. Biochem.* **46**: 573–639.

Winkler C., Modi W., Smith, M.W., *et al.* (1998) Genetic restriction of AIDS pathogenesis by an SDF-1 chemokine gene variant. *Science* **279**: 389–393.

Wolf U., Schempp W., Scherer G. (1992) Molecular biology of human Y chromosome. *Rev. Physiol. Biochem. Pharmacol.* **121**: 148–213.

Wolfe K.H., Shields D.C. (1997) Molecular evidence for an ancient duplication of the entire yeast genome. *Nature* **387**: 708–713.

Wood B. (1992) Origin and evolution of the genus *Homo*. *Nature* **355**: 783–790.

Wood B. (1996) Human evolution. *BioEssays* **18**: 945–954.

Xiong W., Li W.-H., Posner I., Yamamura T., Yamamoto A., Gotto A.M., Chan L. (1991) No severe bottleneck during human evolution: evidence from two apolipoprotein C-II deficiency alleles. *Am. J. Hum. Genet.* **48**: 383–389.

Yen P.H., Marsh B., Allen E., Tsai S.P., Ellison J., Connolly L., Neiswanger. K., Shapiro L.J. (1988) The human X-linked steroid sulfatase gene and a Y-encoded pseudogene: evidence for an inversion of the Y chromosome during primate evolution. *Cell* **55**: 1123–1135.

Young J.Z. (1973) *The Life of Vertebrates*. 2nd edn. Clarendon Press, Oxford.

Yunis J.J., Prakash O. (1982) The origin of man: a chromosomal pictorial legacy. *Science* **215**: 1525–1529.

Zietkiewicz E., Yotova V., Jarnik M., Korab-Laskowska M., Kidd K.K., Modiano D., Scozzari R., Stoneking M., Tishkoff S., Batzer M., Labuda D. (1998) Genetic structure of the ancestral population of modern humans. *J. Mol. Evol.* **47**: 146–155.

Zischler H., Geisert H., Von Haeseler A., Paabo S. (1995) A nuclear 'fossil' of the mitochondrial D-loop and the origin of modern humans. *Nature* **378**: 489–492.

Zivelin A., Rosenberg N., Faier S., Kornbrot N., Peretz H., Mannhalter C., Horellou M.H., Seligsohn U. (1998) A single genetic origin for the common prothrombotic G20210A polymorphism in the prothrombin gene. *Blood* **92**: 1119–1124.

Zlotogora J. (1994) High frequencies of human genetic diseases: founder effect with genetic drift or selection? *Am. J. Med. Genet.* **49**: 10–13.

PART 2
EVOLUTION OF GENE STRUCTURE

Introns, exons, and evolution

3.1 Intron structure, function and evolution

Eukaryotic genes are not usually contiguous entities but are instead interrupted by noncoding sequences termed introns. How were introns first acquired? Although there is at present no clear unambiguous answer to this question, there appears to be a continuous evolutionary line from the archaic self splicing group II introns of eubacteria and simple eukaryotes (Belfort, 1993; Lambowitz and Belfort, 1993) via protein-assisted self-splicing introns to the splicing exhibited by extant eukaryotic nuclear protein-encoding genes. It is therefore reasonable to speculate that group II introns could have been introduced by the invasion of eukaryotic cells by the bacterial endosymbiotic ancestor of the mitochondrion (Cavalier-Smith, 1991).

Whatever their origin, a close correlation exists between intron density and developmental complexity. Thus introns are largely, although not exclusively, confined to the eukaryotes. The archaebacteria, which form a third major taxon distinct from the eukaryotes and eubacteria, possess small introns in their tRNA and rRNA genes whilst eubacteria and a few eukaryotes possess introns that catalyse their own splicing. An average of one intron per kilobase (kb) of coding sequence is found in simple eukaryotes such as *Dictyostelium* and *Plasmodium*, 3–4 per kb in plants and fungi and 6 per kb in vertebrates (Palmer and Logsdon, 1991). The average cellular gene in vertebrates contains about seven introns (Sharp, 1994). Intron size also increases with phylogenetic complexity with intronic sequence accounting for only 10–20% of primary transcripts in the protista but as much as 95% in vertebrates (Cavalier-Smith, 1985).

As far as human genes are concerned, introns are the rule rather than the exception (Chapter 1, section 1.2.1). Some human genes have nevertheless been found which lack introns. These include the sex determining (*SRY*; Yp11.3) gene (O'Neil *et al.*, 1998), the POU domain transcription factor *POU3F2* (6q16) gene (Atanasoski *et al.*, 1995), the thrombomodulin (*THBD*; 20p11.2) gene (Jackman *et al.*, 1987), the β2-adrenergic receptor (*ADRB2*; 5q32-q34) gene (Kobilka *et al.*, 1987), the *JUN* proto-oncogene (1p31-p32; Hattori *et al.*, 1988), the recombination-activating genes *RAG1* and *RAG2* (11p13), the arylamine *N*-acetyltransferase (*AAC1*; 8p21.3-p23.1) gene (Grant *et al.*, 1989), the 13 genes of the interferon-α (*IFNA*) gene cluster and the interferon-β gene (*IFNB1*) at 9p21 (Diaz *et al.*, 1994), the heat shock 70 kDa protein genes (*HSPA2*, 14q22-q24; *HSPA1A*, 6p21.3; Milner and Campbell 1990), the pyruvate dehydrogenase E1α subunit (*PDHA1*; Xp22) gene (Dahl *et al.*, 1990), the formyl peptide receptor (*FPR1*; chromosome 19) gene (De Nardin *et al.*, 1992), the serotonin receptor genes *HTR1D* (1p34.3-p36.3) and *HTR1F* (Demchyshyn *et al.*, 1992; Adham *et al.*, 1993), the casein kinase 2-α1 subunit (*CSNK2A1*; 20p13) gene (Devilat and Carvallo, 1993), the *ID2* gene (2p25) encoding an inhibitor of DNA binding (Kurabayashi *et al.*, 1993), the purinergic receptor (*P2RY1*; chromosome 3) gene (Ayyanathan *et al.*, 1996) and the calmodulin-like genes (*CALML3*, 10p13-pter; *CALML1*, 7p13-pter; Berchtold *et al.*, 1993; Koller and Strehler, 1993). Intriguingly, >90% of the human genes that encode the G-protein-coupled receptor family are intronless, a finding which may reflect a retrotranspositional origin prior to gene duplication (Gentles and Karlin, 1999).

Introns in human genes vary enormously in size, from as little as 24 bp in the case of the parvalbumin (*PVALB*; 22q12-q13) gene, to in excess of 600 kb in the case of the α1(V) collagen (*COL5A1*; 9q34.2-q34.3) gene (Takahara *et al.*, 1995). Minimum intron size may be determined by the need to prevent steric hindrance between splicing factors. Intron size is not usually well conserved between orthologous genes. Thus during the evolution of the higher primates, intron 8 of the lamin B (*LMNB2*; 19p13.3) gene increased very dramatically in size as a result of a repeat expansion (de Stanchina *et al.*, 1997), whilst intron 6 of the Ewing sarcoma breakpoint region 1 (*EWSR1*; 22q12) gene expanded progressively through successive retrotransposition and recombination events involving *Alu* sequences (Zucman-Rossi *et al.*, 1997). Similarly, the paralogous murine phospholipase D1 and D2 genes differ in size as a result of a 20-fold expansion/contraction of intron size in one of the genes (Redina and Frohman, 1998). There are, however, notable exceptions to the rule of lack of conservation; for example, the sizes of the 53 introns of the human and mouse α1(II) collagen (*COL2A1*; 12q12-q13.2) genes differ on average by only 13% (Ala-Kokko *et al.*, 1995).

In the context of intron size, one enigma is the pufferfish (*Fugu rubripes*) whose 400 Mb genome is 1/7 the size of the human genome, is relatively devoid of repetitive DNA, and contains comparatively short introns (75% are less than 120 bp in length) (Brenner *et al.*, 1993). A dramatic example of the economy manifested by the *Fugu* genome is provided by the relative size of intron 7 of the Duchenne muscular dystrophy (*DMD*; Xp21) gene in *Fugu* (2.4 kb) as compared to its human counterpart (109.6 kb) (McNaughton *et al.*, 1997). At least 40% of the human intron is made up of LINE elements, *Alu* sequences, THE-1 and related LTR sequences, interspersed repeat sequences, a *mariner* transposon and other repeats including microsatellites whose insertion served to double the size of the intron over the last 130 Myrs (McNaughton *et al.*, 1997). This example is not unrepresentative of the size differences noted between orthologous human and *Fugu* introns; thus the neurofibromatosis type 1 (*NF1*; 17q11) gene spans only 27 kb in *Fugu* as compared to 335 kb in human (Kehrer-Sawatzki *et al.*, 1998). Whether or not the size of the *Fugu* genome represents the ancestral state of the early vertebrate genome is unclear but its size may well approach the minimum sustainable for a vertebrate. We can only speculate as to the possible reasons why selection has acted on a genome-wide basis to minimize intron size in this species. Perhaps the *Fugu* genome has been resistant to colonization by transposable elements and so has no endogenous source of reverse transcriptase to aid the process of retrotransposition.

Since a high proportion of human genes belong to gene families or super-families which have arisen through a process of duplication and divergence (Chapter 4), it is hardly surprising that the locations of introns are often evolutionarily conserved. Such conservation may be evident in terms of the structures of orthologous genes, for example the phenylalanine hydroxylase (*PAH*; 12q22-q24) genes of *Drosophila* and human (Ruiz-Vazquez *et al.*, 1996) or the myoglobin gene of the abalone *Sulculus diversicolor* and the human gene encoding indoleamine 2,3-dioxygenase (*IDO*; 8p11-p12; Suzuki *et al.*, 1996). It may also be apparent between paralogous genes, members of the same gene family within a given species, for example between the highly homologous human monoamine oxidase A and B genes (*MAOA, MAOB*; Xp11.23; Grimsby *et al.*, 1991) or between the human gene encoding the α-subunit of the granulocyte-macrophage colony stimulating factor receptor (*CSF2RA*; Xp22.32) and other members of the cytokine receptor family (Nakagawa *et al.*, 1994). However, it is also clear that some evolutionarily related genes can possess quite divergent intron distributions, for example those encoding the actin-regulatory proteins, gelsolin (*GSN*; 9q33), the villins (*VIL1*, 2q35-36; *VIL2*, 6q22-q27) and the capping protein Cap G (*CAPG*; 2cen-q24) (Mishra *et al.*, 1994), or the complement proteins *C6* (chromosome 5), *C7* (chromosome 5) and *C9* (5p12-p14) (Hobart *et al.*, 1993) or the fibrinogens (*FGA, FGB, FGG*; 4q31) (Crabtree *et al.*, 1985). A particularly dramatic example of post-duplication structural divergence is provided by the human microfibril-associated glycoprotein genes *MFAP2* (1p35-p36) and *MAGP2* (12p12-p13). These two genes comprise 9 and 10 exons respectively, but sequence and structural conservation as well as conservation of intron location is confined to exons 8 and 9 of the *MFAP2* gene and exons 7 and 8 of the *MAGP2* gene (Hatzinikolas and Gibson, 1998). These exons encode the first six of the seven precisely aligned cysteine residues at the center of both proteins. Divergent exon distribution can arise through intron insertion and deletion (Section 3.5). Alternatively, it can also arise by recombination, an example of this is provided by the human prosaposin (*PSAP*; 10q22.1) gene. The *PSAP* gene is polycistronic, encoding a precursor for four saposins (A, B, C, and D). Saposins A, B, and D are encoded by three exons, saposin C by only two. Analysis of intron locations has indicated that the *PSAP* gene evolved by two duplication events and at least one rearrangement (Rorman *et al.*, 1992). This rearrangement is thought to have involved a double crossover, between the first and second and between the second and third intron positions of the saposin B and C coding regions, after the introduction of introns (Rorman *et al.*, 1992).

Intron sequences themselves are not usually well conserved during evolution (Sharp, 1994), a finding which is not surprising in view of the fact that most intronic sequence is likely to be nonfunctional. However, there are some notable exceptions: the single intron of the oligodendrocyte-myelin glycoprotein (*OMG*; 17q11-q12) gene exhibits 75% overall homology between human and mouse (Mikol *et al.*, 1993) whilst the 53 introns of the *COL2A1* (12q12–13.2) gene differ by 69% between the two species (Ala-Kokko *et al.*, 1995). Sequences in the third intron of the γ-actin (*ACTG1*; 17p11-qter) gene are highly conserved between human and *Xenopus* (Erba *et al.*, 1988). Perhaps the most dramatic example of the conservation of intronic sequence is that evident in a comparison of 97 kb of the human and murine T-cell receptor α/δ (*TCRA, TCRD*; 14q11.2) gene locus which exhibits 66% overall sequence homology even though <6% corresponds to exonic sequence (Koop and Hood, 1994).

There are exceptions to the general rule that intron sequences evolve more rapidly than exons. For instance, the second exons of the human semenogelin genes (*SEMG1, SEMG2*; 20q12-q13.1) have evolved very rapidly when compared to their rat homologues, and more so even than the flanking introns (Lundwall and Lazure, 1995). Similarly, the divergence between the red and green visual pigment (*GCP, RCP*; Xq28) genes in humans, chimpanzees and baboons is lower in intron 4 than in exons 4 and 5 of these genes (Shyue *et al.*, 1994; Zhou and Li, 1996). In this case, homogenization of intron 4 sequences is thought to have been brought about by gene conversion (see Chapter 9, section 9.5) whilst selection has probably acted so as to confine gene conversion to the intron in order to retain the distinct functions of exons 4 and 5 between the two genes.

Evolutionary conservation may imply function but we are only just beginning to elucidate the function of conserved sequences within introns. The introns of some genes contain the coding sequences of other genes (see Chapter 1, section 1.2.1). Thus, the *OMG, EVI2A, EVI2B* genes occur within intron 27b of the human neurofibromatosis type 1 (*NF1*; 17q11.2) gene and are transcribed in the opposite direction to the *NF1* gene. An orthologue of the *EVI2B* gene is present in the corresponding *Fugu* intron but this intron reveals no trace of the *OMG* or *EVI2A* genes (Kehrer-Sawatzki *et al.*, 1998). This indicates that the *EVI2B* gene must have been inserted into the *NF1* gene more than 450 Myrs ago whilst the *OMG* gene must have been a more recent acquisition.

Some intronic sequence motifs perform transcriptional regulatory functions. Positive regulatory (enhancer-like) elements have now been reported from the first introns of a considerable number of human genes, for example the genes encoding type X collagen (*COL10A1*; 6q21-q22.3; Beier *et al.*, 1997), type I 3β-hydroxysteroid dehydrogenase (*HSD3B2*; 1p13.1; Guerin *et al.*, 1995), tissue inhibitor of metalloproteinase 1 (*TIMP1*; Xp11.23-p11.3; Clark *et al.*, 1997), factor IX (*F9*; Xq27; Kurachi *et al.*, 1995), heat shock protein 90 (*HSPCB*; 6p12; Shen *et al.*, 1997), Bruton's tyrosine kinase (*BTK*; Xq21.3-q22; Rohrer and Conley 1998), purine nucleoside phosphorylase (*NP*; 14q13.1; Jonsson *et al.*, 1992), O6-methylguanine DNA methyltransferase (*MGMT*; 10q26; Harris *et al.*, 1994), type I collagen (*COL1A1*; 17q21.3-q22; Hormuzdi *et al.*, 1998), IgE receptor (*FCER1B*; 11q13; Lacy *et al.*, 1994), growth hormone (*GH1*; 17q22-q24; Slater *et al.*, 1985; Kolb *et al.*, 1998), thymidylate synthase (*TYMS*; 18p11.32; Takayanagi *et al.*, 1992) and dystrophin (*DMD*; Xp21.2; Klamut *et al.*, 1996). Although enhancer sequences are most commonly found within the first intron of genes, such sequence elements are also occasionally found in other locations, for example the second intron of the human apolipoprotein B (*APOB*; 2p24) gene (Rosby *et al.*, 1992), the third intron of the human oxytocin receptor (*OXTR*; 3p26) gene (Mizumoto *et al.*, 1997) and intron 8 of the δ-aminolevulinate synthase 2 (*ALAS2*; Xp11.21) gene (Surinya *et al.*, 1998). Negative regulatory elements (repressors) within introns have also been characterized, for example in the first introns of the human Bruton's tyrosine kinase (*BTK*; Xq21.3-q22; Rohrer and Conley, 1998) and acid maltase (*GAA*; 17q25.2-q25.3; Raben *et al.*, 1996) genes. A highly conserved CCAAT element within intron 1 of the proliferating cell nuclear antigen (*PCNA*; 20p12) gene also appears to act as a negative regulatory element (Alder *et al.*, 1992).

Other sequences within introns may affect nucleosome formation (Denisov *et al.*, 1997), or play a regulatory role in the processing of the primary transcript either by modulating mRNA splicing or by influencing mRNA stability through RNA-DNA, RNA-RNA or RNA-protein interactions (Mattick, 1994). For example, sequences within the second intron of the human β-globin (*HBB*; 11p15.5) gene are important in promoting efficient 3' end formation and appear to be essential to the stability of the cytoplasmic *HBB* mRNA (Antoniou *et al.*, 1998). A further fortuitous function of introns may be to act as 'lightning conductors' for the retrotranspositional insertion of mobile elements thereby protecting the nearby coding sequences from inactivation (Ferlini and Muntoni 1998). A glimpse of what surprises introns may have in store for us is to be found within the introns of the human U22 host gene. Seven fibrillarin-associated small nucleolar RNAs (U25-U31), with complementarity to different segments of rRNA, are encoded within different introns of this gene and are highly conserved between human and mouse (Tycowski *et al.*, 1996). The spliced U22 host gene mRNA species has by contrast little coding potential, is short lived and is evolutionarily poorly conserved. The surprise therefore is that in the U22 host gene, it is the 'introns' rather than the 'exons' which appear to specify the functional products.

3.2 The evolution of alternative processing

3.2.1 Alternative splicing

Alternative splicing allows the generation of different mRNAs (and therefore a diverse array of protein isoforms) from the same gene (*Figure 3.1*). It is therefore an important mechanism for the tissue-specific or developmental regulation of gene expression. The potential evolutionary advantages of alternative splicing have been discussed in detail by Smith *et al.* (1989) and hence will now only be summarized briefly. Unlike gene duplication or rearrangement, alternative splicing does not change the gene structure or copy number and need not therefore be irreversible in genetic terms. Since existing splicing pathways need not necessarily be discarded in order to employ new ones, alternative splicing is likely to be particularly useful as a means to generate protein diversity during early development and in very long lived and terminally differentiated cells. The use of alternative splicing may also facilitate the efficient exploitation of intragenic duplications since if the transcript of the newly duplicated gene were to be alternatively spliced, the gene could continue to produce the old gene product as well as the new one. Although alternative splicing implies the existence of cell and developmental stage-specific splicing factors, it is still unclear whether it is a predecessor or a refinement of constitutive splicing. Smith *et al.* (1989) argued that the two processes could have evolved simultaneously. If, indeed, many genes have evolved by exon shuffling (Section 3.6), then novel exon combinations may have been tested by alternative splicing. If these conferred a selective advantage, then they could have been fixed by mutation at the appropriate splice junction rendering them constitutive. Consistent with this view, most alternatively spliced exons appear to have originated by exon duplication.

There are now several examples of alternative splicing being evolutionarily conserved in orthologous genes. For instance, exon 9a of the proto-oncogene MTG8/ETO (*CBFA2T1*; 8q22) is alternatively spliced in both mouse and human (Wolford and Prochazka 1998). A developmental alternative splicing switch, involving exon 16 of the erythroid protein 4.1 (*EPB41*; 1p) gene, occurs during mammalian erythropoiesis and this alternative splice also occurs in *Xenopus* (Winardi *et al.*, 1995). The reason for conservation of this switch over 350 Myrs of evolution probably lies in the fact that it controls the expression of a 21 amino acid peptide required for the protein's high affinity interactions with spectrin and actin that help to regulate erythrocyte membrane stability. The alternative splicing of exons 2 and 27 of the neural cell adhesion molecule L1 (*NCAM1*; 11q23-q24) gene has been conserved in human and the puffer fish (*Fugu rubripes*) demonstrating evolutionary conservation of the alternative splicing mechanism over some 430 Myrs (Coutelle *et al.*, 1998). In the α-tropomyosin (*TPM1*, 15q22) gene, alternative splicing to produce tissue-specific isoforms has been conserved from *Drosophila* to human, corresponding to a timespan of at least 700 Myrs (Wieczorek *et al.*, 1988). Exon 6′ of the chromatin condensation regulator gene (*CHC1*; 1p36.1) is involved in alternative splicing in both humans and hamsters even though this exon encodes 31 amino acids in human but only 13 in hamster

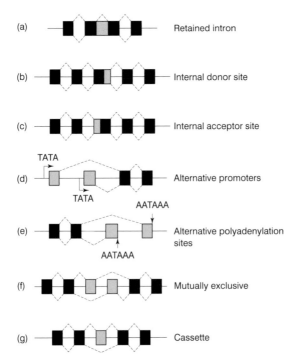

(a) Retained intron

(b) Internal donor site

(c) Internal acceptor site

(d) TATA — Alternative promoters

(e) TATA AATAAA — Alternative polyadenylation sites — AATAAA

(f) Mutually exclusive

(g) Cassette

Figure 3.1. Types of alternative splicing (redrawn from Li, 1997). Constitutively spliced exons are denoted by black boxes and alternatively spliced exons by shaded boxes. Splicing patterns are shown by interrupted diagonal lines. Alternative promoter and polyadenylation sites are denoted by TATA and AATAAA motifs respectively.

(Miyabashira *et al.*, 1994). Alternatively spliced forms of the nuclear factor I/C (*NFIC*) and I/X (*NFIX*) genes, linked on human chromosome 19p13.3, are evolutionarily conserved from chickens to humans (Kruse and Sippel, 1994). The chicken CP49 gene, encoding a 49 kDa cytoskeletal protein, generates two different mRNA transcripts, one containing a novel exon encoding 49 amino acids of helix 1B, the other lacking it (Wallace *et al.*, 1998). The transcript containing this exon is however absent from both the orthologous bovine and human (*BFSP2*; 3q21-q25) genes. Similarly, alternative splicing of the Wilms' tumor (*WT1*; 11p13) gene is evident in various mammals but not in the pufferfish, *Fugu rubripes* (Miles *et al.*, 1998). Finally, in the fibronectin (*FN1*; 2q34) gene, where exon EIIIB is alternatively spliced in a cell-type-specific manner, the pattern of TGCATG repeats in the downstream intron has been evolutionarily conserved across the vertebrates suggesting a possible role for these sequences in splice site selection (Lim and Sharp, 1998).

Alternative processing can however also have had a relatively recent origin. Insertion of a B2 (SINE) element into the 3' untranslated region of the murine leukemia inhibitory factor receptor (*lifr*) gene (human equivalent, *LIFR*; 5p12-p13) encoding the soluble form of the leukemia inhibitory factor receptor (LIFR) has, by potentiating alternative 3' mRNA processing and alternative splicing, given rise to a truncated mRNA species (relative to the mRNA encoding the membrane-anchored LIFR) which encodes soluble LIFR (Michel *et al.*, 1997). In the rat, no such retrotranspositional event has occurred and the soluble form of LIFR is not found. The potential for alternative processing must therefore have arisen in the last 20–30 Myrs. Two species of mRNA encoding the metabotropic glutamate receptor subtype 5 (*GRM5*) gene occur in rat and human which differ in terms of the presence of a 96 bp insertion thought to result from alternative mRNA processing (Minakami *et al.*, 1993). Another example of the relatively recent evolution of alternative splicing has been noted in the lecithin: cholesterol acyltransferase (*LCAT*; 16q22) gene and is present in humans and the great apes but not in gibbons or Old World and New World monkeys (Miller and Zeller, 1997). Known examples of single base-pair substitutions in splice sites of specific orthologous gene pairs that are responsible for the evolutionary emergence of alternative splicing are discussed in Chapter 7, section 7.5.3.

One of the most dramatic inter-specific differences in alternative splicing involves the human *t* complex responder (*TCP10*; 6q27) gene and its murine orthologue (Islam *et al.*, 1993). The human *TCP10* transcript includes two exons not present in mouse transcripts, whilst mouse *tcp10* transcripts include up to four exons that are not present in the human transcript. It is at present unclear whether a given exon has been incorporated into the transcript of one species or lost from the other. The recruitment and removal of exons is potentially explicable by single base-pair substitutions either eliminating splice junctions or converting intronic DNA sequence into novel splice sites. Islam *et al.* (1993) described this mode of evolution as 'punctuated equilibrium', a term that is used to refer to evolutionary situations in which phenotypic characters appear to leap from one equilibrium state to another in a short space of time. These authors suggested that this mode of evolution might have been facilitated by the existence of a gene family in which individual members possessed redundant functions. One family member would therefore have been free to pass through a nonfunctional state in which several base changes accumulated, resulting eventually in a change in splicing phenotype. Whether selection or genetic drift could have been responsible was unclear. Unfortunately in this context, the authors did not see fit to discuss the intron phase of the added/removed exons.

Evidence for intergenic differences in alternative splicing during evolution has also come from the study of paralogous genes. Although the intron–exon distribution in the human annexin VI (*ANX6*; 5q32-q34) gene conforms exactly to that found in the human annexin I (*ANX1*; 9q11-q22) and II (*ANX2*; 15q21-q22) genes, exon 21 of the annexin VI gene (encoding 6 amino acids) is alternatively spliced (Smith *et al.*, 1994). Similarly, the human chloride channel gene *CLCN6* (1p36) contains an alternatively spliced 167 bp exon that is absent in the evolutionarily related *CLCN1* (7q32-qter), *CLCN5* (Xp) and *CLCN7* (16p13) genes (Eggermont 1998). The genetic basis of these examples of gene-specific alternative splicing has however not yet been elucidated. Known examples of single base-pair substitutions in paralogous genes responsible for the emergence of alternative splicing are discussed in Chapter 7, section 7.5.3.

Alternative processing may become redundant after gene duplication as exemplified by the troponin I gene which encodes one of the three subunits of the troponin complex of the thin filaments of vertebrate striated muscle. Invertebrates and ascidians possess a single troponin I gene (Hastings, 1997) and this is alternatively spliced to generate proteins that differ in their N-terminal regions. By contrast, the three troponin I genes in the human genome (*TNNI1*, 1q31; *TNNI2*, 11p15; *TNNI3*, 19q13) and in the genomes of other vertebrates do not undergo alternative splicing although the 'extra' exon in the *TNNI3* gene (as compared with the *TNNI1* and *TNNI2* genes) is thought to correspond to the exon which is alternatively spliced in ascidians and invertebrates (Hastings, 1997).

A database of alternatively spliced genes (ASDB) is available online at **http://cbcg.nersc.gov/asdb** and contains information about alternatively spliced genes in different organisms and in different tissues.

Alternative transcripts may also be generated by the differential utilization of polyadenylation sites (*Figure 3.1*). The differential utilization of two distinct polyadenylation sites in the plasminogen activator inhibitor 1 (*PAI1*; 7q22) gene (yielding two mRNA species of 2.6 kb and 3.6 kb) has been conserved between humans, orangutans and African green monkeys (Cicila *et al.*, 1989). However, only one *PAI1* mRNA species is apparent in lower primates and nonprimate mammals (Cicila *et al.*, 1989) consistent with the acquisition of an extra polyadenylation site during primate evolution. The evolution of alternative processing may have come about through the differential utilization of polyadenylation sites via the newly described mechanism of LINE element-mediated recombination described in Section 3.6.1). Finally, alternative transcripts may also arise through alternative promoter usage. There is very little information on inter-specific differences in promoter site selection: the example of the insulin-like growth factor II gene is cited in Section 3.7.

Herbert and Rich (1999) have stressed the potential importance of RNA processing in evolutionary processes. Their perspective, which deserves close examination, was elegantly summarized in one of their introductory paragraphs:

The outcome of one RNA processing event can affect the outcome of others resulting in regulatory 'networks' that influence both the information content and expression of the 'ribotype' (RNA pool), which in turn influences phenotype. Alterations in RNA processing generate different sets of ribotypes which are subject to natural selection on the basis of the phenotypes they produce. The evolutionary process requires the preservation of successful ribotypes in an inheritable (*sic*) form.

3.2.2 Aberrant transcripts

There are numerous examples in the literature of aberrantly processed mRNA transcripts being produced in addition to correctly processed mRNAs in the cells of normal healthy individuals (Berg *et al.*, 1996). It is possible that some of these 'aberrant transcripts' could be of functional significance but in the majority of cases they are likely to occur simply as a consequence of the absence of any selective pressure to avoid aberrant splicing. One example of this phenomenon is provided by the human glutathione peroxidase type 5 (*GPX5*) gene in which the majority of gene transcripts appear to be incorrectly spliced (Hall *et al.*, 1998). If taken to its extreme, however, and the extent of aberrant splicing increases beyond a certain point, a gene may come to be effectively inactivated as has occurred in the case of the human chorionic somatomammotropin (*CSHL1*; 17q22-q24) 'pseudogene' (Misra-Press *et al.*, 1994).

3.2.3 mRNA surveillance

The cell does however possess a mechanism that is capable of removing those aberrant transcripts that encode 'nonsense mRNAs' which would otherwise be expected to give rise to truncated proteins. Such mRNAs are often inherently unstable as a result of a process known as *mRNA surveillance* which involves nonsense-mediated mRNA decay (Culbertson, 1999). mRNA surveillance appears to be ubiquitous in eukaryotes with homologous proteins encoded by orthologous genes being found in organisms as widely separated as yeast (Upf1, Upf2, Upf3) and human (*RENT1*; 19p13; Culbertson, 1999). Presumably this mechanism has evolved in order to reduce the cost to the cell of producing non-productive transcripts.

3.2.4 Ectopic transcripts

Extremely low levels of correctly spliced mRNA transcripts from tissue-specific genes have been demonstrated in supposedly 'non-expressing' cell types (reviewed by Cooper *et al.*, 1994). Such *ectopic* or *illegitimate* transcripts have been found to occur at levels as low as one mRNA molecule per 500–1000 cells, >1000-fold lower than the level of an average 'low abundance' mRNA. It remains unclear whether every cell is capable of generating ectopic transcripts or alternatively if the occasional cell in an otherwise non-expressing tissue is able to produce comparatively high levels of the transcript in question. Do ectopic transcripts have a biological role? Since the levels involved are often extremely low, it is hard to imagine that such a role involves significant protein synthesis. Perhaps what we observe simply represents a reasonable balance between the cost to the cell of producing nonproductive transcripts from many 'non-expressed' genes and the cost of turning off the transcription of many thousands of 'leaky' genes completely.

3.3 Exon structure and evolution

The average cellular gene in vertebrates contains about eight exons (Sharp, 1994). A comprehensive analysis of 5061 exons from some 2705 intron-containing human genes has yielded a classification scheme for exons according to their transcriptional or translational boundaries (Zhang, 1998). This scheme (depicted in *Figure 3.2*) employs 12 different categories: (1) 5' terminal untranslated exons, (2) 3' terminal untranslated exons, (3) 5' terminal exons having a 5' untranslated sequence (UTR) followed by a coding sequence, (4) 3' terminal exons having a 3' untranslated sequence (UTR) followed by a coding sequence, (5) internal exons having a 5' portion of 5' UTR followed by coding sequence, (6) internal exons having a coding sequence followed by a 3' portion of 3' UTR, (7) internal untranslated exons, (8) internal translated exons, (9) exons containing the complete coding sequence but which do not contain the transcriptional end, (10) exons containing the complete coding sequence but which do not contain the transcriptional start, (11) exons containing the complete coding sequence and both the transcriptional start and the transcriptional end, and (12) exons containing the complete coding sequence but neither the transcriptional start nor the transcriptional end. Although no doubt subject to ascertainment bias, the sample of exons studied by Zhang (1998) exhibited the following frequencies by type for the first 8 categories of exon: (1) 271, (2) 38, (3) 482, (4) 553, (5) 174, (6) 69, (7) 34, (8) 3440. Exons in categories (3) and (7) were in general found to be <100 bp in length whereas exons in categories (2) and (4) were mostly 300–500 bp in length. There appears to be virtually no minimum size for an internal translated exon (category 8) in human genes; the smallest include the 4 bp exon 3 of the skeletal muscle troponin (*TNNI1*; 19q31) gene and two 3 bp exons (10 and 14) in the cardiac myosin binding protein C (*MYBPC3*; 11p11.2) gene. As far as the maximum size of an internal exon is concerned, the human *F8C* (3106 bp; Xq28), *APOB* (7572 bp; 2p23-p24) and *MUC5B* (10,690 bp; 11p15.5) genes are currently league leaders.

Zhang (1998) found that the size of 5' UTR varied between human genes from 0 bp to 2077 bp with a mean of 136 bp. The size of the 3' UTR varies from −2 bp to 3427 bp with a mean of 589 bp (Zhang, 1998). The −2 value, for the human α-D-galactosidase A (*GLA*; Xq) gene, is due to the fact that the coding sequence including the stop codon ends at nucleotide 11268 whilst the poly(A) addition site is located at 11266.

3.4 Introns early or introns late?

Were introns used in the assembly of the first genes or were they added only later to previously contiguous coding sequences? The *'introns early'* theory or simply the *'exon theory of genes'* proposes that the genes encoding complex extant proteins

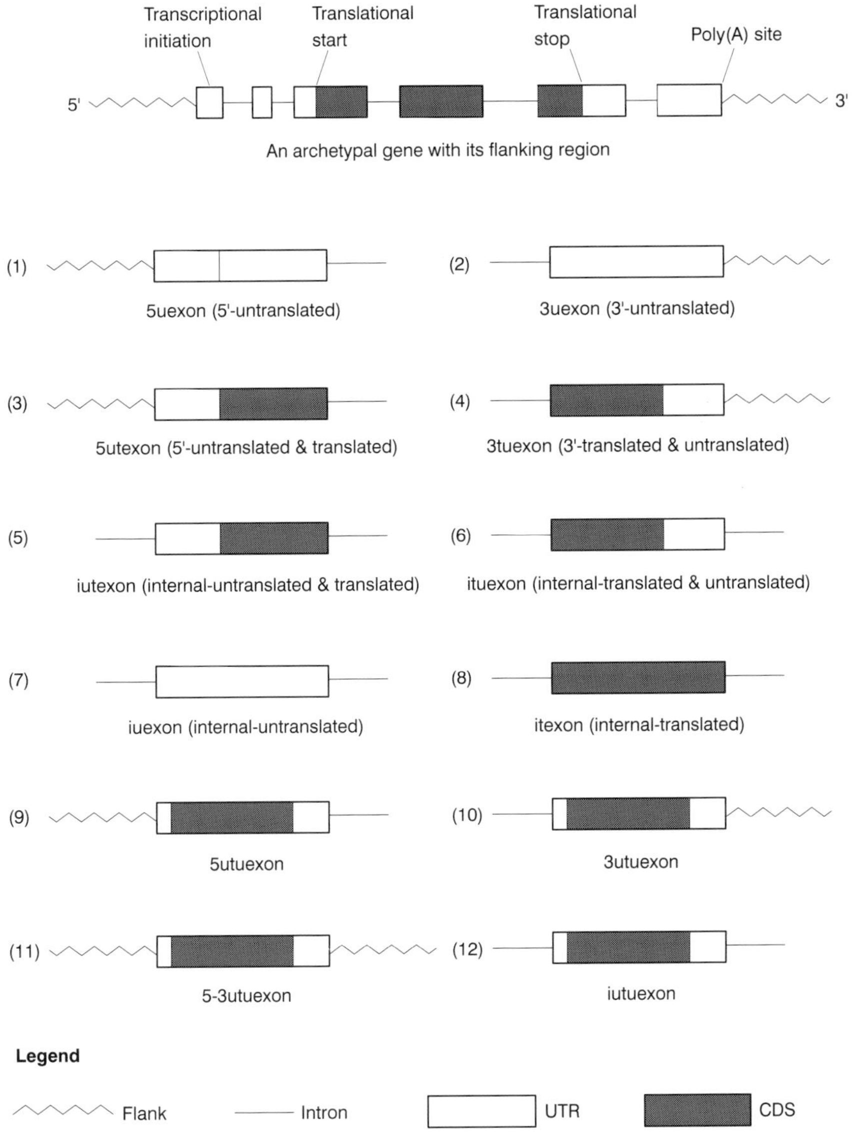

Figure 3.2. Exon classification. All exons can be classified into these 12 mutually exclusive classes. At the top, a schematic gene model is depicted which indicates how some types of exons may be organized in a gene (redrawn from Zhang, 1998). For explanation, see text. UTR: Untranslated region. CDS: Coding sequence.

emerged through the coalescence of primordial minigenes (Gilbert, 1987, 1997). These minigenes are held to have originally encoded protein modules and are now represented as exons whereas the non-coding linker DNA between the minigenes has survived as introns. Introns were then lost and novel exons made by fusing smaller exons together. By contrast, the *'introns late'* theory postulates that fully functional genes had introns inserted into them at different stages in their evolution

(Cavalier-Smith, 1991; Palmer and Logsdon, 1991). At its extremes, the debate between these two viewpoints is reminiscent of the old controversies between selectionists and neutralists. However, perhaps not altogether surprisingly, the debate has generated a significant amount of evidence supporting both theories (Gilbert *et al.*, 1997; Rogers 1990).

One prediction of the introns-early hypothesis is that intron location should tend to demarcate structural or functional domains within proteins (Blake, 1978; Branden *et al.*, 1984; Craik *et al.*, 1983; Go, 1987). This question has proved highly contentious (Traut, 1988). The extreme introns-late view is that no such correlation exists and that introns have been inserted quasi-randomly into the structure of genes. Consistent with this standpoint, several genes encoding evolutionarily ancient proteins (actins, alcohol dehydrogenase, carbonic anhydrase II, pyruvate kinase, globins) appear not to exhibit any correlation between intron position and structural features (Stoltzfus *et al.*, 1994; Weber and Kabsch, 1994). However, more recently, a computer program designed to predict precisely the locations of module boundaries in proteins has helped to demonstrate that there is indeed a strong correlation between intron position and structural elements at least in the sample of 32 ancient proteins examined (de Souza *et al.*, 1996). The introns involved were invariably phase zero (de Souza *et al.*, 1998; see Section 3.6.2).

As we have noted in Section 3.2, the conservation of intron location is a common finding. That intron location is often conserved in evolutionarily ancient genes such as glyceraldehyde-3-phosphate dehydrogenase (*GAPD*; 12p13; Kersanach *et al.*, 1993), carbamoylphosphate synthetase (*CPS1*; 2q35; van den Hoff *et al.*, 1995) and triose phosphate isomerase (*TPI1*; 12p13; Gilbert and Glynias 1993; Marchionni and Gilbert 1986; McKnight *et al.*, 1986; Straus and Gilbert 1985) is certainly compatible with the introns-early hypothesis. Most informative in terms of the introns-early/late debate, however, are discordant cases, examples of evolutionarily related genes from different taxa in which either intron position varies from between several nucleotides to several codons or where the intron is either present or absent (Rogers, 1989). The introns-early theorists have explained these discrepancies by invoking *intron sliding* and deletion (Craik *et al.*, 1983) but despite the occasional convincing example [e.g. intron 8 of the histidyl-tRNA synthetase (*HARS*; chromosome 5) gene; Brenner and Corrochano, 1996; *Figure 3.3*; intron 2 of the glucose-dependent insulinotropic peptide (*GIP*; 17q21.3-q22) gene which results in an 8 amino acid deletion of the prepropeptide of the rat protein as compared to human; Higashimoto and Liddle, 1993], the evidence for the widespread occurrence of intron sliding is still rather weak (Yuasa *et al.*, 1997; Stoltzfus *et al.*, 1997). The introns-late view is that intron sliding is inherently improbable because such a process would almost innevitably involve an intermediate stage that would alter the reading frame thereby leading to the loss or inactivation of the protein product. Therefore introns-late devotees have regarded discordant intron location, as found in the evolutionarily ancient α- and β-tubulin (*TUBA1*, 2q; *TUBA2*, 13q11; *TUBB*, 6p21-pter; Dibb and Newman, 1989), aldehyde dehydrogenase (*ALDH1*, 9p21; *ALDH2*, 12q24; *ALDH3*, 17; *ALDH5*, 9; *ALDH6*, 15q26; *ALDH9*, 1; *ALDH10*, 17p11.2; Rzhetsky *et al.*, 1997) and triose-phosphate isomerase (*TPI1*; 12p13; Kwiatowski *et al.*, 1995; Logsdon *et al.*, 1995) genes as evidence for the occurrence of multiple independent insertional or deletional events.

Several possible relationships may exist between the structural domains and the arrangement of exons in the gene (*Figure 3.4*). In many genes encoding globular proteins (e.g. the globins), the exons correspond closely to the structural domains thereby ensuring that when exon duplication has occurred, it has given rise to domain duplication (Li, 1997). In the majority of cases, however, a more complex relationship exists between domains and exons (Li, 1997). Thus it may be that the correspondence between exons and domains is only approximate. In some cases, an exon may encode two or more domains whilst in others, a single domain may be encoded by two or more exons. Finally, there may be no obvious correspondence between exons and domains.

Taking together the results of all the studies so far performed, the introns-early and introns-late viewpoints can to some extent be reconciled once we accept that they are not necessarily mutually exclusive, even for a single specific gene. A relationship between exon distribution and the domain structure of the encoded protein may therefore exist for some genes or some introns/domains within the same gene/protein. For others, it may be that such a relationship did once exist but has decayed with the passage of evolutionary time (Traut, 1988). Similarly with intron location, some introns were already present in primordial genes, whereas other genes lost or more likely gained (Cho and Doolittle, 1997) their introns at later stages of evolution. De Souza *et al.* (1998) have suggested

(a)   **DNA sequence around intron 8 of the histidyl-tRNA synthetase (HARS) gene**

```
HARS-Fugu      TATGTTGGTATGCAAGgtgagattt---tctgtagGTGGAATGGATTTGGCTGAACGT
HARS-Hamster   TATGTCCAGCAGCACGGTGAGgtaaa-----gctccccagGTGTGTCTGGTAGAGCAG
HARS-Human     TATGTCCAGCAACATGGTGGG             GTATCCCTGGTGGAACAG
```

(b)   **Model for the shift of intron 8**

Taking the *Fugu* intron as the ancestral one, an A to G change (arrowed) creates a cryptic splice site

```
                             ↓
           Y  V  G  M  Q                    G  G  M  D  L  A  E  R
           TATGTTGGTATGCAAGgtgagattt---tctgtagGTGGAATGGATTTGGCTGAACGT
           TATGTTGGTATGCAAGgtgaggttt---tctgtagGTGGAATGGATTTGGCTGAACGT
```

An additional insertion of a G (arrowed) then creates a frameshift (top) which would normally be lethal but is rescued by the cryptic splice site becoming functional (bottom)

```
                                          ↓
           Y  V  G  M  Q                    G  G  D  G  F  G  STOP
           TATGTTGGTATGCAAGgtgaggttt---tctgtagGTGGAGATGGATTTGGCTGAACGT
           TATGTTGGTATGCAAGGTGAGgttt---tctgtaggtggagATGGATTTGGCTGAACGT
           Y  V  G  M  Q  G  E                    M  D  L  A  E  R
```

The result is a 5 nucleotide shift in the position of the intron but only a G to E substitution in the amino acid sequence

```
           Old    Y  V  G  M  Q  G  G  G  M  D  L  A  E  R
           New    Y  V  G  M  Q  G  E  M  D  L  A  E  R
```

Figure 3.3. Intron sliding in the histidyl-tRNA synthetase (*HARS*) gene (after Brenner and Corrochano, 1996). (a) Comparison of the sequence around intron 8 in pufferfish (*Fugu*), human and hamster *HARS* genes. Lower case letters denote intron sequence, upper case letters exon sequence. The segment of the *Fugu* intron identical to hamster exon sequence is given in bold type. (b) Model for the sliding of intron 8. Nucleotide changes are in bold type and marked by arrows.

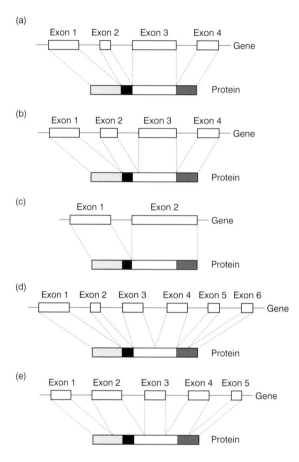

Figure 3.4. Possible relationships between the arrangement of exons in a hypothetical gene and the structural domains of the protein it encodes (from Li 1997), (a) each exon corresponds exactly to a structural domain, (b) the correspondence is only approximate, (c) an exon encodes two or more domains, (d) a single structural domain is encoded by two or more exons, (e) lack of correspondence between exons and domains. The four structural domains of the hypothetical protein are designated by different boxes.

that 30–40% of the present day intron locations correspond to phase zero introns originally present in the progenote. The remainder, they suggest, correspond to introns either added or moved and appear equally in all three phases. In the next section, the evidence for both the insertion or deletion of introns will be examined.

3.5 Mechanisms of intron insertion and deletion

As we have seen, the introns-late theory both assumes and requires that introns have been inserted and deleted during evolution. One example of the gain of an exon during evolution comes from the ten exon human renin (*REN*; 1q32) gene which contains three amino acids encoded by a sixth 9 bp exon not present in the mouse gene (Miyazaki *et al.*, 1984). This exon must have been acquired in the lineage leading to humans rather than lost in the murine lineage because this exon is absent from the evolutionarily related pepsin genes in both human (*PGA3*, *PGA4*, *PGA5*; 11q13) and pig. We may surmise that the sixth exon in the human *REN* gene was acquired by mutational change in intron 5 at some stage in the last

100 Myrs since the divergence of the rodent and human lineages. Another possible example of a late insertional event is provided by the human forkhead FKHL13 (*FKHL13*; 17q22-q25) gene which contains a single intron that has interrrupted the forkhead DNA-binding domain (Murphy *et al.*, 1997).

The evolution of the human *HLA-DRB6* (6p21.3) gene has been characterized not only by the deletion of an exon and the original promoter region but also by the *de novo* creation of an exon. This is thought to have occurred in association with the insertion of a retroviral LTR into intron 1 of the gene >23 Myrs ago, prior to the divergence of Old World monkeys from the human-ape lineage (Mayer *et al.*, 1993). Whether the exon/promoter deletion accompanied or followed the LTR insertion is unclear. Whichever, the gene is still capable of being transcribed. That this is so is due to the creation of an open reading frame for a new exon by the insertion which serendipitously encoded a hydrophobic sequence that was able to function as a leader for the truncated *HLA-DRB6* protein. The new exon also provided a functional donor splice site at its 3′ end which potentiated in-register splicing with exon 2 of the *HLA-DRB6* gene. Finally, since the LTR also provided a substitute promoter region, the downstream *HLA-DRB6* gene could be transcribed.

Retrotransposition of an mRNA intermediate is one mechanism that would serve to erase completely the original exon-intron distribution of a gene. One potential example of this phenomenon is provided by the 68 kDa neurofilament protein (*NEFL*; 8p21) gene, a member of the intermediate filament multigene family that diverged over 600 Myrs ago. Other members of this family include desmin (*DES*; 2q35), vimentin (*VIM*; 10p13), glial fibrillary acidic protein (*GFAP*; 17q21), and the type I and II keratins. Whereas these latter genes possess seven or eight introns, six of which occur at homologous locations, the *NEFL* gene possesses only three, none of which correspond in terms of their location to introns in the other known intermediate filament genes (Lewis and Cowan 1986). Retrotransposition of a cDNA intermediate could account for the present day structure of the *NEFL* gene but three new introns would still have to have been acquired, presumably by insertion. In other cases of putative intron loss, the retrotransposition of semiprocessed mRNAs (Chapter 6, section 6.1.3) should also be considered. Such retrotransposed sequences will, by definition, contain fewer introns than their parent genes but the retention of some introns will often serve to obscure their origin.

The best characterized example of the insertion of an intron into a gene is from the sex-determining gene which in humans (*SRY*; Yp11.3) and other placental mammals is intronless. However, in dasyurid marsupials, an intron was inserted *de novo* into the *Sry* gene about 45 Myrs ago (O'Neill *et al.*, 1998). The 825 bp intron lies within the coding sequence 550 bp from the start codon and was inserted in phase 1 (between the first and second bases of the codon). The intron, which contains a repetitive sequence element specific to marsupials, is correctly spliced out of the primary transcript. Since the *SRY* gene is essential for mammalian sex determination, it is very unlikely that intron insertion could have inactivated the gene even temporarily. We must therefore conclude that the inserted sequence probably contained functional splice junction motifs that ensured its accurate removal from the marsupial *Sry* transcript.

Some gene families have evolved by multiple rounds of gene duplication accompanied by the gain or loss of both introns and exons, for example the human lipoprotein lipase (*LPL*; 8p22), hepatic lipase (*LIPC*; 15q21-q23) and pancreatic lipase (*PNLIP*; 10q24-p26) genes and the homologous yolk protein 1 gene from *Drosophila* (*Figure 3.5*; Kirchgessner *et al.*, 1989). Another example is provided by the human thyroid peroxidase (*TPO*; 2pter-p24) and myeloperoxidase (*MPO*; 17q21.3–q23) genes (Kimura *et al.*, 1989).

The mammalian hepatocyte nuclear factor 1α (HNFα; *TCF1*; 12q24.3) gene possesses nine exons, whereas its avian and amphibian counterparts have ten (Hörlein *et al.*, 1993). This difference is explicable in terms of the loss in mammals of intron 9 which in the chicken gene sub-divides the serine-rich transactivation domain. Interestingly, there is a sequence at the 3′ end of intron 9 in the chicken gene which matches a conserved element used as a joining signal in immunoglobulin and T-cell receptor genes. Similarly, a perfectly conserved heptamer (CACAGTG) box, a recombination signal sequence in immunoglobulin genes, is found at the junction of the fused exon 9–exon 10 in the rat HNFα gene. Hörlein *et al.* (1993) speculated that V(D)J recombinase may have been involved in the excision of intron 9 in the mammalian gene.

The human surfeit 5 (*SURF5*; 9q34.1) gene contains an intron in its 5′ untranslated region which is not present in the mouse or rat *Surf5* genes (Duhig *et al.*, 1998). This additional intron is also present in apes, Old and New World monkeys and the prosimian *Galago*. Duhig *et al.* (1998) speculated that the intron was introduced after the divergence of primates and rodents but before the divergence of the human and prosimian lineages.

A considerable number of examples are therefore now known of introns which have been either gained or lost during evolution as a result of the action of several distinct mechanisms. Such examples are certainly supportive of the introns-late theory.

3.6 Exon shuffling

3.6.1 Exon shuffling in the evolution of human genes

Exon shuffling may be defined as the transfer of exons, encoding specific functional modules, between genes so that the module-associated functions are conferred upon

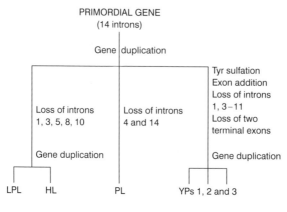

Figure 3.5. A model for the evolution of the lipase gene family. The six known members of this family (lipoprotein lipase (LPL), hepatic lipase (HL), pancreatic lipase (PL) and the *Drosophila* yolk proteins (YPs)) evolved from an ancestral gene containing 14 introns by a series of gene duplications, exon shuffling events together with the gain and loss of introns (after Kirchgessner *et al.*, 1989).

the recipient proteins. One of the first examples of exon shuffling to be recognized was from the human low density lipoprotein receptor (*LDLR*; 19p13.3) gene which contains eight contiguous exons encoding epidermal growth factor (EGF)-like domains as well as seven repeats corresponding to LDL-binding domains (Südhof *et al.*, 1985a,b). Other examples include the human hemopexin (*HPX*; 11p15.4-p15.5) gene which contains 10 exons each of which encodes a 45 amino acid repeat (Altruda *et al.*, 1988) and the human factor XIII b subunit (*F13B*; 1q31-q32; Bottenus *et al.*, 1990) gene which encodes a protein with ten 60 amino acid complement B-type ('sushi') domains each encoded by a single exon. A large number of other human genes possess multiple copies of exons encoding specific domains. These encode proteins of the extracellular matrix (e.g. laminins, collagens), the serine proteases of coagulation (see Section 3.6.3), a variety of receptors (e.g. integrins) and the immunoglobulins among many others (reviewed by Patthy, 1996). Exon shuffling is therefore a widely used evolutionary strategy although it appears to be confined to the genes of higher eukaryotes.

Until recently, exon shuffling was considered to result solely from intron-mediated recombination (Patthy, 1995; Strelets and Lim, 1995). However, an alternative mechanism, *LINE element-mediated recombination*, now appears to be capable of linking previously unlinked genomic DNA segments and may represent a novel means to bring about exon shuffling. Moran *et al.* (1999) have shown that LINE elements are capable of transducing exons from a downstream gene to new genomic locations thereby creating novel genetic combinations. This mechanism requires readthrough of the relatively weak LINE element polyadenylation signal by RNA polymerase to yield a processed mRNA transcript containing the LINE element fused to one or more exons from the neighboring gene. Readthrough is made possible by the presence of more potent polyadenylation signals 3' to the LINE element and Moran *et al.* (1999) have shown that this can occur with high efficiency. The chimeric transcript then serves as a template for reverse transcriptase and if the resulting cDNA is subsequently integrated into the intron of a recipient gene, it may be recruited by that gene. This may help to explain why 3' terminal exons are often much longer than internal exons (Hawkins, 1988). Inefficient mRNA cleavage/polyadenylation might however lead to alternative splicing (Section 3.2). A possible example of this process occurred about 10 Myrs ago during the evolution of the great apes: the co-transduction and integration of a LINE element and a fragment containing exon 9 of the cystic fibrosis transmembrane conductance regulator (*CFTR*; 7q31) gene (Rozmahel *et al.*, 1997). Since some 6% of LINE element insertions occur within genes (Moran *et al.*, 1999), it could be that this mechanism has played an important role in shuffling exons between genes, and may therefore have been central to the creation of the mosaic structures characteristic of so many genes in higher animals.

3.6.2 The phase compatibility of introns

Only a limited proportion of exons can be utilized for exon shuffling since the splice junctions of the shuffled exon must be phase compatible with their flanking exons in order to maintain the reading frame. As we have seen in Chapter 1, section 1.2.1, introns are classified as phase 0 if the intron lies between two codons, phase 1 if the intron lies between the first and second nucleotides of a codon, and phase 2 if it lies between the second and third nucleotides of a codon. The requirement for phase compatibility dictates that only symmetrical exons of class 1–1, class 2–2 and class 0–0 [i.e. exons flanked by introns of the same phase (1, 2, or 0)] are suitable for duplication or insertion. It is therefore not surprising that genes encoding mosaic proteins (proteins which comprise multiple domains and which are likely to have evolved by exon shuffling) contain a disproportionate number of class 1–1 exons encoding a wide variety of different modules, for example EGF-like, calcium-binding, fibronectin-like finger, kringle, complement B, LDL receptor, von Willebrand, thyroglobulin, C-type lectin etc (Patthy 1991a, 1994, 1996). Class 1–1 introns predominate in the immunoglobulin, T-cell receptor and HLA genes, the Thy-1 glycoprotein (*THY1*; 11q22.3-q23) gene and the β2-microglobulin (*B2M*; 15q21-q22) gene, consistent with exon shuffling having been important in their evolution (Patthy *et al.*, 1987). Further examples of this phenomenon include the exons encoding the complement B-type domains in the human C4b-binding protein α gene (*C4BPA*; 1q32; Hillarp *et al.*, 1993), the immunoglobulin-like and fibronectin type III modules in the human Axl oncogene (*AXL*; 19q13.1–13.2; Schulz *et al.*, 1993) gene and four of six thrombospondin modules of the human properdin gene (*BF*; 6p21.3; Nolan *et al.*, 1992). Genes with predominantly class 0–0 exons include those encoding the type III collagens (see Chapter 4) and β-casein (*CSN2*; 4pter-q21) whilst class 2–2 exons are exclusively found in the glucagon (*GCG*; 2q36-q37) gene (Patthy, 1987).

Using a large database of eukaryotic genes, Long *et al.* (1995) demonstrated there to be an excess of *symmetrical* exons over expectation and estimated that at least 19% of exons had been involved in exon shuffling. Why the preponderance of class 1–1 exons? Patthy (1994) suggested that since 92% of introns flanking signal peptide domains were phase 1, modularization of exported proteins with secretory signal peptide domains is likely to have employed class 1–1 introns. According to Patthy (1994), modularization proceeds in different stages: (i) insertion of introns of identical phase at boundaries of the protein fold, (ii) tandem duplication of the symmetrical module by intronic recombination, and (iii) module transfer to a novel location (*Figure 3.6*). Nonsymmetrical exons (those flanked by introns of different phase) are potentially much less versatile and may be expected to have been utilized much less often. Tomita *et al.* (1996) noted that when introns are inserted so as to disrupt codons, the site of insertion occurs much more frequently between the first and second bases than between the second and third bases. The reason for this remains unclear.

There are however various instances which do not conform to the above exon shuffling rules. In some cases, the original exon-intron organization of genes has become eroded with the passage of evolutionary time. Thus, although in the human perlecan (*HSPG2*; 1p35-p36; Cohen *et al.*, 1993) gene, the LDL receptor modules and the immunoglobulin-like modules are still flanked by phase 1 introns, the original introns have been lost from the regions of the gene encoding laminin A, laminin B and epidermal growth factor-like modules. Similarly, in the human complement 6 (*C6*; chromosome 5; Hobart *et al.*, 1993) gene, phase 1 introns are found only at the boundaries of one of the complement B modules having been lost from the boundaries of the class 1–1 thrombospondin,

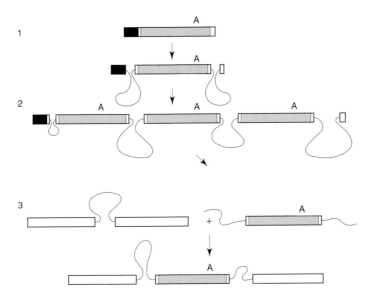

Figure 3.6. Different stages in the conversion of a domain to a module (redrawn from Patthy 1994). The boxes represent exons separated by introns. The exon encoding the signal peptide is solid whilst the exons encoding the protein fold (A) are lightly shaded. Stage 1: Insertion of introns of identical phase at amino and carboxy terminal boundaries of protein fold. Stage 2: Tandem duplications of symmetrical protomodule A via intronic recombination. Stage 3: Module A is transferred to a new location.

LDL receptor, EGF, and C7 modules through multiple occurrences of intron insertion and removal. Finally, the exon shuffling rules do not apply in some cases of genes encoding proteins common to both prokaryotes and eukaryotes (phosphoglycerate kinase, alcohol dehydrogenase, pyruvate kinase, glyceralde-hyde-3-phosphate dehydrogenase, triosephosphate isomerase and dihydrofo-latereductase). This is consistent with the view that these evolutionarily very ancient genes did not evolve by exon shuffling (Patthy, 1987, 1991a). However, Long *et al.* (1995) claimed that there is an excess of symmetrical exons in the ancient conserved regions of eukaryotic genes (regions homologous to prokary-otic genes), a finding that is consistent with at least some introns being of ancient origin (Gilbert *et al.*, 1997).

3.6.3 The serine proteases of coagulation

One of the archetypal examples of the evolution of modular proteins by exon shuf-fling is that of the serine proteases of coagulation. As the completed primary sequences of hemostatic factors became available, it was noticed that certain domains of shared homology recurred many times in diverse proteins. *Figure 3.7* depicts 11 proteins of coagulation and fibrinolysis in such a way as to emphasise their modular composition. The kringle domain is present in prothrombin as well as three proteins of fibrinolysis (tPA, urokinase and plasminogen) and may play a role in fibrin binding. A trypsin-like serine protease domain is common to all the proteolytic enzymes of coagulation. The EGF-like domain was first recognized among clotting proteins in factor X. Two copies of this motif are found in all the vitamin K-dependent proteins of coagulation (factor IX, factor X, factor VII, protein C) except protein S which has four and prothrombin which has none. The action of vitamin K in coaguloprotein synthesis is to promote carboxylation of N-terminal glutamic acid (Gla) residues by a liver microsomal enzyme system. The five proteins which share this N-terminal modification have a homologous propeptide domain that functions as a signal to the carboxylase.

The organization of coagulation factor genes reflects to a considerable extent their functional modular assembly as described above. *Figure 3.8* shows the gene structure of several clotting factors. It is apparent that since factors VII (*F7*; 13q34), X (*F10*; 13q34), IX (*F9*; Xq26.3-q27.1) and protein C (*PROC*; 2q13-q21) have virtually identical protein and gene structures, they must have emerged relatively recently by a process of gene duplication and divergence. Similarly, tissue plasminogen activator (*PLAT*; 8p12-q11.2), factor XI (*F11*; 4q35), factor XII

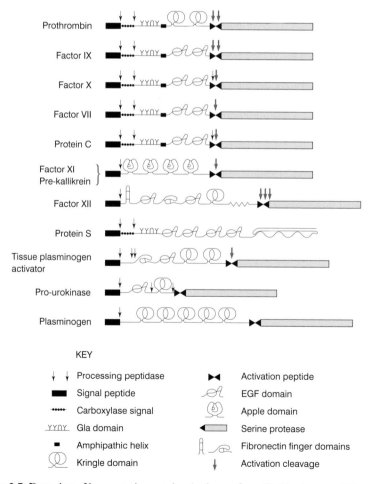

Figure 3.7. Domains of hemostatic proteins (redrawn from Tuddenham and Cooper, 1994).

(*F12*; 5q33-qter) and urokinase (*PLAU*; 10q24-qter) exhibit similar gene organization with respect to their protease domains but differ from prothrombin (*F2*; 11p11-q12) and factors VII, IX, X, and protein C. Kringle domains are encoded by a similar gene structure wherever they occur, as are the calcium-binding Gla domains. EGF-like domains are encoded by a single exon with type 1 splice junctions as are the fibronectin type I (Fn I) domains.

The presence/absence of different modules (Gla and EGF-like domains, kringles) in the different coagulation factor proteins can be used to reconstruct their evolutionary past. The transfer of modules between proteins has arisen via exon shuffling and protein data can be used to construct evolutionary trees (Patthy, 1985) that are consistent with what is already known of the genes in terms of their exon size and distribution. Patthy (1990) has described at length the principles underlying the assembly of present-day coagulation factor genes from their constituent modules. Evolution of these genes has proceeded by repeated insertions, duplications, exchanges and deletions of modules. How has it been possible to produce such a plethora of different proteins/genes by exon shuffling in what has been a comparatively short period of evolutionary time? The answer appears to lie firstly with the close correspondence between exon boundaries and the modular domains of the proteins and secondly with the fact that the ancestors of the kringle, fibronectin, growth factor, protease and Gla modules all had phase 1 introns at both module boundaries.

The phylogenetic tree of the serine proteases, constructed by Patthy (1990) is shown in *Figure 3.9*. In this diagram, the major division between the blood coagulation proteases and those involved in fibrinolysis is very apparent. This division

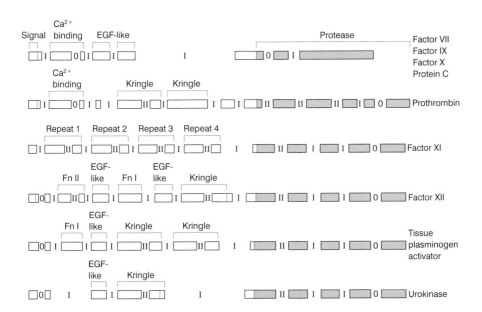

Figure 3.8. Organization of the genes of hemostatic proteins (redrawn from Tuddenham and Cooper, 1994). Fn: Fibronectin-like domain. Intron phase is denoted by O, I or II.

is also apparent when intron/exon distribution and codon usage for the active site serine (Brenner, 1988) are considered: the alternative codons, TCN and AGY, cannot be interconverted by a single nucleotide substitution. The genes encoding factors VII, IX, X, and protein C possess the AGY codon whilst the fibrinolytic enzyme genes possess the TCN codon. Phylogenetic trees for the protease and kringle domains are however not identical (Ikeo *et al.*, 1995) suggesting that these domains may have experienced different evolutionary histories. This implies that in multidomain proteins such as the serine proteases, each domain can represent an independent evolutionary unit.

3.6.4 Protein folds, primordial exons and the emergence of exon shuffling

Proteins are composed of combinations of secondary structural elements, α-helices and β-sheets, which are connected by loop regions at the surface of the molecule. Certain combinations of these elements (termed *motifs* or *folds*) are frequently used

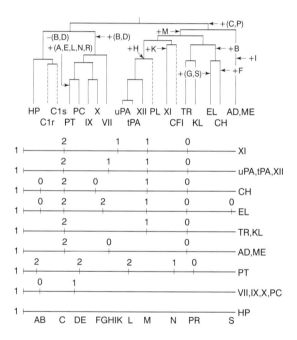

Figure 3.9. Evolution of the serine protease family (redrawn from Patthy, 1990). The genealogy of the serine protease family was elucidated by reference to the known amino acid sequences. Dashed lines indicate proteases whose gene structure has not yet been established. 0, 1, 2 denote the phase class of the introns. Introns are labelled A to S, and putative intron addition and removal events during evolution (consistent with the genealogy based upon amino acid sequence data) are denoted by (+) and (−) respectively. Protein abbreviations used are: HP, haptoglobin; Clr and Cls, complement factors Clr and Cls; PT, prothrombin; PC, protein C; IX, X, VII, factors IX, X, VII; uPA, urokinase; tPA, tissue-type plasminogen activator; XII, factor XII; PL, plasminogen and apolipoprotein(a); XI, factor XI and pre-kallikrein; CFI, complement factor I; TR, trypsin, KL, kallikreins; EL, elastases; CH, chymotrypsin; AD, ME, adipsin, medullasin, mast cell proteases, cytotoxic lymphocyte proteases, cathepsin G.

in proteins and are known as *domains*. Evolutionarily mobile domains are termed *modules*. Motifs can be as simple as in the case of the hexamer repeat unit that forms a left-handed parallel β-helix found in UDP-N-acetylglucosamine acyltransferase or more complex as in the 21–26 amino acid zinc finger motif (Henikoff *et al.*, 1997). For practical purposes, a protein domain may be regarded as a structurally and/or functionally discrete portion of a protein that can fold independently into a stable tertiary structure and may be composed of single or multiple motifs. Since the zinc finger motif can fold independently, it may actually be regarded as a small domain in its own right. Most proteins contain two or more domains although some may just contain a single domain (Doolittle, 1995).

During evolutionary time, the amino acid sequences of proteins are eroded at a much faster rate than are the corresponding 3D structures. In other words, although it may be difficult to discern sequence conservation, structural conservation may still be evident (Creighton, 1993). One example of this is provided by the α-defensin (*DEFA1, DEFA4, DEFA5, DEFA6*; 8p23) and β-defensin (*DEFB1, DEFB2*; 8p23) genes which, despite a complete lack of DNA sequence homology, and major differences between their encoded proteins in terms of their cysteine spacing and disulfide pairing, have nevertheless evolved from a common ancestor (Lei *et al.*, 1997).

It is often quite difficult to ascertain whether structural similarities are *homologous* (i.e. based upon divergent evolution from a common ancestor) or *analogous* (i.e. based upon convergent evolution to a physically favoured secondary or tertiary structure). The α/β barrel domain is common among enzymes and examples of this domain are thought to share a common ancestor rather than to have evolved convergently to a stable fold (Reardon and Farber, 1995). Such examples are a reflection of the evolution of extant protein structures from a small number of basic folds encoded by primordial exons.

Primordial exons encoding functional domains are now widely dispersed among many diverse proteins (Doolittle, 1995). Examples include the ankyrin repeat (Bork, 1993), the spectrin repeat (Pascual *et al.*, 1997), the EGF-like domain (Campbell and Bork 1993), the WD-repeat (Neer *et al.*, 1994), the ATPase domain (Bork *et al.*, 1992) and kringles (Patthy *et al.*, 1984). The EGF-like domain is present in ~1% of human proteins whilst the immunoglobulin domain may be even more prevalent (Henikoff *et al.*, 1997). Other domains such as the Kunitz domain which is less prevalent, may have a more recent origin (Ikeo *et al.*, 1992).

How many primordial exons were required to construct the huge array of extant proteins? Dorit *et al.* (1990) used the frequency with which exons in a 1200 exon database had been 'reused' between genes encoding different proteins to assess the size of the underlying exon pool; they estimated that the number of primordial exons was between 1000 and 7000. A similar study succeeded in distilling 1410 polypeptide chains down to 112 analogous fold families whose members exhibited an average of <18% sequence identity (Orengo *et al.*, 1993). Patthy (1991b) has pointed out, however, that this type of approach may tend to underestimate the true number of exons since it may not adequately have considered the pool of exons that participate in exon shuffling only rarely. Databases of protein folds and their families are available online at **http://www.embl-ebi.ac.uk/dali/** (Homology-derived Structures of Proteins), **http://www.mips.biochem.mpg.de/** (Munich Information

Center for Protein Sequences) and **http://protein.toulouse.inra.fr/prodom.html** (Protein Domain Database).

Mosaic proteins assembled from class 1–1 modules are confined predominantly to multicellular organisms (Patthy 1991a). Exon shuffling probably only came into its own with the radiation of the metazoa when the requirements of multicellularity fuelled the demand for novel and increasingly more complex proteins to potentiate intercellular communication. Perhaps the cellular machinery for exon shuffling only appeared at this time.

3.7 Pseudoexons

Pseudoexons may be defined as gene regions that are functional in a gene from one species and which have been conserved in non-functional form in another species. Good examples of pseudoexons are those found in the human αA-crystallin (*CRYAA*; 21q22.3; Jaworski and Piatigorsky, 1989), glycophorin B (*GYPB*; 4q28-q31; Huang *et al.*, 1995), complement component receptor 2 (*CR2*; 1q32; Holguin *et al.*, 1990) and transketolase-related (*TKT*; Xq28; Coy *et al.*, 1996) genes. In the *CRYAA* example, the pseudoexon is alternatively spliced into the mRNA in rodents and bats but was lost to other mammalian species about 30–40 Myrs ago. Exon III of the glycophorin B (*GYPB*; 4q28-q31) gene is expressed in the chimpanzee and gorilla but has been silenced as a pseudoexon in humans through an exon 3 donor splice site mutation (Huang *et al.*, 1995; Kudo and Fukuda, 1989; Xie *et al.*, 1997). This finding predicts a larger extracellular domain in the chimpanzee protein. Similarly, the human glycophorin E (*GYPE*; 4q28-q31) gene has lost two of its original exons (Kudo and Fukuda, 1990). The recombinational use of these pseudoexons represents a major mechanism for allelic diversification in the glycophorins: depending upon the extent, location and type of recombination, the exchange or transfer of pseudoexons can result in the creation of novel inter-exon or intra-exon hybrid junctions that yield new mRNA splicing patterns (Blumenfeld and Huang, 1995).

Murine apolipoprotein E receptor 2 gene transcripts possess an extra exon as compared with transcripts from the human (*LRP8*; 1p34) gene (Kim *et al.*, 1998). However, the human and marmoset *LRP8* genes possess an intronic pseudoexon corresponding to the murine exon indicating that this exon has been lost in the primate lineage. The incorporation of the pseudoexon into the primate *LRP8* mRNA transcript is prevented by two nucleotide substitutions in the adjacent 5′ donor splice site. The pseudoexon contains a deletion which would have led to a frameshift but it is unclear whether the deletion occurred after inactivation of the exon or if the splice site substitutions served to allow the primate *LRP8* gene to avoid incorporation of the mutation-containing exon.

The human insulin-like growth factor-II (*IGF2*; 11p15.5) gene is transcribed and processed into three different mRNAs under the control of three distinct promoters, two of which have counterparts in the mouse and rat (de Pagter-Holthuisen *et al.*, 1987; 1988). The 5′-most promoter, which is active in adult human tissues, controls the transcription of a cassette of three exons that comprises the 5′ UTR of the human gene. The murine gene does not however contain a structural and functional homologue of exon 1 (and adjacent promoter) of the human gene (Rotwein and Hall, 1990). The absence of this cassette from murine *Igf2* mRNA may account for the disappearance of *Igf2* transcripts from most murine tissues shortly after birth.

The presence of pseudoexons in a number of human genes has thus provided good evidence for the occurrence of exon inactivation (and hence also intron loss) during mammalian evolution. Such events are certainly compatible with the introns early theory. It is however unclear how common pseudoexons are in human genes since they will usually only be detected through the careful comparison of orthologous gene pairs, a procedure that will tend to underestimate their prevalence.

References

Adham N., Kao H.T., Schecter L.E., Bard J., Olsen M., Urquhart D., Durkin M., Hartig P.R., Weinshank R.L., Branchek T.A. (1993) Cloning of another human serotonin receptor (5-HT1F): a fifth 5-HT1 receptor subtype coupled to the inhibition of adenylate cyclase. *Proc. Natl. Acad. Sci. USA* **90**: 408–412.

Ala-Kokko L., Kvist A.P., Metsaranta M., Kivirikko K.I., de Crombrugghe B., Prockop D.J., Vuorio E. (1995) Conservation of the sizes of 53 introns and over 100 intronic sequences for the binding of common transcription factors in the human and mouse genes for type II procollagen (*COL2A1*). *Biochem. J.* **308**: 923–929.

Alder H., Yoshinouchi M, Prystowsky MB, Appasamy P, Baserga R. (1992) A conserved region in intron 1 negatively regulates the expression of the *PCNA* gene. *Nucleic Acids Res.* **20**: 1769–1765.

Altruda F., Poli V., Restagno G., Silengo L. (1988) Structure of the human hemopexin gene and evidence for intron-mediated evolution. *J. Mol. Evol.* **27**: 102–108.

Antoniou M., Geraghty F., Hurst J., Grosveld F. (1998) Efficient 3'-end formation of human β-globin mRNA *in vivo* requires sequences within the last intron but occurs independently of the splicing reaction. *Nucleic Acids Res.* **26**: 721–729.

Atanasoki S., Toldo S.S., Malipiero U., Schreiber E., Fries R., Fontana A. (1995) Isolation of the human genomic brain-2/N-Oct 3 gene (POUF3) and assignment to chromosome 6q16. *Genomics* **26**: 272–280.

Ayyanathan K., Naylor S.L., Kunapuli S.P. (1996) Structural characterization and fine chromosomal mapping of the human P2Y1 purinergic receptor gene (*P2RY1*). *Somat. Cell Molec. Genet.* **22**: 419–424.

Beier F., Vornehm S., Poschl E., von der Mark K., Lammi M.J. (1997) Localization of silencer and enhancer elements in the human type X collagen gene. *J. Cell. Biochem.* **66**: 210–218.

Belfort M. (1993) An expanding universe of introns. *Science* **262**: 1009–1010.

Berchtold M.W., Koller M., Egli R., Rhyner J.A., Hameister H., Strehler E.E. (1993) Localization of the intronless gene coding for calmodulin-like protein CLP to human chromosome 10p13-ter. *Hum. Genet.* **90**: 496–500.

Berg L.-P., Soria J.M., Formstone C.J., Morell M., Kakkar V.V., Estivill X., Sala N., Cooper D.N. (1996) Aberrant RNA splicing of the protein C and protein S genes in healthy individuals. *Blood Coag. Fibrinol.* **7**: 625–631.

Blake C.C.F. (1978) Do genes-in-pieces imply proteins-in-pieces? *Nature* **273**: 267.

Blumenfeld O.O., Huang C.-H. (1995) Molecular genetics of the glycophorin gene family, the antigens for the MNSs blood groups: multiple gene rearrangements and modulation of splice site usage result in extensive diversification. *Hum. Mutation* **6**: 199–209.

Bork P. (1993) Hundreds of ankyrin-like repeats in functionally diverse proteins: mobile modules that cross phyla horizontally? *Proteins Struct. Funct. Genet.* **17**: 363–374.

Bork P., Sander C., Valencia A. (1992) An ATPase domain common to prokaryotic cell cycle proteins, sugar kinases, actin, and hsp70 heat shock proteins. *Proc. Natl. Acad. Sci. USA* **89**: 7290–7294.

Bottenus R.E., Ichinose A., Davie E.W. (1990) Nucleotide sequence of the gene for the b-subunit of human factor XIII. *Biochemistry* **29**: 11195–11209.

Branden C.I., Eklund H., Cambillau C., Pryor A.J. (1984) Correlation of exons with structural domains in alcohol dehydrogenase. *EMBO J.* **3**: 1307–1310.

Brenner S. (1988) The molecular evolution of genes and proteins: a tale of two serines. *Nature* **334**: 528–530.

Brenner S., Corrochano L.M. (1996) Translocation events in the evolution of aminoacyl-tRNA synthetases. *Proc. Natl. Acad. Sci. USA* **93**: 8485–8489.

Brenner S., Elgar G., Sandford R., Macrae A., Venkatesh B., Aparicio S. (1993) Characterization of the pufferfish (*Fugu*) genome as a compact model vertebrate genome. *Nature* **366**: 265–268.

Campbell I.D., Bork P. (1993) EGF-like modules. *Curr. Opin. Struct. Biol.* **2**: 385–392.

Cavalier-Smith T. (1985) Eukaryotic gene numbers, non-coding DNA and genome size. In: *The Evolution of Genome Size*, (Ed. T. Cavalier-Smith). John Wiley, Chichester. pp 70–103.

Cavalier-Smith T. (1991) Intron phylogeny: a new hypothesis. *Trends Genet.* **7**: 145–148.

Cho G., Doolittle R.F. (1997) Intron distribution in ancient paralogs supports random insertion and not random loss. *J. Mol. Evol.* **44**: 573–584.

Cicila G.T., O'Connell T.M., Hahn W.C., Rheinwald J.G. (1989) Cloned cDNA sequence for the human mesothelial protein 'mesosecrin' discloses its identity as a plasminogen activator inhibitor (PAI-1) and a recent evolutionary change in transcript processing. *J. Cell. Sci.* **94**: 1–10.

Clark I.M., Rowan A.D., Edwards D.R., Bech-Hansen T., Mann D.A., Bahr M.J., Cawston T.E. (1997) Transcriptional activity of the human tissue inhibitor of metalloproteinases 1 (TIMP-1) gene in fibroblasts involves elements in the promoter, exon 1 and intron 1. *Biochem. J.* **324**: 611–617.

Cohen I.R., Grassel S., Murdoch A.D., Iozzo R.V. (1993) Structural characterization of the complete human perlecan gene and its promoter. *Proc. Natl. Acad. Sci. USA* **90**: 10404–10408.

Cooper D.N., Berg L.-P., Kakkar V.V., Reiss J. (1994) Ectopic (illegitimate) transcription: new possibilities for the analysis and diagnosis of human genetic disease. *Ann. Med.* **26**: 9–14.

Coutelle O., Nyakatura G., Taudien S., Elgar G., Brenner S., Platzer M., Drescher B., Jouet M., Kenwrick S., Rosenthal A. (1998) The neural cell adhesion molecule L1: genomic organisation and differential splicing is conserved between man and the pufferfish *Fugu*. *Gene* **208**: 7–15.

Coy J.F., Dubel S., Kioschis P., Thomas K., Micklem G., Delius H., Poustka A. (1996) Molecular cloning of tissue-specific transcripts of a transketolase-related gene: implications for the evolution of new vertebrate genes. *Genomics* **32**: 309–316.

Crabtree G.R., Comeau C.M., Fowlkes D.M., Fornace A.J., Malley J.D., Kant J.A. (1985) Evolution and structure of the fibrinogen genes. Random insertion of introns or selective loss? *J. Mol. Biol.* **185**: 1–19.

Craik C.S., Rutter W.J., Fletterick R. (1983) Splice junctions: association with variation in protein structure. *Science* **220**: 1125–1129.

Creighton T.E. (1993) *Proteins. Structures and Molecular Properties*. 2nd Edn. WH Freeman, New York.

Culbertson M.R. (1999) RNA surveillance: unforseen consequences for gene expression, inherited disorders and cancer. *Trends Genet.* **15**: 74–80.

Dahl H.H., Brown R.M., Hutchison W.M., Maragos C., Brown G.K. (1990) A testis-specific form of the human pyruvate dehydrogenase E1 α subunit is coded for by an intronless gene on chromosome 4. *Genomics* **8**: 225–232.

De Nardin E., Radel S.J., Lewis N., Genco R.J., Hammarskjold M. (1992) Identification of a gene encoding for the human formyl peptide receptor. *Biochem. Int.* **26**: 381–387.

De Pagter-Holthuizen P., Jansen M., Van der Kammen R., Van Schaik F.M.A., Sussenbach J.S. (1988) Differential expression of the human insulin-like growth factor II gene. Characterization of the IGF-II mRNAs and an mRNA encoding a putative IGF-II-associated protein. *Biochim. Biophys. Acta* **950**: 282–295.

De Pagter-Holthuizen P., Jansen M., Van Schaik F.M.A., Van der Kammen R., Oosterwijk C., Van der Brande J.L., Sussenbach J.S. (1987) The human insulin-like growth factor II gene contains two development-specific promoters. *FEBS Letts.* **214**: 259–264.

de Souza S., Long M., Schoenbach L., Roy S.W., Gilbert W. (1996) Intron positions correlate with module boundaries in ancient proteins. *Proc. Natl. Acad. Sci. USA* **93**: 14632–14636.

de Souza S., Long M., Klein R.J., Roy S.W., Lin S., Gilbert W. (1998) Towards a resolution of the introns early/late debate: only phase zero introns are correlated with the structure of ancient proteins. *Proc. Natl. Acad. Sci. USA* **95**: 5094–5099.

de Stanchina E., Perini G., Patrone G., Suarez-Covarrubias A., Riva S., Biamonti G. (1997) A repeated element in the human lamin B2 gene covers most of an intron and reiterates the exon/intron junction. *Gene* **196**: 267–277.

Demchyshyn L., Sunahara R.K., Miller K., Teitler M., Hoffman B.J., Kennedy J.L., Seeman P., Van Tol H.H., Niznik H.B. (1992) A human serotonin 1D receptor variant (5HT1D beta) encoded by an intronless gene on chromosome 6. *Proc. Natl. Acad. Sci. USA* **89**: 5522–5526.

Denisov D.A., Shpigelman E.S., Trifonov E.N. (1997) Protective nucleosome centering at splice sites as suggested by sequence-directed mapping of the nucleosomes. *Gene* **205**: 145–149.

Devilat I., Carvallo P. (1993) Structure and sequence of an intronless gene for human casein kinase II-alpha subunit. *FEBS Letts.* **316**: 114–118.

Diaz M.O., Pomykala H.M., Bohlander S.K., Maltepe E., Malik E., Brownstein B., Olopade O.I. (1994) Structure of the human type-I interferon cluster determined from a YAC clone contig. *Genomics* **22**: 540–552.

Dibb N.J., Newman A.J. (1989) Evidence that introns arose at proto-splice sites. *EMBO J.* **8**: 2015–2021.

Doolittle R.F. (1995) The multiplicity of domains in proteins. *Annu. Rev. Biochem.* **64**: 287–314.

Dorit R.L., Schoenbach L., Gilbert W. (1990) How big is the universe of exons? *Science* **250**: 1377–1382.

Duhig T., Ruhrberg C., Mor O., Fried M. (1998) The human surfeit locus. *Genomics* **52**: 72–78.

Eggermont J. (1998) The exon-intron architecture of human chloride channel genes is not conserved. *Biochim. Biophys. Acta* **1397**: 156–160.

Erba H.P., Eddy R., Shows T., Kedes L., Gunning P. (1988) Structure, chromosomal location, and expression of the human γ-actin gene: differential evolution, location, and expression of the cytoskeletal β- and γ-actin genes. *Mol. Cell. Biol.* **8**: 1775–1779.

Ferlini A., Muntoni F. (1998) The 5′ region of intron 11 of the dystrophin gene contains target sequences for mobile elements and three overlapping ORFs. *Biochem. Biophys. Res. Commun.* **242**: 401–406.

Gentles A.J., Karlin S. (1999) Why are human G-protein-coupled receptors predominantly intronless? *Trends Genet.* **15**: 47–49.

Gilbert W. (1987) The exon theory of genes. *Cold Spring Harb. Symp. Quant. Biol.* **52**: 901–905.

Gilbert W., Glynias M. (1993) On the ancient nature of introns. *Gene* **135**: 137–144.

Gilbert W., de Souza S.J., Long M. (1997) Origin of genes. *Proc. Natl. Acad. Sci. USA* **94**: 7698–7703.

Go M. (1987) Protein structure and the origin of introns. *Cold Spring Harb. Symp. Quant. Biol.* **52**: 915–924.

Grant D.M., Blum M., Demierre A., Meyer U.A. (1989) Nucleotide sequence of an intronless gene for a human arylamine N-acetyltransferase related to polymorphic drug acetylation. *Nucleic Acids Res.* **17**: 3978.

Grimsby J., Chen K., Wang L.-J., Lan N.C., Shih J.C. (1991) Human monoamine oxidase A and B genes exhibit identical exon-intron organization. *Proc. Natl. Acad. Sci. USA* **88**: 3637–3641.

Guerin S.L., Leclerc S., Verreault H., Labrie F., Luu-The V. (1995) Overlapping *cis*-acting elements located in the first intron of the gene for type I 3 beta-hydroxysteroid dehydrogenase modulate its transcriptional activity. *Mol. Endocrinol.* **9**: 1583–1597.

Hall L., Williams K., Perry A.C.F., Frayne J., Jury J.A. (1998) The majority of human glutathione peroxidase type 5 (GPX5) transcripts are incorrectly spliced: implications for the role of GPX5 in the male reproductive tract. *Biochem. J.* **333**: 5–9.

Harris L.C., Remack J.S., Brent T.P. (1994) Identification of a 59 bp enhancer located at the first exon/intron boundary of the human O6-methylguanine DNA methyltransferase gene. *Nucleic Acids Res.* **22**: 4614–4619.

Hastings K.E.M. (1997) Molecular evolution of the vertebrate troponin I gene family. *Cell Struct. Func.* **22**: 205–211.

Hattori K., Angel P., Le Beau M.M., Karin M. (1988) Structure and chromosomal localization of the functional intronless human *JUN* protooncogene. *Proc. Natl. Acad. Sci. USA* **85**: 9148–9152.

Hatzinikolas G., Gibson M.A. (1998) The exon structure of the human MAGP-2 gene. *J. Biol. Chem.* **273**: 29 309–29 314.

Hawkins J.D. (1988) A survey on intron and exon lengths. *Nucleic Acids Res.* **16**: 9893–9908.

Henikoff S., Greene E.A., Pietrokovski S., S., Bork P., Attwood T.K., Hood L. (1997) Gene families: the taxonomy of protein paralogs and chimeras. *Science* **278**: 609–614.

Herbert A., Rich A. (1999) RNA processing and the evolution of eukaryotes. *Nature Genet.* 21: 265–269.

Higashimoto Y., Liddle R.A. (1993) Isolation and characterization of the gene encoding rat glucose-dependent insulinotropic peptide. *Biochem. Biophys. Res. Commun.* 193: 182–190.

Hillarp A., Pedro-Manuel F., Ruiz R.R., de Cordoba S.R., Dahlbck B. (1993) The human C4b-binding protein β-chain gene. *J. Biol. Chem.* 268: 15 017–15 023.

Hobart M.J., Fernie B., DiScipio R.G. (1993) Structure of the human C6 gene. *Biochemistry* 32: 6198–6205.

Holguin M.H., Kurtz C.B., Parker C.J., Weis J.J., Weis J.H. (1990) Loss of human CR1- and murine Crry-like exons in human CR2 transcripts due to CR2 gene mutations. *J. Immunol.* 145: 1776–1781.

Hörlein A., Grajer K.-H., Igo-Kemenes T. (1993) Genomic structure of the POU-related hepatic transcription factor HNF-1α. *Biol. Chem. Hoppe-Seyler* 374: 419–425.

Hormuzdi S.G., Penttinen R., Jaenisch R., Bornstein P. (1998) A gene targeting approach identifies a function for the first intron in expression of the α1(I) collagen gene. *Mol. Cell. Biol.* 18: 3368–3375.

Huang C.H., Xie S.S., Socha W., Blumenfeld O.O. (1995) Sequence diversification and exon inactivation in the glycophorin A gene family from chimpanzee to human. *J. Mol. Evol.* 41: 478–486.

Ikeo K., Takahashi K., Gojobori T. (1992) Evolutionary origin of a Kunitz-type trypsin inhibitor domain inserted in the amyloid beta precursor protein of Alzheimer's disease. *J. Mol. Evol.* 34: 536–543.

Ikeo K., Takahashi K., Gojobori T. (1995) Different evolutionary histories of kringle and protease domains in serine proteases: a typical example of domain evolution. *J. Mol. Evol.* 40: 331–336.

Islam S.D., Pilder S.H., Decker C.L., Cebra-Thomas J.A., Silver L.M. (1993) The human homolog of a candidate mouse t complex responder gene: conserved motifs and evolution with punctuated equilibria. *Hum. Molec. Genet.* 2: 2075–2079.

Jackman R.W., Beeler D.L., Fritze L., Soff G., Rosenberg R.D. (1987) Human thrombomodulin gene is intron depleted: nucleic acid sequences of the cDNA and gene predict protein structure and suggest sites of regulatory control. *Proc. Natl. Acad. Sci. USA* 84: 6425–6429.

Jaworski C.J., Piatigorsky J. (1989) A pseudo-exon in the functional human αA-crystallin gene. *Nature* 337: 752–754.

Jonsson J.J., Foresman M.D., Wilson N., McIvor R.S. (1992) Intron requirement for expression of the human purine nucleoside phosphorylase gene. *Nucleic Acids Res.* 20: 3191–3198.

Kehrer-Sawatzki H., Maier C., Moschgath E., Elgar G., Krone W. (1998) Genomic characterization of the neurofibromatosis type 1 gene of *Fugu rubripes*. *Gene* 222: 145–153.

Kersanach R., Brinkmann H., Liaud M.-F., Zhang D.-X., Martin W., Cerff R. (1993) Five identical intron positions in ancient duplicated genes of eubacterial origin. *Nature* 367: 387–389.

Kim H.-J., Kim D.-H., Magoori K., Saeki S., Yamamoto T.T. (1998) Evolution of the apolipoprotein E receptor 2 gene by exon loss. *J. Biochem.* 124: 451–456.

Kimura S., Hong Y.-S., Kotani T., Ohtaki S., Kikkawa F. (1989) Structure of the human thyroid peroxidase gene: comparison and relationship to the human myeloperoxidase gene. *Biochemistry* 28: 4481–4489.

Kirchgessner T.G., Chuat J.-C., Heinzmann C., Etienne J., Guilhot S., Svenson K., Ameis D., Pilon C., D'Auriol L., Andalibi A., Schotz M.C., Galibert F., Lusis A.J. (1989) Organization of the human lipoprotein lipase gene and evolution of the lipase gene family. *Proc. Natl. Acad. Sci. USA* 86: 9647–9651.

Klamut H.J., Bosnoyan-Collins L.O., Worton R.G., Ray P.N., Davis H.L. (1996) Identification of a transcriptional enhancer within muscle intron 1 of the human dystrophin gene. *Hum. Molec. Genet.* 5: 1599–1606.

Kobilka B.K., Frielle T., Collins S., Yang-Feng T., Kobilka T.S., Francke U., Lefkowitz R.J., Caron M.G. (1987) An intronless gene encoding a potential member of the family of receptors coupled to guanine nucleotide regulatory proteins. *Nature* 329: 75–79.

Kolb A.F., Günzberg W.H., Brem G., Erfle V., Salmons B. (1998) A functional eukaryotic promoter is contained within the first intron of the hGH-N coding region. *Biochem. Biophys. Res. Commun.* 247: 332–337.

Koller M., Strehler E.E. (1993) Functional analysis of the promoters of the human CaMIII calmodulin gene and of the intronless gene coding for a calmodulin-like protein. *Biochim. Biophys. Acta* 1163: 1–9.

Koop B.F., Hood L. (1994) Striking sequence similarity over almost 100 kilobases of human and mouse T-cell receptor DNA. *Nature Genet.* **7**: 48–53.

Kruse U., Sippel A.E. (1994) The genes for transcription factor nuclear factor I give rise to corresponding splice variants between vertebrate species. *J. Mol. Biol.* **238**: 860–865.

Kudo S., Fukuda M. (1989) Structural organization of glycophorins A and B genes: glycophorin B gene evolved by homologous recombination at *Alu* repeat sequence. *Proc. Natl. Acad. Sci. USA* **86**: 4619–4623.

Kudo S., Fukuda M. (1990) Identification of a novel human glycophorin, glycophorin E, by isolation of genomic clones and complementary DNA clones utilizing polymerase chain reaction. *J. Biol. Chem.* **265**: 1102–1110.

Kurabayashi M., Jeyaseelan R., Kedes L. (1993) Two distinct cDNA sequences encoding the human helix-loop-helix protein Id2. *Gene* **133**: 305–306.

Kurachi S., Hitomi Y., Furukawa M., Kurachi K. (1995) Role of intron 1 in expression of the human factor IX gene. *J. Biol. Chem.* **270**: 5276–5281.

Kwiatowski J., Krawczyk M., Kornacki M., Bailey K., Ayala F.J. (1995) Evidence against the exon theory of genes derived from the triose-phosphate isomerase gene. *Proc. Natl. Acad. Sci. USA* **92**: 8503–8506.

Lacy J., Roth G., Shieh B. (1994) Regulation of the human IgE receptor (Fc epsilon RII/CD23) by EBV. Localization of an intron EBV-responsive enhancer and characterization of its cognate GC-box binding factors. *J. Immunol.* **153**: 5537–5548.

Lambowitz A.M., Belfort M. (1993) Introns as mobile genetic elements. *Annu. Rev. Biochem.* **62**: 587–622.

Lei L., Zhao C., Heng H.H.Q., Ganz T. (1997) The human β-defensin-1 and α-defensins are encoded by adjacent genes: two peptide families with differing disulfide topology share a common ancestry. *Genomics* **43**: 316–320.

Lewis S.A., Cowan N.J. (1986) Anomalous placement of introns in a member of the intermediate filament multigene family: an evolutionary conundrum. *Mol. Cell. Biol.* **6**: 1529–1534.

Li W.-H. (1997) *Molecular Evolution*. Sinauer Associates, Sunderland, Mass.

Lim L.P., Sharp P.A. (1998) Alternative splicing of the fibronectin EIIIB exon depends on specific TGCATG repeats. *Mol. Cell. Biol.* **18**: 3900–3906.

Logsdon J.M., Tyshenko M.G., Dixon C., Jafari J.D., Walker V.K., Palmer J.D. (1995) Seven newly discovered intron positions in the triose-phosphate isomerase gene: evidence for the introns-late theory. *Proc. Natl. Acad. Sci. USA* **92**: 8507–8511.

Long M., Rosenberg C., Gilbert W. (1995) Intron phase correlations and the evolution of the intron/exon structure of genes. *Proc. Natl. Acad. Sci. USA* **92**: 12 495–12 499.

Lundwall A., Lazure L. (1995) A novel gene family encoding proteins with highly differing structure because of a rapidly evolving exon. *FEBS Letts.* **374**: 53–56.

McKnight G.L., O'Hara P.J., Parker M.L. (1986) Nucleotide sequence of the triosephosphate isomerase gene from *Aspergillus nidulans*: implication for a differential loss of introns. *Cell* **46**: 143–149.

McNaughton J.C., Hughes G., Jones W.A., Stockwell P.A., Klamut H.J., Petersen G.B. (1997) The evolution of an intron: analysis of a long, deletion-prone intron in the human dystrophin gene. *Genomics* **40**: 294–304.

Marchionni M., Gilbert W. (1986) The triosephosphate isomerase gene from maize: introns antedate the plant-animal divergence. *Cell* **46**: 133–142.

Mattick J.S. (1994) Introns: evolution and function. *Curr. Opin. Genet. Devel.* **4**: 823–831.

Mayer W.E., O'hUigin C., Klein J. (1993) Resolution of the *HLA-DRB6* puzzle: a case of grafting a *de novo*-generated exon on an existing gene. *Proc. Natl. Acad. Sci. USA* **90**: 10 720–10 724.

Michel D., Chatelain G., Mauduit C., Benahmed M., Brun G. (1997) Recent evolutionary acquisition of alternative pre-mRNA splicing and 3′ processing regulations induced by intronic B2 SINE insertion. *Nucleic Acids Res.* **25**: 3228–3234.

Mikol D.D., Rongnoparut P., Allwardt B.A., Marton L.S., Stefansson K. (1993) The oligodendrocyte-myelin glycoprotein of mouse: primary structure and gene structure. *Genomics* **17**: 604–610.

Miles C., Elgar G., Coles E., Kleinjan D.-J., Van Heyningen V., Hastie N. (1998) Complete sequencing of the *Fugu* WAGR region from WT1 to PAX6: dramatic compaction and conservation of synteny with human chromosome 11p13. *Proc. Natl. Acad. Sci. USA* **95**: 13 068–13 072.

Miller M., Zeller K. (1997) Alternative splicing in lecithin: cholesterol acyltransferase mRNA: an evolutionary paradigm in humans and great apes. *Gene* **190**: 309–313.

Milner C.M., Campbell R.D. (1990) Structure and expression of the three MHC-linked HSP70 genes. *Immunogenetics* **32**: 242–251.

Minakami R., Katsuki F., Sugiyama H. (1993) A variant of metabotropic glutamate receptor subtype 5: an evolutionarily conserved insertion with no termination codon. *Biochem. Biophys. Res. Commun.* **194**: 622–627.

Mishra VS., Henske E.P., Kwiatkowski D.J., Southwick F.S. (1994) The human actin-regulatory protein Cap G: gene structure and chromosome location. *Genomics* **23**: 560–565.

Misra-Press A., Cooke N.E., Liebhaber S.A. (1994) Complex alternative splicing partially inactivates the human chorionic somatomammotropin-like (*hCS-L*) gene. *J. Biol. Chem.* **269**: 23 220–23 229.

Miyabashira J., Sekiguchi T., Nishimoto T. (1994) Mammalian cells have two functional RCC1 proteins produced by alternative splicing. *J. Cell Sci.* **107**: 2203–2208.

Miyazaki H., Fukamizu A., Hirose S., Hayashi T., Hori H., Ohkubo H., Nakanishi S., Murakami K. (1984) Structure of the human renin gene. *Proc. Natl. Acad. Sci. USA* **81**: 5999–6003.

Mizumoto Y., Kimura T., Ivell R. (1997) A genomic element within the third intron of the human oxytocin receptor gene may be involved in transcriptional suppression. *Mol. Cell. Endocrinol.* **135**: 129–138.

Moran J.V., DeBerardinis R.J., Kazazian H.H. (1999) Exon shuffling by L1 retrotransposition. *Science* **283**: 1530–1534.

Murphy D.B., Seemann S., Wiese S., Kirschner R., Grzeschik K.H., Thies U. (1997) The human hepatocyte nuclear factor 3/fork head gene *FKHL13*: genomic structure and pattern of expression. *Genomics* **40**: 462–469.

Nakagawa Y., Kosugi H., Miyajima A., Arai K., Yokota T. (1994) Structure of the gene encoding the α subunit of the human granulocyte-macrophage colony stimulating factor receptor. *J. Biol. Chem.* **269**: 10 905–10 912.

Neer E.J., Schmidt C.J., Nambudripad R., Smith T.F. (1994) The ancient regulatory-protein family of WD-repeat proteins. *Nature* **371**: 297–300.

Nolan K.F., Kaluz S., Higgins J.M.G., Goundis D., Reid K.B.M. (1992) Characterization of the human properdin gene. *Biochem. J.* **287**: 291–297.

O'Neil R.J.W., Brennan F.E., Delbridge M.L., Croier R.H., Marshall Graves J.A. (1998) *De novo* insertion of an intron into the mammalian sex determining gene, *SRY. Proc. Natl. Acad. Sci. USA* **95**: 1653–1657.

Orengo C.A., Flores T.P., Taylor W.R., Thornton J.M. (1993) Identification and classification of protein fold families. *Protein Eng.* **6**: 485–500.

Pascual J., Castresana J., Saraste M. (1997) Evolution of the spectrin repeat. *BioEssays* **19**: 811–817.

Patthy L. (1985) Evolution of the proteases of blood coagulation and fibrinolysis by assembly from modules. *Cell* **41**: 657–663.

Patthy L. (1987) Intron-dependent evolution: preferred types of exons and introns. *FEBS Letts.* **214**: 1–7.

Patthy L. (1990) Evolutionary assembly of blood coagulation proteins. *Sem. Thromb. Hemostas.* **16**: 245–259.

Patthy L. (1991a) Modular exchange principles in proteins. *Curr. Opin. Struct. Biol.* **1**: 351–361.

Patthy L. (1991b) Exons – original building blocks of proteins? *BioEssays* **13**: 187–192.

Patthy L. (1994) Introns and exons. *Curr. Opin. Struct. Biol.* **4**: 383–392.

Patthy L. (1995) *Protein Evolution by Exon-Shuffling.* Springer-Verlag, New York.

Patthy L., Trexler M., Vali Z., Banyai L., Varadi A. (1984) Kringles: modules specialized for protein binding. *FEBS Letts.* **171**: 131–136.

Plamer J.D., Logsdon J.M. (1991) The recent origins of introns. *Curr. Opin. Genet. Devel.* **1**: 470–477.

Raben N., Nichols R.C., Martiniuk F., Plotz P.H. (1996) A model of mRNA splicing in adult lysosomal storage disease (glycogenosis type II). *Hum. Molec. Genet.* **5**: 995–1000.

Reardon D., Farber G.K. (1995) The structure and evolution of α/β barrel proteins. *FASEB J.* **9**: 497–503.

Redina O.E., Frohman M.A. (1998) Genomic analysis of murine phospholipase D1 and comparison to phospholipase D2 reveals an anusual difference in gene size. *Gene* **222**: 53–60.

Rogers J.H. (1989) How were introns inserted into nuclear genes? *Trends Genet.* **5**: 213–216.

Rogers J.H. (1990) The role of introns in evolution. *FEBS Letts.* **268**: 339–343.

Rohrer J., Conley M.E. (1998) Transcriptional regulatory elements within the first intron of Bruton's tyrosine kinase gene. *Blood* **91**: 214–221.

Rorman E.G., Scheinker V., Grabowski G.A. (1992) Structure and evolution of the human prosaposin chromosomal gene. *Genomics 13: 312–318.*

Rosby O., Poledne R., Hjermann I., Tonstad S., Berg K., Leren T.P. (1992) *Sty*I polymorphism in an enhancer region of the second intron of the apolipoprotein B gene in hyper- and hypocholesterolemic subjects. *Clin. Genet.* **42**: 217–223.

Rotwein P., Hall L.J. (1990) Evolution of insulin-like growth factor II: characterization of the mouse IGF-II gene and identification of two pseudo-exons. *DNA Cell Biol.* **9**: 725–735.

Rozmahel R., Heng H.H., Duncan A.M., Shi X.M., Rommens J.M., Tsui L.C. (1997) Amplification of *CFTR* exon 9 sequences to multiple locations in the human genome. *Genomics* **45**: 554–561.

Ruiz-Vazquez P., Moulard M., Silva F.J. (1996) Structure of the phenylalanine hydroxylase gene in *Drosophila melanogaster* and evidence of alternative promoter usage. *Biochem. Biophys. Res. Commun.* **225**: 238–242.

Rzhetsky A., Ayala F.J., Hsu L.C., Chang C., Yoshida A. (1997) Exon/intron structure of aldehyde dehydrogenase genes supports the 'introns-late' theory. *Proc. Natl. Acad. Sci. USA* **94**: 6820–6825.

Schulz A.S., Schleithof L., Faust M., Bartram C.R., Janssen J.W.G. (1993) The genomic structure of the human UFO receptor. *Oncogene* **8**: 509–513.

Sharp P.A. (1994) Split genes and RNA splicing. *Cell* **77**: 805–815.

Shen Y., Liu J., Wang X., Cheng X., Wang Y., Wu N. (1997) Essential role of the first intron in the transcription of hsp90beta gene. *FEBS Letts.* **413**: 92–98.

Shyue S.-K., Chang B.H.-J., Li W.-H. (1994) Intronic gene conversion in the evolution of human X-linked color vision genes. *Mol. Biol. Evol.* **11**: 548–551.

Slater E.P., Rabenau O., Karin M., Baxter J.D., Beato M. (1985) Glucocorticoid receptor binding and activation of a heterologous promoter by dexamethasone by the first intron of the human growth hormone gene. *Mol. Cell. Biol.* **5**: 2984–2992.

Smith C.W.J., Patton J.G., Nadal-Ginard B. (1989) Alternative splicing in the control of gene expression. *Annu. Rev. Genet.* **23**: 527–577.

Smith P.D., Davies A., Crumpton M.J., Moss S.E. (1994) Structure of the human annexin VI gene. *Proc. Natl. Acad. Sci. USA* **91**: 2713–2717.

Stoltzfus A., Spencer D.F., Zuker M., Logsdon J.M., Doolittle W.F. (1994) Testing the exon theory of genes: the evidence from protein structure. *Science* **265**: 202–207.

Stoltzfus A., Logsdon J.M., Palmer J.D., Doolittle W.F. (1997) Intron 'sliding' and the diversity of intron positions. *Proc. Natl. Acad. Sci. USA* **94**: 10 739–10 744.

Straus D., Gilbert W. (1985) Genetic engineering in the precambrian: structure of the chicken triosephosphate isomerase gene. *Mol. Cell. Biol.* **5**: 3497–3506.

Strelets V.B., Lim H.A. (1995) Ancient splice junction shadows with relation to blocks in protein structure. *Biosystems* **36**: 37–41.

Südhof T.C., Goldstein J.L., Brown M.S., Russell D.W. (1985a) The LDL receptor gene: a mosaic of exons shared with different proteins. *Science* **228**: 815–819.

Südhof T.C., Russell D.W., Goldstein J.L., Brown M.S., Sanchez-Pescador R., Bell G.I. (1985b) Cassette of eight exons shared by genes for LDL receptor and EGF precursor. *Science* **228**: 893–897.

Surinya K.H., Cox T.C., May B.K. (1998) Identification and characterization of a conserved erythroid-specific enhancer located in intron 8 of the human 5-aminolevulinate synthase 2 gene. *J. Biol. Chem.* **273**: 16 798–16 809.

Suzuki T., Yuasa H., Imai K. (1996) Convergent evolution. The gene structure of *Sulculus* 41 kDa myoglobin is homologous with that of human indoleamine dioxygenase. *Biochim. Biophys. Acta* **1308**: 41–48.

Takahara K., Hoffman G.G., Greenspan D.S. (1995) Complete structural organization of the human α1(V) collagen gene (*COL5A1*): divergence from the conserved organization of the other characterized fibrillar collagen genes. *Genomics* **29**: 588–597.

Takayanagi A., Kaneda S., Ayusawa D., Seno T. (1992) Intron 1 and the 5′-flanking region of the human thymidylate synthase gene as a regulatory determinant of growth-dependent expression. *Nucleic Acids Res.* **20**: 4021–4025.

Tomita M., Shimizu N., Brutlag D.L. (1996) Introns and reading frames: correlation between splicing sites and their codon positions. *Mol. Biol. Evol.* **13**: 1219–1223.

Traut T.W. (1988) Do exons code for structural or functional units in proteins? *Proc. Natl. Acad. Sci. USA* **85**: 2944–2948.

Tuddenham E.G.D., Cooper D.N. (1994) *The Molecular Genetics of Haemostasis and its Inherited Disorders.* Oxford University Press, Oxford.

Tycowski K.T., Shu M.-D., Steitz J.A. (1996) A mammalian gene with introns instead of exons generating stable RNA products. *Nature* **379**: 464–466.

van den Hoff M.J.B., Jonker A., Beintema J.J., Lamers W.H. (1995) Evolutionary relationships of the carbamoylphosphate synthetase genes. *J. Mol. Evol.* **41**: 813–832.

Venta P.J., Montgomery J.C., Hewett-Emmett H., Wiebauer K., Tashian R.E. (1985) Structure and exon-to-protein relationship of the mouse carbonic anhydrase II gene. *J. Biol. Chem.* **260**: 12130–12135.

Wallace P., Signer E., Paton I.R., Burt D., Quinlan R. (1998) The chicken CP49 gene contains an extra exon compared to the human CP49 gene which identifies an important step in the evolution of the eye lens intermediate filament proteins. *Gene* **211**: 19–27.

Weber K., Kabsch W. (1994) Intron positions in actin genes seem unrelated to the secondary structure of the protein. *EMBO J.* **13**: 1280–1286.

Wieczorek D.F., Smith C.W.J., Nadal-Ginard B. (1988) The rat α-tropomyosin gene generates a minimum of six different mRNAs coding for striated, smooth, and nonmuscle isoforms by alternative splicing. *Mol. Cell. Biol.* **8**: 679–694.

Winardi R., Discher D., Kelley C., Zon L., Mays K., Mohandas N., Conboy J.G. (1995) Evolutionarily conserved alternative pre-mRNA splicing regulates structure and function of the spectrin-actin binding domain of erythroid protein 4.1. *Blood* **86**: 4315–4322.

Wolford J.K., Prochazka M. (1998) Structure and expression of the human MTG8/ETO gene. *Gene* **212**: 103–109.

Xie S.S., Huang C.H., Reid M.E., Blancher A., Blumenfeld O.O. (1997) The glycophorin A gene family in gorillas: structure, expression, and comparison with the human and chimpanzee homologues. *Biochem. Genet.* **35**: 59–76.

Yuasa H.J., Cox J.A., Takagi T. (1997) Diversity of the troponin C genes during chordate evolution. *J. Biochem.* **123**: 1180–1190.

Zhang M.Q. (1998) Statistical features of human exons and their flanking sequences. *Hum. Molec. Genet.* **7**: 919–932.

Zucman-Rossi J., Batzer MA., Stoneking M., Delattre O., Thomas G. (1997) Interethnic polymorphism of EWS intron 6: genome plasticity mediated by *Alu* retroposition and recombination. *Hum. Genet.* **99**: 357–363.

Genes and gene families

4.1 The origins of human genes

Certainly the growth of the forebrain has been a success:
He has not got lost in a backwater like the lampshell
Or the limpet; he has not died out like the super-lizards.
His boneless worm-like ancestors would be amazed
At the upright position, the breasts, the four-chambered heart,
The clandestine evolution in the mother's shadow.
W.H. Auden. In Time of War, Commentary from *Journey to a War* (1939)

Some human genes appear to have originated comparatively recently. Indeed a very few have emerged in the relatively short space of time that has elapsed since divergence from chimpanzee some 7 Myrs ago. Other human genes appear to have homologues in simple organisms implying a very ancient origin, in some cases even predating the divergence of prokaryotes and eukaryotes. In the following sections, representative examples of genes are cited in order to illustrate the enormous variability in the antiquity of extant human genes. To some extent of course, the origin of a given gene is a matter merely of semantic distinction. Thus Doolittle's (1992) truism stated: 'new proteins come from old proteins as a result of gene duplications followed by base substitutions.' This notwithstanding, it has still been possible to draw up a working classificatory scheme for proteins based on their 'apparent invention time' (*Table 4.1*).

4.1.1 Genes with a specifically human origin

There are relatively few known examples of human genes which originated in the last 7 Myrs after the divergence of the human and chimpanzee lineages. One such gene is the potentially functional immunoglobulin Vκ gene located 1.5 Mb telomeric to the human Cκ gene which is putatively human-specific (Huber *et al.*, 1994). Further studies of the immunoglobulin κ locus (comprising one Cκ, 5 Jκ, and 76 Vκ gene segments on the short arm of human chromosome 2), have provided evidence for the human-specific partial duplication of the locus (Ermert *et al.*, 1995). The duplicated portion of the immunoglobulin κ locus, which contains 36 Vκ genes and pseudogenes, is not found in the chimpanzee or gorilla. Thus, the duplication event must have occurred after the branchpoint of humans from the great apes. Huber *et al.* (1994) estimated that this duplication event occurred 1 Myrs ago.

Table 4.1. Classification of human proteins by 'invention period' (after Doolittle, 1992)

Ancient proteins
> *First editions*: direct descent to both human and contemporary prokaryotes. Mostly mainstream metabolic enzymes (e.g. triosephosphate isomerase).
> *Second editions*: homologous sequences in humans and prokaryotes but apparently serve different functions (e.g. 27% identity between glutathione reductase (human red blood cells) and mercury reductase (*Pseudomonas*).

Middle age proteins
> Proteins found in most eukaryotes but prokaryotic counterparts are as yet unknown (e.g. actin).

Modern proteins
> *Recent vintage*: proteins found in animals or plants but not both. Not found in prokaryotes (e.g. collagen).
> *Very recent inventions*: Proteins confined to vertebrates (e.g. albumin).
> *Recent mosaics*: Modern proteins resulting from exon shuffling (e.g. low density lipoprotein receptor).

Another example of human specificity is provided by the salivary amylase (*AMY1*) genes clustered on human chromosome 1p21: three such genes are to be found in the human genome whereas only one is present in chimpanzee (Samuelson *et al.*, 1990). The most parsimonious explanation is that amplification of the *AMY1* gene sequences has occurred during the last 7 Myrs, the time since human and chimpanzee last shared a common ancestor. Similarly, the high-affinity immunoglobulin receptor genes (*FCGR1A*, 1q21; *FCGR1B*, 1p12; *FCGR1C*, 1q21) were triplicated from a single primate ancestral *Fcgr1* gene about 3 Myrs ago, after the divergence of chimpanzees from the human lineage (Maresco *et al.*, 1996). Finally, evidence for substantial deletions/translocations in the chimpanzee genome as compared to human is apparent from pulsed field gel electrophoretic studies of the region between the *HLA-B* and *TNF* genes on chimpanzee chromosome 5p21.3 (Leelayuwat *et al.*, 1993).

The gene coding regions of humans and chimpanzees are typically of the order of 99% homologous (e.g. *BRCA1*; Hacia *et al.*, 1998). As yet, however, there are no known differences between the human and chimpanzee genomes in terms of either the presence/absence of genes or gene copy number that could account for the considerable anatomical and behavioral differences between the two species. We may surmise that subtle differences in the expression of regulatory genes or alternatively differences in the responsiveness of those genes which serve as targets for the action of regulatory proteins, may help to explain why humans are not chimpanzees. Candidate genes could include not only those involved in developmental regulation but possibly also those situated adjacent to human-specific karyotypic rearrangements.

4.1.2 Human genes which originated after the divergence of Old World monkeys and New World monkeys

The haptoglobin gene appears to have become amplified after the separation of Old World monkeys from New World monkeys. The gorilla, chimpanzee, orangutan and rhesus monkey have three haptoglobin genes, humans have two (*HP, HPR*; 16q22), whilst the spider monkey and cebus monkey have only one (McEvoy and Maeda 1988). The apolipoprotein(a) (*LPA*; 6q27; Lawn *et al.*, 1997; Pesole *et al.*, 1994) gene probably also emerged during this time since its presence has only been detected in humans, apes, and Old World monkeys.

The number of rhesus blood group antigen genes varies between primate species, humans having two (*RHD, RHCE*; 1p34-p36.2), gorillas two, chimpanzees three, whilst orangutan, gibbons and all Old World and New World monkeys possess only a single *RH*-like gene (Salvignol *et al.*, 1995). Duplication of the *RH*-like genes probably therefore occurred between 8 and 11 Myrs ago before divergence of the human and gorilla lineages.

Eosinophil-derived neurotoxin (EDN) and eosinophil cationic protein (ECP) are host defense proteins which are members of the ribonuclease family. They are encoded respectively by the *RNASE2* and *RNASE3* genes located on human chromosome 14q24-q31. Divergence of the nucleotide sequences of the two related genes was found to be consistent with a duplication event occurring some 25–40 Myrs ago after the divergence of Old World monkeys from New World monkeys (Hamann *et al.*, 1990). In agreement with this prediction, *RNASE2* and *RNASE3* orthologues are evident in chimpanzee, gorilla, orangutan and macaque but not in the marmoset, a New World monkey (Rosenberg *et al.*, 1995; Zhang *et al.*, 1998). Since their duplication, the two proteins have diverged very rapidly, with evolution promoting the acquisition of novel functions viz. increased cationicity/toxicity (ECP) and enhanced ribonuclease activity (Rosenberg and Dyer, 1995).

4.1.3 Human genes which originated during primate evolution

The *ZNF91* Kruppel-associated box-containing zinc finger gene family on human chromosome 19p12-p13 is present in the great apes, the gibbons, Old World and New World monkeys but is not found in prosimians or rodents (Bellefroid *et al.*, 1995). The origin of this gene family is unclear. Presumably, the *ZNF91* gene cluster arose by duplication/amplification at least 55 Myrs ago in the common ancestor of simians.

The three alcohol dehydrogenase genes (*ADH1, ADH2,* and *ADH3*), linked on human chromosome 4q22, also originated during the adaptive radiation of the primates; two successive duplications are estimated to have occurred 45 ± 8 Myrs and 60 ± 8 Myrs ago (Duester *et al.*, 1986; Ikuta *et al.*, 1986; Trezise *et al.*, 1989; 1991; Yokoyama *et al.*, 1987).

Other examples of human genes with a primate origin include the interferon-α gene (*IFNA*) and growth hormone (*GH*) gene clusters. The *IFNA* cluster, containing 13 members (see Section 4.2.1) and located at chromosome 9p21, has emerged by a process of duplication and divergence over the last 26 Myrs (Miyata and Hayashida 1982). The *GH* gene cluster is also essentially a primate creation with several genes (*GH1, GH2, CSH1,* and *CSH2*; 17q23) having emerged over the last 25–50 Myrs by a process of duplication and divergence (Chen *et al.*, 1989; Golos *et al.*, 1993; Miller and Eberhardt 1983). see Section 4.2.1).

Some genes emerged in the last 100 Myrs during the adaptive radiation of the mammals. For example, the fucosyltransferase gene family (*FUT1, FUT2,*

19q13.1-qter; *FUT3, FUT5, FUT6*, 19p13.3; *FUT4*, 11q21; *FUT7*, 9; *FUT8*, 14q23) is thought to have emerged during this period by a process of gene duplication, translocation and divergence (Costache *et al.*, 1997).

α-Lactalbumin is a mammary gland-expressed calcium-binding protein which interacts with galactosyl transferase to promote lactose synthesis. Although this function is clearly confined to mammals, the α-lactalbumin gene (*LALBA*; 12q13) is thought to have arisen by duplication of the gene encoding lysozyme *c* which encodes a bacteriolytic enzyme present in the tissues and secretions of mammals, birds, reptiles and even insects. This gene duplication probably occurred 300–400 Myrs ago (Nitta and Sugai, 1989; Prager, 1996; Prager and Wilson, 1988; Qasba and Kumar, 1997) well before the evolutionary emergence of the mammals and the process of lactation.

4.1.4 Human genes whose origin preceded the divergence of mammals

Other genes were specifically vertebrate creations, their emergence through duplication and divergence having perhaps been made possible by the genome duplications thought to have occurred prior to the adaptive radiation of this subphylum (Chapter 2, section 2.1). A new wave of gene creation may have accompanied the emergence of the amphibians and early reptiles thereby equipping them for terrrestrial life.

One example of a vertebrate creation appears to be the vitamin K-dependent serine proteases of blood coagulation (factors VII, IX, X, protein C and prothrombin). Prothrombin is present in bony fish (trout), cartilaginous fish (dogfish) and the hagfish, one of the modern representatives of the jawless *Agnatha* (Banfield and MacGillivray, 1992; Doolittle, 1993). There is no evidence, however, for the existence of thrombin or a thrombin-like protein in either protochordates or echinoderms. Whether the other four vitamin K-dependent factors of coagulation are present in fish is as yet unclear (Doolittle, 1993). If they are, the adaptive radiation of the vitamin K-dependent factors of hemostasis must have occurred during the space of some 50 Myrs between the divergence of the protochordates and the appearance of the *Agnatha*, some 450 Myrs ago (Doolittle, 1993). The subsequent evolution of the vitamin K-dependent factors of coagulation is explored in Section 4.2.3, *Serine protease genes*.

Another specifically vertebrate invention was pulmonary surfactant which comprises a series of proteins which serve to reduce surface tension at the air–liquid interface in the lung. This was probably a prerequisite for air breathing. The human genome contains a number of clustered pulmonary surfactant genes (Kölble *et al.*, 1993), among them *SFTPA1* (10q21-q24) encoding pulmonary surfactant protein A. Orthologues of this gene have been found in airbreathing lungfish and surfactant protein has been detected both in the swimbladder of goldfish and in the lungs of lungfish (Sullivan *et al.*, 1998).

Transthyretin is a thyroid-binding protein which in human is synthesized mainly in the liver but also in the choroid plexus and retina. The transthyretin gene (*TTR*; 18q11-q12) is a vertebrate invention having made its first appearance in the stem reptiles some 300 Myrs ago (Schreiber and Richardson, 1997). Transthyretin is expressed in the choroid plexus in reptiles. Liver synthesis of the protein only evolved later (and independently) in birds, eutherian mammals and in some marsupials (Schreiber and Richardson, 1997).

Troponin I, together with troponins C and T comprise the three subunits of the troponin complex of the thin filaments of vertebrate striated muscle. Since invertebrates and ascidians possess a single troponin I gene (Hastings, 1997), at least two duplications must have taken place during the vertebrate lineage to generate the three troponin I genes evident in the human genome (*TNNI1*, 1q31; *TNNI2*, 11p15; *TNNI3*, 19q13).

The human ABO blood group gene (*ABO*; 9q34) has a common evolutionary origin with the chromosomally linked and now inactive α-1,3-galactosyltransferase (*GGTA1*; 9q33-q34; see Chapter 6, Section 6.2) gene; the two genes emerged by duplication and divergence about 400 Myrs ago during the evolution of the early vertebrates (Saitou and Yamamoto, 1997).

Olfactory receptors are extremely important to mammals; dogs, rats and mice may have as many as 1000 genes encoding them. Although the human genome may contain fewer functional genes than other mammals, pseudogenes abound. Indeed, it has been estimated that >0.1% of the human genome is composed of olfactory receptor genes and pseudogenes dispersed between at least 13 different chromosomes (Trask *et al.*, 1998). [The classification of this gene family is still in its infancy but known members include *OR1A1*, 17p13; *OR1D2*, *OR1D4*, *OR1D5*, 17p13; *OR1E1*, *OR1E2*, 17p13; *OR1F1*, 16p13; *OR1G1*, 17p13; *OR2D2*, 11p15; *OR3A1*, *OR3A2*, *OR3A3*, 17p13; *OR5D3*, *OR5D4*, 11q12; *OR5F1*, 11q12; *OR6A1*, 11p15; *OR10A1*, 11p15]. The origin and emergence of the present-day size of the olfactory receptor gene family probably preceded the divergence of the mammals (Ben-Arie *et al.*, 1993; Buettner *et al.*, 1998; Issel-Tarver and Rine, 1997). However, at least in the primates, the olfactory receptor genes appear to have still been in a considerable state of flux with numerous translocations, duplications and deletions having occurred during the evolution of the great apes (Trask *et al.*, 1998; see section 4.2.3, *Olfactory receptor genes*).

4.1.5 Human genes whose origin preceded the divergence of the vertebrates

One example of a human gene whose origin preceded the advent of the vertebrates ~500 Myrs ago is the c-*myc* proto-oncogene (Atchley and Fitch, 1995). Present in all vertebrates, a homologue of the human gene (*MYC*; 8q24) is detectable in echinoderms but not in *Caenorhabditis* or *Drosophila* (Walker *et al.*, 1992). Other examples include the insulin-like growth factor genes (*IGF1* and *IGF2*; 12q22-q24 and 11p15.5 respectively) which emerged during the evolution of the protochordates more than 600 Myrs ago (McRory and Sherwood, 1997; see Section 4.2.3).

4.1.6 Human genes whose origin preceded the divergence of the metazoa

Human genes with homologues in the fruitfly, *Drosophila*, must have originated prior to the divergence of the deuterostomes from the protostomes ~700 Myrs ago (Doolittle *et al.*, 1996; Ayala *et al.*, 1998). A large number of *Drosophila* genes have been shown to have human homologues and *vice versa* (see FlyBase at

http://flybase.bio.indiana.edu/). Examples (with human chromosomal locations where known) include proto-oncogenes *jun* (*JUN*; 1p31-p32; Perkins *et al.*, 1988) and *fos* (*FOS*; 14q24; Perkins *et al.*, 1988), *flightless I* (*FLI1*; 17p11.2; Campbell *et al.*, 1997), *minibrain* (*MNBH*; 21q22.2; Guimera *et al.*, 1996), *dorsal* (*REL*; 2p12-p13; Steward 1987), *atonal* (*ATOH1*; 4q22; Ben-Arie *et al.*, 1996), *archain* (*ARCN1*; 11q23; Radice *et al.*, 1995), tumor suppressor gene lethal 2 giant larvae (*LLGL1*; 17p11-p12; Strand *et al.*, 1995), transient receptor potential channel-related protein 1 (*TRPC1*; Wes *et al.*, 1995), succinate dehydrogenase, iron-sulfur protein subunit (*SDH1*; 1p22.1-qter; Au and Scheffler 1994), *slowpoke* (*SLO*, chromosome 10; Pallanck and Ganetzky 1994), *enhancer of split* (*TLE1*; 19p13.3; Stifani *et al.*, 1992), *dodo* (*PIN1*, Maleszka *et al.*, 1996), EGF receptor (*EGFR*; 7p12.1-p12.3; Livneh *et al.*, 1985), *disheveled* (*DVL1*; 1p36; Pizzuti *et al.*, 1996), *son of sevenless* (*SOS1*, 2p21-p22; *SOS2*, 14q21; Della *et al.*, 1995), *awd* (*NME1* (17q21.3; Zinyk *et al.*, 1993), *Ddx1* DEAD box polypeptide (*DDX1*; 2p24; Rafti *et al.*, 1996), *cut* (*CUTL*; 7q22; Neufeld *et al.*, 1992), and the paired box domain family of transcription factors that play an important role in development (*PAX1*, 20p11.2; *PAX2*, 10q25; *PAX3*, 2q36; *PAX4*, 7q22-qter; *PAX5*, 9p13; *PAX6*, 11p13; *PAX7*, 1p36; *PAX8*, 2q12-q14; *PAX9*, 14q12-q14; Quiring *et al.*, 1994; Balczarek *et al.*, 1997; Czerny *et al.*, 1997).

Homologs of the *ETS1* proto-oncogene, located on human chromosome 11q23, have been found not only in *Drosophila* (Laudet *et al.*, 1993) but throughout the metazoa in organisms as evolutionarily distant as sponges, anenomes, flatworms and nematodes (Degnan *et al.*, 1993; Laudet *et al.*, 1999). The origin of the *ETS* genes therefore appears to predate the divergence of the schizocoeles (arthropods, annelids etc) from the pseudocoeles (nematodes) more than 750 Myrs ago (Doolittle *et al.*, 1996). STAT (Signal Transducers and Activators of Transcription) proteins represent an example of proteins present in *Drosophila* and *Dictyostelium* but not in yeast (Kawata *et al.*, 1997) and so must have emerged at some stage during the adaptive radiation of the metazoa. Homology is also evident between members of the human lipase gene family [lipoprotein lipase (*LPL*; 8p22), hepatic lipase (*LIPC*; 15q21-q23) and pancreatic lipase (*PNLIP*; 10q26)] and the yolk proteins of *Drosophila* (Hide *et al.*, 1992; Kirchgessner *et al.*, 1989).

Another example of a gene with an ancestry stretching back at least as far as 750 Myrs, is a human retina-expressed gene, *CAGR1* (13q13) which is homologous to the *Caenorhabditis elegans* cell fate-determining gene, *mab*-21 (Margolis *et al.*, 1996). Ahringer (1997) presented evidence that at least 50% of *Caenorhabditis* genes are likely to have counterparts in the human genome. Similarly, the human neurotrophic tyrosine kinase receptor (*NTRK3*; 15q25) gene has a homologue in the snail, *Lymnea stagnalis* (van Kesteren *et al.*, 1998) although not apparently in *C. elegans*.

Ancient conserved regions, deemed to be regions of the greatest structural or functional importance on account of their evolutionary conservation are also evident in various other proteins with homologues in both human and *C. elegans*, for example adenylate cyclases, epidermal growth factor-like domains, gelsolin, intermediate filament proteins, kinesins, neurotransmitter transporters and the ubiquitins (Green *et al.*, 1993).

4.1.7 Human genes whose origin preceded the divergence of animals and fungi

> We must acknowledge that man with all his noble qualities still bears in his bodily frame the indelible stamp of his lowly origin.
> C. Darwin *The Descent of Man* (1871)

The 12 068 kb sequence of the 16 chromosomes of the yeast *Saccharomyces cerevisiae* genome has now been established (Goffeau *et al.*, 1996). A considerable proportion of the organism's 5885 genes are significantly related to sequences in the human genome. Tugendreich *et al.* (1994) used the BLASTX P program to measure the extent of this homology: 29% of human cDNAs found in GenBank matched a yeast protein with a P value of less than 10^{-5}. More specifically, these authors showed that some important human disease genes manifest considerable sequence similarity to a yeast counterpart e.g. adrenoleukodystrophy (*ALD*, Xq28) with a yeast 70 kD peroxisomal membrane protein, the myotonic dystrophy protein (*DMPK*, 19q13) with a yeast cAMP-dependent protein kinase, the Wilms' tumor protein (*WT1*, 11p13) with a yeast zinc finger protein and the Ret (*RET*, 10q11.2) protein underlying multiple endocrine neoplasia type 2A with a yeast cell division control protein (Tugendreich *et al.*, 1994). Similarly, human genes *QM* (Xq28) encoding a c-*jun*-associated transcription factor (Farmer *et al.*, 1994), *SEC13L1* (3p24–25) encoding a protein putatively involved in vesicle biosynthesis from the endoplasmic reticulum during protein transport (Swaroop *et al.*, 1994), *DNECL* (14q32) encoding cytoplasmic dynein (Gibbons, 1995), the MCM family (*MCM2*, 3q21; *MCM3*, 6p12; *MCM4*, 8q12-q13; *MCM5*, 22q13; *MCM6*, 2q21; *MCM7*, 7q21-q22) of DNA replication proteins (Kearsey and Labib, 1998), the *PTEN* (10q23.3) tumor suppressor gene (Li *et al.*, 1997) and the G protein α-subunit gene family (Wilkie *et al.*, 1992; see Section 4.2.1), have yeast homologues.

Since *S. cerevisiae* diverged from higher eukaryotes some 1000 Myrs ago (Doolittle *et al.*, 1996), homologies between human and yeast genes are clearly very ancient. Such conservation at the amino acid level is likely to reflect the conservation of fairly basic biological functions. Thus, it comes as no surprise to find that the yeast genes encoding cytochrome *c* (Wu *et al.*, 1986), histone deacetylase (Leipe and Landsman, 1997), the origin recognition complex (Gavin *et al.*, 1995), the recombination protein *Rad51* (Brendel *et al.*, 1997; Shinohara *et al.*, 1993) and the nucleotide excision repair gene *Rad23* (Masutani *et al.*, 1994) have highly homologous human counterparts viz. *CYC1* (8q24.3), *HDAC1* (1p34), *ORC1L* (1p32), *RECA* (15q15.1) and *RAD23A* and *RAD23B* (19p13 and 3p25, respectively).

Sometimes the homology is regionally localized, for example the GTPase-activating protein-related domain of neurofibromin (encoded by the *NF1* gene on human chromosome 17q11.2) exhibits extensive homology to the *S. cerevisiae* proteins IRA1 and IRA2 (Xu *et al.*, 1990). Such *ancient conserved regions* (ACRs) represent regions of the greatest structural or functional importance. Eukaryotic-specific ACRs are also evident in various other proteins with homologues in both human and yeast, for example hexokinases, β-transducins, protein kinase catalytic domains and the Src homology domain (Green *et al.*, 1993).

4.1.8 Human genes whose origin preceded the divergence of plants and animals

Some human genes have their counterparts in both plants and fungi implying that they originated before the divergence of the three kingdoms (Doolittle *et al.*, 1996). Members of the *myb* gene family (*MYB*; 6q23.3-q24) are found in animals, plants, fungi and even slime moulds (Lipsick, 1996; Rosinski and Atchley, 1998). Myb proteins function as regulators of cell growth and differentiation by binding to DNA and regulating gene expression. Since slime moulds such as *Dictyostelium* are thought to have diverged before the main eukaryotic radiation which gave rise to animals, plants and fungi ~1000 Myrs ago (Doolittle *et al.*, 1996), the presence of Myb proteins in this organism serves to date the emergence of this ancient regulator of gene expression. Remarkably, despite this ancient origin, the Myb-related proteins of *Drosophila* and *Dictyostelium* are still able to recognize and interact with the same cognate DNA sequence as their vertebrate counterparts (Lipsick, 1996).

The High Mobility Group (HMG) proteins are another human gene family represented in both plants and fungi (Laudet *et al.*, 1993). Two members of this family, *HMG1* and *HMG2*, are present on human chromosomes 13q12 and 4q31 respectively. Members of the HMG protein superfamily are characterized by the possession of one or more HMG boxes, each of which comprises ~80 amino acid residues and is capable of interacting with DNA. Both the actins (Hennessey *et al.*, 1993) and the CCAAT-specific transcription factor, NF-Y (Li *et al.*, 1992) have their counterparts in plants and fungi whilst the annexins (Morgan *et al.*, 1998) and the cystatins (Rawlings and Barrett, 1990) are present in plants but not fungi.

4.1.9 Human genes whose origin preceded the divergence of prokaryotes and eukaryotes

> Would it be too bold to imagine that in the great lengths of time, since the earth began to exist, perhaps millions of ages before the commencement of the history of mankind, would it be too bold to imagine that all the warmblooded animals have arisen from one living filament which the Great First Cause endued with animality and thus possessing the faculty of continuing to improve by its own inherent activity and of delivering down those improvements by generation to its posterity, world without end.
> Erasmus Darwin *Zoonomia* (1794)

Genes that are common to prokaryotes and eukaryotes must have arisen more than 2000 Myrs ago before the divergence of the two groups (Doolittle *et al.*, 1996). A number of human genes fit into this category. Hexokinase is one of the best characterized examples with sequences available from bacteria, yeast, plants and vertebrates (three hexokinase genes *HK1*, *HK2*, and *HK3* exist in humans on chromosomes 10q22, 2p12 and 5q35 respectively). A proposed multi-kingdom phylogeny of the hexokinase gene is shown in *Figure 4.1* and depicts a series of duplication and fusion events that must have occurred during its evolution (Cardenas *et al.*, 1998). The matrix metalloproteinases that play an important role in tissue remodelling and wound healing have homologues in plants, animals and bacteria (Massova *et al.*, 1998).

Universal stress protein UspA of *E. coli* is homologous to the MADS-box transcription regulators of eukaryotes (Mushegian and Koonin, 1996) whilst the FUS6 family of eukaryotic proteins contains a putative DNA-binding domain related to bacterial helix-turn-helix transcription regulators (Mushegian and Koonin, 1996). Other human genes with homologues in *E. coli* are the Y-box family (*YB1*; 1p34) which encode nucleic acid-binding proteins which serve as transcriptional repressors in humans and the cold shock response in bacteria (Wolffe *et al.*, 1992; Wolffe, 1994), the *SIR2L* gene family which are involved in cell cycle progression and chromosome stability and chromatin silencing (Brachmann *et al.*, 1995), *CYP51* (7q21), the most conserved member of the P450 monooxygenase family involved in the demethylation of sterol precursors (Yoshida *et al.*, 1997), the *MLH1* (3p21), *MSH2* (2p21-p22), *GTBP* (2p16), *PMS1* (2p31-q33) and *PMS2* (7p22) mismatch repair genes (Fishel *et al.*, 1993; Kolodner, 1996), the *RAD51A* (15q15) DNA repair protein gene homologous to the RecA protein of *E. coli* (Brendel *et al.*, 1997), and *HES1* (21q22.3) with homology to the σ cross-reacting protein 27A of *E. coli* (Scott *et al.*, 1997). A considerable degree of homology has also been noted between prokaryotic and eukaryotic RNA polymerases (Iwabe *et al.*, 1991; Klenk *et al.*, 1992; Sweetser *et al.*, 1987), prokaryotic σ factors and eukaryotic transcription factors TBP (TFIIB) and TFIIE (Malik *et al.*, 1991; Ohkuma *et al.*, 1991; Rowlands *et al.*, 1994; Sumimoto *et al.*, 1991), eukaryotic translation initiation factors eIF-1A and eIF-5A and their prokaryotic counterparts (Hashimoto and Hasegawa, 1996;

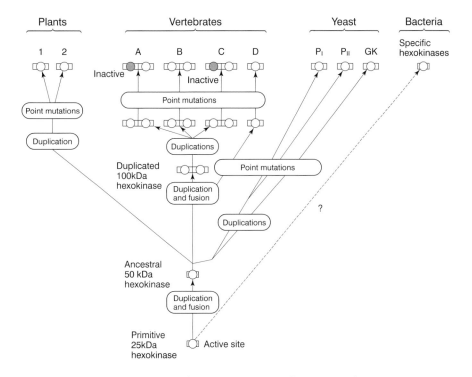

Figure 4.1. A proposed phylogeny for the hexokinases (Cardenas *et al.*, 1998).

Kyrpides and Woese, 1998), between prokaryotic and eukaryotic RNA-binding proteins (Mulligan *et al.*, 1994) and between prokaryotic and eukaryotic type II DNA topoisomerases and DNA polymerases (Forterre *et al.*, 1994). A considerable degree of homology has also been noted between prokaryotic and eukaryotic aminoacyl-tRNA synthetases (Francklyn *et al.* 1997; Nagel and Doolittle, 1991; see Section 4.2.1, *Genes encoding tRNAs and aminoacyl-tRNA synthetase*).

The eukaryotic cell cytoskeleton contains a number of proteins, the most abundant being actins, tubulins, myosins and cytokeratins (Little and Seehaus, 1988). The evolutionary origin of these proteins is intimately bound up with the origins of the eukaryotic cell itself (Doolittle, 1992). Thus γ-tubulin is thought to have evolved before the α- and β-tubulins diverged from each other and since these three tubulin isotypes are found in all eukaryotes, their origins must have preceded the eukaryotic divergence (Ludueña 1998). Consistent with this postulate, sequence similarities exist between eukaryotic actins and tubulins and bacterial *ftsA* and *ftsZ* proteins (Doolittle, 1995).

There are many instances of genes present in *E. coli* or yeast having multiple homologues in the human genome indicating that the gene family in question must have expanded during evolution. One example of this are the helicase genes of the RecQ family of which there are five known in the human genome (*WRN*, 8p; *BLM*, 15; *RECQL*, 12p12; *RECQLA*, 8q24.3; *RECQL5*, 17q25) (Kitao *et al.*, 1998).

As one travels further back in evolutionary time, the similarity between gene sequences tends to decay to an extent that only specific portions may be recognizably homologous. These *ancient conserved regions* (ACRs) represent those regions of greatest structural or functional importance and often correspond to specific domains. ACRs common to both prokaryotes and eukaryotes have been noted in a diverse array of proteins, for example enolase, glyceraldehyde 3-phosphate dehydrogenase, cytochrome *c* oxidase subunit I, aminoacyl-transfer RNA (II) synthetases, HSP70 and HSP90 (see section 4.2.3), phosphoglycerate kinase, pyruvate dehydrogenase E1α, pyruvate kinase, ribosomal proteins L3 and P0 and triosephosphate isomerase (Green *et al.*, 1993). Ceruloplasmin (*CP*; 3q23-q25) and coagulation factors V (*F5*; 1q23) and VIII (*F8C*; Xq28) manifest homologies to the small blue proteins of bacteria which have a role in electron transfer (Rydén and Hunt, 1993). A 60 amino acid domain found in cystathion-ine-β-synthase (*CBS*; 21q22.3) is also found in a bacterial ABC transporter protein and in a putative protein found in archaebacteria (Bateman, 1997).

Phylogenetic studies have suggested that the emergence of cytochrome oxidase (a key enzyme in aerobic metabolism) (Castresana *et al.*, 1994), the aldehyde dehydrogenases (e.g. *ALDH1*, 9q21; *ALDH3*, 17; *ALDH5*, 9; *ALDH6*, 15q26; *ALDH9*, 1; *ALDH10*, 17p11.2; Yoshida *et al.*, 1998), carbamoyl phosphate synthetase (a key enzyme of arginine and pyrimidine biosynthesis) (*CPS1*; 2q35; Schofield 1993; Lawson *et al.*, 1996), the protein synthesis elongation factors Tu (*TUFM*; 16p11) and G (*EEF1G*) (Baldauf *et al.*, 1996), tRNA splicing endonucle-ase (Trotta *et al.*, 1997) and the glucose-6-phosphate isomerase gene family (*GPI*; 19q13; Hattori *et al.*, 1995) all predated the divergence of prokaryotes and eukaryotes. The RadA protein that catalyzes DNA pairing and strand exchange is present not only in all eukaryotes including human (*RECA*; 15q15.1) but also has homologues in both prokaryotes (RecA) and the archaea (Seitz *et al.*, 1998).

Phylogenetic studies have provided evidence for the extreme antiquity of the glutamine synthetase gene (Kumada *et al.*, 1993). Glutamine synthetase is an essential enzyme of nitrogen metabolism with vital functions in both glutamine biosynthesis and ammonia assimilation. It is thought therefore to be indispensible to all living organisms. Comparison of the glutamine synthetase genes of extant organisms as diverse as bacteria and vertebrates allowed Kumada *et al.* (1993) to estimate the time of duplication of the ancestral glutamine synthetase gene to ~3500 Myrs ago, that is >1000 Myrs before the divergence of the prokaryotes and eukaryotes.

4.1.10 The emergence of genes and gene families has paralleled organismal evolution

> Probably all the organic beings which have ever lived on this earth have descended from some one primordial form.
> C. Darwin *The Origin of Species* (1859)

Eukaryotic genomes have been largely constructed by a continual process of gene duplication and divergence utilizing a set of basic gene types (Chapter 3, section 3.6.4) many of which possess counterparts in the genomes of primitive organisms. Much of this gene diversification may have taken place in the 'Cambrian explosion' which is thought to have begun ~570 Myrs ago.

Genes which have emerged during primate evolution are relatively few in number and tend to involve the addition of new members to pre-existing multigene families (e.g. genes involved in host defence such as the immunoglobulins; the alcohol dehydrogenases whose expansion may have been associated with a move to a fruit diet). During mammalian evolution, certain genes emerged which were intimately associated with mammalian physiology such as lactation. The emergence of genes with novel functions (e.g. pulmonary surfactant) and the expansion of other gene families (e.g. coagulation factors and olfactory receptors) occurred in parallel with the emergence of the first terrestrial vertebrates. By contrast, genes whose origin preceded the divergence of animals and fungi, or animals and plants, encode proteins involved in fairly basic cellular processes. Finally, the truly primordial genes shared by both prokaryotes and eukaryotes encode protein products which are absolutely required for fundamental cellular processes such as DNA repair and replication, transcription and translation, as well as certain very basic enzymes of metabolism such as glutamine synthetase.

Iwabe *et al.* (1996) performed their own analysis of gene duplication during organismal evolution. They concluded that most gene duplications giving rise to novel functions predated the divergence of the vertebrate and arthropod lineages. However, genes encoding products that perform virtually identical functions but which differ in their tissue distribution (tissue-specific isoforms) underwent duplications independently in vertebrates and arthropods after divergence of their respective lineages. Finally, genes which encode proteins that are localized to cell compartments (compartmentalized isoforms) emerged by duplications which predated the separation of animals and fungi. Iwabe *et al.* (1996) concluded that there was a good correspondence between molecular evolution at the level of the gene, and tissue and organismal evolution.

4.2 Multigene families

4.2.1 Gene families

Dayhoff *et al.* (1978) defined a gene family as a group of related genes encoding proteins differing at fewer than half their amino acid positions. Owing to complex evolutionary histories, both the size of gene families and the relationship between their constituent members are very variable (Henikoff *et al.*, 1997). In principle, the members of multigene families can evolve in three different ways—by concerted evolution, divergent evolution or by a birth-and-death process (Ota and Nei, 1994; *Figure 4.2*). *Concerted evolution* may operate through either unequal crossing over or gene conversion (Chapter 9, section 9.5) both of which serve to homogenize multigene family members. Examples of this process include the rRNA and histone genes (Sections 4.2.2, *Histone genes* and 4.2.2, *Ribosomal RNA genes*). By contrast, some distinct groups of genes within multigene families appear to have been maintained over long periods of evolutionary time. An example of this *divergent evolution* is provided by the immunoglobulin V_H genes (Section 4.2.4, *Immunoglobulin genes*), the diversification of which may have conferred a selective advantage. In other multigene families, gene duplication serves to create new genes whilst other genes are inactivated by deleterious mutation or are eliminated by unequal crossing over ('*birth-and-death*'). Once again, the immunoglobulin V_H genes provide an example of this process. It can be seen therefore that some multigene families possess members that serve to illustrate the action of all three of the above processes. No one multigene family can be regarded as being archetypal. Each illustrates certain principles and so a variety of examples will be discussed.

Actin genes. A total of six actin genes are found in mammals although a considerably larger number of pseudogenes may be found (Engel *et al.*, 1981; Erba *et al.*,

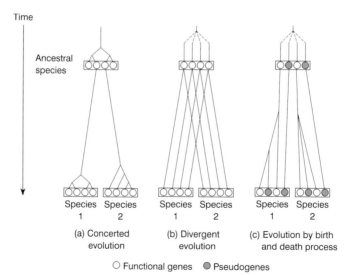

Figure 4.2. Three different modes of evolution of multigene families (after Ota and Nei 1994).

1988; Gunning *et al.*, 1994; Kedes *et al.*, 1985; Ng *et al.*, 1985; Ponte *et al.*, 1993). The actin genes are highly conserved evolutionarily and encode protein isoforms that differ from each other by only a few amino acids, mostly at their amino terminals. Those characterized in human are α-smooth muscle aortic actin (*ACTA2*; 10q22-q24), α-skeletal actin (*ACTA1*; 1q42.1), α-cardiac actin (*ACTC*; 15q14), β-nonmuscle cytoplasmic (*ACTB*; 7p12-p15), γ-nonmuscle cytoplasmic (*ACTG1*; 17q25) and γ-smooth muscle enteric (*ACTG2*; 2p13.1) actins. Comparison of the structures of these six human actin genes (*Figure 4.3*) indicates that comparable regions are identically interrupted by introns although the sizes of the introns differ (Miwa *et al.*, 1991). Comparison of nucleotide and amino acid sequences as well as gene structures also allowed Miwa *et al.* (1991) to propose a possible phylogenetic tree for the actin gene family (*Figure 4.4*). Since all four muscle actins differ from the cytoplasmic actins by substitutions at 19 amino acid residues, the muscle actins must have shared a common ancestral actin gene. Duplication and divergence then led to the emergence of the smooth muscle and striated muscle actin genes. Introns were gained or lost at different stages in this process. The ancestral smooth muscle actin gene acquired an amino acid substitution at residue 89 before being duplicated; the smooth muscle γ-actin subsequently lost an amino acid residue at position 4.

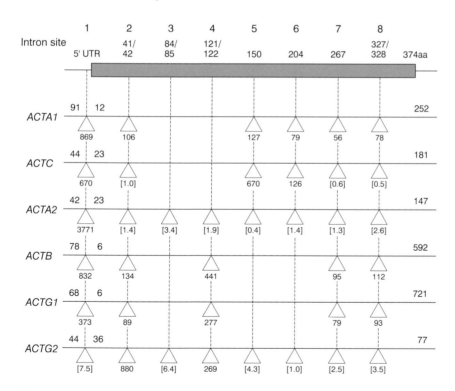

Figure 4.3. Comparison of gene structures of the six human actin genes (after Miwa *et al.*, 1991). Triangles indicate intron positions. The numbers below indicate intron sizes in base-pairs or kilobases (in square brackets). The numbers above the lines indicate the sizes of the 5′ UTR in exons 1 and 2 and those of the 3′ UTR in the last exon.

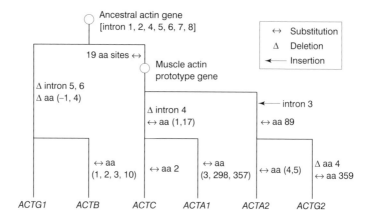

Figure 4.4. Hypothetical phylogenetic tree for the human actin gene family (after Miwa *et al.*, 1991). The vertical scale is non-linear and merely represents relative evolutionary time.

Albumin genes. The human albumin gene family comprises four genes, encoding albumin (*ALB*), α-fetoprotein (*AFP*), α-albumin/afamin (*AFM*) and group-specific component/vitamin D-binding globulin (*GC*) that are closely linked on chromosome 4q11-q13 (Nishio *et al.*, 1996). These genes are similar in terms of their structure and encode homologous proteins (Nishio and Dugaiczyk, 1996). A proposed phylogeny for this gene family is presented in *Figure 4.5*. The *GC* gene appears to be the oldest member of the family; it has lost two of its original 15 exons and with them four of the original 18 disulfide bridges characteristic of the putative ancestral protein. The *AFP* and *ALB* genes have been evolving at a particularly rapid rate (Minghetti *et al.*, 1985). However, the *ALB* gene still displays a high degree of evolutionary conservation in terms of its sequence which is perhaps a little surprising on account of its apparent nonessential nature evidenced by the absence of overt clinical signs in analbuminemic humans and rats (Ohno, 1981).

Albumin is synthesized in the adult liver whereas α-fetoprotein, being the fetal counterpart of albumin, is produced in the fetal liver and yolk sac. The mechanism of the developmental switch bringing about *AFP* gene repression and *ALB* gene activation is not yet understood (Nakata *et al.*, 1992).

The chromosomal region containing the albumin gene cluster has been involved in a number of pericentric inversions during the evolution of higher primates. As a result of one such inversion, the *ALB* and *AFP* genes were translocated to the short arm of chimpanzee chromosome 3 (analogous to human chromosome 4) (Magenis *et al.*, 1987). Similar inversions in the gorilla and orangutan however left the *ALB* and *AFP* genes on the long arm of chromosome 3 in these species (Magenis *et al.*, 1989).

Apolipoprotein genes. The lipoproteins are the major carriers of cholesterol, triglycerides and other lipids in human plasma (Breslow 1985). They are encoded by a multigene family which is dispersed but not fully dispersed in the human genome: *APOC1, APOC2, APOC4, APOE* (19q13.2) *APOC3, APOA1, APOA4* (11q23), and *APOA2* (1q21-q23). The genes evolved from a primordial *APOC1*-like gene, which is thought to have existed ~680 Myrs ago, via a series of internal and complete gene duplications (*Figure 4.6*; Luo *et al.*, 1986; Boguski *et al.*, 1986). Since both *apoA1* and

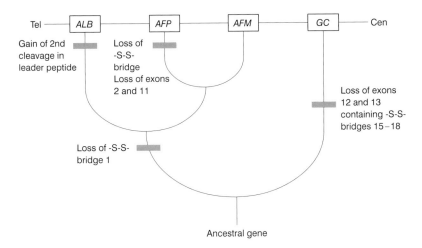

Figure 4.5. Proposed model for the evolution of the human albumin gene family (after Nishio *et al.*, 1996). Tel: telomere. Cen: centromere. -S-S- : disulfide bridge. The *ALB* and *GC* genes diverged before the emergence of amphibians 400 Myrs ago whilst the *AFP* gene emerged after the separation of amphibians and reptiles some 350 Myrs ago.

apoE genes have been found in fish, the duplication event from which they both arose must have occurred before the divergence of tetrapods and teleost fish (Babin *et al.*, 1997). The structures of the genes show remarkable similarities: all (with the exception of *APOA4*) possess four exons with introns interrupting the coding sequence at very similar locations (*Figure 4.7*). The major difference between the genes is in the length of the last exon which encodes a variable number of lipid-binding domains that contain multiple repeats of 22 amino acids each of which represents a tandem array of two 11-mers (Li *et al.*, 1988).

Two of the major known apolipoproteins do not fit easily into the above scheme: the 29 exon *APOB* (2p24) gene encodes a protein which may be distantly related to the other apolipoproteins (Li *et al.*, 1988) whilst the *APOD* (3q26.2-qter) gene encodes a protein with a high degree of homology to retinol-binding protein but little similarity to the other apolipoproteins (Drayna *et al.*, 1987).

Complement genes. The vertebrate complement system may be regarded as a primitive immune system which lacks the capability to recognize foreign antigens made possible by the evolution of the MHC complex, the T cell receptors and the immunoglobulins. Gene duplication has played a major role in the evolution of the complement components (Farries and Atkinson, 1991). This is still evident from the close linkage exhibited by the human complement genes: *C1QA*, *C1QB*, *C1QG* (1p34-p36), *C1S*, *C1R* (12p13), B factor (*BF*), *C2*, *C4A*, *C4B* within the HLA locus at chromosome 6p21, *C6*, *C7*, *C9* (5p13), *C8A*, *C8B* (1p32), *C5*, *C8G* (9q34), membrane cofactor protein (*MCP*), H factor (*HF1*) and decay accelerating factor (*DAF*) on chromosome 1q32. Some loci however appear to be isolated, for example I factor (*IF*; 4q25) and *C3* (19p13). Thus, some complement components exhibit extensive structural homology even though their genes are not linked (e.g. C3, C4, and C5).

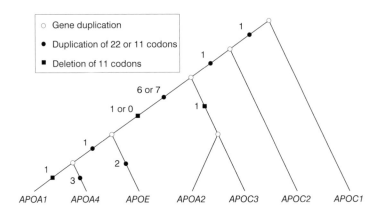

Figure 4.6. A hypothetical scheme for the evolution of the apolipoprotein genes (after Luo *et al.*, 1986). Numerals refer to the number of specified mutational events.

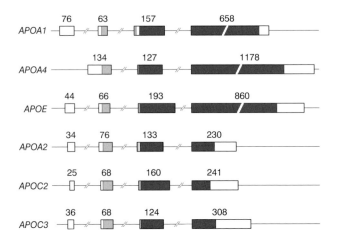

Figure 4.7. Structural organization of the human apolipoprotein genes (after Li *et al.*, 1988). Exons are denoted by bars, the 5′ and 3′ flanking regions and introns by lines. Open bars represent the 5′ and 3′ untranslated regions, hatched bars the signal peptide regions and solid bars the mature peptide regions. The numbers above the exons indicate their length in base-pairs.

The genes encoding the different components of the complement system belong to several evolutionarily unrelated gene families. Phylogenetic analysis of the *C3*, *C4* and *C5* genes, together with the more distantly related α2-macroglobulin (*A2M*; 12p12-p13) gene, supports the view that C5 diverged first with C3 and C4 subsequently diverging before the separation of the jawed and jawless fishes (Hughes, 1994). Similar analyses have been performed for the complement C1 components (Dodds and Petry, 1993).

Exon shuffling has been extremely important in diversifying the structure of the complement genes thereby conferring novel functions upon the complement proteins. Thus for example, C1r, C1s, and C2 possess serine protease domains, C6,

C7, C8α, C8β, and C9 possess thrombospondin and low density lipoprotein receptor domains, and C1r, C1s, C6, C7, C8a, C8b, and C9 possess homologies with epidermal growth factor (Farries and Atkinson, 1991).

Crystallin genes. The α-, β-, and γ-crystallins of the lens are ubiquitous in vertebrates. These proteins have ancient origins as evidenced by the relationship of the α-crystallins to small heat shock proteins (Caspers *et al.*, 1995; Lu *et al.*, 1995) and the relationship between the β- and γ-crystallins to proteins found in the bacterium *Myxococcus* and the primitive eukaryote *Physarum* respectively (Wistow, 1993). In humans, there appear to be two active α-crystallin genes (*CRYAA*, 21q22.3; *CRYAB*, 11q22.3-q23.1), six active β-crystallin genes (*CRYBB2*, *CRYBB3*, *CRYBA4*, *CRYBB1*, 22q11.2-q12.1; *CRYBA2*, 2q34-q36; *CRYBA1*, 17q11.1-q12) and five active γ-crystallin genes (*CRYGS*, 3; *CRYGA*, *CRYGB*, *CRYGC*, *CRYGD*, 2q33-q35).

Vertebrate lens crystallins have undergone dramatic changes during evolution including gene duplication, gene inactivation, gene recruitment and changes in gene expression (Lu *et al.*, 1996; Lubsen *et al.*, 1988; Wistow, 1993). The vertebrate lenses with the highest refractive index are found in fish and in the predominantly nocturnal rodents; in these lenses, γ-crystallin predominates (Graw *et al.*, 1993). Most vertebrates have however reduced the refractive index of their lenses in order to aid accommodation and focusing at a distance; this process has been accompanied by the loss of γ-crystallins (for example, replaced by δ-crystallins in birds). Humans have reduced γ-crystallin levels by partial or complete inactivation of several members of the γ-crystallin gene family once found in the common ancestor of primates and rodents (Brakenhoff *et al.*, 1990; den Dunnen *et al.*, 1989; Lubsen *et al.*, 1988; Meakin *et al.*, 1985, 1987). Thus, of the human γ-crystallin genes, only *CRYGC* and *CRYGD* (2q33-q35) are still fully active. Several γ-crystallin genes have been inactivated by deletion or point mutation whilst the *CRYGA* and *CRYGB* (2q33-q35) genes have been down-regulated by promoter mutation and increased mRNA turnover respectively (Brakenhoff *et al.*, 1990).

In addition to the ubiquitous crystallins mentioned above, there is a group of *taxon-specific* crystallins that are restricted to certain evolutionary lineages. These crystallins are pre-existing metabolic enzymes which have been recruited to a novel lens-specific role, not on account of their enzymatic activity but rather for structural reasons in that they are able to modify the refractive index of the lens. Thus, in most reptiles and birds, δ2-crystallin is the enzyme argininosuccinate lyase (Piatigorsky *et al.*, 1988) whilst δ1 crystallin has arisen by duplication of the δ2 crystallin gene but lacks its enzymatic activity (Mori *et al.*, 1990). Various other examples of *gene recruitment* (also termed *gene sharing*) have been documented among the crystallins, for example lactate dehydrogenase B (crystallin ε; crocodiles and some birds), NADPH: quinone oxidoreductase (crystallin ζ; guinea pigs and camelids) and aldehyde dehydrogenase 1 (crystallin η; elephant shrews). Recruitment of these gene products to a new lens function has been achieved by the acquisition of lens expression for these genes through the emergence of novel promoter elements (Wistow, 1993). In the case of ζ-crystallin, recruitment appears to have occurred independently in guinea pigs and camels; different regions of the same nonfunctional intronic sequence have been altered by single base-pair substitutions and micro-deletions and insertions to perform a role in lens-specific expression (Gonzalez *et al.*, 1995). By contrast, in birds, recruitment of the δ1-, δ2- and ε-crystallins appears to have come about through the modification of pre-existing promoters utilized for nonlens tissue expression (Hodin and Wistow, 1993).

Collagen genes. The collagens are multi-domain proteins which serve as structural molecules of the extracellular matrix. They have a very ancient origin as evidenced by their presence in sea urchins, annelids, *Drosophila* and even jellyfish and sponges (Exposito and Garrone, 1990; Exposito *et al.*, 1991; reviewed by Garrone, 1998). These proteins contain one or more domains with a triple helical conformation characterised by a repeating Gly-X-Y amino acid motif.

A total of 19 types of collagen (I to XIX) have so far been defined in humans; these comprise homo- or heteromeric assemblies of specific polypeptide chains (*Table 4.2*). A minimum of 34 genes on 13 different human chromosomes are required to encode these chains (*Table 4.2*). Based on their structures, the collagens may be divided into several distinct classes: fibrillar (I, II, III, V, XI), basement membrane (IV), fibril-associated with interrupted triple helices (IX, XII, XIV, XIX), filament-producing (VI), network forming (VIII, X) and anchoring fibril (VII). The class structure of the proteins and the homologies between the members of each class is reflected in the evolutionary relationships between the genes that encode them, their structure and often their chromosomal location (reviewed by Prockop and Kivirikko, 1995; van der Rest and Garrone, 1991; Vuorio and de Crombrugghe, 1990).

The fibrillar collagens are proteins with a continuous triple helical domain containing uninterrupted Gly-X-Y motifs. Their genes are thought to have diverged some 800–900 Myrs ago (Runnegar 1985) but still share a very similar exon-intron organization comprising 52–54 exons of which 42 have specific lengths: 45 bp, 54 bp, 99 bp, 108 bp, and 162 bp representing multiples of the 9 bp encoding the Gly-X-Y motif. The latter three lengths are combinations of the first two suggesting that they may have been derived from an ancestral 54 bp exon by duplication and recombination events (*Figure 4.8*). These events may have been mediated by recombination between introns (Butticé *et al.*, 1990; Chu *et al.*, 1984; Yamada *et al.*, 1980). Partial gene duplication events are unlikely to have disrupted gene function because multiples of 9 bp would not have altered the reading frame of the proteins. The 66 exon *COL5A1* and *COL11A2* genes appear to have increased in size through an increase in the number of 54 bp exons (Takahara *et al.*, 1995; Vuoristo *et al.*, 1995). The *COL3A1* and *COL5A2* genes are both located at chromosome 2q31, indicative of a close evolutionary relationship.

Although the genes encoding the basement membrane collagens possess 46–52 exons, they differ from the fibrillar collagen genes in that they possess many fewer exons of length 45 bp and 54 bp. Further, the exons vary widely in size, do not always begin with a Gly codon and often split codons. Their similar exon–intron organization testifies however to a common evolutionary origin. Moreover, the six type IV genes are arranged in syntenic pairs [*COL4A1* and *COL4A2* (13q34), *COL4A3* and *COL4A4* (2q36-q37), *COL4A5* and *COL4A6* (Xq22)] with head-to-head arrangement (sometimes sharing promoter elements), consistent with a model of gene duplication and divergence.

Table 4.2. Types of human collagen and the genes that encode them

Collagen type	Constituent chains	Gene symbol	Chromosomal localization
I	α1(I)	COL1A1	17q21.3-q22
	α2(I)	COL1A2	7q22.1
II	α1(II)	COL2A1	12q13
III	α1(III)	COL3A1	2q31
IV	α1(IV)	COL4A1	13q34
	α2(IV)	COL4A2	13q34
	α3(IV)	COL4A3	2q36-q37
	α4(IV)	COL4A4	2q36-q37
	α5(IV)	COL4A5	Xq22
	α6(IV)	COL4A6	Xq22
V	α1(V)	COL5A1	9q34.2-q34.3
	α2(V)	COL5A2	2q31
VI	α1(VI)	COL6A1	21q22.3
	α2(VI)	COL6A2	21q22.3
	α3(VI)	COL6A3	2q37
VII	α1(VII)	COL7A1	3p21.3
VIII	α1(VIII)	COL8A1	3q12-q13
	α2(VIII)	COL8A2	1p32.3-p34.2
IX	α1(IX)	COL9A1	6q13
	α2(IX)	COL9A2	1p32.2-p33
	α3(IX)	COL9A3	20q13.3
X	α1(X)	COL10A1	6q21-q22.3
XI	α1(XI)	COL11A1	1p21
	α2(XI)	COL11A2	6p21.3
XII	α1(XII)	COL12A1	6q12-q13
XIII	α1(XIII)	COL13A1	10q22
XIV	α1(XIV)	COL14A1	8q23
XV	α1(XV)	COL15A1	9q21-q22
XVI	α1(XVI)	COL16A1	1q34
XVII	α1(XVII)	COL17A1	10q24.3
XVIII	α1(XVIII)	COL18A1	21q22.3
XIX	α1(XIX)	COL19A1	6q12-q14

The fibril-associated collagens with interrupted triple helices are distinct from both the fibrillar and basement membrane collagens. The COL9A1, COL9A2, COL12A1, and COL19A1 genes are similar in structure indicative of a close evolutionary relationship (Kaleduzzaman et al., 1997): all contain some exons of length 54 bp. The COL9A1, COL12A1, and COL19A1 genes are also closely linked on chromosome 6q12-q14.

The genes encoding the filament-producing collagens, COL6A1 and COL6A2, possess a similar structure and are closely linked in head-to-tail orientation (Trikka et al., 1997). Although they contain exons of length 45 bp and 54 bp, most are 63 bp in length. The type VII collagen gene, COL7A1, contains 118 exons some of which are of 45 bp or 54 bp in length but rather more are of 36 bp, the remainder invariably being multiples of 9 bp (Christiano et al., 1994). By contrast to the extremely fragmented organization of the other collagen genes, the COL10A1 gene is remarkably compact, containing only three exons.

The evolution of the collagens is thus broadly consistent with a model of gene duplication and divergence coupled with intragenic exon duplications, deletions

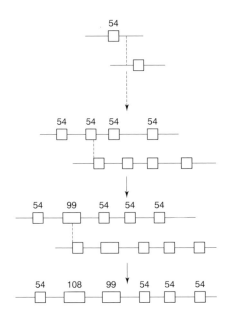

Figure 4.8. A hypothetical scheme for the evolution of the exons in a region of the *a2*(I) collagen (*COL1A2*; 7q22) gene coding for the helical region of the polypeptide (Li, 1997). The number of base-pairs in each exon is given. The dotted line denotes the position of an unequal crossover. An unequal crossover between two 54 bp exons could give rise to an exon of 99 bp and an exon of 9 bp whereas an unequal crossover between an exon of 99 bp and exon of 54 bp could give rise to an exon of 108 bp and an exon of 45 bp.

and fusions. The recurring sizes of many of the exons that encode the extant human collagens provides ample evidence for the exon amplification and contraction processes that must have fashioned them. Although the evolution of several families of related collagen genes has been studied in some detail, the evolution of the gene family in its entirety has not yet been pieced together. Since both the fibrillar and nonfibrillar collagens have a very ancient origin, this task is one of some very considerable complexity.

Genes for the fibroblast growth factors and their receptors. The fibroblast growth factors (FGF) constitute a family of polypeptide growth factors that have multiple functions in mitogenesis, angiogenesis and wound healing. They contain an extracellular portion with either 2 or 3 immunoglobulin-like domains, a transmembrane domain and a cytoplasmic tyrosine kinase domain. The FGF family of proteins interacts with the membrane-associated FGF receptors (FGFR) and these interactions are important for a variety of developmental processes such as the formation of the mesoderm during gastrulation, the integration of growth, budding and patterning during early post-implantation, and the development of various tissues including the skeletal system.

A total of 14 FGF genes have been identified in the human genome: *FGF1* (5q31), *FGF2* (4q25-q27), *FGF3* (11q13), *FGF4* (11q13), *FGF5* (4q21), *FGF6* (12p13), *FGF7* (15q15-q21), *FGF8* (10q24), *FGF9* (13q12), *FGF10* (5p12-p13), *FGF11* (17q21), *FGF12* (3q28), *FGF13* (Xq26) and *FGF14* (13q34). Thus, with the exception of the closely linked and tandemly repeated *FGF3* and *FGF4* genes, the FGF genes are dispersed in the human genome. FGF genes are present in invertebrate genomes and probably expanded in number after the divergence of protostomes and deuterostomes. The mammalian FGF gene family has emerged by a process of phased duplication and divergence (*Figure 4.9*; Coulier *et al.*, 1997; Johnson and Williams, 1993). Some of the paralogous mammalian FGF genes may be derived from genome duplications (Chapter 2, section 2.2) that occurred early on in the evolution of the vertebrate genome, for example *FGF4* and *FGF6* on chromosomes 11 and 12, respectively, and *FGF1* and *FGF2* on chromosomes 4 and 5, respectively.

The generation of FGF diversity may have played a role in the various innovations of the vertebrate skeletal system and the novel FGF family members are likely to have co-evolved with the FGFRs. FGFR genes are also found in invertebrate genomes (Coulier *et al.*, 1997) and four are found in the human genome: *FGFR1* (8p11), *FGFR2* (10q26), *FGFR3* (4p16) and *FGFR4* (5q35-qter). However, by contrast with the FGF family expansion, the FGFR gene family appears to have undergone only a single phase of expansion (Coulier *et al.*, 1997).

GABA receptor genes. The γ-aminobutyric acid (GABA) receptor is a pentameric ion channel complex which mediates fast inhibitory synaptic transmission in the central nervous system. The receptor exists as many different isoforms assembled from different combinations of subunit subtypes, α_{1-6}, β_{1-4}, γ_{1-4}, δ, ε, and ρ_{1-2}. The human GABA receptor genes are clustered at different chromosomal locations: *GABRA2, GABRA4, GABRB1, GABRG1* (4p12-p13), *GABRA1, GABRA6, GABRB2, GABRG2* (5q34-q35), *GABRA5, GABRB3, GABRG3* (15q11-q13), *GABRA3, GABRB4, GABRE* (Xq28), *GABRR1, GABRR2* (6q14-q21), and *GABRD* (1p) (McLean *et al.*, 1995; Russek and Farb, 1994). The organization of this gene family is consistent with an evolutionary model of intracluster gene duplication

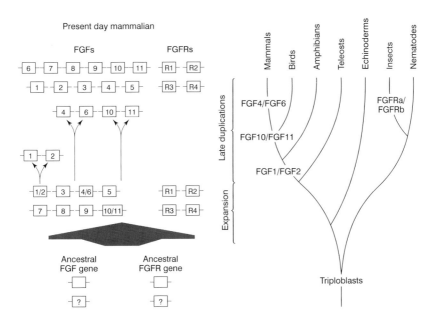

Figure 4.9. A hypothetical scheme for the evolution of the fibroblast growth factors and their receptors. The putative phases of gene duplications (on the left) are tentatively related to a phylogenetic tree of the metazoa (on the right). After Coulier *et al.* (1997). Since this phylogenetic study was performed, several further *FGF* genes have been characterized in the human genome (see text).

as well as the duplication of a primordial gene cluster. The tandem duplication of the ancestral α subunit gene appears to have occurred before the duplication and translocation of at least one of the original gene clusters.

G protein α subunit genes. G proteins are ubiquitous in eukaryotes and serve to mediate signal transduction via appropriate receptors (see Section 4.2.3, *G-protein-coupled receptor genes*) by coupling extracellular signals to intracellular effectors such as adenylyl cyclase, phospholipases and ion channels. More than a dozen G protein α subunit genes have been identified in the human genome (*Figure 4.10*) and their evolutionary history appears to stretch back over 1500 Myrs (Wilkie *et al.*, 1992). Whilst some members of the human gene family are chromosomally linked [e.g. *GNAI2* and *GNAT1* (3p21), *GNAI3* and *GNAT2* (1p13), *GNA11* and *GNA15* (19p13)], the remainder are scattered around the genome. This organization is potential explicable by a combination of successive genome duplications, tandem duplication, and duplication/translocation events (Wilkie *et al.*, 1992).

Globin genes. Proteins homologous to the globins are ubiquitous in eukaryotes and even have counterparts among the prokaryotes (Hardison, 1998; Riggs, 1991). The myoglobin (*MB*; 22q11.2-q13) gene and the ancestor of the extant globin genes are thought to have arisen from a common ancestral gene encoding a heme-containing protein ~700 Myrs ago before the advent of the vertebrates (Czelusniak *et al.*, 1982; Suzuki and Imai, 1998). The α- and β-globin genes diverged from each other about 500 Myrs ago and at some stage became chromosomally separated (Efstratiadis *et al.*, 1980). The α-globin cluster subsequently evolved by a process of successive duplication and divergence (*Figure 4.11*): the ζ/α gene divergence occurring about 400 Myrs ago (Czelusniak *et al.*, 1982; Proudfoot *et al.*, 1982) and the θ/α gene divergence about 260 Myrs ago (Hsu *et al.*, 1988). The α1- and α2-globin genes arose from a further duplication event between 50 Myrs and 60 Myrs ago whilst the ψα1 pseudogene was inactivated ~45 Myrs ago (Proudfoot and Maniatis, 1980; *Figure 4.11*). The 40 kb human

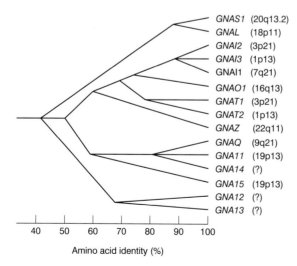

Figure 4.10. Human G protein α subunit multigene family.

α-globin gene cluster (16p13.3) therefore comprises the embryonically expressed ζ (*HBZ*) gene, the post-natally expressed α2 (*HBA2*) and α1 (*HBA1*) genes, the θ-globin (*HBQ1*) gene (probably a transcribed pseudogene) and three conventional pseudogenes (*HBZP*, *HBAP1*, *HBAP2*) (*Figure 4.12*). The genes in the α-globin gene cluster are therefore arranged in the order of their activation during ontogeny. The human *HBA1* and *HBA2* genes have remained virtually identical to each other as a consequence of crossing over and gene conversion (Bailey *et al.*, 1992; Hess *et al.*, 1984; Michelson and Orkin 1983).

The ancestral β-globin gene is thought to have duplicated about 200 Myrs ago to yield a δ/β ancestral gene and a ε/γ ancestral gene (Hardies *et al.*, 1984; *Figure 4.11*). The δ- and β-globin genes diverged from the δ/β ancestral gene about 40 Myrs ago.

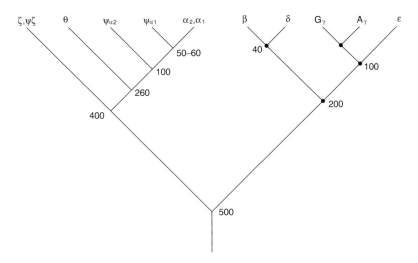

Figure 4.11. Evolution of the vertebrate globin genes (redrawn from Efstratiadis *et al.*, 1980 and Higgs *et al.*, 1989). Numbers denote approximate estimated times of divergence in Myrs before present.

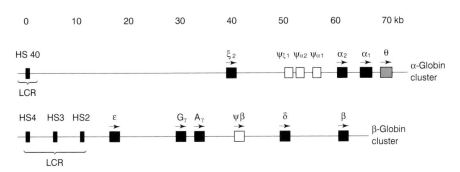

Figure 4.12. Organization of the human α- and β-globin clusters. Arrows denote the direction of transcription. Solid boxes denote genes, empty boxes pseudogenes and the hatched box the θ-globin (*HBQ1*) gene which is transcribed but probably not expressed. The relative positions of the locus control regions (LCRs) are denoted together with the constituent DNase I hypersensitive sites (HS).

The ε- and γ-globin genes diverged from the ε/γ ancestral gene about 100 Myrs ago (*Figure 4.11*) followed by the duplication of the γ-globin gene to form the Gγ and Aγ genes, an event which occurred before the divergence of the Old World and New World monkeys (Chiu *et al.*, 1997; Slightom *et al.*, 1987). The human β-globin gene cluster (11p15.5) therefore comprises the embryonically expressed ε (*HBE1*) gene, the fetal Gγ (*HBG2*) and Aγ (*HBG1*) genes, the post-natally expressed β (*HBB*) and δ (*HBD*) genes plus a single ψη pseudogene (*HBBP1*) (*Figure 4.12*). Thus, as with the α-globin genes, the genes in the 75 kb β-globin gene cluster are arranged in the order of their activation during ontogeny.

The *HBD* gene is present in humans, apes and New World monkeys but has been inactivated in Old World monkeys (Kimura and Takagi, 1983; Martin *et al.*, 1980). During evolution, gene conversion has operated on the *HBD* gene allowing it to acquire sequence characteristics of the *HBB* gene (Koop *et al.*, 1989; Tagle *et al.*, 1991). Gene conversion events have however been even more frequent between the *HBG2* and *HBG1* genes (Chiu *et al.*, 1997; Slightom *et al.*, 1987; 1988). In addition to gene conversion, a number of other events have occurred during evolution which have altered the β-globin cluster in specific ways in different mammalian orders; these include insertion of repeat sequences, change in expression profile, gene duplication, gene fusion and gene loss or inactivation (*Figure 4.13*). Thus, the lagomorphs and rodents lack the η-globin locus whilst the artiodactyls lack the γ-globin locus. As a consequence of some of these changes, the β-globin cluster in mammals varies from as little as 20 kb in lemurs to about 90 kb in goats (*Figure 4.13*).

The ε and γ globin genes were originally embryonically expressed and fetally inactive and this early expression pattern is still found in the galago and lemurs (Tagle *et al.*, 1988). In higher primates, the γ-globin gene was duplicated before the divergence of Old World and New World monkeys (perhaps by an unequal homologous crossing over event mediated by LINE elements; Fitch *et al.*, 1991) and became fetally expressed as a direct result of the accumulation of sequence changes in the 5' flanking region (Johnson *et al.*, 1996; TomHon *et al.*, 1997). In Old World monkeys, apes and humans, both the γ1- and γ2-globin genes are functional but the expression of the γ1 gene is three-fold higher than that of the γ2 gene. In New World monkeys, only one γ-globin gene is functional, usually γ2 (Chiu *et al.*, 1996; 1997; Johnson *et al.*, 1996; Meireles *et al.*, 1995).

Growth hormone and somatomammotropin genes. The growth hormone and prolactin genes are thought to have emerged as a result of a duplication event some 470 Myrs ago. In nonprimates, with the exception of caprine ruminants (Wallis *et al.*, 1998), GH is encoded by a single gene whilst in primates, the gene has expanded to a gene cluster. Thus, the human gene encoding pituitary-expressed growth hormone (*GH1*) is located on chromosome 17q23 within a cluster of five related genes (Chen *et al.*, 1989). The other loci present in the growth hormone gene cluster are two chorionic somatomammotropin genes (*CSH1* and *CSH2*), a chorionic somatomammotropin pseudogene (*CSHP1*) and a second growth hormone gene (*GH2*). These genes are separated by intergenic regions of 6 kb to 13 kb in length, lie in the same transcriptional orientation, are placentally expressed and are under the control of a downstream tissue-specific

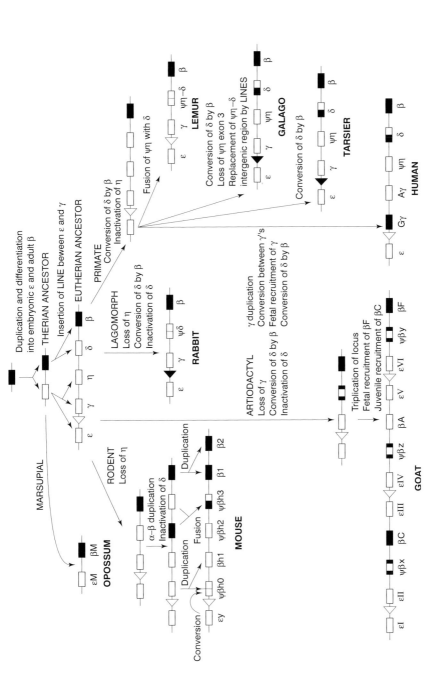

Figure 4.13. Evolution of the β-globin gene cluster in mammals (redrawn from Tagle *et al.*, 1992). Genes represented by boxes composed of two patterns denote the occurrence of specific recombinational events (gene conversions or crossovers). The insertion of a LINE element between the ε and γ globin genes is shown as an open triangle. The solid triangles denote orthology verified by sequence comparisons.

enhancer (Jacquemin *et al.*, 1994). The *GH2* locus encodes a protein that differs from the *GH1*-derived growth hormone at 13 amino acid residues. All five genes share a very similar structure with five exons interrupted at identical positions by short introns (Hirt *et al.*, 1987).

On the basis of sequence data, it was initially thought that the evolution of the GH gene cluster had taken place comparatively recently over the last 15 Myrs by a process of gene duplication and divergence (Chen *et al.*, 1989; Miller and Eberhardt, 1983). The first event is thought to have been the duplication of a single ancestral GH gene to generate pre-GH and pre-CSH genes (Barsh *et al.*, 1983) followed by the duplication of the newly created gene pair to yield *GH1*, *CSH1*, *GH2*, and *CSH2* (*Figure 4.14*). Finally, the *CSH1* gene was duplicated to form two genes, one of which (*CSHP1*) became inactivated through mutation. Although this sequence of events is probably correct, the estimated timings now appear to be seriously inaccurate because a *GH2* gene has subsequently been reported in macaque (Golos *et al.*, 1993). The duplication event creating the *GH1* and *GH2* genes must therefore have occurred before the divergence of Old World monkeys and the anthropoid apes ~25 Myrs ago. No study has been performed on either prosimians or New World monkeys and so the timing of this duplication event may have to be revised still further as new data emerge. The initially misleading conclusion as to the timing of the duplication events was probably due to a failure to consider the effect of gene conversion in minimizing sequence differences between the different GH loci (Giordano *et al.*, 1997).

The 70 kb human GH gene cluster contains some 48 *Alu* sequences (Chen *et al.*, 1989) some of which may have mediated the unequal recombination events responsible for the gene duplications (Barsh *et al.*, 1983). On the other hand, some *Alu* sequences have become duplicated along with their associated genes during the duplication process. One consequence of the relatively recent evolutionary changes in the GH gene cluster is that multiple sequence homologies and internal repetitions are still evident within it.

Glycophorin genes. The human glycophorins are encoded by a multigene family that has evolved from an ancestral gene that ceased to be functional at some stage during primate evolution. Glycophorins A and B are the major sialoglycoproteins of the human erythrocyte membrane and carry the MN and Ss blood group antigens respectively. They are encoded by the *GYPA* and *GYPB* genes which occur in a 330 kb cluster together with the glycophorin E (*GYPE*) gene on chromosome 4q28-q31. This cluster is thought to have arisen by two successive duplications, the first creating the *GYPA* gene by duplication of an ancestral gene (between 9 Myrs and 35 Myrs ago) and the second (5–21 Myrs ago) generating the *GYPB* and *GYPE* genes (Kudo and Fukuda, 1989; Onda and Fukuda, 1995; *Figure 4.15*). The *GYPB* gene differs from the *GYPA* gene by virtue of the presence of, (i) a G→T transversion at the +1 position of the intron 3 donor splice site which serves to inactivate the expression of exon III and (ii) a 9 bp insertion at the 5′ end of exon V (Blumenfeld *et al.*, 1997). The *GYPE* gene differs from the *GYPB* gene in that exon IV has been inactivated by a splice site mutation, exon V contains a 24 bp insertion and the encoded protein has been shortened by 5 amino acids through the introduction of a premature Stop codon (Blumenfeld *et al.*,

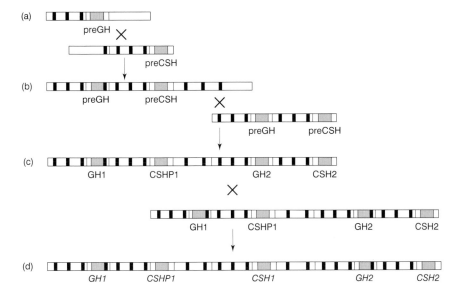

Figure 4.14. Evolution of the human growth hormone gene cluster by serial unequal crossing over events (after Phillips, 1995). (a) Duplication to yield pre-growth hormone (*GH*) and pre-chorionic somatomammotropin (*CSH*) genes. (b) Unequal crossing over between pre-*GH* and pre-*CSH* genes to yield (c) the *GH1*, *GH2*, and *CSH2* genes plus a *CSHP1* locus which was to become inactivated. (d) Subsequently, the *CSH1* gene emerged by duplication of the *CSHP1* gene.

1997). Both the *GYPB* and *GYPE* genes lack exons 6 and 7 present in the *GYPA* gene that encode the cytoplasmic domain.

The ancestral gene from which the *GYPA* gene was derived is thought to have been inactivated but its remnants may still be present on chromosome 4, just downstream of the *GPA* gene (Onda *et al.*, 1993, 1994). The ancestor of the human *GYPB* and *GYPE* genes underwent a rearrangement in its 3′ region by recombination between *Alu* sequences and in so doing, acquired 3′ sequence from another gene. This unequal crossing over event served to generate short last exons in both *GYPB* and *GYPE* genes; that of *GYPB* comprises a single amino acid whilst that of *GYPE* encodes an untranslated region.

At least one *GYPA*-like gene occurs in all hominoid primates whereas one *GYPB*-like gene is present in man, both species of chimpanzee and the gorilla but not orangutan and gibbon (Rearden *et al.*, 1993). The *GYPE* gene is present in all hominoid primates that possess a *GYPB* gene, but is polymorphically present/absent in gorillas (Rearden *et al.*, 1993). Thus duplication of the *GYPB*/*GYPE* progenitor sequence to yield the *GYPB* and *GYPE* genes must have occurred prior to the divergence of the human-gorilla-chimpanzee clade ~10 Myrs ago. Subsequently, the *GYPE* gene has acquired DNA sequence from the *GYPA* gene by gene conversion (Kudo and Fukuda, 1994; Rearden *et al.*, 1993).

Homeobox genes. Homeobox (*HOX*) genes represent a major class of transcription factors whose origin preceded the radiation of the triploblastic metazoa (Finnerty

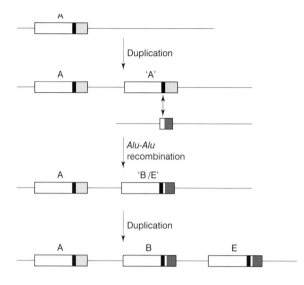

Figure 4.15. Evolution of the human glycophorin genes, *GYPA*, *GYPB* and *GYPE* (after Ondo *et al.*, 1993).

and Martindale, 1998) and which are directly involved in the control of embryogenesis in animals as diverse as nematodes, insects and vertebrates. The expression of these genes is spatially and temporally regulated during embryonic development and plays an important role in establishing the system of positioning along the anterior-posterior axis of the embryo. HOX proteins share a common homeobox domain and influence gene transcription by binding to a 4 bp core sequence in the promoters of their target genes.

The *HOX* genes appear to have been duplicated on several occasions during evolution, perhaps even before the origin of angiosperms, fungi and the metazoa (Bharathan *et al.*, 1997). Early in vertebrate evolution, an ancestral *HOX* gene cluster present in invertebrates and the cephalochordate *Amphioxus* was duplicated twice probably by whole genome duplication (see Chapter 2, section 2.1; Bailey *et al.*, 1997; Garcia-Fernàndez and Holland, 1994; Holland and Garcia-Fernàndez, 1996; Schughart *et al.*, 1989) to give rise to the four linkage groups (HOX A, HOX B, HOX C, HOX D). From studies of *HOX* gene number, this duplication event is likely to have occurred after the divergence of ray-finned and lobe-finned fishes but before the radiation of the teleosts (Amores *et al.*, 1998). A scheme for the evolution of mammalian *HOX* genes is presented in *Figure 4.16*. It can be seen that cluster duplication must have been followed by the loss of some specific *HOX* genes and indeed this is evident from comparison of *HOX* gene number between clusters (*Figure 4.17*). The physical order of genes within the clusters has however been conserved during vertebrate evolution (Schughart *et al.*, 1989) perhaps as a result of enhancer sharing between different *HOX* genes (Mann, 1997).

In human, most of the *HOX* genes are located in four clusters containing between them 39 *HOX* genes (Acampora *et al.*, 1989). The HOX A cluster is located at 7p14-p15 and contains 11 genes: *HOXA1, HOXA2, HOXA3, HOXA4, HOXA5, HOXA6, HOXA7, HOXA9, HOXA10, HOXA11* and *HOXA13*. The HOX B cluster is located at 17q21-q22 and contains 10 genes: *HOXB1, HOXB2, HOXB3, HOXB4, HOXB5, HOXB6, HOXB7, HOXB8, HOXB9* and *HOXB13*. The HOX C cluster is located at 12q13 and contains 9 genes *HOXC4, HOXC5,*

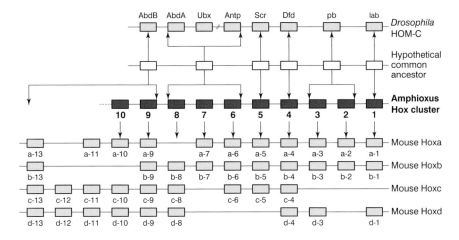

Figure 4.16. Model for the evolution of the homeobox gene clusters of eukaryotes (redrawn from Holland and Garcia-Fernandez, 1996). In *Drosophila*, the eight *Hox* genes are found in a single complex which is split between two clusters. AbdB: Abdominal-B. AbdA: Abdominal-A. Ubx: Ultrabithorax. Antp: Antennapedia. Scr: Sex combs reduced. Dfd: Deformed. Pb: Proboscipedia. Lab: Labial. In mammals, four *Hox* gene clusters are found on separate chromosomes.

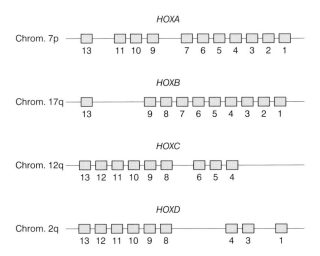

Figure 4.17. Human *HOX* gene clusters. Physical distances are not to scale.

HOXC6, HOXC8, HOXC9, HOXC10, HOXC11, HOXC12, and *HOXC13.* The HOX D cluster is located at 2q31-q32 and contains 9 genes: *HOXD1, HOXD3, HOXD4, HOXD8, HOXD9, HOXD10, HOXD11, HOXD12,* and *HOXD13.* In addition to the *HOX* gene clusters, other dispersed paralogous *HOX* genes are found in the human genome (e.g. *MSX1,* 4p16; *HMX2,* 10q25-q26) and may also be of functional importance (Brooke *et al.,* 1998).

The development of the vertebrate body plan both by elaboration of primitive chordate characters and the development of novel morphological characters (e.g. cranial ganglia, teeth and bone) must have required substantial reprogramming of gene networks (Shubin *et al.*, 1997). There are several different ways in which changes in the *HOX* genes have probably influenced morphological evolution: (i) expansion in the structural diversity of *HOX* genes within a given complex or class (Sharkey *et al.*, 1997), (ii) expansion of the number of HOX complexes (Mayer *et al.*, 1998), (iii) the loss of specific *HOX* genes (Aparicio *et al.*, 1997), (iv) changes in the location, timing or level of *HOX* gene expression (Burke, 1995; Gellon and McGinnis, 1998), and (v) changes in the interactions between HOX proteins and their target genes, possibly as a result of mutational changes in *cis*-acting regulatory elements (Carroll, 1995).

Integrin genes. The integrins are a family of heterodimeric membrane glycoproteins that participate in cell adhesion and are involved in a wide range of cell-cell and cell-matrix interactions. The diversity and specificity of integrin function is paralleled by the structural diversity potentiated by the existence of at least 16 different alpha chains and 8 different beta chains. Although, in principle, the alpha and beta chains could associate in a multitude of ways, in practice diversity is limited to certain combinations. These chains are encoded by a family of genes which are widely dispersed in the human genome (*Table 4.3*). The alpha chain integrins can themselves be divided into two subgroups by virtue of the insertion of an I domain of about 180 amino acids in the extracellular region. The I-integrin alpha chain genes (*ITGA1*, *ITGA2*, *ITGAD*, *ITGAM*, *ITGAL*, and *ITGAE*) are thought to have arisen as a result of an early insertion into a non-I gene followed by gene duplication and divergence. The clustering of these genes on chromosomes 5 and 16 (*Table 4.3*) is thought to have resulted from relatively recent gene duplications (Wang *et al.*, 1995). The non-I alpha chain genes (*ITGA2*, *ITGA3*, *ITGA4*, *ITGA5*, *ITGA6*, *ITGA7*, *ITGA8*, *ITGA9*) are largely confined to clusters on chromosomes 2, 12, and 17 (Wang *et al.*, 1995; *Table 4.3*), locations which coincide closely with the homeobox (*HOX*) gene clusters (see Chapter 2, section 2.1 and chapter 4 section 4.2.1, *G protein α subunit genes*). This is suggestive of the occurrence of genomic or chromosomal duplications involving both types of gene cluster. Some of the beta chain (*ITGB*) genes are also located on chromosomes 2, 12, and 17 (*Table 4.3*) indicating common ancestry with the non-I alpha chain genes.

Keratin genes. The keratins, cytoskeletal proteins of the epithelium, belong to two families: type I (acidic) and type II (basic) (Fuchs *et al.*, 1982). Since keratins are obligate heteropolymers, keratin intermediate filaments are composed of one type I and one type II polypeptide. This has led to consistent co-expression of type I-type II keratin pairs in different types of epithelial cells. The keratins are evolutionarily related to the family of intermediate filament proteins which includes vimentin and desmin (Hanukoglu and Fuchs, 1982). Type II keratins are no more homologous to the type I keratins than they are to other intermediate proteins (Klinge *et al.*, 1987) indicating that the type I and type II keratins diverged from the common ancestor of all intermediate filaments at about the same time. This common ancestral gene probably had its origins among the lower eukaryotes (Fuchs and Marchuk, 1983; Krieg *et al.*, 1985). By contrast to

Table 4.3. Types of human integrin chain and the genes that encode them

Integrin chain	Gene symbol	Chromosomal location
α1	*ITGA1*	5
α2	*ITGA2*	5q23-q31
α3	*ITGA3*	17
α4	*ITGA4*	2q31-q32
α5	*ITGA5*	12q11-q13
α6	*ITGA6*	2
α7	*ITGA7*	12q13
α8	*ITGA8*	?
α9	*ITGA9*	3p21.3
αllb	*ITGA2B*	17q21.3
αE	*ITGAE*	?
αL	*ITGAL*	16p11.2
αM	*ITGAM*	16p11.2
αV	*ITGAV*	2q31-q32
αX	*ITGAX*	16p11-p13
αD	*ITGAD*	16p11.2
β1	*ITGB1*	10p11.2
β2	*ITGB2*	21q22.3
β3	*ITGB3*	17q21.3
β4	*ITGB4*	17q11-qter
β5	*ITGB5*	?
β6	*ITGB6*	2
β7	*ITGB7*	12q13.1
β8	*ITGB8*	7p

the conserved central alpha-helical domains, the variable terminal glycine-rich domains of keratins 1 and 10 (comprising short 4–10 amino acid segments repeated 3–15 times) appear to have evolved by a series of tandem duplications brought about by unequal crossing over (Klinge *et al.*, 1987).

In the human genome, the genes encoding these keratin families have been found to be tightly clustered; at 17q12-q21 for the acidic keratins (e.g. *KRT9*, *KRT10, KRT14, KRT15, KRT16, KRT17,* and *KRT19*), and chromosome 12 for the basic keratins (e.g. *KRT1, KRT2A, KRT5, KRT6A, KRT6B,* and *KRT8*) (Milisavljevic *et al.*, 1996; *Table 4.4*). There is one exception to this rule: *KRT18* is a type I keratin gene but is located on chromosome 12q11-q13. There is some evidence for gene conversion between type I and type II keratin genes which could have facilitated their coevolution (Klinge *et al.*, 1987). Concerted gene duplications also provide evidence for coevolution of type I and type II keratin genes (Blumenberg, 1988). Thus, in both families, the genes expressed in the embryo duplicated and diverged first, followed by the genes expressed in various differentiated cells. Further gene duplications gave rise to the hair keratin genes (Blumenberg, 1988; Powell *et al.*, 1992; Rogers *et al.*, 1998; *Table 4.4*). This parallelism of gene duplication cannot be explained by gene conversion but as yet the underlying mechanism which apparently allows duplications in one family to influence duplications in the other is still unclear. Coevolution may have been driven by the obligate heteropolymer status of the proteins. The tight regulation of the coordinate expression of the type I and type II keratin genes implies that unbalanced production is likely to be deleterious. It may therefore follow that the duplication and functional divergence of a type I keratin gene might lead to a specific type II gene duplication being selectively favored.

Genes of the major histocompatibility complex. The major histocompatibility complex (MHC) region on chromosome 6p21.3 spans 4 Mb DNA and contains the genes encoding the human leukocyte antigens (HLA), a large family of cell surface glycoproteins (*Figure 4.18*). The function of the MHC proteins, members of the immunoglobulin superfamily (Section 4.2.4, *Immunoglobulin genes*), is to present peptides to T cells. The MHC family is divided into classes I and II whose members differ in terms of their structure, function, expression and polymorphism (Hughes, 1996). Class I molecules (the classical transplantation antigens) are expressed on most cells whereas the expression of class II molecules is confined to antigen-presenting cells. Both class I and II molecules are heterodimers with four extracellular domains; α1, α2 and α3 associated with β2-microglobulin (class I) and α1, α2, β1 and β2 (class II). These molecules bind both self- and pathogen-derived peptides in a surface groove (*peptide-binding region*) and present these to cytotoxic T cells that lyse the pathogen-infected cell, thereby limiting host infection.

Table 4.4. Types of human keratin and the genes that encode them

Keratin type	Gene	Chromosomal location
Keratin 1	*KRT1*	12q11-q13
Keratin 2A	*KRT2A*	12q11-q13
Keratin 3	*KRT3*	12q12-q13
Keratin 4	*KRT4*	12p12-q11
Keratin 5	*KRT5*	12q
Keratin 6A	*KRT6A*	12q12-q21
Keratin 6B	*KRT6B*	12
Keratin 7	*KRT7*	12q12-q21
Keratin 8	*KRT8*	12
Keratin 9	*KRT9*	17q21
Keratin 10	*KRT10*	17q21-q23
Keratin 12	*KRT12*	17q11-q12
Keratin 13	*KRT13*	17q21-q23
Keratin 14	*KRT14*	17q12-q21
Keratin 15	*KRT15*	17q21-q23
Keratin 16	*KRT16*	17q12-q21
Keratin 17	*KRT17*	17q12-q21
Keratin 18	*KRT18*	12q11-q13
Keratin 19	*KRT19*	17q21-q23
Keratin, hair, acidic 1	*KRTHA1*	17q12-q21
Keratin, hair, acidic 2	*KRTHA2*	17q12-q21
Keratin, hair, acidic 3A	*KRTHA3A*	17q12-q21
Keratin, hair, acidic 3B	*KRTHA3B*	17q12-q21
Keratin, hair, acidic 4	*KRTHA4*	?
Keratin, hair, acidic 5	*KRTHA5*	17q12-q21
Keratin, hair, basic 1	*KRTHB1*	12q13
Keratin, hair, basic 2	*KRTHB2*	?
Keratin, hair, basic 3	*KRTHB3*	12q12-q13
Keratin, hair, basic 4	*KRTHB4*	?
Keratin, hair, basic 5	*KRTHB5*	12q12-q13
Keratin, hair, basic 6	*KRTHB6*	12q13

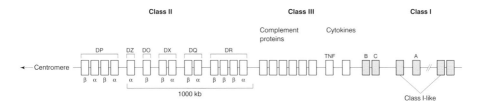

Figure 4.18. Map of the human histocompatibility complex loci on 6p21.3.

In humans, the class I loci are *HLA-A, HLA-B*, and *HLA-C* (classical or type Ia) and *HLA-E, HLA-F*, and *HLA-G* (non-classical or type Ib) (*Figure 4.18*). The class I genes are interspersed with five MHC class I-related (MIC) loci (*MICA, MICB, MICC, MICD,* and *MICE*), five full-length pseudogenes (*HLA-H, HLA-J, HLA-K, HLA-L,* and *HLA-X*) plus several truncated pseudogenes and remnants (*Figure 4.19*). The 5 MIC genes are distantly related to the hemochromatosis (*HFE*; 6p21.3) gene. The class II loci are clustered into three regions, HLA-DR (*HLA-DRA, HLA-DRB1, HLA-DRB2, HLA-DRB3, HLA-DRB4,* and *HLA-DRB5;* NB. *HLA-DRB3, HLA-DRB4,* and *HLA-DRB5* may be variably present or absent depending upon the haplotype), HLA-DQ (*HLA-DQA1, HLA-DQA2, HLA-DQB1, HLA-DQB2, HLA-DQB3*), and HLA-DP (*HLA-DPA1, HLA-DPB1,* and *HLA-DNA*) (*Figure 4.18*). The class I and II regions are separated by a gene-dense 1100 kb region containing a number of so-called type III genes including *BF, C2, C4A, C4B,* and *TNF*.

The ancestral MHC molecule appears to have been assembled by combining its three constituent domains (peptide-binding domain, immunoglobulin-like domain and membrane-anchoring domain) by exon shuffling (*Figure 4.20*). Hughes and Nei (1993) estimated that the class II A and B genes diverged more than 500 Myrs ago. Class I genes then emerged by duplication and divergence during the primate radiation of the last 60 Myrs (Hughes and Yeager, 1997; Kulski *et al.*, 1997). The putative ancestral duplication(s) early in vertebrate evolution (Chapter 2, section 2.1) may well have provided an impetus to the emergence of the HLA complex in that the consequent redundancy could have created opportunities for the emergence of a variety of accessory and effector molecules (Kasahara *et al.*, 1997). A scheme for the subsequent evolution of these genes is presented in *Figure 4.21*.

Figure 4.19. Map of the class I region of the human histocompatibility complex (redrawn from Wells and Parham, 1996).

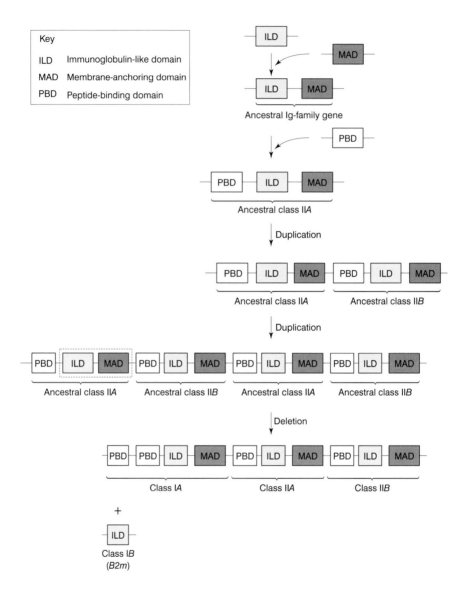

Figure 4.20. Postulated origin of MHC class I and class II genes (redrawn from Klein and O'hUigen, 1993). An exon encoding a soluble immunoglobulin-like domain (ILD) is joined with an exon encoding a membrane-anchoring domain (MAD) to produce an ancestral immunoglobulin-family gene. This is joined by an exon encoding a peptide-binding domain (PBD) to produce an ancestral class II-like gene, assumed to be of the A variety (coding for the α-chain). Duplication and deletion events, along with sequence divergence would produce the present day array of class I A and class II A and B genes. The class I B gene is in fact the β2-microglobin (*B2M*; 15q21-q22) gene, which is not linked to the main MHC complex.

So far, the MHC appears to be confined to vertebrates (Klein *et al.*, 1993; Klein and Sato 1998). In a phylogenetic analysis of class I genes from different mammalian orders, Hughes and Nei (1989b) found that the class Ib genes clustered with the class Ia genes. This is explicable either by postulating an independent origin for the class Ib genes in different orders of mammals (Hughes, 1991) or by invoking the homogenizing influence of gene conversion between class I loci within each mammalian order (Rada *et al.*, 1990). Class I genes evolved by a process of repeated gene duplication followed by a reduction in the level of expression of some genes, and the transcriptional silencing or deletion of others (Watkins, 1995). This explains why orthologous relationships are found among class I loci of mammals of the same order but not among mammals of different orders. By contrast to the class I loci, orthologous relationships between mammalian class II loci are the rule rather than the exception, a consequence of the early origin of the class II MHC loci prior to the mammalian radiation (Hughes and Nei, 1990).

The MHC class I genes are conserved in the great apes and Old World monkeys. Thus orthologues of *HLA-A, HLA-B, HLA-E, HLA-F,* and *HLA-G* are present in macaques although the *HLA-C* locus has been found only in chimpanzees and gorillas (Watkins 1995; Lienert and Parham, 1996). The MHC class I genes of New World monkeys are similar to the *HLA-G* and *HLA-F* genes (Watkins, 1995). The subfamily Callitrichinae (tamarins and marmosets) manifests an unusually high rate of turnover of class I MHC loci with different sets of nonorthologous MHC

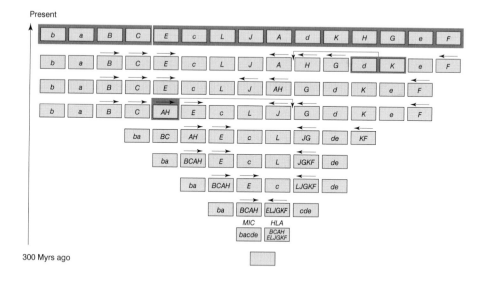

Figure 4.21. Evolution of the class I MHC genes by serial duplication (after Klein *et al.*, 1998). Capital letters denote HLA loci, small letters MIC loci. Each rectangle represents one locus. Multiple letters within a rectangle denote ancestors of genes specified by individual letters. Short arrows indicate transcriptional orientation of loci whilst long arrows denote transpositions of loci.

class I genes being expressed (Cadavid *et al.*, 1997). It would thus appear that class I genes have been differentially duplicated, deleted and amplified in different primate lineages (Klein *et al.*, 1998).

There is an extremely high level of polymorphism at certain MHC loci, some 20 times higher than the average level of nucleotide polymorphism found in the human genome. New MHC sequence variants have arisen by single base-pair substitution, reciprocal recombination or gene conversion (Ohta 1997; Watkins *et al.*, 1991; Yeager and Hughes, 1999). Gene conversion at the MHC locus can occur at relatively high frequency (10^{-4} per locus per generation; Zangenberg *et al.*, 1995) but, on its own, it is insufficient to account for the level of polymorphism observed. The polymorphisms have gradually accumulated during human evolution and have probably been maintained at high frequency by a variety of means, one of the most important being *overdominant selection* in which the heterozygote responds better to challenge than either homozygote (Li, 1997; Wells and Parham, 1996).

Certain MHC alleles are very ancient. Indeed, it would appear that some of the class I MHC allelic lineages are shared by human, chimpanzee and gorilla, implying that these polymorphisms were present in the common ancestor of the three species (Fan *et al.*, 1989; Gyllensten *et al.*, 1991; Kupfermann *et al.*, 1992; Lawlor *et al.*, 1988; 1991; Mayer *et al.*, 1988; 1992). Similar findings have been reported in the Cercopithecinae (Castro *et al.*, 1996). The long-term persistence of such *trans-species polymorphism* is potentially explicable by overdominant selection because neutral alleles are unlikely to survive the process of speciation (Klein *et al.*, 1993). Selection would be expected to act on MHC alleles that differed in their ability to bind and present foreign peptides. In a population exposed to pathogens, an individual heterozygous at many MHC loci would be able to present a larger number of foreign peptides than a homozygote and might therefore be resistant to a wider range of pathogens (Parham and Ohta, 1996). Support for the overdominant selection hypothesis has come from studies (Hughes and Nei, 1988, 1989a; Ohta, 1991) that noted a significantly higher rate of nonsynonymous than synonymous nucleotide substitution in the peptide-binding region (antigen recognition site), strong evidence for the action of positive selection (Ayala *et al.*, 1994). It should however be noted that negative or purifying selection may also operate on the HLA system so as to reduce diversity. One example of this is found in the peptide-binding region of the *HLA-E* gene in New World monkeys (Knapp *et al.*, 1998).

The high frequency of polymorphism has probably also been maintained by other means such as temporal variation in selection pressure driven by changes in pathogens. Rare allele advantage (frequency-dependent selection) may play a role in that individuals with a rare MHC allele may respond better to challenge from new pathogen variants that have evolved in such a way as to evade the products of the more common MHC alleles. Population size and structure may also be important since diversity may be promoted by high effective population size or as a result of the agglomeration of smaller populations each bearing a few different alleles (Wells and Parham, 1996).

A high level of polymorphism can also be maintained by a 'genetic hitch-hiker' effect. Thus the occurrence of at least 11 olfactory receptor genes (*OR2C1*) within the MHC complex (Fan *et al.*, 1995) may potentiate the selection of specific receptor alleles matched to a subset of MHC-determined odorants, such that those haplotypes that carry certain matched alleles would have a selective advantage in the population (Gruen and Weissman, 1997). If odorant discrimination is learned, this could have implications for the population genetics of species in which odorant-mediated kin recognition is important. Intriguingly, Wedekind *et al.* (1995) have claimed there to be evidence in humans for a preference for particular MHC-linked variations in sweat odors.

The common marmoset (*Callithrix jacchus*), a New World primate, possesses limited MHC class II variability as a result of the inactivation of the MHC-DP region and limited polymorphism at the MHC-DR and -DQ loci (Antunes *et al.*, 1998). This limited MHC class II repertoire could play a role in the apparently increased susceptibility of this species to viral, bacterial, protozoan and helminth infections.

Mucin genes. Many epithelial tissues such as trachea, mammary gland, pancreas, stomach, cervix and intestine produce high molecular weight glycoproteins known as mucins which are the major proteins of mucus. The mucins display only limited homologies with one another (Desseyn *et al.*, 1997a) but do share the property of containing extensive tandemly repetitive regions. These regions can vary in length from 8 amino acid residues in MUC5AC to 169 residues in MUC6 and can be highly polymorphic. Several of the human mucin genes are located within a 400 kb cluster on chromosome 11p15 (*MUC2, MUC5AC, MUC5B, MUC6*), consistent with a series of successive gene duplications (Gum 1992), whereas others are solitary (*MUC1*, 1q21; *MUC3*, 7q22; *MUC4*, 3q29; *MUC7*, 4q13-q21; *MUC8*, 12q24) (Pigny *et al.*, 1996). Although the human mucins display only a limited degree of homology with one another, some mucin genes possess exons of similar length and distribution consistent with their having evolved from a common ancestor (Buisine *et al.*, 1998; Desseyn *et al.*, 1997a, 1998).

As noted above, the mucin genes often contain internal tandemly repetitive domains. Thus the *MUC2* gene contains two regions with a high degree of internal homology but no homology with each other. The first region comprises multiple 48 bp repeats interrupted by 21–24 bp segments whilst the second region is composed of 69 bp repeats arranged in a tandem array of up to 115 copies (Toribara *et al.*, 1991). The *MUC5B* gene contains an extremely large (10.7 kb) exon which encodes a 3570 amino acid protein that contains 19 subdomains which can be grouped into four larger composite units of 528 amino acids ('superrepeats') (Desseyn *et al.*, 1997b). Similarly, the *MUC4* gene contains an uninterrupted 18 kb exon encoding about 380 units of length 48 bp (Nollet *et al.*, 1998). Presumably these enormous exons have gradually become extended by serial internal duplication events that have not altered the splicing pattern, merely the length of the exon to be spliced. A model for the evolution of these complex and highly variable genes is presented in *Figure 4.22*.

Genes encoding RNA-binding proteins. RNA-binding proteins are involved in a wide range of biological functions including mRNA splicing, processing and translation. These proteins constitute a family insofar as they contain RNA-binding domains (*Figure 4.23*) which share a common evolutionary origin that predates the divergence of prokaryotes and eukaryotes (Fukami-Kobayashi *et al.*,

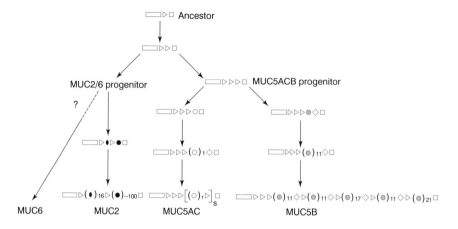

Figure 4.22. Hypothetical scheme for the evolution of the *MUC6, MUC2, MUC5AC,* and *MUC5B* mucin genes from a common ancestor (redrawn from Desseyn *et al.,* 1998). Unique domains flanking the tandem repeat arrays, conserved in each protein, are represented by rectangular or square boxes. A triangle represents the Cys sub-domain. The repetitive central domains are represented by the following symbols: circles (closed, empty and hatched) for the 23 amino acid tandem repeats of *MUC2,* the 8 amino acid tandem repeats of *MUC5AC,* and the 29 amino acid tandem repeats of *MUC5B.* The oval represents the irregular 16 amino acid tandem repeat polypeptide of *MUC2.* Diamonds represent the *MUC5B* R-end sub-domain type. t, r, and s denote unknown numbers of repeats.

1993). Examples of human genes encoding RNA-binding proteins include the 70 kDa nuclear ribonucleoprotein (*SNRP70*; 19q13.3) gene, the heterogeneous nuclear ribonuclear riboprotein A1 (*HNRPA1*; 12q13.1) gene, the Arg/Ser-rich splicing factor (*SFRS2*; chromosome 17) gene, the poly(A) binding protein (*PABPL1*; 3q22-q25, 12q13-q14, 13q12-q13) genes, and the *NCL* (2q12-qter) gene which encodes nucleolin, a protein involved in the control of transcription of rRNA genes and in the nucleocytoplasmic transport of ribosomal components.

Sulfatase genes. Sulfatases catalyze the hydrolysis of sulfate ester bonds in a variety of substrates. Sulfatase genes have been described in lower eukaryotes and comprise an evolutionarily conserved multigene family in human (Parenti *et al.,* 1997): arylsulfatase A (*ARSA*; 22q13), arylsulfatase B (*ARSB*; 5q11-q14), arylsulfatase D (*ARSD*; Xp22.3), arylsulfatase E (*ARSE*; Xp22.3), arylsulfatase F (*ARSF*; Xp22.3) proximal to the pseudoautosomal boundary, steroid sulfatase (*STS*; Xp22.3) 4 Mb distant from the *ARSD/ARSE/ARSF* cluster, iduronate-2-sulfatase (*IDS*; Xq27–28), galactose 6-sulfatase (*GALNS*; 16q24) and glucosamine 6-sulfatase (*GNS*; 12q14). The chromosomally dispersed sulfatase genes exhibit a relatively low degree of sequence identity consistent with these genes having emerged comparatively early during evolution. By contrast, the four genes located at Xp22.3, within the pseudoautosomal region, are more similar in sequence, share a very similar exon-intron organization and possess homologues on the Y chromosome (Meroni *et al.,* 1996), consistent with duplication events that occurred before the X and Y copies of these genes started to diverge (i.e. while they were still pseudoautosomal).

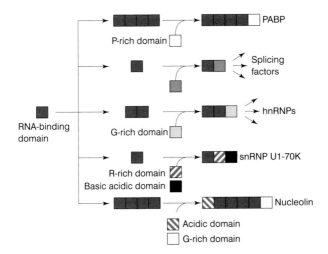

Figure 4.23. Evolutionary schema for various types of RNA-binding protein (redrawn from Fukami-Kobayashi *et al.*, 1993). PABP: poly(A) binding protein. hnRNP: heterogeneous nuclear ribonucleoprotein. snRNP: small nuclear ribonucleoprotein.

Genes encoding tRNAs and aminoacyl-tRNA synthetases. tRNA molecules perform a central role in protein biosynthesis by potentiating the transfer of amino acids to the ribosome. Each amino acid is bound to the tRNA by an aminoacyl-tRNA synthetase. Each tRNA contains sequence elements (located in the anticodon, acceptor stem or the discriminator base at position 73) that are unambiguously recognized by its cognate synthetase. The tRNA-tRNA synthetase complex, which also includes an elongation factor and GTP, enters the ribosome where the anticodon of the tRNA interacts specifically with one of the codon triplets of the mRNA. Subsequently, the amino acid is covalently bound to the C-terminal of the nascent polypeptide chain. tRNAs have similar sizes and tertiary structures (Mans *et al.*, 1991) but are subdivided into *isoaccepting groups* which, despite having different sequences, recognize the same amino acid.

It is unclear how many functional tRNA genes are present in the human genome. An early saturation hybridization study put the total number of tRNA genes at about 1300 per haploid human genome, an average of 65 copies for each tRNA species (Hatlen and Attardi, 1971). Chromosomally allocated human tRNA genes include *TRAN* (alanine, 6p21-p22), *TRE* (glutamic acid, 1p36), *TRG1* (glycine, chromosome 1), *TRK1* (lysine, 17p13), *TRL1* (leucine 1, 14q11-q12), *TRL2* (leucine 2, 17p13), *TRMI1* and *TRMI2* (methionine, 6p23-q12), *TRN* (asparagine, 1p36), *TRP1* and *TRP2* (proline 1 and 2, 14q11-q12), *TRP3* (proline 3, chromosome 5), *TRQ1* (glutamine 1, 17p13), *TRR1* (arginine 1, 17p13), *TRR3* (arginine 3, 6p21-p22), *TRR4* (arginine 4, 6p21-p22), *TRT1* (threonine 1, chromosome 5), and *TRT2* (threonine 2, 14q11-q12). Thus tRNAs from different isoaccepting groups often cluster together, for example 6p22 (*TRAN, TRMI1, TRMI2, TRR3, TRR4*; Buckland *et al.*, 1996), 14q11-q12 (*TRL1, TRP1, TRP2, TRT2*; Chang *et al.*, 1986) and 17p13 (*TRK1, TRL2, TRQ1, TRR1*; Morrison *et al.*, 1991). Other tRNA genes are dispersed and, since these genes are often flanked by short 8–12 bp direct terminal repeats, they may have arisen by RNA-mediated transposition (McBride *et al.*, 1989).

A human gene (*TRSP*) encoding an opal suppressor phosphoserine tRNA has been characterized on chromosome 19q13 (O'Neill *et al.*, 1985). This tRNA is potentially capable of suppressing nonsense mutations since it is able to recognize a different codon than that corresponding to the amino acid it carries. This ability is conferred upon it by dint of a base change in its anticodon which allows it to translate a UGA Stop codon and insert a serine at this position.

Phylogenetic studies of tRNA genes have shown that members of isoaccepting groups do not invariably cluster together (Saks and Sampson, 1995). It would appear that during evolution, tRNAs have been recruited from one isoaccepting group to another by mutations in the anticodon (Saks and Sampson, 1995).

The aminoacyl-tRNA synthetases constitute a large gene family in the human genome including glutaminyl/prolyl (*EPRS*; 1q32-q42), lysyl (*KARS*; 16q23-q24), alanyl (*AARS*; 16q22), arginyl (*RARS*; 5pter-q11), valyl (*VARS1*; 9), histidyl (*HARS*; chromosome 5), asparginyl (*NARS*; chromosome 18), threonyl (*TARS*; 5p13-cen), methionyl (*MARS*; chromosome 12), isoleucyl (*IARS*; 9q21), glycyl (*GARS*; 7p15), tryptophanyl (*WARS*; 14q32), cysteinyl (*CARS*; 11p15.5) and leucyl (*LARS*; 5cen-q11). Whilst many of the members of this gene family are chromosomally widely dispersed, similar chromosomal locations for some members are suggestive of linkage. The aminoacyl-tRNA synthetases may be grouped into two distinct classes, each with ten members, based upon sequence data and structure (Cusack *et al.*, 1991; Eriani *et al.*, 1990). These groups are thought to have arisen from two progenitor aminoacyl-tRNA synthetases by gene duplication and divergence (Nagel and Doolittle, 1995). Despite their ancient origin and their presence in both prokaryotes and eukaryotes, sequence similarity between aminoacyl-tRNA synthetases from eukaryotes and prokaryotes may be as low as 15% (Nagel and Doolittle, 1995).

Interestingly, there is a relationship between synthetase class and the nucleotide that is conserved at position 73. Eight of the ten tRNAs that are aminoacylated by class I synthetases have an alanine residue at this position whereas those that are aminoacylated by class II synthetases manifest greater nucleotide diversity (Saks and Sampson, 1995). It is clear that the genes for both tRNAs and aminoacyl-tRNA synthetases have a very long evolutionary history. Their origins and the means by which they may have coevolved are as yet unclear (Saks and Sampson, 1995) although recent work has suggested that the tRNA synthetases may have been preceded by their tRNAs (Ribas de Pouplana *et al.*, 1998).

Ubiquitin genes. The ubiquitin genes are extremely highly conserved 76 amino acid proteins which are required for ATP-dependent nonlysosomal intracellular protein degradation of defective proteins and proteins with a rapid turnover (Schlesinger and Bond 1987). Although found in all eukaryotes, they have not so far been found in prokaryotes.

The human genome contains multiple ubiquitin-related sequences most of which are processed pseudogenes (Schlesinger and Bond, 1987). There are however at least four functional ubiquitin genes which together have provided an interesting challenge to carefully crafted general definitions of the gene (Chapter 1, section 1.2.1). These genes belong to one of two distinct classes. The first is exemplified by the ubiquitin C (*UBC*; 12q24) gene which contains 9 direct repeats of the ubiquitin amino acid sequence with no spacer regions and no introns (Wiborg *et al.*, 1985). The *UBC* locus has a single termination codon and yields a 2500 nucleotide mRNA which upon translation generates a polyubiquitin precursor molecule which is post-translationally cleaved to active ubiquitin. Unequal crossing over events at the *UBC* locus have been responsible for polymorphism in ubiquitin repeat unit number (Baker and Board, 1989). A second human polyubiquitin gene (*UBB*, 17p11-p12) with 3 repeat units yields a 1000 nucleotide mRNA (Webb *et al.*, 1990). The second class comprises ubiquitin fusion genes which encode a single ubiquitin fused in-frame to a ribosomal protein of either 52 (*UBA52*; 19p13; Baker and Board, 1991; 1992) or 76–80 (*UBA80*) amino acids (Lund *et al.*, 1985).

The number of nucleotide differences between ubiquitin repeats within a given species is very low as compared to those between species (Sharp and Li, 1987). Thus, repeats within a locus share a more recent common ancestor than any two repeats in different species. This finding is explicable in terms of concerted evolution, probably mediated by unequal crossing over or gene conversion (Sharp and Li, 1987; Vrana and Wheeler, 1996). In human, concerted evolution is evident both within and between ubiquitin loci but appears to occur at a higher rate for some repeats than others (Tan *et al.*, 1993).

An overview of the evolution of multigene families in eukaryotes. During eukaryotic evolution, gene families have been created by sequential, phased rounds of gene duplication with specific genes becoming duplicated as a result of whole genome duplications, subchromosomal regional duplications, and through the rather more discrete duplication of individual gene loci. Since the number of genes in a gene family varies quite widely, we may surmise that some sequences are more predisposed to duplicate than others. As a consequence of their common origin, multigene family members usually have a similar structure both in terms of their nucleotide sequences and exon-intron organization (although the gain or loss of some introns in some members can serve to obscure their common ancestry). Whereas some genes have retained their syntenic relationship with each other after duplication and have remained chromosomally linked for relatively long periods of evolutionary time, others have been translocated to another chromosomal location. In some gene families, new rounds of gene duplication have then led to the formation of gene clusters. Once diversification of gene clusters, sub-clusters and individual genes had occurred through the acquisition of comparatively subtle mutational changes, the duplication of entire gene clusters as well as portions of clusters would then have led to the emergence of a sub-family structure within the gene family.

Diversification of individual genes within multigene families has occurred in a host of different ways in different lineages. Amino acid substitutions may appear to constitute relatively subtle structural changes but these changes can be quite dramatic in functional terms if, for example, substrate specificity is altered. Diversification can also proceed by internal duplication and deletion (e.g. gain or loss of individual exons), the insertion or removal of individual amino acid residues, repeat expansion and by the more complex processes of gene conversion, fusion and recombination. Family members have sometimes been inactivated or lost so that in some cases, the immediate ancestors of extant genes may now no longer exist (molecular 'missing links'). In others, promoter changes have led to the emergence of differences in gene expression either in terms of tissue specificity or level of expression or in responsiveness to inductive stimuli whether of environmental (e.g. temperature), systemic (e.g. hormonal) or intra-cellular origin (e.g. transcription factors). Some multigene family members have evolved more quickly than others whether as a result of selection pressure or the stochastic processes of neutralist evolution. What is common to all the gene families cited above is the principle that gene duplication has created redundancy which has then allowed evolutionary experimentation through diversification, ultimately leading to the recruitment of a new generation of genes encoding proteins with novel properties.

4.2.2 Highly repetitive multigene families

Histone genes. Histones are basic nuclear proteins which make up the nucleosome within the chromatin fibre. Pairs of H2A, H2B, H3, and H4 form the octamer with H1 being responsible for linking the nucleosomes and potentiating the formation of higher order chromosome structure. In mammals, the histones may be sub-divided into three types: (i) main-type replication-dependent histones, (ii) replication-independent 'replacement' histones, and (iii) tissue-specific histones. The genes encoding the replication-dependent and tissue-specific histones lack introns, give rise to non-polyadenylated mRNAs, contain 3' elements with dyadic symmetry essential for mRNA processing, and are chromosomally clustered. By contrast, the genes encoding the replacement histones can contain introns, give rise to polyadenylated mRNAs and are solitary rather than clustered (Brush *et al.*, 1985; Doenecke *et al.*, 1994).

Three main clusters of histone genes are apparent in the human genome and contain between them about 60 genes. Two clusters, located at 6p21.3, are separated by ~2 Mb and contain all replication-dependent H1 histone (*H1F1, H1F2, H1F3, H1F4, H1F5; Figure 4.24*) genes and surrounding core histone (*H2A, H2B, H3, H4*) genes (Albig *et al.*, 1993; Albig and Doenecke, 1997; Albig *et al.*, 1997a, 1997b). The other cluster at 1q21 is smaller consisting of at least four core histone genes. Various solitary replacement histone genes have been located on different chromosomes, for example H1° (*H1F0*; 22q13; Albig *et al.*, 1993), H2A.X (*H2AX*; 11q23; Ivanova *et al.*, 1994), H2A.Z (*H2AZ*; 4q24; Popescu *et al.*, 1994) and H3.3B (*H3F3B*; 17q25; Albig *et al.*, 1995). Finally, testis-specific histone genes *H3F3A* (Albig *et al.*, 1996) and *H1FT* (Albig *et al.*, 1997) have been localized to chromosomes 1q42 and 6p21, respectively.

Histones are very ancient proteins as evidenced by homologies between prokaryotic and eukaryotic histones H2A/B, H3, and H4 (Slesarev *et al.*, 1998; Ouzounis and Kyrpides, 1996). Interestingly, homology also exists between the core histones and the CCAAT-binding factor (Ouzounis and Kyrpides, 1996) suggesting that transcriptional regulation and nucleosomal packing may have been intimately related for a very considerable period of evolutionary time.

In lower eukaryotes, histone genes usually occur in long tandemly repetitive arrays but in mammals, although the genes are clustered, they are less ordered and

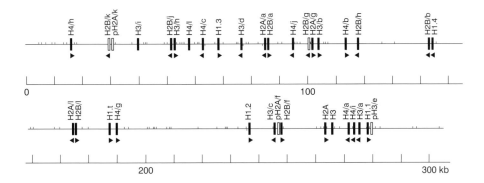

Figure 4.24. Organization of the human histone gene cluster on chromosome 6 (after Albig *et al.*, 1997a).Solid rectangles: histone genes. Open rectangles: pseudogenes. Orientations of histone genes are denoted by arrowheads. Restriction sites are shown as short vertical bars above (*Eco*RI) or below (*Mlu*I) the line.

do not occur as tandem repeats (Heintz *et al.*, 1981). A map of part of the major histone cluster on human chromosome 6 is shown in *Figure 4.24* and it is evident that the arrangement of the histone genes is fairly random; the only regularity is the arrangement of the divergently orientated *H2A* and *H2B* gene pairs. For both main-type and replacement histone types, chromosomal localization, gene number and gene organization appear to be fairly well conserved between human and mouse (Albig and Doenecke, 1997; Albig *et al.*, 1997). This indicates that the separation of these two groups of histone genes must have occurred at an early stage in mammalian evolution. Chromosomal clustering may be a prerequisite for coordinate regulation. An unexpectedly high degree of within-species homogeneity of tandemly repeated histone genes is explicable in terms of unequal crossing over and/or gene conversion (Maxson *et al.*, 1983).

The residual sequence similarity (pairwise 15–20%) exhibited by the four core histones together with their structural similarity argues strongly for descent from a common ancestor, albeit a very ancient one (Ramakrishnan, 1995). A phylogenetic analysis of the core histones has suggested that the histones that form dimers (viz. H2A/H2B and H3/H4) have very similar trees and appear to have co-evolved (Thatcher and Gorovsky, 1994). H3 and H4 have evolved ~10-fold more slowly than H2A and H2B and are very highly conserved (Thatcher and Gorovsky, 1994). The H2A.Z variant arose early and appears to be more highly conserved than the main-type 2A histone whilst H3.3 variants have arisen independently on many occasions (Thatcher and Gorovsky, 1994). By contrast, H2A.X arose comparatively recently during the evolution of the vertebrates (Thatcher and Gorovsky, 1994).

Ribosomal RNA genes. The ribosomal RNA (rRNA) genes occur as 300–400 copies in the human genome. They are organized in tandemly repeated blocks which are located on the acrocentric chromosomes: *RNR1* (13p12), *RNR2* (14p12), *RNR3* (15p12), *RNR4* (21p12), and *RNR5* (22p12). The 44 kb rDNA repeat unit contains a 13.3 kb RNA polymerase I-transcribed portion and a 31 kb nontranscribed spacer region (*Figure 4.25*; Gonzalez *et al.*, 1992). The 13 kb (45S) transcript is then processed to generate the 28S, 18S and 5S mature rRNA molecules with respective lengths of 1.8 kb, 0.15 kb, and 5.8 kb.

The 28S rRNA contains both conserved and variable regions. The former are constant in size and sequence whereas the latter are variable (Gorski *et al.*, 1987) and exhibit species-specific differences between primates (Gonzalez *et al.*, 1988; 1990). The variable regions potentiate hairpin loop formation and are essential for rRNA secondary structure (Gorski *et al.*, 1987).

The number of rRNA genes on a given chromosome varies in the population, while in any one individual, rRNA gene number varies between clusters (Arnheim *et al.*, 1980). The rRNA clusters appear to have evolved in concerted fashion. Thus rRNA genes on non-homologous chromosomes are far more similar to one another than would be expected if these genes had evolved independently. Several mechanisms have been proposed to account for this relative homogeneity of rDNA including unequal crossing over between rDNA sequences on nonhomologous chromosomes (Worton *et al.*, 1988). Gene conversion may then act so as to homogenize rDNA repeats within each cluster. On the basis of linkage disequilibrium data, Seperack *et al.* (1988) have suggested that gene conversion and unequal crossing over may also occur within a chromosome (i.e. by sister chromatid exchanges).

5S ribosomal RNA genes. The human 5S rRNA (*RN5S1*) genes occur in clusters of 2.3 kb and 1.6 kb tandem repeat units which differ from each other by virtue of a deletion in the 3′ flanking region (Sorensen and Frederiksen, 1991). There appear to be 300–400 copies of 5S rRNA genes per haploid human genome. Between 100 and 150 genes are derived from the 2.3 kb cluster (probably located at chromosome 1q42), 5–10 from the 1.6 kb cluster whereas the remaining 200–300 genes/gene variants are not found in repeat structures and may be dispersed around the genome. In addition to these genes, a large number of 5S rRNA pseudogenes exist which brings the total number of 5S rRNA homologous sequences per haploid genome to around 2000.

Small nuclear RNA genes. The human U2 snRNA (*RNU2*) genes are clustered in tandem arrays of between 6 and 30 copies at 17q21-q22 (Van Arsdell and Weiner, 1984; Westin *et al.*, 1984). The primate U2 snRNA arrays have evolved in concerted fashion with each repeat being essentially homogeneous within a species although somewhat different between species, an observation consistent with the action of gene conversion (Liao and Weiner, 1995; Liao *et al.*, 1997; Matera *et al.*, 1990). By contrast to the situation in higher primates, the U2 snRNA genes of both the mouse and the prosimian, *Galago crassicaudatus*, are dispersed rather than clustered suggesting that the arrays characteristic of higher primates may have resulted from amplification of a common ancestral gene (Matera *et al.*, 1990). Once established in the simian lineage, however, the U2 tandem repeat array has remained at the same chromosomal locus through multiple speciation events over a period of >35 Myrs (Pavelitz *et al.*, 1995).

Some 30 copies of the human U1 snRNA (*RNU1*) genes are also clustered at 1p36.3. This site is distinct from the cluster of U1 snRNA pseudogenes at 1q12-q22 that outnumber their functional counterparts by some 15–30 fold (Lindgren *et al.*, 1985). This contrasts with the situation found for the U2 snRNA genes, where the active genes outnumber the pseudogenes.

4.2.3 Gene superfamilies

The superfamily is used to describe a group of gene families whose individual members share a common evolutionary origin, possess common features within a family but differ with respect to certain other features between families. A selection of some of the most important human gene superfamilies will be presented.

Cadherin genes. The cadherins are membrane-associated glycoproteins which act as calcium-dependent cell adhesion molecules but which may also be involved in signal transduction. Cadherins may be classified into four groups (classical, desmosomal, protocadherins, cadherin-related proteins) which are structurally similar, each containing an extracellular domain consisting of between 4 and 30+ repeats of an 110 amino acid cadherin-specific motif. Although classical cadherins appear to be confined to the vertebrates, the superfamily has ancient origins with members present in organisms as diverse as nematodes and humans. The classical and desmosomal cadherins are thought to have evolved more recently by a process of duplication and divergence from the primordial protocadherin-like proteins (Suzuki, 1996).

A cluster of human classical cadherin genes is present at chromosome 16q22.1 (*CDH1*, *CDH5*, *CDH3*) with *CDH13* and *CDH15* located at 16q24 and *CDH11* not yet regionally localized on chromosome 16. Other genes encoding classical cadherins are present on 5p13-p14 (*CDH12*) and 18q11 (*CDH2*) whilst a protocadherin (*PCDH7*) gene has been mapped to 4p15.

Cytochrome P450 genes. The cytochrome P450 enzymes comprise a large liver-expressed family of heme-containing electron transport molecules which are involved in the oxidative metabolism of a wide range of substrates including steroids, drugs and xenobiotics. This family can be subdivided into sub-families on the basis of structural and functional criteria and many are extremely polymorphic, for example CYP2D6 (Marez *et al.*, 1997). With nearly 40 CYP P450 genes identified (and perhaps between 20 and 150 yet to be identified), this superfamily is one of the larger superfamilies represented in the human genome: *CYP1A1* and *CYP1A2* (15q22), *CYP1B1* (2p21), *CYP2A6*, *CYP2A7*, *CYP2A13* (19q13.1), *CYP2B6* and *CYP2B7* (19q13.2), *CYP2C8*, *CYP2C9*, *CYP2C10*, *CYP2C18*, *CYP2C19* (10q24), *CYP2D6* (22q13.1), *CYP2E* (10q24-qter), *CYP2F1* (19q13.2), *CYP2J2* (1p31), *CYP3A4* (7q22.1), *CYP4A11* (1), *CYP4B1* (1p12-p34), *CYP7A1* (8q11-q12), *CYP11A* (15q23-q24), *CYP11B1* and *CYP11B2* (8q21), *CYP17* (10q24.3), *CYP19* (15q21), *CYP21* (6p21.3), *CYP24* (20q13), *CYP26A1* (10q23-q24), *CYP27A1* (2q33-qter), *CYP27B1* (12q13.3-q14) and *CYP51* (7q21). Several gene clusters are apparent, for example on 10q24 and 19q13, and this is likely to reflect a history of gene duplication and divergence (Hoffman *et al.*, 1995; Nelson *et al.*, 1996). A common origin may however also be reflected by the possession of a similar exon/intron arrangement as in the case of the *CYP17* and *CYP21* genes or the *CYP11*, *CYP24*, and *CYP27* genes. The phylogeny of the cytochrome P450 superfamily has been explored to some extent and various attempts have been made to date the various gene duplication events which have given rise to extant genes (Degtyarenko and Archakov, 1993; Nelson and Strobel, 1987; Nebert et al., 1989). Much of the increase in CYP P450 gene number took place around 400 Myrs ago. This was the time when tetrapods first began to colonize the land and feed upon the plants that had become established in the late Silurian and early Devonian periods. Since these terrestrial plants probably contained toxic compounds, expansion of the CYP P450 gene family to provide a large set of detoxifying enzymes was probably an adaptive response to this chemical challenge.

The *CYP51* gene encodes sterol 14-α-demethylase which plays an important role in sterol biosynthesis in fungi and plants as well as animals. As such, *CYP51* is the only P450 family member which is recognizable across all eukaryotic phyla (Rozman *et al.*, 1996). The *CYP51* gene also appears to have its counterparts in prokaryotes (Aoyama *et al.*, 1998) and the gene family may have originated before the divergence of the eukaryotic and prokaryotic kingdoms. When members of the various mammalian and fungal P450 gene families were aligned and compared (Rozman *et al.*, 1996), more than 80 intron locations were identified. Since it is unlikely that all of these introns were present in the primordial eukaryotic P450 gene, it may be concluded that P450 gene structures have evolved very considerably over the last 2 billion years either by intron insertion (Chapter 3, section 3.5) or intron sliding (Chapter 3, section 3.4).

Cystatin genes. The cystatin superfamily comprises a number of proteins, many of which are cysteine protease inhibitors, but which have evolved to take on a variety of different physiological functions (Rawlings and Barrett, 1990). The family has emerged through a process of duplication and divergence from a primordial gene which is thought to have existed more than 1200 Myrs ago (Rawlings and Barrett, 1990). Human cystatin superfamily genes belong to family 1 [cystatins A and B (*CSTA*, 3cen-q21; *CSTB*, 21q22.3)], family 2 cystatins C, S, SA and SN (*CST3*, *CST4*, *CST2*, and *CST1*; 20p11.2; Thiesse et al., 1994), family 3 kininogen (*KNG*, 3q21-qter) or family 4 histidine-rich glycoprotein and α_2HS-glycoprotein (*HRG*, 3q27; *AHSG*, 3q27-q29). The evolutionary relationship of these genes remains to be unravelled (Brown and Dziegielewska, 1997; Müller-Esterl *et al.*, 1985).

G-protein-coupled receptor genes. The G-protein-coupled receptor (GPCR) superfamily can be separated into five different groups on the basis of their natural ligands: (i) peptides and peptide hormones, for example endothelin receptors (*EDNRB*, 13q22; *EDNRA*, chromosome 4), adrenocorticotropic hormone receptor (*MC2R*; 18p11.2), angiotensin receptor (*AGTR1*; 3q21-q25), thyrotropin receptor (*TSHR*; 14q31), (ii) neurotransmitters, for example dopamine receptors (*DRD1*, 5q35; *DRD2*, 11q23; *DRD3*, 3q13; *DRD4*, 11p15; *DRD5*, 4p15-p16), (iii) other regulatory factors, for example thrombin receptor (*F2R*; 5q13), (iv) sensory stimuli, for example opsins (see Chapter 7, section 7.5.2, *The visual pigments*), and (v) unknown ligands. The GPCRs are evolutionarily and phylogenetically related, most of them possessing a common motif of 7 transmembrane domains (Bockaert and Pin, 1999; Yokoyama and Starmer, 1996). A multitude of human genes encoding G-protein-coupled receptor with unknown ligands are also known and these are widely scattered around the genome, for example *GPR1* (15q21.6), *GPR2* (17q21), *GPR3* (1p34-p36), *GPR4* (19q13), *GPR5* (3p21), *GPR6* (6q21-q22), *GPR7* (10q11-q21), *GPR8* (20q13), *GPR9* (8p11-p12), *GPR10* (10q25-q26), *GPR12* (13q12), *GPR13* (3p21-pter), *GPR15* (3q11-q13), *GPR18* (13q32), *GPR20* (8q24).

Heat shock genes. The heat shock proteins are evolutionarily ubiquitous ('universal') proteins that function as molecular chaperones under both physiological and stress conditions. These proteins recognize and stabilize partially folded proteins during the processes of translation, translocation across membranes and multimer assembly. *Escherichia coli* possesses one heat shock protein gene (*dnaK*), yeast possesses nine as does *Drosophila*. This superfamily can be subdivided into three major classes based upon the molecular weight and degree of homology of the proteins.

The most highly conserved class is the HSP70 family represented in human by at least eleven genes: *HSPA1A*, *HSPA1B* and *HSPA1L* linked on chromosome 6p21, *HSPA2* (14q24), *HSPA3* (chromosome 21), *HSPA4* and *HSPA9* linked on chromosome 5q31, *HSPA5* (9q34), *HSPA6* and *HSPA7* linked on chromosome 1q, and *HSPA8* (11q23-q25). Some members of the HSP70 family are ubiquitously expressed, others are tissue-specific, some are constitutively expressed, others are expressed only in response to stress (Günther and Walter, 1994). The HSP70 proteins also differ in their subcellular localization, for example cytosol, nucleus, nucleolus, endoplasmic reticulum, mitochondrion (Günther and Walter, 1994). The HSP70 genes are extremely ancient having arisen before the diversification of cellular life into bacteria, archaebacteria and eukaryotes. Indeed, sequence data from these proteins have been used to argue for the origin of eukaryotes via a fusion between archaebacteria and gram-negative bacteria (Gupta and Golding, 1993; Gupta and Singh, 1994). Not surprisingly in view of their universality, they exhibit extreme evolutionary conservation (Boorstein *et al.*, 1994). Chromosomal localization has also been conserved evolutionarily as evidenced by the linkage of three HSP70 genes to the HLA/MHC complex in both humans and rodents.

The 90 kDa heat shock proteins comprise a second family of evolutionarily highly conserved proteins which have counterparts in bacteria as well as in all eukaryotes. Phylogenetic analysis has suggested that the first of the HSP90 gene duplications occurred very early on in the evolution of the eukaryotes (Gupta, 1995). Human homologues of this gene family are dispersed in the genome and include *HSPCA* (1q21-q22) and *HSPCB* (6p12). The human genome also contains a group of heat shock proteins of molecular weight 15–30 kDa which are highly conserved and evolutionarily related to the α-crystallins (de Jong *et al.*, 1993; see Section 4.2.1). Interestingly, one of the human genes encoding heat shock protein 27 (*HSPB2*) is closely linked to the α-B-crystallin (*CRYAB*) gene on chromosome 11q22-q23.

Insulin and insulin-like growth factor genes. The insulin gene family has ancient origins among the primitive eukaryotes (Le Roith *et al.*, 1980; Smit *et al.*, 1993). Human representatives of this superfamily include insulin (*INS*; 11p15.5), insulin receptor (*INSR*; 19p13), insulin-like growth factor I (*IGF1*; 12q22-q24), insulin-like growth factor II (*IGF2*; 11p15.5) and the relaxins (*RLN1* and *RLN2*; 9pter-q12). The origin of the *IGF1* and *IGF2* genes, which encode important regulators of growth and development, antedates the emergence of the first vertebrates, since they appeared early on in the evolution of the protochordates more than 600 Myrs ago (Chan *et al.*, 1990; McRory and Sherwood, 1997; Nagamatsu *et al.*, 1991; Upton *et al.*, 1997). It is likely that in primitive organisms, insulin-like peptides functioned so as to promote the uptake and utilization of nutrients, and as a consequence, stimulated growth. With increasing complexity, nutrition and growth became uncoupled and the insulin family diversified into proteins and receptors with differing capacities for regulating metabolism and growth (Steiner *et al.*, 1985).

Interferon genes. The interferon superfamily of viral defence proteins comprises two main classes of gene (type I and type II). The type II interferons have only one member, interferon-γ, encoded by a gene (*IFNG*) on chromosome 12q14. By contrast, type I interferons comprise several sub-families of genes most of which are clustered in a 400 kb region of chromosome 9p21. This cluster contains 13 interferon-α genes (*IFNA1, IFNA2, IFNA4, IFNA5, IFNA6, IFNA7, IFNA8, IFNA10, IFNA13, IFNA14, IFNA16, IFNA17* and *IFNA21*), a single interferon-Ω gene (*IFNW1*), a single interferon-β gene (*IFNB1*) and several pseudogenes (Diaz *et al.*, 1994; *Figure 4.25*). The genes are arranged in tandem and most of the functional genes are oriented with their 3′ ends pointing in a telomeric direction. There are two exceptions, *IFNA1* and *IFNA8*, which together with four pseudogenes orient towards the centromere (*Figure 4.26*). This is consistent with the occurrence at some stage of an inverted duplication within the gene cluster with its breakpoint between *IFNP12* and *IFNP11*.

The evolutionary relationships of the various *IFNA* family members have been determined by phylogenetic analysis (Golding and Glickman, 1985) and are consistent with the emergence of this gene family by a process of gene duplication

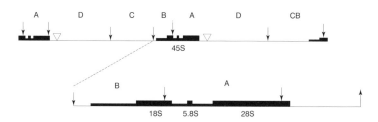

Figure 4.25. Organization of the human ribosomal gene clusters (after Erickson and Schmickel, 1985). The repeat pattern consists of four *Eco*RI fragments: A (7.3 kb), B (6.1 kb), C (11.7 kb), and D (16–19.6 kb) which together comprise a total repeat length of 41.1–44.7 kb. The inverted triangle denotes an area of length variability. *Eco*RI fragments are indicated by arrows. The 45S precursor rRNA transcript is processed to yield the mature 18S, 5.8S, and 28S rRNAs.

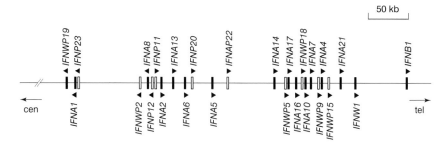

Figure 4.26. Map of the human type-I interferon gene cluster on chromosome 9p22 (after Diaz *et al.*, 1994). Solid rectangles: genes; open rectangles: pseudogenes.

and divergence (Miyata *et al.*, 1985). The proximal genes, *IFNA1*, *IFNA2*, *IFNA5*, *IFNA6*, *IFNA13*, appear more closely related to each other than they are to the distal genes, *IFNA4*, *IFNA7*, *IFNA10*, *IFNA16*, *IFNA17*, and *IFNA21*, whilst *IFNA8* is equally divergent from both groups (Gillespie and Carter, 1983; Henco *et al.*, 1985; Miyata and Hayashida, 1982). Since these two groups are located at opposite ends of the gene cluster, this may reflect an early multi-gene duplication. The gene cluster appears to have evolved by gene duplication as a result of unequal crossing over involving tandem units of *IFNA* and *IFNW1* genes (Diaz *et al.*, 1994) but divergence between gene sequences as well as subsequent gene deletions and duplications have served to erase the evolutionary history of parts of the gene cluster (Diaz, 1995). The *IFNA14* gene, located midway between the two groups, is similar to the distal group in its 5′ half and the proximal group in its 3′ half (Diaz *et al.*, 1994). This is explicable either in terms of a unequal crossing over event between distal and proximal interferon genes or a gene conversion event which has corrected the 5′ half of the gene against a donor proximal gene. Several of the human *IFNA* genes differ from the other family members by one or more relatively subtle mutations. Golding and Glickman have shown in their insightful and prescient (1985) study that these changes are explicable in terms of their templation by the local DNA sequence environment. Thus, the in-frame deletion of a GAT codon in the *IFNA2* gene may have been templated by a 5 bp inverted repeat 9 bp 5′ to the observed deletion (*Figure 4.27a*). Similarly, a 9 bp inverted repeat (separated by 25 bp DNA) may have templated an AA to GT change in the 5′ flanking region of the interferon "α9" gene (*Figure 4.27b*) whilst a 16 bp direct repeat could have templated five distinct sequence changes (two transitions, two transversions and a guanine insertion) as the product of one mutational event (*Figure 4.27c*).

The type I interferon genes not only lack introns but also exhibit sequence homology indicating that they share a common ancestry (Miyata and Hayashida, 1982). Since the spacer regions between primate *IFNA* genes still retain some sequence similarity, we may surmise that some of the gene duplication events have been relatively recent (Ullrich *et al.*, 1982). Indeed, Miyata and Hayashida (1982) estimated that they arose within the last 26 Myrs. The duplicational expansion of the *IFNA* gene cluster may have been driven by the need to produce large quantities of interferon rapidly, by selection for a novel temporal or spatial pattern of expression, or by selection for a specialized novel function, perhaps associated with a variant receptor with lower affinity for its ligand (Diaz, 1995).

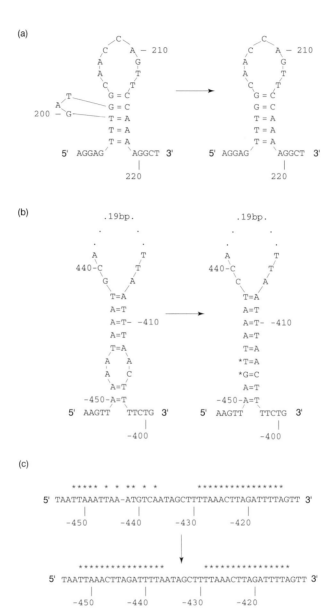

Figure 4.27. Evolution of the human α-interferon gene cluster; the role of template-mediated mutational changes (redrawn from Golding and Glickman, 1985). (a) An inverted repeat may have formed a template for the deletion of a GAT codon in the *IFNA2* gene. (b) An inverted repeat may have formed a template for an AA to GT change within the 5′ flanking region of the interferon "α9" gene. (c) A direct repeat may have directed five changes in the 5′ flanking region of the *IFNA10* gene. The 16 bp repeat is marked by asterisks. The putative ancestral sequence is shown above, the descendent sequence below.

The absence of introns in the *IFNA* and *IFNB* genes probably reflects the organization of the ancestral gene which could conceivably have been a retrotransposed copy of an intron-containing interferon gene such as that encoding interferon-γ (*IFNG*; 12q14) which contains three introns. The *IFNA* and *IFNB* genes also exhibit an axis of internal symmetry, presumably the result of an internal duplication which must have occurred prior to the divergence of the two gene families (Erickson *et al.*, 1984; Miyata *et al.*, 1985).

Nuclear receptor genes. Nuclear receptors are ligand-activated transcription factors that regulate the expression of their target genes by binding to specific *cis*-acting sequences in their promoter regions. This superfamily may be divided into a minimum of three distinct groups, one containing the receptors for steroid hormones (glucocorticoids, androgens, estrogens, progesterone, etc), a second containing receptors for vitamin D, thyroid hormone and retinoic acid and a third containing various 'orphan' receptors which putatively interact with ligands that still remain to be identified. Nuclear receptors exhibit a modular organization and contain at least four domains: an A/B domain involved in transactivation, a highly conserved zinc finger-containing domain (C) involved in DNA binding, a hinge (D) domain and a carboxy terminal (E) domain that is required for ligand binding, dimerization and transcriptional regulation.

Human genes belonging to this superfamily include the androgen receptor (*AR*; Xq11-q12), estrogen receptor (*ESR1*; 6q25), glucocorticoid receptor (*GRL*; 5q31), mineralocorticoid receptor (*MLR*; 4q31), progesterone receptor (*PGR*; 11q22), retinoic acid receptors α (*RARA*; 17q21), β (*RARB*; 3p24) and γ (*RARG*; 12q13), thyroid hormone receptors α (*THRA*; 17q21) and β (*THRB*; 3p24) and vitamin D receptor (*VDR*; 12q12-q14). Genes encoding the 'orphan' receptors include hepatocyte nuclear factor 4 (*HNF4A*; 20q12-q13), the COUP transcription factors (*TFCOUP1*, 5q14; *TFCOUP2*; 15q26), the retinoid X receptors (*RXRA*, 9q34; *RXRB*, 6p21.3; *RXRG*, 1q22-q23) and the peroxisome proliferator-activated receptors α (*PPARA*; 22q12-q13), γ (*PPARG*) and δ (*PPARD*; 6p21). It can be seen that the retinoic acid and thyroid hormone receptor genes are arranged in two syntenic groups.

The ancestral nuclear receptor gene may have originated very early by fusion of DNA-binding and steroid-binding domains (Amero *et al.*, 1992; Moore, 1990; O'Malley, 1989). Indeed, Escriva *et al.* (1997) have proposed that the ancestral nuclear receptor was an orphan receptor which subsequently acquired its ligand-binding potential. Since numerous vertebrate nuclear receptors have *Drosophila* homologues, the nuclear receptor superfamily must have diversified before the divergence of arthropods and vertebrates more than 500 Myrs ago. This diversification may have proceeded along the lines of the schema laid out in *Figure 4.28*. Two waves of gene duplication occurred, one very early on giving rise to the different receptor groups and a second in the vertebrate lineage. This was accompanied by domain shuffling between genes (Escriva *et al.*, 1997; Laudet, 1997; Laudet *et al.*, 1992). Sequence from the thyroid hormone receptor α-related gene, *THRAL*, appears to have been translocated to the *THRA* locus thereby creating the final *THRA* exon. This must have occurred early on in mammalian evolution since this organization is present in rat and human but not in chicken (Laudet *et*

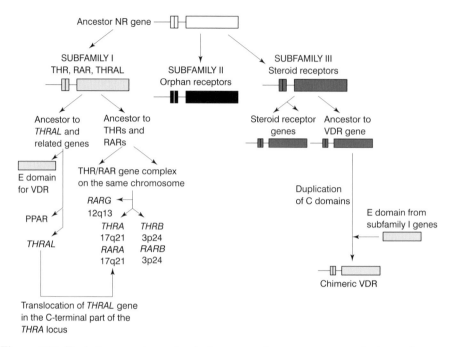

Figure 4.28. Evolutionary schema for the human nuclear receptor genes (redrawn from Laudet *et al.*, 1992).

al., 1992; *Figure 4.28*). An E domain from subfamily I has been been acquired by the *VDR* gene which distinguishes it from the other genes in the steroid hormone receptor family (Laudet *et al.*, 1992; *Figure 4.28*). There appears however to be no relationship between the phylogenetic position of a liganded receptor and the chemical nature of its ligand (Escriva *et al.*, 1997).

Protein kinase C genes. The members of the mammalian protein kinase C superfamily are involved in a number of signalling systems underlying a range of cellular processes. The characterized members of the human superfamily can be divided into three subfamilies, class I: α (*PRKCA*; 17q22-q23), β (*PRKCB1*; 16p11) and γ (*PRKCG*; 19q13), class II: δ (*PRKCD*; 3p) and θ (*PRKCQ*; 10p15), class III: ε (*PRKCE*), class IV: ι (*PRKCI*; Xq21.3), and ζ (*PRKCZ*). These genes have an ancient origin, their homologues having been found in both nematode and yeast (Mellor and Parker, 1998). The deduced amino acid sequences are highly conserved among mammals (>95% homology between humans and rodents) although homology is lower (45–65%) between nematodes and mammals. The expression pattern of the different proteins can however vary markedly. Structurally, each subfamily is characterized by a different arrangement of regulatory domains which serve to determine signalling specificity (Mellor and Parker, 1998).

Serine protease genes. The serine protease superfamily provide an archetypal example of evolution by gene duplication and divergence coupled with exon shuffling. Thus, mast cell chymase (*CMA1*; 14q11.2; Huang and Hellman, 1994), the digestive pancreatic proteases [trypsin (*PRSS1* and *PRSS2*; 7q35), chymotrypsin (*CTRB1*; 16q23) and elastase (*ELA1*; 12q13)] and the hepatic proteases of coagulation (e.g. thrombin; *F2*; 11p11-q12) have all evolved from the same common ancestral precursor gene (Greer, 1990). This ancestral gene appears to predate the divergence of eukaryotes and prokaryotes (Rypniewski *et al.*, 1994) and may itself have been the product of an internal duplication (McLachlan, 1979). Enormous structural and functional diversity has been generated both by the addition of non-proteolytic domains and by changes in the enzyme active sites which account for differences in their specificity.

The vitamin K-dependent serine proteases of coagulation exhibit substantial sequence and structural homology and their evolution is described in some detail in Section 4.3 and Chapter 3, section 3.6.3 and Chapter 10, section 10.2. Human representatives of this serine protease family include factor VII (*F7*; 13q34), factor IX (*F9*; Xq27), factor X (*F10*; 13q34), prothrombin (*F2*; 11p11-q12), protein C (*PROC*; 2q13-q14) and protein S (*PROS1*; 3p11). Various serine proteases have also been recruited to functions in fibrinolysis e.g. plasminogen (*PLG*; 6q26-q27), tissue-type plasminogen activator (*PLAT*; 8p11-q12) and urokinase (*PLAU*; 10q24-qter) and, phylogenetically, these are closely related to the contact factors: factor XI (*F11*; 4q34), factor XII (*F12*; 5q33-qter) and plasma kallikrein (*KLK3*; 4q34-q35) (*Figure 3.7*).

The division between the proteases of coagulation and fibrinolysis is apparent at the level of exon/intron organization but also in terms of codon usage for the active site serine residue (Brenner, 1988). The genes encoding the vitamin K-dependent factors of coagulation possess an AGY codon whereas the fibrinolytic enzyme genes exhibit a TCN codon also found in the serine protease genes of eubacteria and invertebrates. These alternative codons cannot be interconverted by a single nucleotide substitution. Brenner (1988) proposed that the AGY codon could have been derived from an active cysteine residue encoded by TGY in a cysteine protease that existed billions of years ago. Irwin (1988) has however argued that the AGY codon evolved from a TCN codon on at least two separate occasions, once on the lineage leading to the vitamin K-dependent factors of coagulation and once on the lineage leading to plasminogen and apolipoprotein(a).

Hypervariability of the active site regions is apparent in some serine proteases indicating that rapid evolution of the reactive center may have driven the functional divergence of these enzymes (Creighton and Darby, 1989; Huang and Hellman, 1994; Lesk and Fordham, 1996; Ohta, 1994). Since a similar phenomenon is apparent for the serine protease inhibitors (see Section 4.2.3, *Serpin genes*), it may be that the proteolytic enzymes and their inhibitors have coevolved.

Serpin genes. The serine protease inhibitors (Serpins) are a superfamily of proteins with over 100 members in mammals, and counterparts in invertebrates, plants and even some viruses (Marshall, 1993). Serpins interact with the substrate-binding sites of their cognate proteases via an exposed binding site of canonical conformation (Bode and Huber, 1992). The protease and its inhibitor rapidly form a tightly bound 1 : 1 stoichiometric complex with the reactive center of the serpin acting as a bait for the appropriate serine protease. Mammalian serpins share a common ancestor some 500 Myrs ago (Bao *et al.*, 1987) and include α1-proteinase inhibitor (*PI*; 14q32), α1-antichymotrypsin (*AACT*; 14q32), heparin cofactor II (*HCF2*; 22q11), antithrombin (*AT3*; 1q23), α2-antiplasmin (*PLI*; 17p13), pigment epithelium-derived factor (*PEDF*; 17p13), C1-inhibitor (*CII*; 11q11-q13), protein C inhibitor (*PCI*; 14q32) and plasminogen activator inhibitors 1 and 2 (*PAI1*, 7q21-q22; *PAI2*, 18q21.3). Other less well characterized mammalian serpins include cytoplasmic antiproteinase 2 (*PI8*; 18q21.3), maspin (*PI5*; 18q21.3), nexin (*PI7*; 2q33-q35), kallistatin (*PI4*; 14), neuroserpin (*PI12*; 3q26). Thus, serpin gene clusters are found at 14q32, 17p13 and 18q21.3 in the human genome.

Not all serpins act as protease inhibitors. Thus, angiotensinogen (*AGT*; 1q41) and corticosteroid-binding globulin (*CBG*; 14q32) act as blood pressure regulatory hormones whilst ovalbumin functions as an avian egg storage protein. Although there is no ovalbumin gene in the human genome, there are several genes encoding ovalbumin-like serpins (ov-serpins) which are clustered either on 18q21.3 (*PI10, SCCA1, SCCA2*) or 6p25 (*PI2, PI6, PI9*). The ov-serpins on chromosome 6 share more amino acid sequence identity with one another than they do with their chromosome 18 counterparts with the exception of PI8 (Bartuski *et al.*, 1997). By contrast, most of the chromosome 18 ov-serpins share greater sequence identity with a chromosome 6 ov-serpin than with each other. To account for these observations, Bartuski *et al.* (1997) proposed that the 6p25 loci arose first and then gave rise to the chromosome 18 cluster.

Dendrograms derived from comparisons of serpin nucleotide and amino acid sequences and of intron positions differ significantly (Wright, 1993). The exon/intron organization of extant serpin genes is therefore unlikely to be explicable simply in terms of the loss of introns from a large primordial gene. Wright (1993) proposed that the serpin gene family arose from an early recombination event which fused the amino and carboxyl domains. The subsequent insertion of sequence (possibly intronic) served to create β-sheet A and stabilized the new structure. Few of the introns demarcate regions of secondary or tertiary structure and further insertions, deletions and migrations of introns must have occurred.

Substrate specificity manifested by the serpin is determined, at least in part, by the P1 residue of the bait loop which contains Met or Val for elastase, Lys for trypsin, Leu for chymotrypsin and Arg for thrombin. After gene duplication, divergence in terms of substrate specificity has been very rapid and has resulted from 'accelerated evolution' of the reactive center region (Hill and Hastie, 1987). This rapid evolution is thought to have taken place by a combination of mechanisms, from genetic drift to gene conversion to positive Darwinian selection (Graur and Li, 1988; Hill and Hastie, 1987; Ohta, 1994).

Zinc finger genes. The term 'zinc finger' refers to a 28 amino acid sequence motif which binds zinc ions thereby stabilizing the structure of a small DNA-binding domain (El-Baradi and Pieler, 1991). It has been estimated that between 300 and 700 human genes encode zinc finger-containing proteins (Hoovers *et al.*, 1992). Most of these proteins belong to the *Kruppel*-type family and act as transcription factors (El-Baradi and Pieler, 1991). Zinc finger proteins often contain multiple zinc finger motifs, the number varying from between 2 and 37 (Klug and Schwabe, 1995).

To date, some 250 human zinc finger genes have been allocated a symbol and some have been chromosomally localized. Inspection of the available mapping data reveals at least 9 distinct clusters of zinc finger genes in the human genome: 3p21-p22 (*ZNF52, ZNF64, ZNF35, ZNF166, ZNF167, ZNF168*), 6p21 (*ZNF76, ZNF165, ZNF173, ZNF204*), 8q24 (*ZNF7, ZNF16, ZNF34*), 10p11 (*ZNF11A, ZNF25, ZNF33A, ZNF37A*), 10q11 (*ZNF11B, ZNF22, ZNF33B, ZNF37B*), 11q23 (*ZNF123, ZNF125, ZNF128, ZNF129, ZNF145*), 19p12-p13 (*ZNF43, ZNF56, ZNF58, ZNF66, ZNF67, ZNF85, ZNF90, ZNF91, ZNF92, ZNF208*), 19q13 (*ZNF42, ZNF45, ZNF83, ZNF93, ZNF132, ZNF134, ZNF135, ZNF136, ZNF137, ZNF146, ZNF154, ZNF155, ZNF160, ZNF175*), 22q11 (*ZNF69, ZNF70, ZNF71, ZNF74*). This superfamily has presumably evolved through cycles of gene transposition and duplication. Duplications in the ZNF91 family are known to have occurred some 55 Myrs ago in the common ancestor of the simians (Bellefroid *et al.*, 1995).

The human *ZNF45* and *ZNF93* genes are very similar to their murine orthologues although the human genes encode more zinc finger repeats than their murine counterparts, consistent with the occurrence of intragenic deletions/duplications (Shannon and Stubbs, 1998). Similarly, the human transcription factor MOK2 (*MOK2*; 19q13.2-q13.3) contains 10 zinc finger motifs in comparison to 7 in the murine homologue (Ernoult-Lange *et al.*, 1995).

Olfactory receptor genes. The olfactory system is thought to be capable of distinguishing several thousand odorant molecules. This is potentiated by olfactory receptors (ORs) which are responsible for the recognition and G protein-mediated transduction of specific odorant signals. With possibly 1000 members in the human genome, the OR gene superfamily constitutes by far the largest family encoding G protein-coupled receptors. OR genes are intronless and occur in clusters that are present at more than 25 chromosomal locations in the human genome. However, more than 70% of human OR-homologous sequences are probably pseudogenes (Rouquier *et al.*, 1998; Trask *et al.*, 1998).

Human OR gene clusters appear to be disproportionately located in subtelomeric regions and have been subject to frequent duplications and inter-chromosomal rearrangements (Trask *et al.*, 1998) some of which appear to have been mediated by recombination between repetitive sequence elements (Glusman *et al.*, 1996). OR genes within the clusters belong to at least four different subfamilies which display as much sequence variability within clusters as between clusters (Ben-Arie *et al.*, 1993). The classification of human OR genes is still in its infancy but chromosomally localized members include *OR1A1*, 17p13; *OR1D2, OR1D4, OR1D5*, 17p13; *OR1E1, OR1E2*, 17p13; *OR1F1*, 16p13; *OR1G1*, 17p13; *OR2D2*, 11p15; *OR3A1, OR3A2, OR3A3*, 17p13; *OR5D3, OR5D4*, 11q12; *OR5F1*, 11q12; *OR6A1*, 11p15; *OR10A1*, 11p15.

The olfactory system is combinatorial in that one OR can recognize multiple odorant molecules, and one odorant molecule may be recognized by multiple ORs, whilst different odorants are recognized by different combinations of ORs (Malnic *et al.*, 1999). Each nasal olfactory sensory neuron expresses only one allele of a single OR gene (Chess *et al.*, 1994; Sullivan *et al.*, 1996). In the olfactory epithelium, different sets of ORs are expressed in distinct spatial zones; neurons expressing a given OR gene are located in the same zone but in that zone they are interspersed with neurons expressing other ORs (Sullivan *et al.*, 1996). However, OR genes expressed in the same zone map to numerous different loci whereas a single OR locus may contain genes expressed in different zones (Sullivan *et al.*, 1996). Since some of the OR pseudogenes have been found to be expressed, it is possible that some of those neurons that only express a single receptor type may be nonfunctional (Crowe *et al.*, 1996).

The origin and emergence of the olfactory receptor gene family probably preceded the divergence of the mammals (Ben-Arie *et al.*, 1993; Issel-Tarver and Rine, 1997). In the primates, the olfactory receptor genes appear to have been in a considerable state of flux with numerous translocations, duplications and deletions occurring during the evolution of the great apes (Trask *et al.*, 1998). The origin of ORs is unclear but their expression during spermatogenesis and their presence in mature sperm cells suggests that their original function could have been in sperm physiology (Vanderhaeghen *et al.*, 1997a, 1997b).

4.2.4 Genes which undergo programmed rearrangement

The immunoglobulin and T-cell receptor genes are members of the immunoglobulin superfamily and share the common property of being assembled from multiple gene coding segments in the lymphocyte. In the germline, the genes encoding the variable portions of these receptors are usually split into variable (V), J (joining) and D (diversity) segments that are joined together by a somatic process of site-specific recombination that involves the recombination-activating proteins RAG1 and RAG2 (see Chapter 9, section 9.4.2) to generate exons that encode the antigen-binding portion of the polypeptide.

Immunoglobulin genes. A diverse repertoire of human antibodies is generated by the combinatorial somatic rearrangement of a relatively small number of gene segments, variable (V_H), diversity (D) and joining (J_H) segments for the heavy chain V region, and variable (V_L) and joining (J_L) segments for the light chain V region. The J_H and J_L segments are spliced to the constant region genes of the heavy (C_H) and light (C_L) chains following mRNA transcription.

The human immunoglobulin genes are mainly distributed between three chromosomal locations: 14q32 for the heavy chain loci, 2p12 for the κ light chain loci and 22q11 for the λ light chain loci. The 1100 kb human heavy chain locus contains 51 functional V_H gene (*IGHV*) segments, ~30 D segments (*IGHDY*), 6 functional J_H segments (*IGHJ*) and 9 functional C_H genes: μ (*IGHM*), δ (*IGHD*), γ3 (*IGHG3*), γ1 (*IGHG1*), α1 (*IGHA1*), γ2 (*IGHG2*), γ4 (*IGHG4*), ε1 (*IGHE*) and α2 (*IGHA2*) (*Figure 4.29*; Cook and Tomlinson, 1995). The human C_H gene locus clearly exhibits dyadic symmetry as a result of a duplication event that included the Cγ-Cγ-Cε-Cα gene cluster in the ancestor of the great apes (Kawamura and Ueda, 1992). The symmetry is imperfect in extant species because the Cε gene was lost through independent deletion events in humans, chimpanzees, gorilla and the white-handed gibbon (*Hylobates lar*) whereas in the orangutan, the Cα gene was deleted as well as the Cε gene (Kawamura and Ueda, 1992).

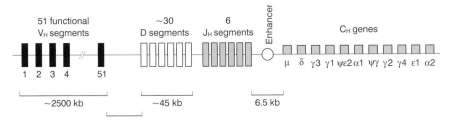

Figure 4.29. Schematic organization of the human immunoglobulin heavy chain (*IGHV*) locus.

The κ light chain locus contains 76 V_κ genes (*IGKV*) and pseudogenes, 5 functional J_κ segments (*IGKJ*) and a single C_κ gene (*IGKC*). This locus consists of two copies of a DNA region arranged with opposite polarity (Weichhold *et al.*, 1993), the result of an intra-locus duplication which has occurred since the divergence of human and chimpanzee (Ermert *et al.*, 1995). The λ light chain locus contains contains 36 functional V_λ genes (*IGLV*), 33 V_λ pseudogenes, 34 V_λ 'relics' containing large deletions or insertions, 6 J_λ segments (*IGLJ*) and 6 C_λ genes (*IGLC*) (Kawasaki *et al.*, 1997). In addition to the above, 8 V_H segments (*IGHV2*) and a cluster of D segments (*IGHDY2*) have been identified on chromosome 15q11.2 with a further 16 V_H segments (*IGHV3*) on chromosome 16p11.2 (Tomlinson *et al.*, 1994). These 'orphon' V_H sequences are thought to have been translocated to chromosomes 15 and 16 some 20 Myrs ago (Matsuda and Honjo, 1996) and ~40% may be functional. A small cluster of orphon Vκ sequences has also been found on chromosome 22, and flanking direct and inverted repeats have been invoked to account for their transposition (Borden *et al.*, 1990).

It has been apparent for some time that the heavy and light immunoglobulin chains are homologous and that these proteins must have been assembled by extensive duplication of a short ancestral 100 amino acid polypeptide chain (Hill *et al.*, 1966). This domain, the immunoglobulin fold, has now been found in a very wide array of different proteins throughout the animal kingdom. The immunoglobulins of the immune system appear however to be a vertebrate creation (Schluter *et al.*, 1997).

The immunoglobulin genes are thought to have originated very early in the evolution of the jawed vertebrates (Rast *et al.*, 1997). Phylogenetic analyses of the human V_H heavy chain genes (Haino *et al.*, 1994; Vargas-Madrazo *et al.*, 1997), C_H genes (Takahashi *et al.*, 1982), V_λ light chain genes and pseudogenes (Kawasaki *et al.*, 1997) and Vκ light chain genes (Kurth *et al.*, 1993; Sitnikova and Nei 1998; Sitnikova and Su 1998; Vargas-Madrazo *et al.*, 1997) are consistent with a history of multiple, successive duplication events, both extensive and of more limited extent, during the evolution of all branches of the immunoglobulin gene family. This mode of evolution is reminiscent of a 'birth and death process' rather than any mechanism of concerted evolution (Section 4.2.1). Some immunoglobulin genes have have been present in vertebrate genomes for >400 Myrs, others are more recent creations that have emerged during the mammalian radiation (reviewed by Andersson and Matsunaga, 1995; Matsuda and Honjo, 1996; Schluter *et al.*, 1997). The expansion of the immunoglobulin gene family was almost certainly an adaptive response to the wide range of novel pathogens that challenged early tetrapods in their new terrestrial environment.

The evolution of the immunoglobulin genes is subject to at least two distinct types of constraint: existing gene segments must be conserved in order to meet recurring immunological challenges but diversity must be continually created to meet new immunological challenges. The duplication of gene segments provides a means to achieve this by ensuring that old structures/functions are maintained at the same time as new ones are being created.

T-cell receptor genes. The T-cell receptor plays an important role in antigen recognition during the immune response. The four T-cell receptor chains (α, β, γ, and δ), members of the immunoglobulin superfamily, dimerize to form two types of receptor (α/β, γ/δ). In the human genome, these chains are encoded by four genes: *TCRA* and *TCRD* (14q11.2), *TCRB* (7q35), and *TCRG* (7p14-p15). The location of the *TCRD* gene is highly unusual in that it lies within the *TCRA* gene: the multiple gene units are organized as Vα-Vδ-Dδ-Jδ-Cδ-Jα-Cα (Hockett *et al.*, 1988). Each locus comprises a large number of tandemly arrayed variable (V) genes, diversity (D) segments (*TCRB* and *TCRD*), a number of clustered joining (J) elements and one or two constant (C) region genes. These gene segments undergo somatic rearrangement during lymphocyte differentiation to yield either V/J or V/D/J genes. As with the immunoglobulin genes (Section 4.2.4, *Immunoglobin genes*), the origin of the T-cell receptor genes probably occurred very early in the evolution of the jawed vertebrates (Rast *et al.*, 1997). The tandem arrays of V genes seen in extant vertebrate genomes have arisen subsequently by serial duplication, generating the sizeable repertoire of T-cell receptor V genes essential for creating diversity of antigen recognition specificity.

The *TCRB* gene has been the best characterized of the four loci with the sequencing of 685 kb from the region (Rowen *et al.*, 1996). This region contains 46 variable gene segments, 19 pseudogenes and two clusters of D, J, and C segments. A further 22 additional sequences, termed 'relics' by the authors, were also identified. These sequences exhibited limited homology to V gene segments and represent partial pseudogenes extensively altered by insertions and deletions. Some 30% of the *TCRB* locus is composed of genome-wide interspersed repeats but these sequences do not appear to have facilitated the duplication of locus-specific repeats. Some of these repeats within the *TCRB* locus have been shown to harbor trypsinogen genes (*PRSS1*, *PRSS2*). These genes must have been conserved for at least 350 Myrs because a trypsinogen gene cluster is present at this location in both mouse and chicken. Higher primates contain similar numbers of V gene segments at the *TCRB* locus: human (47), gorilla (45), orangutan (57), macaque (57) with the greater number of V genes in orangutan and macaques being due to amplification of the TCRBV7, 9 and 23 subfamilies (Charmley *et al.*, 1995).

4.3 Convergent evolution

Most members of the gene and protein families discussed in Section 4.2 have been subject to *divergent evolution*, the process by which homologous proteins or domains develop from a common origin but gradually acquire their own unique identity in terms of structure and function. By contrast, *convergent evolution* refers to the similarity between two protein structures, amino acid sequences or nucleotide sequences due to their independent evolution from different origins rather than through their possession of a common ancestor. Evolutionary convergence therefore implies 'adaptive change in which lesser related entities come to appear more related than they are' (Doolittle, 1994). The term convergence has however been used in many different contexts with very different meanings. Doolittle (1994) distinguished between *functional* convergence, *mechanistic* convergence and *structural* convergence.

Functional convergence is exemplified by various pairs of enzymes that have evolved independently to catalyse the same biochemical reactions, for example superoxide dismutases, aldolases, alcohol dehydrogenases and topoisomerases (Doolittle, 1994). Myoglobin from the abalone *Sulculus diversicolor* exhibits functional convergence with vertebrate myoglobin although it is not in any way homologous to it (Suzuki *et al.*, 1996). Instead, the *Sulculus* myoglobin gene is evolutionarily related to the human indoleamine 2,3-dioxygenase (*IDO*; 8p11-p12) gene (Suzuki *et al.*, 1996). The original *Sulculus* myoglobin gene may have been lost and a modified *Ido* gene could have evolved as a substitute.

Perhaps the best example of mechanistic convergence is provided by the serine proteases subtilisin and chymotrypsin which, although unrelated evolutionarily, have independently evolved similar enzymatic mechanisms. Chymotrypsin has a catalytic triad with a histidine at residue 57, an aspartate at residue 102 and a serine at residue 195. The bacterial protein subtilisin possesses a completely different structure but also has a catalytic triad comprising an aspartate at residue 32, a histidine at residue 64 and a serine at residue 221. The human genome contains representatives of both families. Thus, the chymotrypsinogen B (*CTRB1*; 16q23), chymotrypsin-like protease (*CTRL*; 16q22), trypsin 1 (*PRSS1*; 7q35), trypsin 2 (*PRSS2*; 7q35), elastase 1 (*ELA1*; 12q13), plasminogen (*PLG*; 6q26) and prothrombin (*F2*; 11p11-q12) genes are members of the chymotrypsin family of serine proteases. Human genes encoding subtilisin-like serine proteases include the proprotein convertases (*PCSK1*, 5q15-q21; *PCSK2*, 20p11; *PCSK4*, chromosome 19; *PCSK5*, 9q21.3), furin (*PACE*; 15q25-q26) and paired basic amino acid cleaving enzyme 4 (*PACE4*; 15q26). Another example of mechanistic convergence is provided by the cold shock domain protein family and the family of RNA-binding proteins that contain an RNA-binding domain. Both of these domains contain conserved ribonucleoprotein motifs on similar single stranded nucleic acid-binding surfaces (Graumann and Marahiel, 1996).

Structural convergence, on the other hand, may reflect the tendency of specific amino acid sequences to fold into certain favored conformations. Thus, structurally dissimilar families of transport proteins have been found to exhibit similar structural units consisting of six tightly packed α-helices which may comprise all or part of a transmembrane channel (Saier, 1994).

In none of the above examples is there any evidence for *sequence convergence*. For two sequences to be shown to display convergence in this strict sense, they would not only have to be shown to be evolutionarily unrelated but chance would also have to be excluded as a reason for their similarity. Indeed, sequence convergence would imply the occurrence of adaptive amino acid replacements that have been positively selected during evolution. Although sequence convergence has been invoked in many different situations, there are still no convincing examples of unrelated sequences which adhere to these criteria such that they warrant the use of the term *sequence convergence*.

4.4 Coevolution

Coevolution can be said to be operating in cases where the process of evolutionary change experienced at one locus is influenced by changes that have occurred at another locus. At its simplest, coevolution can occur when two genes are intimately associated as in the case of shared bidrectional promoter elements (Chapter 5, section 5.1.5). More complex situations involve unlinked genes. Fryxell (1996) proposed that 'the acquisition of a novel function by a duplicated gene could be facilitated by pre-existing heterogeneity in proteins that interact directly with the product of the duplicated gene'. Thus, the duplication and functional divergence of one gene might serve to create an altered genetic environment that could promote the divergence of duplicate copies of genes encoding proteins that interact with the protein products of the first gene.

In a study of the interspecies diversity manifested by a series of 48 ligand-receptor pairs, Murphy (1993) demonstrated there to be a linear relationship between receptor divergence and ligand divergence. Interestingly, the inter-species differences in receptor structure were nonrandomly distributed and largely confined to the extracellular domains that interact with the ligand. Since this study employed ligand-receptor pairs from diverse systems (host defence proteins, neurotransmitters, hormones, growth factors and cell adhesion proteins), it is reasonable to suppose that the coevolution of genes encoding interacting proteins is not an uncommon phenomenon.

The coevolution of families of receptors and their ligands is perhaps best exemplified by the insulin-nerve growth factor family and their receptors (Section 4.2.3, *Insulin and insulin-like growth factor genes*; Fryxell, 1996). These ligand-receptor pairs include insulin (*INS*; 11p15.5) and insulin receptor (*INSR*; 19p13), insulin-like growth factor 1 (*IGF1*; 12q22-q23) and its receptor (*IGF1R*; 15q25-qter), brain-derived neurotrophic factor (*BDNF*; 11p13) and neurotrophin 5 (*NTF5*; 19q13.3) and their cognate receptor neurotrophic tyrosine kinase receptor type 2 (*NTRK2*; 9q22.1), neurotrophin 3 (*NTF3*; 12p13) and neurotrophic tyrosine kinase receptor type 3 (*NTRK3*; 15q25), nerve growth factor (*NGFB*; 1p13.1) and neurotrophic tyrosine kinase receptor type 1 (*NTRK1*; 1q21-q22). The ligand-encoding genes share a common ancestry as do the genes encoding their receptors (Fryxell, 1996). As new trophic factors emerged by duplication and divergence, so their cognate receptors evolved by a similar parallel process. Other examples of the coevolution of ligands and their receptors include interleukin 8 (*IL8*; 4q13–21) and its receptors (*IL8RA, IL8RB*; 2q35; Ahuja *et al.*, 1992), interleukin 4 (*IL4*; 5q23-q31), and interleukin 4 receptor (*IL4R*; 16p11-p12; Richter *et al.*, 1995), the gonadotropins and their receptors (Moyle *et al.*, 1994) and the nuclear receptors and their ligands (Escriva *et al.*, 1997; section 4.2.3, *Nuclear receptor genes*).

Coevolution occurs between genes encoding different subunits of heterodimeric proteins, for example the *myc* proto-oncogene (*MYC*; 8q24) and its dimerization partner *max* (*MAX*; 14q23) (Atchley and Fitch 1995), the genes encoding the type I and II keratins (Section 4.2.1, *Keratin genes*) and the genes encoding the α- and β-integrin chains (Hughes 1992; Section 4.2.1, *Integrin genes*). Coevolution can also occur between agonists and antagonists, for example interleukins 1α (*IL1A*; 2q13) and 1β (*IL1B*; 2q13-q21), and the interleukin 1 receptor antagonist (*IL1RN*; 2q14.2) which binds to the IL1 receptor and blocks IL1α and IL1β binding without inducing a signal of its own; all three genes share a common ancestry (Eisenberg *et al.*, 1991). The cytokines and their receptors have both arisen by a process of gene duplication and divergence and although the relative timing of these duplicative events is still unclear, evidence is emerging for ligand-receptor coevolution (He and Wu, 1993; Kosugi *et al.*, 1995; Shields *et al.*, 1995).

One prediction of Fryxell's (1996) hypothesis is that duplication/divergence events in functionally related gene families may be temporally correlated. This prediction appears to be bourne out at least for the α- and β-integrin chain genes (Hughes, 1992) and the fibroblast growth factor/fibroblast growth factor receptor genes (Coulier *et al.*, 1997).

Whilst numerous examples of 'ligand promiscuity' have been recognized e.g. the type I interferon genes clustered on 9p21 (see Section 4.2.3, *Interferon genes*), 'receptor promiscuity' appears to be much rarer (Ahuja *et al.*, 1992). Examples of receptor promiscuity include the interleukin 8 receptors (*IL8RA, IL8RB*; 2q35; Ahuja *et al.*, 1992) and the interferon receptors α, β, and ω, 1 (*IFNAR1*) and 2 (*IFNAR2*) and interferon receptor γ2 (*IFNGR2*) whose genes are closely linked to each other on 21q22.1.

The coevolution of ligand-receptor pairs by parallel pathways of gene duplication and functional divergence has probably been facilitated in two quite distinct ways. Firstly, the close linkage of multiple genes encoding either ligands or their receptors will have served to promote gene duplication. Secondly, whole genome duplications early on in vertebrate evolution (Chapter 2, section 2.1) may have, by simultaneously increasing both ligand and ligand receptor diversity, provided the raw material for selection to recruit novel receptor-ligand interactions thereby potentiating the dramatic increase in the biochemical and physiological complexity characteristic of the vertebrates.

References

Acampora D., D'Esposito M., Faiella A., Pannese M., Migliaccio E., Morelli F., Stornaiuolo A., Nigro V., Simeone A., Boncinelli E. (1989) The human HOX gene family. *Nucleic Acids Res.* **17**: 10385–10402.

Ahringer J. (1997) Turn to the worm! *Curr. Op. Genet. Devel.* **7**: 410–415.

Ahuja S.K., Özçelik T., Milatovitch A, Francke U, Murphy PM. (1992) Molecular evolution of the human interleukin-8 receptor gene cluster. *Nature Genet.* **2**: 31–36.

Albig W., Doenecke D. (1997) The human histone gene cluster at the D6S105 locus. *Hum. Genet.* **101**: 284–294.

Albig W., Drabent B., Kunz J., Kalff-Suske M., Grzeschik K.H., Doenecke D. (1993) All known human H1 histone genes except the H1° gene are clustered on chromosome 6. *Genomics* **16**: 649–654.

Albig W., Bramlage B., Gruber K., Klobeck H.G., Kunz J., Doenecke D. (1995) The human replacement histone gene variant H3.3B (*H3F3B*). *Genomics* **30**: 264–272.

Albig W., Ebentheuer J., Klobeck G., Kunz J., Doenecke D. (1996) A solitary human H3 histone gene on chromosome 1. *Hum. Genet.* **97**: 486–491.

Albig W., Kioschis P., Poustka A., Meergans K., Doenecke D. (1997a) Human histone gene organization: nonregular arrangement within a large cluster. *Genomics* **40**: 314–322.

Albig W., Meergans K., Doenecke D. (1997b) Characterization of the H1.5 gene completes the set of human H1 subtype genes. *Gene* **184**: 141–148.

Amero S.A., Kretsinger R.H., Moncrief N.D., Yamamoto KR., Pearson W.R. (1992) The origin of nuclear receptor proteins: a single precursor distinct from other transcription factors. *Molec. Endocrinol.* **6**: 3–7.

Amores A., Force A., Yan Y.-L., Joly L., Amemiya C., Fritz A., Ho R.K., Langeland J., Prince V., Wang Y.-L., Westerfield M., Ekker M., Postlethwait J.H. (1998) Zebrafish *hox* clusters and vertebrate genome evolution. *Science* **282**: 1711–1714.

Andersson E., Matsunaga T. (1995) Evolution of immunoglobulin heavy chain variable region genes: a V$_H$ family can last for 150–200 million years or longer. *Immunogenetics* **41**: 18–28.

Antunes S.G., De Groot N.G., Brok H., Doxiadis G., Menezes A.A.L., Otting N., Bontrop R.E. (1998) The common marmoset: a new world primate species with limited Mhc class II variability. *Proc. Natl. Acad. Sci. USA* **95**: 11 745–11 750.

Aoyama Y., Horiuchi T., Gotoh O., Noshiro M., Yoshida Y. (1998) *CYP51*-like gene of *Mycobacterium tuberculosis* actually encodes a P450 similar to eukaryotic CYP51. *J. Biochem.* **124**: 694–696.

Aparicio S., Hawker K., Cottage A., Mikawa Y., Zuo L., Venkatesh B., Chen E., Krumlauf R., Brenner S. (1997) Organization of the *Fugu rubripes Hox* clusters: evidence for continuing evolution of vertebrate *Hox* complexes. *Nature Genet.* **16**: 79–83.

Arnheim N., Krystal M., Schmickel R., Wilson G., Ryder O., Zimmer E. (1980) Molecular evidence for genetic exchanges among ribosomal genes on non-homologous chromosomes in man and apes. *Proc. Natl. Acad. Sci. USA* **77**: 7323–7327.

Atchley W.R., Fitch W.M. (1995) Myc and Max: molecular evolution of a family of proto-oncogene products and their dimerization partner. *Proc. Natl. Acad. Sci. USA* **92**: 10 217–10 221.

Au H.C., Scheffler I.E. (1994) Characterization of the gene encoding the iron-sulfur protein subunit of succinate dehydrogenase from *Drosophila melanogaster*. *Gene* **149**: 261–265.

Ayala F.J., Escalante A., O'hUigen C., Klein J. (1994) Molecular genetics of speciation and human origins. *Proc. Natl. Acad. Sci. USA* **91**: 6787–6794.

Ayala F.J., Rzhetsky A., Ayala F.J. (1998) Origin of the metazoan phyla: molecular clocks confirm paleontological estimates. *Proc. Natl. Acad. Sci. USA* **95**: 606–611.

Babin P.J., Thisse C., Durliat M., Andre M., Akimenko M.-A., Thisse B. (1997) Both apolipoprotein E and A-I genes are present in a nonmammalian vertebrate and are highly expressed during embryonic development. *Proc. Natl. Acad. Sci. USA* **94**: 8622–8627.

Bailey A.D., Stanhope M., Slightom J.L., Goodman M., Shen C.C., Shen C.-K.J. (1992) Tandemly duplicated α globin genes of gibbon. *J. Biol. Chem.* **267**: 18 398–18 406.

Bailey W.J., Kim J., Wagner G.P., Ruddle F.H. (1997) Phylogenetic reconstruction of vertebrate Hox cluster duplications. *Mol. Biol. Evol.* **14**: 843–853.

Baker R.T., Board P.G. (1989) Unequal crossover generates variation in ubiquitin coding unit number at the human UbC polyubiquitin locus. *Am. J. Hum. Genet.* **44**: 534–542.

Baker R.T., Board P.G. (1991) The human ubiquitin 52 amino acid fusion protein gene shares several structural features with mammalian ribosomal protein genes. *Nucleic Acids Res.* **19**: 1035–1040.

Baker R.T., Board P.G. (1992) The human ubiquitin/52-residue ribosomal protein fusion gene subfamily (UbA$_{52}$) is composed primarily of processed pseudogenes. *Genomics* **14**: 520–522.

Balczarek K.A., Lai Z.-C., Kumar S. (1997) Evolution and functional diversification of the paired box (*Pax*) DNA-binding domains. *Mol. Biol. Evol.* **14**: 829–842.

Baldauf S.L., Palmer J.D., Doolittle W.F. (1996) The root of the universal tree and the origin of eukaryotes based on elongation factor phylogeny. *Proc. Natl. Acad. Sci. USA* **93**: 7749–7754.

Bao J.-J., Sifers R.N., Kidd V.J., Ledley F.D., Woo S.L.C. (1987) Molecular evolution of serpins: homologous structure of the human α1-antichymotrypsin and α1-antitrypsin genes. *Biochemistry* **26**: 7755–7759.

Barsh G.S., Seeburg P.H., Gelinas R.E. (1983) The human growth hormone gene family: structure and evolution of the chromosomal locus. *Nucleic Acids Res.* **11**: 3939–3958.

Bateman A. (1997) The structure of a domain common to archaebacteria and the homocystinuria disease protein. *Trends Biochem. Sci.* **22**: 12–13.

Batuski A.J., Kamachi Y., Schick C., Overhauser J., Silverman G.A. (1997) Cytoplasmic antiproteinase 2 (PI8) and bomapin (PI10) map to the serpin cluster at 18q21.3. *Genomics* **43**: 321–328.

Bellefroid E.J., Marine J.C., Matera A.G., Bourguignon C., Desai T., Healy K.C., Bray-Ward P., Martial J.A., Ihle J.N., Ward D.C. (1995) Emergence of the *ZNF91* Kruppel-associated box-containing zinc finger gene family in the last common ancestor of anthropoidea. *Proc. Natl. Acad. Sci. USA* **92**: 10 757–10 761.

Ben-Arie N., McCall A.E., Berkman S., Eichele G., Bellen H.J., Zoghbi H.Y. (1996) Evolutionary conservation of sequence and expression of the bHLH protein atonal suggests a conserved role in neurogenesis. *Hum. Molec. Genet.* **5**: 1207–1216.

Ben-Arie N., Lancet D., Taylor C., Khen M., Walker N., Ledbetter D.H., Carrozzo R., Paten K., Sheer D., Lehrach H., North M.A. (1993) Olfactory receptor gene cluster on human chromosome 17: possible duplication of an ancestral receptor repertoire. *Hum. Molec. Genet.* **3**: 229–235.

Bernstein T.F., Koetzle T.F., Williams G.J.B., Meyer E.F. Jr, Brice M.D., Rodgers J.R., Kennard O., Shimanouchi T., Tasumi M. (1977) The Protein Data Bank: a computer-based archival file for macromolecular structures. *J. Mol. Biol.* **112**: 535–542.

Bharathan G., Janssen B.-J., Kellogg E.A., Sinha N. (1997) Did homeodomain proteins duplicate before the origin of angiosperms, fungi and metazoa? *Proc. Natl. Acad. Sci. USA* **94**: 13 749–13 753.

Blumenberg M. (1988) Concerted gene duplications in the two keratin gene families. *J. Mol. Evol.* **27**: 203–211.

Blumenfeld O.O., Huang C.-H., Xie S.S., Blancher A. (1997) Molecular biology of glycophorins of human and nonhuman primates. In *Molecular Biology and Evolution of Blood Group and MHC Antigens in Primates.* (eds A Blancher, J Klein, WW Socha). Springer-Verlag, Berlin. pp 113–146.

Bockaert J., Pin J.P. (1999) Molecular tinkering of G protein-coupled receptors: an evolutionary success. *EMBO J.* **18**: 1723–1729.

Bode W., Huber R. (1992) Natural protein proteinase inhibitors and their interaction with proteinases. *Eur. J. Biochem.* **204**: 433–451.

Bode W., Turk D., Karshikov A. (1992) The refined 1.9Å X-ray crystal structure of D-Phe-Pro-Arg chloromethylketone inhibited human α-thrombin: structure analysis, overall structure, electrostatic properties, detailed active-site geometry, and structure-function relationships. *Protein Sci.* **1**: 426–471.

Boguski M.S., Berkenmeier E.H., Elshourbagy N.A., Taylor J.M., Gordon J.I. (1986) Evolution of the apolipoproteins. Structure of the rat apoA-IV gene and its relationship to the human genes for apoA-I, C-III and E. *J. Biol. Chem.* **261**: 6398–6407.

Boorstein W.R., Ziegelhoffer T., Craig E.A. (1994) Molecular evolution of the HSP70 multigene family. *J. Mol. Evol.* **38**: 1–17.

Borden P., Jaenichen R., Zachau H.G. (1990) Structural features of transposed human Vk genes and implications for the mechanism of their transpositions. *Nucleic Acids Res.* **18**: 2101–2107.

Brachmann C.B., Sherman J.M., Devine S.E., Cameron E.E., Pilus L., Boeke J.D. (1995) The SIR2 gene families, conserved from bacteria to humans, functions in silencing, cell cycle progression, and chromosome stability. *Genes Devel.* **9**: 2888–2902.

Brakenhoff R.H., Aaarts H.J.M., Reek F.H., Lubsen N.H., Schoenmakers J.G.G. (1990) Human γ-crystallin genes; a gene family on its way to extinction. *J. Mol. Biol.* **216**: 519–532.

Brendel V., Brocchieri L., Sandler S.J., Clark A.J., Karlin S. (1997) Evolutionary comparisons of RecA-like proteins across all major kingdoms of living organisms. *J. Mol. Evol.* **44**: 528–541.

Brenner S. (1988) The molecular evolution of genes and proteins: a tale of two serines. *Nature* **334**: 528–530.

Breslow J.L. (1985) Human apolipoprotein molecular biology and genetic variation. *Ann. Rev. Biochem.* **54**: 699–727.

Brooke N.M., Garcia-Fernandez J., Holland P.W.H. (1998) The ParaHox gene cluster is an evolutionary sister of the Hox gene cluster. *Nature* **392**: 920–922.

Brown W.M., Dziegielewska K.M. (1997) Friends and relations of the cystatin superfamily – new members and their evolution. *Protein Sci.* **6**: 5–12.

Brush D., Dodgson J.B., Choi O.R., Stevens P.W., Engel J.D. (1985) Replacement variant genes contain intervening sequences. *Mol. Cell. Biol.* **5**: 1307–1317.

Buckland R.A., Maule J.C., Sealey P.G. (1996) A cluster of transfer RNA genes (TRM1, TRR3 and TRAN) on the short arm of human chromosome 6. *Genomics* **35**: 164–171.

Buettner J.A., Glusman G., Ben-Arie N., Ramos P., Lancet D., Evans G.A. (1998) Organization and evolution of olfactory receptor genes on human chromosome 11. *Genomics* **53**: 56–68.

Burke A.C., Nelson C.E., Morgan B.A., Tabin C. (1995) *Hox* genes and the evolution of vertebrate axial morphology. *Development* **121**: 333–346.

Butticé G., Kaytes P., D'Armiento J., Vogeli G., Kurkinen M. (1990) Evolution of collagen IV genes from a 54-base exon: a role for introns in gene evolution. *J. Mol. Evol.* **30**: 479–488.

Cadavid L.F., Shufflebotham C., Ruiz F.J., Yeager M., Hughes A.L., Watkins D.I. (1997) Evolutionary instability of the major histocompatibility complex class I loci in New World primates. *Proc. Natl. Acad. Sci. USA* **94**: 14 536–14 541.

Campbell H.D., Fountain S., Young I.G., Claudianos C., Hoheisel J.D., Chen K.-S., Lupski J.R. (1997) Genomic structure, evolution, and expression of human FLII, a gelsolin and leucine-rich-repeat family member: overlap with LLGL. *Genomics* **42**: 46–54.

Cardenas M.L., Cornish-Bowden A., Ureta T. (1998) Evolution and regulatory role of the hexokinases. *Biochim. Biophys. Acta* **1401**: 242–264.

Carroll S.B. (1995) Homeotic genes and the evolution of arthropods and chordates. *Nature* **376**: 479–485.

Caspers G.-J., Leunissen J.A.M., de Jong W.W. (1995) The expanding small heat-shock protein family, and structure predictions of the conserved 'α-crystallin domain'. *J. Mol. Evol.* **40**: 238–248.

Castresana J., Lubben M., Saraste M., Higgins D.G. (1994) Evolution of cytochrome oxidase, an enzyme older than atmospheric oxygen. *EMBO J.* **13**: 2516–2525.

Castro M.J., Morales P., Fernandez-Soria V., Suarez B., Recio M.J., Alvarez M., Martin-Villa M., Arnaiz-Villena A. (1996) Allelic diversity at the primate Mhc-G locus: exon 3 bears stop codons in all Cercopithecinae sequences. *Immunogenetics* **43**: 327–336.

Chan S.J., Cao Q.P., Steiner D.F. (1990) Evolution of the insulin superfamily: cloning of a hybrid insulin/insulin-like growth factor cDNA from *Amphioxus*. *Proc. Natl. Acad. Sci. USA* **87**: 9319–9323.

Chang Y.N., Pirtle I.L., Pirtle R.M. (1986) Nucleotide sequence and transcription of a human tRNA gene cluster with four genes. *Gene* **48**: 165–174.

Charmley P., Keretan E., Snyder K., Clark E.A., Concannon P. (1995) Relative size and evolution of the germline repertoire of T-cell receptor β-chain gene segments in nonhuman primates. *Genomics* **25**: 150–156.

Chen E.Y., Liao Y.-C., Smith D.H., Barrera-Saldana H.A., Gelinas R.E., Seeburg P.H. (1989) The human growth hormone locus: nucleotide sequence, biology and evolution. *Genomics* **4**: 479–497.

Chess A., Simon I., Cedar H., Axel R. (1994) Allelic inactivation regulates olfactory receptor gene expression. *Cell* **78**: 823–834.

Chiu C.H., Schneider H., Schneider M.P.C., Sampaio I., Meireles C.M.M., Slightom J.L., Gumucio D.L., Goodman M. (1996) Reduction of two functional γ-globin genes to one: an evolutionary trend in New World monkeys (Infraorder *Platyrrhini*). *Proc. Natl. Acad. Sci. USA* **93**: 6510–6515.

Chiu C.H., Schneider H., Slightom J.L., Gumucio D.L., Goodman M. (1997) Dynamics of regulatory evolution in primate β-globin gene clusters: *cis*-mediated acquisition of simian γ fetal expression patterns. *Gene* **205**: 47–57.

Christiano A.M., Hoffman G.G., Chung-Honet L.C., Lee S., Cheng W., Uitto J., Greenspan D.S. (1994) Structural organization of the human type VII collagen gene (*COL7A1*), composed of more exons than any previously characterized gene. *Genomics* **21**: 169–179.

Chu M.-L., de Wet W., Bernard M., Ding J.-F., Morabito M., Myers J., Williams C., Ramirez F. (1984) Human proα1(I) collagen gene structure reveals evolutionary conservation of a pattern of introns and exons. *Nature* **310**: 337–340.

Cook G.P., Tomlinson I.M. (1995) The human immunoglobulin V$_H$ repertoire. *Immunology Today* **16**: 237–242.

Costache M., Apoil P.A., Cailleau A., Elmgren A., Larson G., Henry S., Blancher A., Iordachescu D., Oriol R., Mollicone R. (1997) Evolution of the fucosyltransferase genes in vertebrates. *J. Biol. Chem.* **272**: 29 721–29 728.

Coulier F., Pontarotti P., Roubin R., Hartung H., Goldfarb M., Birnbaum D. (1997) Of worms and men: an evolutionary perspective on the fibroblast growth factor (FGF) and FGF receptor families. *J. Mol. Evol.* **44**: 43–56.

Crowe M.L., Perry B.N., Connerton I.F. (1996) Olfactory receptor-encoding genes and pseudogenes are expressed in humans. *Gene* **169**: 247–249.

Cusack S., Härtlein M., Leberman R. (1991) Sequence, structural and evolutionary relationships between class 2 aminoacyl-tRNA synthetases. *Nucleic Acids Res.* **19**: 3489–3498.

Czelusniak J., Goodman M., Hewett-Emmett D., Weiss M.L., Venta P.J., Tashian R.E. (1982) Phylogenetic origins and adaptive evolution of avian and mammalian haemoglobin genes. *Nature* **298**: 297–300.

Czerny T., Bouchard M., Kozmik Z., Busslinger M. (1997) The characterization of novel Pax genes of the sea urchin and *Drosophila* reveal an ancient evolutionary origin of the Pax2/5/8 subfamily. *Mech. Devel.* **67**: 179–192.

Dayhoff M.O., Barker W.C., Hunt L.T. (1978) Protein superfamilies. In *Atlas of Protein Sequence and Structure* Vol. 5, Suppl. 3. (ed. MO Dayhoff). National Biomedical Research Foundation, Washington DC.

de Jong W.W., Leunissen J.A.M., Voorter C.E.M. (1993) Evolution of the α-crystallin/ small heat shock protein family. *Mol. Biol. Evol.* **10**: 103–126.

Degnan B.M., Degnan S.M., Naganuma T., Morse D.E. (1993) The *ets* multigene family is conserved throughout the metazoa. *Nucleic Acids Res.* **21**: 3479–3484.

Degtyarenko K.N., Archakov A.I. (1993) Molecular evolution of P450 superfamily and P450-containing monooxygenase systems. *FEBS Letts* **332**: 1–8.

Della N.G., Hu Y., Holloway A.J., Wang D., Bowtell D.D.L. (1995) A combined genetic and biochemical approach to mammalian signal transduction. *Aust. NZ J. Med.* **25**: 845–851.

den Dunnen J.T., van Neck J.W., Cremers F.P.M., Lubsen N.H, Schoenmakers J.G.G. (1989) Nucleotide sequence of the rat γ-crystallin gene region and comparison with an orthologous human region. *Gene* **78**: 201–213.

Desseyn J.-L., Aubert J.-P., Laine A. (1997a) Genomic organization of the 3′ region of the human mucin gene *MUC5B*. *J. Biol. Chem.* **272**: 16 873–16 883.

Desseyn J.-L., Guyonnet-Dupérat V., Porchet N., Aubert J.-P., Laine A. (1997b) Human mucin gene *MUC5B*, the 10.7 kb large central exon encodes various alternate subdomains resulting in a super-repeat: structural evidence for a 11p15.5 gene family. *J. Biol. Chem.* **272**: 3168–3178.

Desseyn J.-L., Buisine M.-P., Porchet N., Aubert J.-P., Laine A. (1998) Evolutionary history of the 11p15 human mucin gene family. *J. Mol. Evol.* **46**: 102–106.

Diaz M.O. (1995) The human type I interferon gene cluster. *Semin. Virol.* **6**: 143–149.

Diaz M.O., Pomykala H.M., Bohlander S.K., Maltepe E., Malik E., Brownstein B., Olopade O.I. (1994) Structure of the human type-I interferon cluster determined from a YAC clone contig. *Genomics* **22**: 540–552.

Dodds A.W., Petry F. (1993) The phylogeny and evolution of the first component of complement, C1. *Behring Inst. Mitt.* **93**: 87–102.

Doenecke D., Albig W., Bouterfa H., Drabent B. (1994) Organization and expression of H1 histone and H1 replacement histone genes. *J. Cell. Biochem.* **54**: 423–431.

Doolittle R.F. (1992) Reconstructing history with amino acid sequences. *Protein Sci.* **1**: 191–200.

Doolittle R.F. (1994) Convergent evolution: the need to be explicit. *Trends Biochem. Sci.* **19**: 15–18.

Doolittle R.F. (1995) The origins and evolution of eukaryotic proteins. *Phil. Trans. R. Soc. Lond. B* **349**: 235–240.

Doolittle R.F., Feng D.-F., Tsang S., Cho G., Little E. (1996) Determining divergence times of the major kingdoms of living organisms with a protein clock. *Science* **271**: 470–477.

Drayna D.T., McLean J.W., Wion K.L., Trent J.M., Drabkin H.A., Lawn R.M. (1987) Human apolipoprotein D gene: gene sequence, chromosome localization, and homology to the α2μ-globulin superfamily. *DNA* **6**: 199–204.

Duester G., Smith M., Bilanchone V., Hatfield G.W. (1986) Molecular analysis of the human class I alcohol dehydrogenase gene family and nucleotide sequence of the gene encoding the β subunit. *J. Biol. Chem.* **261**: 2027–2033.

Efstratiadis A., Posakony J.W., Maniatis T., Lawn R.M., O'Connor C., Spritz R.A., DeRiel J.K., Forget B.G., Weissman S.M., Slightom J.L., Blechl A.E., Smithies O., Baralle F.E., Shoulders

C.C., Proudfoot N.J. (1980) The structure and evolution of the human β-globin gene family. *Cell* **21**: 653–668.

Eisenberg S.P., Brewer M.T., Verderber E., Heimdal P., Brandhuber B.J., Thompson R.C. (1991) Interleukin 1 receptor antagonist is a member of the interleukin 1 gene family: evolution of a cytokine control mechanism. *Proc. Natl. Acad. Sci. USA* **88**: 5232–5236.

El-Baradi T., Pieler T. (1991) Zinc finger proteins: what we know and what we would like to know. *Mech. Devel.* **35**: 155–169.

Engel J.N., Gunning P.W., Kedes L.H. (1981) Isolation and characterization of human actin genes. *Proc. Natl. Acad. Sci. USA* **78**: 4674–4678.

Erba H.P., Eddy R., Shows T., Kedes L., Gunning P. (1988) Structure, chromosome location, and expression of the human γ-actin gene: differential evolution, location, and expression of the cytoskeletal β- and γ-actin genes. *Mol. Cell. Biol.* **8**: 1775–1789.

Eriani G., Delarue M., Poch O., Gangloff J., Moras D. (1990) Partition of tRNA synthetases into two classes based on mutually exclusive sets of sequence motifs. *Nature* **347**: *203–206.*

Erickson J.M., Schmickel R.D. (1985) A molecular basis for discrete size variation in human ribosomal *DNA. Am. J. Hum. Genet.* **37**: 311–325.

Erickson B.W., May L.T., Sehgal P.B. (1984) Internal duplication in human α1 and β1 interferons. *Proc. Natl. Acad. Sci. USA* **81**: 7171–7175.

Ermert K., Mitlohner H., Schempp W., Zachau H.G. (1995) The immunoglobulin κ locus of primates. *Genomics* **25**: 623–629.

Ernoult-Lange M., Arranz V., Le Coniat M., Berger R., Kress M. (1995) Human and mouse Kruppel-like (MOK2) orthologue genes encode two different zinc finger proteins. *J. Mol. Evol.* **41**: 784–794.

Escriva H., Safi R., Hnni C., Langlois M.C., Saumitou-Laprade P., Stéhelin D., Capron A., Pierce R., Laudet V. (1997) Ligand binding was acquired during evolution of nuclear receptors. *Proc. Natl. Acad. Sci. USA* **94**: 6803–6808.

Exposito J.-V., Garrone R. (1990) Characterization of a fibrillar collagen gene in sponges reveals the early evolutionary appearance of two collagen families. *Proc. Natl. Acad. Sci. USA* **87**: 6669–6673.

Exposito J.-V., Le Guellec D., Lu Q., Garrone R. (1991) Short chain collagens in sponges are encoded by a family of closely related genes. *J. Biol. Chem.* **266**: 21 923–21 928.

Fan W., Kasahara M., Gutknecht J., Klein D., Mayer W.E., Jonker M., Klein J. (1989) Shared class II polymorphisms between human and chimpanzees. *Hum. Immunol.* **26**: 107–112.

Fan W., Liu Y.-C., Parimoo S., Weissman S.M. (1995) Olfactory receptor-like genes are located in the human major histocompatibility complex. *Genomics* **27**: 119–123.

Farmer A.A., Loftus T.M., Mills A.A., Sata M.Y., Neill J.D., Tron T., Yang M., Trumpower B.L., Stanbridge E.J. (1994) Extreme evolutionary conservation of QM, a novel c-*jun* associated transcription factor. *Hum. Molec. Genet.* **3**: 723–728.

Farries T.C., Atkinson J.P. (1991) Evolution of the complement system. *Immunology Today* **12**: 295–300.

Finnerty J.R., Martindale M.Q. (1998) The evolution of the Hox cluster: insights from outgroups. *Curr. Opin. Genet. Devel.* **8**: 681–687.

Fishel R., Lescoe M.K., Rao M.R.S., Copeland N.G., Jenkins N.A., Garber J., Kane M., Kolodner R. (1993) The human mutator gene homolog *MSH2* and its association with hereditary nonpolyposis colon cancer. *Cell* **75**: 1027–1038.

Fitch D.H., Bailey W.J., Tagle D.A., Goodman M., Sien L., Slightom J.L. (1991) Duplication of the γ-globin gene mediated by L1 long interspersed repetitive elements in an early ancestor of simian primates. *Proc. Natl. Acad. Sci. USA* **88**: 7396–7400.

Forterre P., Bergerat A., Gadella D., Elie C., Lottspeich F., Confalonieri F., Duguet M., Holmes M., Dyall-Smith M. (1994) Evolution of DNA topoisomerases and DNA polymerases: a perspective from archaea. *System Appl. Microbiol.* **16**: 746–758.

Francklyn C., Musier-Forsyth, Martinis S.A. (1997) Aminoacyl-tRNA synthetases in biology and disease: new evidence for structural and functional deversity in an ancient family of enzymes. *RNA* **3**: 954–960.

Fryxell K.J. (1996) The coevolution of gene family trees. *Trends Genet.* **12**: 364–369.

Fuchs E., Marchuk D. (1983) Type I and type II keratins have evolved from lower eukaryotes to form the epidermal intermediate filaments in mammalian skin. *Proc. Natl. Acad. Sci. USA* **80**: 5857–5861.

Fuchs E.V., Coppock S.M., Green H., Cleveland D.W. (1981) Two distinct classes of keratin genes and their evolutionary significance. *Cell* **27**: 75–84.

Fukami-Kobayashi K., Tomoda S., Go M. (1993) Evolutionary clustering and functional similarity of RNA-binding proteins. *FEBS Letts.* **335**: 289–293.

Garcia-Fernàndez J., Holland P.W.H. (1994) Archetypal organization of the amphioxus *Hox* gene cluster. *Nature* **370**: 563–566.

Garrone R. (1998) Evolution of metazoan collagens. In *Molecular Evolution: Towards the Origin of Metazoa.* (ed., WEG Muller) Springer-Verlag, Berlin. pp 119–139.

Gavin K.A., Hidaka M., Stillman B. (1995) Conserved initiator proteins in eukaryotes. *Science* **270**: 1667–1670.

Gellon G., McGinnis W. (1998) Shaping animal body plans in development and evolution by modulation of *Hox* expression patterns. *BioEssays* **20**: 116–125.

Gibbons I.R. (1995) Dynein family of motor proteins: present status and future questions. *Cell Motil. Cytoskel.* **32**: 136–144.

Gillespie D., Carter W. (1983) Concerted evolution of human interferon alpha genes. *J. Interferon Res.* **3**: 83–88.

Giordano M., Marchetti C., Chiorboli E., Bona G., Richiardi P.M. (1997) Evidence for gene conversion in the generation of extensive polymorphism in the promoter of the growth hormone gene. *Hum. Genet.* **100**: 249–255.

Glusman G., Clifton S., Roe B., Lancet D. (1996) Sequence analysis in the olfactory receptor gene cluster on human chromosome 17: recombinatorial events affecting receptor diversity. *Genomics* **37**: 147–160.

Goffeau A., Barrell B.G., Bussey H., Davis R.W., Dujon B., Feldmann H., Galibert F., Hoheisel J.D., Jacq C., Johnston M., Louis E.J., Mewes H.W., Murakami Y., Philippsen P., Tettelin H., Oliver S.G. (1996) Life with 6000 genes. *Science* **274**: 546–567.

Golding G.B., Glickman B.W. (1985) Sequence-directed mutagenesis: evidence from a phylogenetic history of human α-interferon genes. *Proc. Natl. Acad. Sci. USA* **82**: 8577–8581.

Golos T.G., Durning M., Fisher J.M., Fowler P.D. (1993) Cloning of four GH/chorionic somatomammotropin-related cDNAs differentially expressed during pregnancy in the rhesus monkey placenta. *Endocrinology* **133**: 1744–1752.

Gonzalez I.L., Sylvester J.E., Schmickel R.D. (1988) Human 28S ribosomal RNA sequence heterogeneity. *Nucleic Acids Res.* **16**: 10213–10224.

Gonzalez I.L., Sylvester J.E., Smith T.F., Stambolian D., Schmickel R.D. (1990) Ribosomal RNA gene sequences and hominoid phylogeny. *Mol. Biol. Evol.* **7**: 203–219.

Gonzalez I.L., Wu S., Li W.-M., Kuo B.A., Sylvester J.E. (1992) Human ribosomal RNA intergenic spacer sequence. *Nucleic Acids Res.* **20**: 5846.

Gonzalez P., Rao P.V., Nuñez S.B., Zigler J.S. (1995) Evidence for independent recruitment of ζ-crystallin/quinone reductase (*CRYZ*) as a crystallin in camelids and hystricomorph rodents. *Mol. Biol. Evol.* **12**: 773–781.

Gorski J.L., Gonzalez I.L., Schmickel R.D. (1987) The secondary structure of human 28S rRNA: the structure and evolution of a mosaic rRNA gene. *J. Mol. Evol.* **24**: 236–251.

Graumann P., Marahiel M.A. (1996) A case of convergent evolution of nucleic acid binding modules. *BioEssays* **18**: 309–315.

Graur D., Li W.-H. (1988) Evolution of protein inhibitors of serine proteinases: positive Darwinian selection or compositional effects? *J. Mol. Evol.* **28**: 131–135.

Green P., Lipman D., Hillier L., Waterson R., States D., Claverie J.-M. (1993) Ancient conserved regions in new gene sequences and the protein databases. *Science* **259**: 1711–1716

Guimera J., Casas C., Pucharcos C., Solans S., Domenech A., Planas A.M., Ashley J., Lovett M., Estivill X., Pritchard M.A. (1996) A human homologue of *Drosophila minibrain* (*MNB*) is expressed in the neuronal regional affected in Down Syndrome and maps to the critical region. *Hum. Molec. Genet.* **5**: 1305–1310.

Gum J.R. (1992) Mucin genes and the proteins they encode: structure, diversity and regulation. *Am. J. Respir. Cell Mol. Biol.* **7**: 557–564.

Gunning P., Ponte P., Kedes L., Eddy R., Shows T. (1984) Chromosomal location of the co-expressed human skeletal and cardiac actin genes. *Proc. Natl. Acad. Sci. USA* **81**: 1813–1817.

Günther E., Walter L. (1994) Genetic aspects of the hsp70 multigene family in vertebrates. *Experientia* **50**: 987–1002.

Gupta R.S. (1995) Phylogenetic analysis of the 90 kD heat shock family of protein sequences and an examination of the relationship among animals, plants, and fungi species. *Mol. Biol. Evol.* **12**: 1063–1073.

Gupta R.S., Golding G.B. (1993) Evolution of HSP70 gene and its implications regarding relationships between archaebacteria, eubacteria and eukaryotes. *J. Mol. Evol.* **37**: 573–582.

Gupta R.S., Singh B. (1994) Phylogenetic analysis of 70 kD heat shock protein sequences suggest a chimeric origin for the eukaryotic cell nucleus. *Curr. Biol.* **4**: 1104–1114.

Gyllensten U.B., Erlich H.A. (1989) Ancient roots for polymorphism at the *HLA-DQα* locus in primates. *Proc. Natl. Acad. Sci. USA* **86**: 9986–9992.

Hacia J.G., Makalowski W., Edgemon K., Erdos M.R., Robbins C.M., Fodor S.P.A., Brody L.C., Collins F.S. (1998) Evolutionary sequence comparisons using high-density oligonucleotide arrays. *Nature Genet.* **18**: 155–158.

Hamann K.J., Ten R.M., Loefering D.A., Jenkins R.B., Heise M.T., Schad C.R., Pease L.R., Gleich G.J., Barker R.L. (1990) Structure and chromosome localization of the human eosinophil derived neurotoxin and eosinophil cationic protein genes: evidence for intronless coding sequences in the ribonuclease gene superfamily. *Genomics* **7**: 535–546.

Hanukoglu I., Fuchs E. (1982) The cDNA sequence of a human epidermal keratin: divergence of sequence but conservation of structure among intermediate filament proteins. *Cell* **31**: 243–252.

Hardies S.C., Edgell M.H., Hutchinson C.A. (1984) Evolution of the mammalian β-globin cluster. *J. Biol. Chem.* **259**: 3748–3756.

Hardison R. (1998) Hemoglobins from bacteria to man: evolution of different patterns of gene expression. *J. Exp. Biol.* **201**: 1099–1117.

Hashimoto T., Hasegawa M. (1996) Origin and early evolution of eukaryotes inferred from the amino acid sequences of translation elongation factors 1α/Tu and 2/G. *Adv. Biophys.* **32**: 73–120.

Hastings K.E.M. (1997) Molecular evolution of the vertebrate troponin I gene family. *Cell Struct. Func.* **22**: 205–211.

Hatlen L., Attardi G. (1971) Proportion of HeLa cell genome complementary to transfer RNA and 5S RNA. *J. Mol. Biol.* **56**: 535–553.

Hattori J., Baum B.R., Miki B.L. (1995) Ancient diversity of the glucose-6-phosphate isomerase genes. *Biochem. Systemat. Ecol.* **23**: 33–38.

He F.-C., Wu C.-T. (1993) Molecular evolution of cytokines and their receptors. *Exp. Hematol.* **21**: 521–524.

Heintz N., Zernik M., Roeder RG. (1981) The structure of the human histone genes: clustered but not tandemly repeated. *Cell* **24**: 661–668.

Henco K., Brosius J., Fujisawa A., Fujisawa J.-I., Haynes J.R., Hochstadt J., Kovacic T., Pasek M., Schambōck A., Schmid J., Todokoro K., Wlchli M., Nagata S., Weissmann C. (1985) Structural relationship of human alpha genes and pseudogenes. *J. Mol. Biol.* **185**: 227–260.

Henikoff S., Greene E.A., Pietrokovski S., Bork P., Attwood T.K., Hood L. (1997) Gene families: the taxonomy of protein paralogs and chimeras. *Science* **278**: 609–614.

Hennessey E.S., Drummond D.R., Sparrow J.C. (1993) Molecular genetics of actin function. *Biochem. J.* **282**: 657–671.

Hess J.F., Schmid C.W., Shen C.-K.J. (1984) A gradient of sequence divergence in the human adult α-globin duplication units. *Science* **226**: 67–70.

Hide W.A., Chan L., Li W.-H. (1992) Structure and evolution of the lipase superfamily. *J. Lipid Res.* **33**: 167–178.

Higgs D.R., Vickers M.A., Wilkie A.O.M., Pretorius I.-M., Jarman A.P., Weatherall D.J. (1989) A review of the molecular genetics of the human α-globin gene cluster. *Blood* **73**: 1081–1104.

Hill R.L., Delaney R., Fellows R.E., Lebowitz H.E. (1996) The evolutionary origins of the immunoglobulins. *Proc. Natl. Acad. Sci. USA* **56**: 1762–1769.

Hirt K., Kimelman J., Birnbaum M.J., Chen E.Y., Seeburg P.H., Eberhardt N.L., Barta A. (1987) The human growth hormone gene locus: structure, evolution and allelic variations. *DNA* **6**: 59–70.

Hockett R.D., de Villartay J.-P., Pollock K., Poplack D.G., Cohen D.I., Korsmeyer S.J. (1988) Human T-cell antigen receptor (TCR) delta-chain locus and elements responsible for its deletion are within the TCR alpha-chain locus. *Proc. Natl. Acad. Sci. USA* **85**: 9694–9698.

Hodin J., Wistow G. (1993) 5'-RACE PCR of mRNA for three taxon-specific crystallins: for each gene one promoter controls both lens and non-lens expression. *Biochem. Biophys. Res. Commun.* **190**: 391–396.

Hoffman S.M., Fernandez-Salguero P., Gonzalez F.J., Mohrenweiser H.W. (1995) Organization and evolution of the cytochrome P450 CYP2A-2B-2F subfamily gene cluster on human chromosome 19. *J. Mol. Evol.* **41**: 894–900.

Holland P.W.H., Garcia-Fernàndez J. (1996) *Hox* genes and chordate evolution. *Dev. Biol.* **173**: 382–395.

Hoovers J.M.N., Mannens M., John R., Bliek J., van Heyningen V., Porteous D.J., Leschot N.J., Westerveld A., Little P.F.R. (1992) High-resolution localization of 69 potential human zinc finger protein genes: a number are clustered. *Genomics* **12**: 254–263.

Hsu S.-L., Marks J., Shaw J.-P., Tam M., Higgs D.R., Shen C.C., Shen C.-K.J. (1988) Structure and expression of the human θ1-globin gene. *Nature* **331**: 94–96.

Huang R., Hellman L. (1994) Genes for mast-cell serine protease and their molecular evolution. *Immunogenetics* **40**: 397–414.

Huber C., Thiebe R., Zachau H.G. (1994) A potentially functional V gene at a distance of 1.5Mb from the immunoglobulin κ locus. *Genomics* **22**: 213–215.

Hughes A.L. (1991) Independent gene duplications, not concerted evolution, explain relationships among class I MHC genes of murine rodents. *Immunogenetics* **33**: 367–373.

Hughes A.L. (1992) Coevolution of the vertebrate integrin α- and β-chain genes. *Mol. Biol. Evol.* **9**: 216–234.

Hughes A.L. (1994) Phylogeny of the C3/C4/C5 complement-component gene family indicates that C5 diverged first. *Mol. Biol. Evol.* **11**: 417–425.

Hughes A.L. (1996) Evolution of the HLA complex. In *Human Genome Evolution.* (eds. M. Jackson, T. Strachan, G. Dover). BIOS Scientific Publishers, Oxford, pp.73–92.

Hughes A.L, Nei M. (1988) Pattern of nucleotide substitution at major histocompatibility complex class I loci reveals overdominant selection. *Nature* **335**: 167–170.

Hughes A.L., Nei M. (1989a) Nucleotide substitution at major histocompatibility complex class II loci: evidence for overdominant selection. *Proc. Natl. Acad. Sci. USA* **86**: 958–962.

Hughes A.L., Nei M. (1989b) Evolution of the major histocompatibility complex: independent origin of non-classical class I genes in different groups of mammals. *Mol. Biol. Evol.* **6**: 559–579.

Hughes A.L., Nei M. (1990) Evolutionary relationships of class II major histocompatibility complex genes in mammals. *Mol. Biol. Evol.* **7**: 491–514.

Hughes A.L., Nei M. (1993) Evolutionary relationships of the classes of major histocompatibility complex genes. *Immunogenetics* **37**: 337–346.

Hughes A.L., Yeager M. (1997) Molecular evolution of the vertebrate immune system. *BioEssays* **19**: 777–786.

Ikuta T., Szeto S., Yoshida A. (1986) Three human alcohol dehydrogenase subunits: cDNA structure and molecular and evolutionary divergence. *Proc. Natl. Acad. Sci. USA* **83**: 634–638.

Irwin D.M. (1988) Evolution of an active-site codon in serine proteases. *Nature* **336**: 429–430.

Issel-Tarver L., Rine J. (1997) The evolution of mammalian olfactory receptor genes. *Genetics* **145**: 185–195.

Ivanova V.S., Zimonjic D., Popescu N., Bonner W.M. (1994) Chromosomal localization of the human H2A.X gene to 11q23.2-q23.3 by fluorescence *in situ* hybridization. *Hum. Genet.* **94**: 303–306.

Iwabe N., Kuma K.-I., Kishino H., Hasegawa M., Miyata T. (1991) Evolution of RNA polymerases and branching patterns of the three major groups of archaebacteria. *J. Mol. Evol.* **32**: 70–78.

Iwabe N., Kuma K., Miyata T. (1996) Evolution of gene families and relationship with organismal evolution: rapid divergence of tissue-specific genes in the early evolution of chordates. *Mol. Biol. Evol.* **13**: 483–493.

Jacquemin P., Oury C., Peers B., Morin A., Belayew A., Martial J.A. (1994) Characterization of a single strong tissue-specific enhancer downstream from the three human genes encoding placental lactogen. *Mol. Cell. Biol.* **14**: 93–103.

Johnson D.E., Williams L.T. (1993) Structural and functional diversity in the FGF receptor multigene family. *Adv. Cancer Res.* **60**: 1–41.

Johnson R.M., Buck S., Chiu C.-H., Schneider H., Sampaio I., Gage D.A., Shen T.-L., Schneider M.P.C., Muniz J.A., Gumucio D.L., Goodman M. (1996) Fetal globin expression in New World monkeys. *J. Biol. Chem.* **271**: 14684–14691.

Kasahara M., Nakaya J., Satta Y., Takahata N. (1997) Chromosomal duplication and the emergence of the adaptive immune system. *Trends Genet.* **13**: 90–92.

Kawamura S., Ueda S. (1992) Immunoglobulin C_H gene family in hominoids and its evolutionary history. *Genomics* **13**: 194–200.

Kawasaki K., Minoshima S., Nakato E., Shibuya K., Shintani A., Schmeits J.L., Wang J., Shimizu N. (1997) One-megabase sequence analysis of the human immunoglobulin λ gene locus. *Genome Res.* **7**: 250–261.

Kawata T., Shevchenko A., Fukuzawa M., Jermyn K.A., Totty N.F., Zhukovskaya N.V., Sterling A.E., Mann M., Williams J.G. (1997) SH2 signaling in a lower eukaryote: a STAT protein that regulates stalk cell differentiation in *Dictyostelium*. *Cell* **89**: 909–916.

Kearsey S.E., Labib K. (1998) MCM proteins: evolution, properties, and role in DNA replication. *Biochim. Biophys. Acta* **1398**: 113–136.

Kedes L., Ng S.-Y., Lin C.-S., Gunning P., Eddy R., Shows T., Leavitt J. (1985) The human beta-actin multigene family. *Trans. Assoc. Am. Phys.* **98**: 42–46.

Khaleduzzaman M., Sumiyoshi H., Ueki Y., Inoguchi K., Ninomiya Y., Yoshioka H. (1997) Structure of the human type XIX collagen (*COL19A1*) gene, which suggests it has arisen from an ancestor gene of the FACIT family. *Genomics* **45**: 304–312.

Kimura A., Takagi Y. (1983) A frameshift addition causes silencing of the δ-globin gene in an Old World monkey, an anubis (*Papio doguera*). *Nucleic Acids Res.* **11**: 2541–2549.

Kirchgessner T.G., Chuat J.-C., Heinzmamm C.A., Etienne J., Guilhot S., Svenson K., Ameis D., Pilon C., D'Auriol L., Andalibi A., Schotz MC., Galibert F., Lusis A.J. (1989) Organization of the human lipoprotein lipase gene and evolution of the lipase gene family. *Proc. Natl. Acad. Sci. USA* **86**: 9647–9651.

Kitao S., Ohsugi I., Ichikawa K., Goto M., Furuichi Y., Shimamoto A. (1998) Cloning of two new human helicase genes of the RecQ family: biological significance of multiple species in higher eukaryotes. *Genomics* **54**: 443–452.

Klein J., O'hUigin C. (1993) Composite origin of major histocompatibility complex genes. *Curr. Opin. Genet. Devel.* **3**: 923–930.

Klein J., Sato A. (1998) Birth of the major histocompatibility complex. *Scand. J. Immunol.* **47**: 199–209.

Klein J., Stao A., O'hUigin C. (1998) Evolution by gene duplication in the major histocompatibility complex. *Cytogenet. Cell Genet.* **80**: 123–127.

Klein J., Satta Y., O'hUigin C., Takahata N. (1993) The molecular descent of the major histocompatibility complex. *Annu. Rev. Immunol.* **11**: 269–295.

Klenk H.-P., Palm P., Lottspeich F., Zillig W. (1992) Component H of the DNA-dependent RNA polymerases of Archaea is homologous to a subunit shared by the three eucaryal nuclear RNA polymerases. *Proc. Natl. Acad. Sci. USA* **89**: 407–410.

Klinge E.M., Sylvestre Y.R., Freedberg I.M., Blumenberg M. (1987) Evolution of keratin genes: different protein domains evolve by different pathways. *J. Mol. Evol.* **24**: 319–329.

Klug A., Schwabe J.W.R. (1995) Zinc fingers. *FASEB J.* **9**: 597–604.

Knapp L.A., Cadvid L.F., Watkins D.I. (1998) The *MHC-E* locus is the most well conserved of all known primate class I histocompatibilty genes. *J. Immunol.* **160**: 189–196.

Kölble K., Lu J., Mole S.E., Kaluz S., Reid K.B.M. (1993) Assignment of the human pulmonary surfactant protein D gene (*SFTP4*) to 10q22-q23 close to the surfactant protein A gene cluster. *Genomics* **17**: 294–298.

Kolodner R. (1996) Biochemistry and genetics of eukaryotic mismatch repair. *Genes Dev.* **10**: 1433–1442.

Koop B.F., Siemieniak D., Slightom J.L., Goodman M., Dunbar J., Wright P.C., Simons E.L. (1989) Tarsius δ- and β-globin genes: conversions, evolution, and systematic implications. *J. Biol. Chem.* **264**: 68–79.

Kosugi H., Nakagawa Y., Arai K.-I., Yokota T. (1995) Gene structures of the α subunits of human IL-3 and granulocyte-macrophage colony-stimulating factor receptors: comparison with the cytokine receptor superfamily. *J. Allergy Clin. Immunol.* **96**: 1115–1125.

Krieg T.M., Schafer M.P., Cheng C.K., Filpula D., Flaherty P., Steinert P.M., Roop D.R. (1985) Organization of a type 1 keratin gene. Evidence for evolution of intermediate filaments from a common ancestral gene. *J. Biol. Chem.* **260**: 5867–5870.

Kudo S., Fukuda M. (1989) Structural organization of glycophorins A and B genes: glycophorin B gene evolved by homologous recombination at *Alu* repeat sequence. *Proc. Natl. Acad. Sci. USA* **86**: 4619–4623.

Kudo S., Fukuda M. (1994) Contribution of gene conversion to the retention of the sequence for M blood group type determinant in glycophorin E gene. *J. Biol. Chem.* **269**: 22969–22974.

Kulski J.K., Gaudieri S., Bellgard M., Balmer L., Giles K., Inoko H., Dawkins R.L. (1997) The evolution of MHC diversity by segmental duplication and transposition of retroelements. *J. Mol. Evol.* **45**: 599–609.

Kumada Y., Benson D.R., Hillemann D., Hosted T.J., Rochefort D.A., Thompson C.J., Wohlleben W., Tateno Y. (1993) Evolution of the glutamine synthetase gene, one of the oldest existing and functioning genes. *Proc. Natl. Acad. Sci. USA* **90**: 3009–3013.

Kupfermann H., Mayer W.E., O'hUigin C., Klein D., Klein J. (1992) Shared polymorphism between gorilla and human major histocompatibility complex *DRB* loci. *Hum. Immunol.* **34**: 267–272.

Kurth J.H., Mountain J.L., Cavalli-Sforza L.L. (1993) Subclustering of human immunoglobulin kappa light chain variable region genes. *Genomics* **16**: 69–77.

Kyrpides N.C., Woese C.R. (1998) Universally conserved translation initiation factors. *Proc. Natl. Acad. Sci. USA* **95**: 224–228.

Laudet V. (1997) Evolution of the nuclear receptor superfamily: early diversification from an ancestral orphan receptor. *J. Molec. Endocrinol.* **19**: 207–226.

Laudet V., Hnni C., Coll J., Catzeflis F., Stéhelin D. (1992) Evolution of the nuclear receptor gene superfamily. *EMBO J.* **11**: 1003–1013.

Laudet V., Stehelin D., Clevers H. (1993) Ancestry and diversity of the HMG box superfamily. *Nucleic Acids Res.* **21**: 2493–2501.

Laudet V., Hnni C., Stéhelin D., Duterque-Coquillaud M. (1999) Molecular phylogeny of the ETS gene family. *Oncogene* **18**: 1351–1359.

Lawlor D.A., Ward F.E., Ennis P.D., Jackson A.P., Parham P. (1988) HLA-A, B polymorphisms predate the divergence of humans and chimpanzees. *Nature* **335**: 268–271.

Lawlor D.A., Warren E., Taylor P., Parham P. (1991) Gorilla class I major histocompatibility complex alleles: comparison to human and chimpanzee class I. *J. Exp. Med.* **174**: 1491–1509.

Lawn R.M., Schwartz K., Patthy L. (1997) Convergent evolution of apolipoprotein(a) in primates and hedgehog. *Proc. Natl. Acad. Sci. USA* **94**: 11992–11997.

Lawson F.S., Charlebois R.L., Dillon J.A. (1996) Phylogenetic analysis of carbamoylphosphate synthetase genes: complex evolutionary history includes an internal duplication within a gene which can root the tree of life. *Mol. Biol. Evol.* **13**: 970–977.

Le Roith D., Shiloach J., Roth J., Lesniak M.A. (1980) Evolutionary origins of vertebrate hormones: substances similar to mammalian insulins are native to unicellular eukaryotes. *Proc. Natl. Acad. Sci. USA* **77**: 6184–6188.

Lee P., Kuhl W., Gelbart., Kamimura T., West C., Beutler E. (1994) Homology between a human protein and a protein of the green garden pea. *Genomics* **21**: 371–378.

Leelayuwat C., Zhang W.J., Townend D.C., Gaudieri S., Dawkins R.L. (1993) Differences in the central MHC between man and chimpanzee: implications for autoimmunity and acquired immune deficiency development. *Hum. Immunol.* **38**: 30–41.

Leipe D.D., Landsman D. (1997) Histone deacetylases, acetoin utilization proteins and acetylpolyamine amidohydrolases are members of an ancient protein superfamily. *Nucleic Acids Res.* **25**: 3693–3697.

Lesk A.M., Fordham W.D. (1996) Conservation and variability in the structures of serine proteinases of the chymotrypsin family. *J. Mol. Biol.* **258**: 501–537.

Li L., Ernsting B.R., Wishart M.J., Lohse D.L., Dixon J.E. (1997) A family of putative tumor suppressors is structurally and functionally conserved in humans and yeast. *J. Biol. Chem.* **272**: 29403–29406.

Li W.-H. (1997) *Molecular Evolution*. Sinauer Associates, Sunderland.

Li W.-H., Tanimura M., Luo C.-C., Datta S., Chan L. (1988) The apolipoprotein multigene family: biosynthesis, structure, structure-function relationships and evolution. *J. Lipid Res.* **29**: 245–271.

Li X.Y., Mantovani R., Hooft van Huijsduijnen R., Andre I., Benoist C., Mathis D. (1992) Evolutionary variation of the CCAAT-binding transcription factor NF-Y. *Nucleic Acids Res.* **20**: 1087–1091.

Liao D., Weiner A.M. (1995) Concerted evolution of the tandemly repeated genes encoding primate U2 small nuclear RNA (the *RNU2* locus) does not prevent rapid diversification of the $(CT)_n \cdot (GA)_n$ microsatellite embedded within the U2 repeat unit. *Genomics* **30**: 583–593.

Liao D., Paveliz T., Kidd J.R., Kidd K.K., Weiner A.M. (1997) Concerted evolution of the tandemly repeated genes encoding human U2 snRNA (the *RNU2* locus) involves rapid intrachromosomal homogenization and rare interchromosomal gene conversion. *EMBO J.* **16**: 588–598.

Lienert K., Parham P. (1996) Evolution of MHC class I genes in higher primates. *Immunol. Cell Biol.* **74**: 349–356.

Lindgren V., Bernstein L.B., Weiner A.M., Francke U. (1985) Human U1 small nuclear RNA pseudogenes do not map to the site of the U1 genes in 1p36 but are clustered in 1q12-q22. *Molec. Cell. Biol.* **5**: 2172–2180.

Lipsick J.S. (1996) One billion years of Myb. *Oncogene* **13**: 223–235.

Little M., Seehaus T. (1988) Comparative analysis of tubulin sequences. *Comp. Biochem. Physiol.* **90B**: 655–670.

Livneh E., Glazer L., Segal D., Schiessinger J., Shilo B.-Z. (1985) The *Drosophila* EGF receptor gene homolog: conservation of both hormone binding and kinase domains. *Cell* **40**: 599–607.

Lu S.-F., Pan F.-M., Chiou S.-H. (1995) Sequence analysis of frog αB-crystallin cDNA: sequence homology and evolutionary comparison of αA, αB and heat shock proteins. *Biochem. Biophys. Res. Commun.* **216**: 881–891.

Lu S.F., Pan F.M, Chiou S.H. (1996) Sequence analysis of four acidic beta-crystallin subunits of amphibian lenses: phylogenetic comparison between β- and γ-crystallins. *Biochem. Biophys. Res. Commun.* **221**: 219–228.

Lubsen N.H., Aarts H.J.M., Schoenmakers J.G.G. (1988) The evolution of lenticular proteins: the β- and γ-crystallin super family. *Prog. Biophys. Mol. Biol.* **51**: 47–76.

Ludueña R.F. (1998) Multiple forms of tubulin: different gene products and covalent modifications. *Int. Rev. Cytol.* **178**: 207–266.

Lund P.K., Moats-Staats B.M., Simmons J.G., Hoyt E., D'Ercole A.J., Martin F., Van Wyk J.J. (1985) Nucleotide sequence analysis of a cDNA encoding human ubiquitin reveals that ubiquitin is synthesized as a precursor. *J. Biol. Chem.* **260**: 7609–7613.

Luo C.-C., Li W.-H., Moore M.N., Chan L. (1986) Structure and evolution of the apolipoprotein multigene family. *J. Mol. Biol.* **187**: 325–340.

McBride O.W., Pirtle I.L., Pirtle R.M. (1989) Localization of three DNA segments encompassing tRNA genes to human chromosomes 1, 5 and 16: proposed mechanism and significance of tRNA gene dispersion. *Genomics* **5**: 561–573.

McEvoy S.M., Maeda N. (1988) Complex events in the evolution of the haptoglobin gene cluster in primates. *J. Biol. Chem.* **263**: 15740–15747.

McLachlan A.D. (1979) Gene duplications in the structural evolution of chymotrypsin. *J. Mol. Biol.* **128**: 49–79.

McLean P.J., Farb D.H., Russek S.J. (1995) Mapping of the α₄ subunit gene (*GABRA4*) to human chromosome 4 defines an α2-α4-β1-γ1 gene cluster: further evidence that modern GABA$_A$ receptor gene clusters are derived from an ancestral cluster. *Genomics* **26**: 580–586.

McRory J.E., Sherwood N.M. (1997) Ancient divergence of insulin and insulin-like growth factor. *DNA Cell Biol.* **16**: 939–949.

Magenis R.E., Sheehy R., Dugaiczyk A., Gibbs P.E. (1987) Chromosomal localization of the albumin and α-fetoprotein genes in the chimpanzee (*Pan troglodytes*). *Cytogenet. Cell Genet.* **46**: 654–658.

Magenis R.E., Luo X.Y., Ryan S.C., Dugaiczyk A., Oosterhuis J.E. (1989) Chromosomal localization of the albumin and α-fetoprotein genes in the orangutan (*Pongo pymaeus*) and gorilla (*Gorilla gorilla*). *Cytogenet. Cell Genet.* **51**: 1037–1041.

Maleszka R., Hanes S.D., Hackett R.D., de Couet H.G. (1996) The *Drosophila melanogaster* dodo (*dod*) gene, conserved in humans, is functionally interchangeable with the ESS1 cell division gene of *Saccharomyces cerevisiae*. *Proc. Natl. Acad. Sci. USA* **93**: 447–451.

Malik S., Hisatake K., Sumimoto H., Horikoshi M., Roeder R.G. (1991) Sequence of general transcription factor TFIIB and relationships to other initiation factors. *Proc. Natl. Acad. Sci. USA* **88**: 9553–9557.

Malnic B., Hirono J., Sato T., Buck L.B. (1999) Combinatorial receptor codes for odors. *Cell* **96**: 713–723.

Mann R.S. (1997) Why are *Hox* genes clustered? *BioEssays* **19**: 661–664.

Mans R.M.W., Pleij C.W.A., Bosch L. (1991) tRNA-like structures. Structure, function and evolutionary significance. *Eur. J. Biochem.* **201**: 303–324.

Maresco D.L., Chang E., Theil K.S., Francke U., Anderson C.L. (1996) The three genes of the *FCGR1* gene family encoding FcγR1 flank the centromere of chromosome 1 at 1p12 and 1q21. *Cytogenet. Cell Genet.* **73**: 157–163.

Marez D., Legrand M., Sabbagh N., Guidice J.M., Spire C., Lafitte J.J., Meyer U.A., Broly F. (1997) Polymorphism of the cytochrome P450 *CYP2D6* gene in a European population: characterization of 48 mutations and 53 alleles, their frequency and evolution. *Pharmacogenetics* **7**: 193–202.

Margolis R.L., Stine O.C., McInnis M.G., Ranen N.G., Rubinsztein D.C., Leggo J., Jones Brando L.V., Kidwai A.S. (1996) cDNA cloning of a human homologue of the *Caenorhabditis elegans* cell fate-determining gene *mab-21*: expression, chromosomal localization and analysis of a highly polymorphic (CAG)$_n$ trinucleotide repeat. *Hum. Molec. Genet.* **5**: 607–616.

Marshall C.J. (1993) Evolutionary relationships among the serpins. *Phil. Trans. R. Soc.Lond. B* **342**: 101–119.

Martin S.L., Zimmer E.A., Kan Y.W., Wilson A.C. (1980) Silent δ-globin gene in Old World monkeys. *Proc. Natl. Acad. Sci. USA* **77**: 3563–3567.

Massova I., Kotra L.P., Fridman R., Mobashery S. (1998) Matrix metalloproteinases: structures, evolution and diversification. *FASEB J.* **12**: 1075–1095.

Masutani C., Sugasawa K., Yanagisawa J., Sonoyama T., Ui M., Enemoto T., Takio K., Tanaka K., van der Spek P.J., Bootsma D., Hoeijmakers J.H.J., Hanaoka F. (1994) Purification and cloning of a nucleotide excision repair complex involving the xeroderma pigmentosum group C protein and a human homolog of yeast RAD23. *EMBO J.* **13**: 1831–1843.

Matera A.G., Weiner A.M., Schmid C.W. (1990) Structure and evolution of the U2 small nuclear RNA multigene family in primates: gene amplification under natural selection? *Molec. Cell. Biol.* **10**: 5876–5882.

Matsuda F., Honjo T. (1996) Organization of the human immunoglobulin heavy-chain locus. *Adv. Immunol.* **62**: 1–29.

Maxson R., Cohn R., Kedes L., Mohun T. (1983) Expression and organization of histone genes. *Ann. Rev. Genet.* **17**: 239–277.

Mayer W.E., Jonker D., Klein D., Ivanyl P., van Seventer G., Klein J. (1988) Nucleotide sequence of chimpanzee MHC class I alleles: evidence for trans-species mode of evolution. *EMBO J.* **7**: 2765–2774.

Mayer W.E., O'hUigin C., Zaleska-Rutczynska Z., Klein J. (1992) Trans-species origin of *Mhc-DRB* polymorphism in the chimpanzee. *Immunogenetics* **37**: 12–18.

Meakin S.O., Breitman M.L., Tsui L.-C. (1985) Structural and evolutionary relationships among five members of the human γ-crystallin gene family. *Mol. Cell. Biol.* **5**: 1408–1414.

Meakin S.O., Du R.P., Tsui L.-C., Breitman M.L. (1987) γ-Crystallins of the human eye lens: expression analysis of five members of the gene family. *Mol. Cell. Biol.* **7**: 2671–2679.

Meireles C.M.M., Schneider M.P.C., Sampaio M.I.C., Schneider H., Slightom J.L., Chiu C.-H., Neiswanger K., Gumucio D.L., Czelusniak J., Goodman M.. (1995) Fate of a redundant γ-globin gene in the atelid clade of New World monkeys: implications concerning fetal globin gene expression. *Proc. Natl. Acad. Sci. USA* **92**: 2607–2611.

Mellor H., Parker P.J. (1998) The extended protein kinase C superfamily. *Biochem. J.* **332**: 281–292.

Meroni G., Franco B., Archdiacono N., Messali S., Andolfi G., Rocchi M., Ballabio A. (1996) Characterization of a cluster of sulfatase genes on Xp22.3 suggests gene duplications in an ancestral pseudoautosomal region. *Hum. Mol. Genet.* **5**: 423–431.

Meyer A. (1998) Hox gene variation and evolution. *Nature* **391**: 226–228.

Michelson A.M., Orkin S.H. (1983) Boundaries of gene conversion within the duplicated human α-globin genes. Concerted evolution by segmental recombination. *J. Biol. Chem.* **258**: 15245–15253.

Milisavljevic V., Freedberg I.M., Blumenberg M. (1996) Close linkage of the two keratin gene clusters in the human genome. *Genomics* **34**: 134–138.

Miller W.L., Eberhardt N.L. (1983) Structure and evolution of the growth hormone gene family. *Endocr. Rev.* **4**: 97–130.

Minghetti P.P., Law S.W., Dugaiczyk A. (1985) The rate of molecular evolution of α-fetoprotein approaches that of pseudogenes. *Mol. Biol. Evol.* **2**: 347–358.

Miwa T., Manabe Y., Kurokawa K., Kamada S., Kanda N., Bruns G., Ueyama H., Kakunaga T. (1991) Structure, chromosome location, and expression of the human smooth muscle (enteric type) γ-actin gene: evolution of six human actin genes. *Mol. Cell. Biol.* **11**: 3296–3306.

Miyata T., Hayashida H. (1982) Recent divergence from a common ancestor of human IFN-α genes. *Nature* **295**: 165–168.

Miyata T., Hayashida H., Kikuno R., Toh H., Kawade Y. (1985) Evolution of interferon genes. *Interferon* **6**: 1–30.

Moore D.D. (1990) Diversity and unity in the nuclear hormone receptors: a terpenoid receptor superfamily. *New Biology* **2**: 100–105.

Morgan R.O., Bell D.W., Testa J.R., Fernandez M.P. (1998) Genomic locations of *ANX11* and *ANX13* and the evolutionary genetics of human annexins. *Genomics* **48**: 100–110.

Mori M., Matsubara T., Amaya Y., Takiguchi M. (1990) Molecular evolution from arginosuccinate lyase to δ-crystallin. In *Isozymes: Structure, Function, and Use in Biology and Medicine*. Wiley-Liss, New York. pp 683–699.

Morrison N., Goddard J.P., Ledbetter D.H., Boyd E., Bourn D., Connor J.M. (1991) Chromosomal assignment of a large tRNA cluster (tRNA-leu, tRNA-gln, tRNA-lys, tRNA-arg, tRNA-gly) to 17p13.1. *Hum. Genet.* **87**: 226–230.

Moyle W.R., Campbell R.K., Myers R.V., Bernard M.P., Han Y., Wang X. (1994) Co-evolution of ligand-receptor pairs. *Nature* **368**: 251–255.

Müller-Esterl W., Fritz H., Kellermann J., Lottspeich F., Machleidt W., Turk V. (1985) Genealogy of mammalian cysteine proteinase inhibitors. Common evolutionary origin of stefins, cystatins and kininogens. *FEBS Letts.* **191**: 221–226.

Mulligan M.E., Jackman D.M., Murphy S.T. (1994) Heterocyst-forming filamentous cyanobacteria encode proteins that resemble eukaryotic RNA-binding proteins of the RNP family. *J. Mol. Biol.* **235**: 1162–1170.

Murphy P.M. (1993) Molecular mimicry and the generation of host defense protein diversity. *Cell* **72**: 823–826.

Mushegian A.R., Koonin E.V. (1996) Sequence analysis of eukaryotic development proteins: ancient and novel domains. *Genetics* **144**: 817–828.

Nagamatsu S., Chan S.J., Falkmer S., Steiner D.F. (1991) Evolution of the insulin gene superfamily. *J. Biol. Chem.* **266**: 2397–2402.

Nagel G., Doolittle R.F (1991) Evolution and relatedness in two aminoacyl-tRNA synthetase families. *Proc. Natl. Acad. Sci. USA* **88**: 8121–8125.

Nagel G.M., Doolittle R.F. (1995) Phylogenetic analysis of the aminoacyl-tRNA synthetases. *J. Mol. Evol.* **40**: 487–498.

Nakata K., Motomura M., Nakabayashi H., Ido A., Tamaoki T. (1992) A possible mechanism of inverse developmental regulation of α-fetoprotein and albumin genes. Studies with epidermal growth factor and phorbol ester. *J. Biol. Chem.* **267**: 1331–1334.

Nebert D.W., Nelson D.R. Feyereisen R. (1989) Evolution of the cytochrome P450 genes. *Xenobiotica* **19**: 1149–1160.

Nelson D.R., Strobel H.W. (1987) Evolution of cytochrome P-450 proteins. *Mol. Biol. Evol.* **4**: 572–593.

Nelson D.R., Koymans L., Kamataki T., Stegeman J.J., Feyereisen R., Waxman D.J., Waterman M.R., Gotoh O., Coon M.J., Estabrook R.W., Gunsalus I.C., Nebert D.W. (1996) Cytochrome P450 superfamily: update on new sequences, gene mapping, accession numbers, and nomenclature. *Pharmacogenetics* **6**: 1–42.

Neufeld E.J., Skalnik G., Lievens P.M.-J., Orkin S.H. (1992) Human CCAAT displacement protein is homologous to the *Drosophila* homeoprotein, *cut*. *Nature Genet.* **1**: 50–51.

Ng S.-Y., Gunning P., Eddy R., Ponte P., Leavitt J., Shows T., Kedes L. (1985) Evolution of the functional human beta-actin gene and its multi-pseudogene family: conservation of the noncoding regions and chromosomal dispersion of pseudogenes. *Mol. Cell. Biol.* **5**: 2720–2732.

Nishio H., Dugaiczyk A. (1996) Complete structure of the α-albumin gene, a new member of the serum albumin multigene family. *Proc. Natl. Acad. Sci. USA* **93**: 7557–7561.

Nishio H., Heiskanen M., Palotie A., Bélanger L., Dugaiczyk A. (1996) Tandem arrangement of the human serum albumin multigene family in the sub-centromeric region of 4q: evolution and chromosomal direction of transcription. *J. Mol. Biol.* **259**: 113–119.

Nitta K., Sugai S. (1989) The evolution of lysozyme and α-lactalbumin. *Eur. J. Biochem.* **182**: 111–118.

Nollet S., Moniaux N., Maury J., Petitprez D., Degand P., Laine A., Porchet N., Aubert J-P. (1998) Human mucin gene *MUC4*: organization of its 5′-region and polymorphism of its central tandem repeat array. *Biochem. J.* **332**: 739–748.

O'Malley B.W. (1989) Did eukaryotic steroid receptors evolve from intracrine gene regulators? *Endocrinology* **125**: 1119–1120.

O'Neill V.A., Eden FC., Pratt K., Hatfield D.L. (1985) A human opal suppressor tRNA gene and pseudogene. *J. Biol. Chem.* **260**: 2501–2508.

Ohkuma Y., Sumimoto H., Hoffmann A., Shimasaki S., Horikoshi M., Roeder R.G. (1991) Structural motifs and potential sigma homologies in the large subunit of human general transcription factor TFIIE. *Nature* **354**: 398–401.

Ohno S. (1970) *Evolution by Gene Duplication*. George Allen & Unwin, London.

Ohno S. (1981) Original domain for the serum albumin family arose from repeated sequences. *Proc. Natl. Acad. Sci. USA* **78**: 7657–7661.

Ohta T. (1991) Role of diversifying selection and gene conversion in evolution of major histocompatibility complex loci. *Proc. Natl. Acad. Sci. USA* **88**: 6716–6720.

Ohta T. (1997) Role of gene conversion in generating polymorphisms at major histocompatibility complex loci. *Hereditas* **127**: 97–103.

Onda M., Kudo S., Rearden A., Mattei M.-G., Fukuda M. (1993) Identification of a precursor genomic segment that provided a sequence unique to glycophorin B and E genes. *Proc. Natl. Acad. Sci. USA* **90**: 7220–7224.

Onda M., Kudo S., Fukuda M. (1994) Genomic organization of glycophorin A gene family revealed by yeast artificial chromosomes containing human genomic DNA. *J. Biol. Chem.* **269**: 13013–13020.

Onda M., Fukuda M. (1995) Detailed physical mapping of the genes encoding glycophorins A, B and E, as revealed by P1 plasmids containing human genomic DNA. *Gene* **159**: 225–230.

Ota T., Nei M. (1994) Divergent evolution and evolution by the birth-and-death process in the immunoglobulin V_H gene family. *Mol. Biol. Evol.* **11**: 469–482.

Ouzounis C.A., Kyrpides N.C. (1996) Parallel origins of the nucleosome core and eukaryotic transcription from Archaea. *J. Mol. Evol.* **42**: 234–239.

Padmanabhan K., Padmanabhan K.P., Tulinsky A., Park CH., Bode W., Huber R., Blankenship D.T., Cardin A.D., Kisiel W. (1993) Structure of human des(1–45) factor Xa at 2.2 Å resolution. *J. Mol. Biol.* **232**: 947–966.

Pallanck L., Ganetzky B. (1994) Cloning and characterization of human and mouse homologs of the *Drosophila* calcium-activated potassium channel gene, *slowpoke*. *Hum. Molec. Genet.* **3**: 1239–1243.

Parenti G., Meroni G., Ballabio A. (1997) The sulfatase gene family. *Curr. Opin. Genet. Devel.* **7**: 386–391.

Parham P., Ohta T. (1996) Population biology of antigen presentation by MHC class I molecules. *Science* **272**: 67–72.

Pavelitz T., Rusché L., Matera A.G., Scharf J.M., Weiner A.M. (1995) Concerted evolution of the tandem array encoding primate U2 snRNA occurs *in situ*, without changing the cytological context of the *RNU2* locus. *EMBO J.* **14**: 169–177.

Pemble S.E., Taylor J.B. (1992) An evolutionary perspective on glutathione transferases inferred from class-theta glutathione transferase cDNA sequences. *Biochem. J.* **287**: 957–963.

Perkins K.K., Dailey G.M., Tjian R. (1988) Novel Jun- and Fos-related proteins in *Drosophila* are functionally homologous to enhancer factor AP-1. *EMBO J.* **7**: 4265–4273.

Pesole G., Gerardi A., di Jeso F., Saccone C. (1994) The peculiar evolution of apolipoprotein(a) in human and rhesus macaque. *Genetics* **136**: 255–260.

Phillips J.A. (1995) Inherited defects in growth hormone synthesis and action. In *The Metabolic and Molecular Bases of Inherited Disease*. 7th Ed. (eds. CR Scriver, AL Beaudet, WS Sly, D Valle). McGraw-Hill, New York. pp 3023–3044.

Piatigorsky J., O'Brien W.E., Norman B.L., Kalumuck K., Wistow G.J., Borras T., Nickerson J.M., Wawrousek E.F. (1988) Gene sharing by δ-crystallin and argininosuccinate lyase. *Proc. Natl. Acad. Sci. USA* **85**: 3479–3483.

Pigny P., Guyonnet-Duperat V., Hill A.S., Pratt W.S., Galiegue-Zouitina S., d'Hooge M.C., Laine A., van Seuningen I., Degand P., Gum J.R., Kim Y.S., Swallow D.M., Aubert J-P, Porchet N. (1996) Human mucin genes assigned to 11p15.5: identification and organization of a cluster of genes. *Genomics* **38**: 340–352.

Pizzuti A., Amati F., Calabrese G., Mari A., Closimo A., Silani V., Giardino L., Ratti A., Penso D., Calza L., Palka G., Scarlato G., Novelli G., Dallapiccola B. (1996) cDNA characterization and chromosomal mapping of two human homologues of the *Drosophila dishevelled* polarity gene. *Hum. Molec. Genet.* 5: 953–958.

Ponte P., Gunning P., Blau H., Kedes L. (1983) Human actin genes are single copy for α-skeletal and α-cardiac actin but multicopy for β- and γ-cytoskeletal genes: 3' untranslated regions are isotype specific but are conserved in evolution. *Mol. Cell. Biol.* 3: 1783–1791.

Popescu N., Zimonjic D., Hatch C., Bonner W.M. (1994) Chromosomal mapping of the human histone gene H2A.Z to 4q24 by fluorescence *in situ* hybridization. *Genomics* 20: 333–335.

Powell B., Crocker L., Rogers G. (1992) Hair follicle differentiation: expression, structure and evolutionary conservation of the hair type II keratin intermediate filament gene family. *Development* 114: 417–433.

Prager E.M. (1996) Adaptive evolution of lysozyme: changes in amino acid sequence, regulation of expression and gene number. In *Lysozmes: Model Enzymes in Biochemistry and Biology.* (ed., P. Jollés). Birkhäuser, Basel. pp 323–345.

Prager E.M, Wilson A.C. (1988) Ancient origin of lactalbumin from lysozyme: analysis of DNA and amino acid sequences. *J. Mol. Evol.* 27: 326–335.

Prockop D.J., Kivirikko K.I. (1995) Collagens: molecular biology, diseases, and potentials for therapy. *Annu. Rev. Biochem.* 64: 403–434.

Proudfoot N.J., Maniatis T. (1980) The structure of a human α-globin pseudogene and its relationship to α-globin gene duplication. *Cell* 21: 537–548.

Proudfoot N.J., Gil A., Maniatis T. (1982) The structure of the human ζ-globin gene and a closely linked, nearly identical pseudogene. *Cell* 31: 553–562.

Qasba P.K., Kumar S. (1997) Molecular divergence of lysozymes and α-lactalbumin. *Curr. Rev. Biochem. Mol. Biol.* 32: 255–306.

Quiring R., Waldorf U., Kloter U., Gehring W. (1994) Homology of the *eyeless* gene of *Drosophila* to the small eye gene in mice and aniridia in humans. *Science* 265: 785–789.

Rada C., Lorenzi R., Powis S.J., van den Bogaerde J., Parham P., Howard J.C. (1990) Concerted evolution of class I genes in the major histocompatibility complex of murine rodents. *Proc. Natl. Acad. Sci. USA* 87: 2167–2171.

Radice P., Pensotti V., Jones C., Perry H., Pierotti M.A., Tunnacliffe A. (1995) The human archain gene, *ARCN1*, has highly conserved homologs in rice and *Drosophila*. *Genomics* 26: 101–106.

Rafti F., Scarvelis D., Lasko P.F. (1996) A *Drosophila melanogaster* homologue of the human DEAD-box gene *DDX1*. *Gene* 171: 225–229.

Ramakrishnan V. (1995) The histone fold: evolutionary questions. *Proc. Natl. Acad. Sci. USA* 92: 11328–11330.

Rast J.P., Anderson M.K., Strong S.J., Luer C., Litman R.T., Litman G.W. (1997) α, β, γ, and δ T cell antigen receptor genes arose early in vertebrate phylogeny. *Immunity* 6: 1–11.

Rawlings N.D., Barrett A.J. (1990) Evolution of proteins of the cystatin family. *J. Mol. Evol.* 30: 60–71.

Rearden A., Magnet A., Kudo S., Fukuda M. (1993) Glycophorin B and glycophorin E genes arose from the glycophorin A ancestral gene via two duplications during primate evolution. *J. Biol. Chem.* 268: 2260–2267.

Ribas de Pouplana L., Turner R.J., Steer B.A., Schimmel P. (1998) Genetic code origins: tRNAs older than their synthetases? *Proc. Natl. Acad. Sci. USA* 95: 11295–11300.

Richter G., Hein G., Blankenstein T., Diamantstein T. (1995) The rat interleukin 4 receptor: coevolution of ligand and receptor. *Cytokine* 7: 237–241.

Riggs A.F. (1991) Aspects of the origin and evolution of non-vertebrate hemoglobins. *Amer. Zool.* 31: 535–545.

Rocques P.J, Clark J., Ball S., Crew J., Gill S., Christodoulou Z., Borts R.H., Louis E.J., Davies K.E., Cooper C.S. (1995) The human *SB1.8* gene (*DXS423E*) encodes a putative chromosome segregation protein conserved in lower eukaryotes and prokaryotes. *Hum. Molec. Genet.* 4: 243–249.

Rogers M.A., Winter H., Wolf C., Heck M., Schweizer J. (1998) Characterization of a 190-kilobase pair domain of human type I hair keratin genes. *J. Biol. Chem.* 273: 26683–26691.

Rosenberg H.F, Dyer K.D. (1995) Eosinophil cationic protein and eosinophil-derived neurotoxin. *J. Biol. Chem.* 270: 21539–21544.

Rosenberg H.F., Dyer K.D., Tiffany L., Gonzalez M. (1995) Rapid evolution of a unique family of primate ribonuclease genes. *Nature Genet.* **10**: 219–223.

Rosinski J.A., Atchley W.R. (1998) Molecular evolution of the Myb family of transcription factors: evidence for polyphyletic origin. *J. Mol. Evol.* **46**: 74–83.

Rouquier S., Taviaux S., Trask B.J., Brand-Arpon V., van den Engh G., Demaille J., Giorgi D. (1998) Distribution of olfactory receptor genes in the human genome. *Nature Genet.* **18**: 243–250.

Rowen L., Koop B.F, Hood L. (1996) The complete 685-kilobase DNA sequence of the human β T cell receptor locus. *Science* **272**: 1755–1762.

Rowlands T., Baumann P., Jackson S.P. (1994) The TATA-binding protein: a general transcription factor in eukaryotes and archaebacteria. *Science* **264**: 1326–1328.

Rozman D., Stromstedt M., Tsui L.-C., Scherer S.W., Waterman M.R. (1996) Structure and mapping of the human lanosterol 14α-demethylase gene (*CYP51*) encoding the cytochrome P450 involved in cholesterol biosynthesis; comparison of exon/intron organization with other mammalian and fungal CYP genes. *Genomics* **38**: 371–381.

Runnegar B. (1985) Collagen gene construction and evolution. *J. Mol. Evol.* **22**: 141–149.

Russek S.J., Farb D.H. (1994) Mapping of the β_2 subunit gene (*GABRB2*) to microdissected human chromosome 5q34-q35 defines a gene cluster for the most abundant $GABA_A$ receptor isoform. *Genomics* **23**: 528–533.

Ryden L.G, Hunt L.T. (1993) Evolution of protein complexity: the blue copper-containing oxidases and related proteins. *J. Mol. Evol.* **35**: 41–66.

Rypniewski W.R, Perrakis A, Vorgias C.E, Wilson K.S. (1994) Evolutionary divergence and conservation of trypsin. *Protein Eng.* **7**: 57–64.

Saier M.H. (1994) Convergence and divergence in the evolution of transport proteins. *BioEssays* **16**: 23–29.

Saitou N., Yamamoto F. (1997) Evolution of primate ABO blood group genes and their homologous genes. *Mol. Biol. Evol.* **14**: 399–411.

Saks M.E., Sampson J.R. (1995) Evolution of tRNA recognition systems and tRNA gene sequences. *J. Mol. Evol.* **40**: 509–518.

Salvignol I., Calvas P., Socha W.W., Coli Y., Le Van Kim C., Bailly P., Ruffie J., Cartron J.-P., Blancher A. (1995) Structural analysis of the RH-like blood group gene products in nonhuman primates. *Immunogenetics* **41**: 271–281.

Samuelson L.C., Wiebauer K., Snow C.M., Meisler M.H. (1990) Retroviral and pseudogene insertion sites reveal the lineage of human salivary and pancreatic amylase genes from a single gene during primate evolution. *Molec. Cell. Biol.* **10**: 2513–2520.

Schlesinger M.J., Bond U. (1987) Ubiquitin genes. In *Oxford Surveys of Eukaryotic Genes*. Vol. 4. Oxford University Press, Oxford. pp. 77–91.

Schluter S.F., Bernstein R.M., Marchalonis J.J. (1997) Molecular origins and evolution of immunoglobulin heavy-chain genes of jawed vertebrates. *Immunology Today* **18**: 543–549.

Schofield J.P. (1993) Molecular studies on an ancient gene encoding carbamoyl-phosphate synthetase. *Clin. Sci.* **84**: 119–128.

Schreiber G., Richardson S.J. (1997) The evolution of gene expression, structure and function of transthyretin. *Comp. Biochem. Physiol.* **116B**: 137–160.

Schughart K., Kappen C., Ruddle F.H. (1989) Duplication of large genomic regions during the evolution of vertebrate homeobox genes. *Proc. Natl. Acad. Sci. USA* **86**: 7067–7071.

Scott S., Chen H., Rossier C., Lalioti M.D., Antonarakis S.E. (1997) Isolation of a human gene (*HES1*) with homology to an *Escherichia coli* and a zebrafish protein that maps to chromosome 21q22.3. *Hum. Genet.* **99**: 616–623.

Seitz E.M., Brockman J.P., Sandler S.J., Clark A.J., Kowalczykowski S.C. (1998) RadA protein is an archael RecA protein homolog that catalyzes DNA strand exchange. *Genes Devel.* **12**: 1248–1253.

Sepepack P., Slatkin M., Arnheim N. (1988) Linkage disequilibrium in human ribosomal genes: implications for multigene family evolution. *Genetics* **119**: 943–949.

Shannon M., Stubbs L. (1998) Analysis of homologous XRCC1-linked zinc-finger gene families in human and mouse: evidence for orthologous genes. *Genomics* **49**: 112–121.

Sharkey M., Graba Y., Scott M.P. (1997) *Hox* genes in evolution: protein surfaces and paralog groups. *Trends Genet.* **13**: 145–151.

Sharp P.M., Li W.-H. (1987) Ubiquitin genes as a paradigm of concerted evolution of tandem repeats. *J. Mol. Evol.* **25**: 58–64.

Shields D.C., Harmon D.L., Nunez F., Whitehead A.S. (1995) The evolution of haematopoietic cytokine/receptor complexes. *Cytokine* **7**: 679–688.

Shindoh N., Kodoh J., Maeda H., Yamaki A., Minoshima S., Simizu Y., Shimizu N. (1996) Cloning of a human homolog of the *Drosophila Minibrain*/Rat *Dyrk* gene from 'the Down Syndrome Critical Region' of chromosome 21. *Biochem. Biophys. Res. Commun.* **225**: 92–99.

Shinohara A., Ogawa H., Matsuda Y., Ushio N., Ikeo K., Ogawa T. (1993) Cloning of human, mouse and fission yeast recombination genes homologous to *RAD51* and *recA*. *Nature Genet.* **4**: 239–243.

Shubin N., Tabin C., Carroll S. (1997) Fossils, genes and the evolution of animal limbs. *Nature* **388**: 639–648.

Sitnikova T., Nei M. (1998) Evolution of immunoglobulin kappa chain variable region genes in vertebrates. *Mol. Biol. Evol.* **15**: 50–60.

Sitnikova T., Su C. (1998) Coevolution of immunoglobulin heavy- and light-chain variable-region gene families. *Mol. Biol. Evol.* **15**: 617–625.

Slesarev A.I., Belova G.I., Kozyavkin S.A., Lake J.A. (1998) Evidence for an early prokaryotic origin of histones H2A and H4 prior to the emergence of eukaryotes. *Nucleic Acids Res.* **26**: 427–430.

Slightom J.L., Theisen T.W., Koop B.F., Goodman M. (1987) Orangutan fetal globin genes. Nucleotide sequences reveal multiple gene conversions during hominid phylogeny. *J. Biol. Chem.* **262**: 7472–7483.

Slightom J.L., Koop B.F., Xu P.L., Goodman M. (1988) Rhesus fetal globin genes. Concerted gene evolution in the descent of higher primates. *J. Biol. Chem.* **263**: 12 427–12 438.

Smit A.B., van Marle A., van Elk R., Bogerd J., van Heerikhuizen H., Geraerts W.P. (1993) Evolutionary conservation of the insulin gene structure in invertebrates: cloning of the gene encoding molluscan insulin-related peptide III from *Lymnea stagnalis*. *J. Molec. Endocrinol.* **11**: 103–113.

Sorensen P.D., Frederiksen S. (1991) Characterization of human 5S rRNA genes. *Nucleic Acids Res.* **19**: 4147–4151.

Steiner D.F., Chan S.J., Welsh J.M., Kwok S.C.M. (1985) Structure and evolution of the insulin gene. *Ann. Rev. Genet.* **19**: 463–484.

Steward R. (1987) *Dorsal*, an embryonic polarity gene in *Drosophila*, is homologous to the vertebrate proto-oncogene, c-*rel*. *Science* **238**: 692–694.

Stifani S., Blaumueller C.M., Redhead N.J., Hill R.E., Artavanis-Tsakonas S. (1992) Human homologs of a *Drosophila* enhancer of split gene product define a novel family of nuclear proteins. *Nature Genet.* **2**: 119–120.

Strand D., Unger S., Corvi R., Hartenstein K., Schenkel H., Kalmes A., Merdes G., Neumann B., Kreig-Schneider F., Coy J.F. et al. (1995) A human homologue of the *Drosophila* tumour suppressor gene 1(2)gl maps to 17p11.2–12 and codes for a cytoskeletal protein that associates with nonmuscle myosin II heavy chain. *Oncogene* **11**: 291–310.

Sullivan L.C., Daniels C.B., Phillips I.D., Orgeig S., Whitsett J.A. (1998) Conservation of surfactant protein A: evidence for a single origin for vertebrate pulmonary surfactant. *J. Mol. Evol.* **46**: 131–138.

Sullivan S.L., Adamson M.C., Ressler K.J., Kozak C.A., Buck L.B. (1996) The chromosomal distribution of mouse odorant receptor genes. *Proc. Natl. Acad. Sci. USA* **93**: 884–888.

Sumimoto H., Ohkuma Y., Sinn E., Kato H., Shimasaki S., Horikoshi M., Roeder R.G. (1991) Conserved sequence motifs in the small subunit of human general transcription factor TFIIE. *Nature* **354**: 401–404.

Suzuki S.T. (1996) Structural and functional diversity of cadherin superfamily: are new members of cadherin superfamily involved in signal transduction pathway? *J. Cell. Biochem.* **61**: 531–542.

Suzuki T., Imai K. (1998) Evolution of myoglobin. *Cell. Mol. Life Sci.* **54**: 979–1004.

Suzuki T., Yuasa H., Imai K. (1996) Convergent evolution. The gene structure of *Sulculus* 41 kDa myoglobin is homologous with that of human indoleamine dioxygenase. *Biochim. Biophys. Acta* **1308**: 41–48.

Swaroop A., Yang-Feng T.L., Liu W., Gieser L., Barrow L.L., Chen K.-C., Agarwal N., Meisler M.H., Smith D.I. (1994) Molecular characterization of a novel human gene, *SEC13R*, related to the yeast secretory pathway gene *SEC13*, and mapping to a conserved linkage group on human chromosome 3p24-p25 and mouse chromosome 6. *Hum. Molec. Genet.* **3**: 1281–1286

Sweetser D., Nonet M., Young R.A. (1987) Prokaryotic and eukaryotic RNA polymerases have homologous core subunits. *Proc. Natl. Acad. Sci. USA* **84**: 1192–1196.

Tagle D.A., Koop B.F., Goodman M., Slightom J.L., Hess D.L., Jones R.T. (1988) Embryonic ε and γ globin genes of a prosimian primate (*Galago crassicaudatus*). *J. Mol. Biol.* **203**: 439–455.

Tagle D.A., Slightom J.L., Jones R.T., Goodman M. (1991) Concerted evolution led to high expression of a prosimian primate δ-globin gene locus. *J. Biol. Chem.* **266**: 7469–7475.

Tagle D.A., Stanhope M.J., Siemieniak D.R., Benson P., Goodman M., Slightom J.L. (1992) The β globin gene cluster of the prosimian primate *Galago crassicaudatus*: nucleotide sequence determination of the 41-kb cluster and comparative sequence analyses. *Genomics* **13**: 741–760.

Takahara K., Hoffman G.G., Greenspan D.S. (1995) Complete structural organization of the human α1(V) collagen gene (*COL5A1*): divergence from the conserved organization of the other characterized fibrillar collagen genes. *Genomics* **29**: 588–597.

Takahashi N., Ueda S., Obata M., Nikaido T., Nakai S., Honjo T. (1982) Structure of human immunoglobulin gamma genes: implications for evolution of a gene family. *Cell* **29**: 671–679.

Tan Y., Sishoff S.T., Riley M.A. (1993) Ubiquitins revisited: further examples of within- and between-locus concerted evolution. *Molec. Phylogenet. Evol.* **2**: 351–360.

Thatcher T.H., Gorovsky M.A. (1994) Phylogenetic analysis of the core histones H2A, H2B, H3 and H4. *Nucleic Acids Res.* **22**: 174–179.

Thiesse M., Millar S.J., Dickinson D.P. (1994) The human type 2 cystatin gene family consists of eight to nine members, with at least seven genes clustered at a single locus on human chromosome 20. *DNA Cell Biol.* **13**: 97–116.

TomHon C., Zhu W., Millinoff D., Hayasaka K., Slightom J.L., Goodman M., Gumucio D.L. (1997) Evolution of a fetal expression pattern via changes near the γ-globin gene. *J. Biol. Chem.* **272**: 14062–14066.

Tomlinson I.M., Cook G.P., Carter N.P., Elaswarapu R., Smith S., Walter G., Buluwela L., Rabbitts T.H., Winter G. (1994) Human immunoglobulin V_H and D segments on chromosomes 15q11.2 and 16p11.2. *Hum. Molec. Genet.* **3**: 853–860.

Toribara N.W., Gum J.R., Culhane P.J., Lagace R.E., Hicks J.W., Petersen G.M., Kim Y.S. (1991) MUC-2 human small intestinal mucin gene structure: repeated arrays and polymorphism. *J. Clin. Invest.* **88**: 1005–1013.

Trask B.J., Massa M., Brand-Arpon V., Chan K., Friedman C., Nguyen O.T.H., Eichler E., van den Engh G., Rouquier S., Shizuya H., Giorgi D. (1998) Large multi-chromosomal duplications encompass many members of the olfactory receptor gene family in the human genome. *Hum. Molec. Genet.* **7**: 2007–2020.

Trezise A.E., Godfrey E.A., Holmes R.S., Beacham I.R. (1989) Cloning and sequencing of cDNA encoding baboon liver alcohol dehydrogenase: evidence for a common ancestral lineage with the human alcohol dehydrogenase β subunit and for class I *ADH* gene duplications predating primate radiation. *Proc. Natl. Acad. Sci. USA* **86**: 5454–5458.

Trezise A.E., Cheung B., Holmes R.S., Beacham I.R. (1991) Evidence for three genes encoding class-I alcohol dehydrogenase subunits in baboon and analysis of the 5′ region of the gene encoding the ADHβ subunit. *Gene* **103**: 211–218.

Trikka D., Davis T., Lapenta V., Brahe C., Kessling A.M. (1997) Human *COL6A1*: genomic characterization of the globular domains, structural and evolutionary comparison with *COL6A2*. *Mamm. Genome* **8**: 342–345.

Trotta C.R., Miao F., Arn E.A., Stevens S.W., Ho C.K., Rauhut R., Abelson J.N. (1997) The yeast tRNA splicing endonuclease: a tetrameric enzyme with two active site subunits homologous to the archaeal tRNA endonuclases. *Cell* **89**: 849–858.

Tugendreich S., Bassett D.E., McKusick V., Boguski M.S., Hieter P. (1994) Genes conserved in yeast and humans. *Hum. Molec. Genet.* **3**: 1509–1517.

Ullrich A., GrayA., Goedell D.V., Dull T.J. (1982) Nucleotide sequence of a portion of human chromosome 9 containing a leukocyte interferon gene cluster. *J. Mol. Biol.* **156**: 467–486.

Upton Z., Francis G.L., Chan S.J., Steiner D.F., Wallace J.C., Ballard F.J. (1997) Evolution of insulin-like growth factor (IGF) function: production and characterization of recombinant hagfish IGF. *Gen. Comp. Endocrinol.* **105**: 79–90.

Van Arsdell S.W., Weiner A.M. (1984) Human genes for U2 small nuclear RNA are tandemly repeated. *Molec. Cell. Biol.* **4**: 492–499.

Vanderhaeghen P., Schurmans S., Vassart G., Parmentier M. (1997a) Specific repertoire of olfactory receptor genes in the male germ cells of several mammalian species. *Genomics* **39**: 239–246.

Vanderhaeghen P., Schurmans S., Vassart G., Parmentier M. (1997b) Molecular cloning and chromosomal mapping of olfactory receptor genes expressed in the male germ line: evidence for their wide distribution in the human genome. *Biochem. Biophys. Res. Commun.* **237**: 283–287.

Van der Rest M., Garrone R. (1991) Collagen family of proteins. *FASEB J.* 5: 2814–2823.

Van Kesteren R.E., Fainzilber M., Hauser G., van Minnen J., Vreugdenhil E., Smit A.B., Ibanez C.F., Geraerts W.P.M., Bulloch A.G.M. (1998) Early evolutionary origin of the neurotrophin receptor family. *EMBO J.* **17**: 2534–2542.

Vargas-Madrazo E., Lara-Ochoa F., Ramirez-Benites M.C., Almagro J.C. (1997) Evolution of the structural repertoire of the human V(H) and Vkappa germline genes. *Int. Immunol.* **9**: 1801–1815.

Vrana P.B., Wheeler W.C. (1996) Molecular evolution and phylogenetic utility of the polyubiquitin locus in mammals and higher vertebrates. *Molec. Phylogenet. Evol.* **6**: 259–269.

Vulpe C., Levinson B., Whitney S., Packman S., Gitschier J. (1993) Isolation of a candidate gene for Menkes disease and evidence that it encodes a copper-transporting ATPase. *Nature Genet.* **3**: 7–14.

Vuorio E., de Crombrugghe B. (1990) The family of collagen genes. *Annu. Rev. Biochem.* **59**: 837–872.

Vuoristo M.M., Pihlajamaa T., Vandenberg P., Prockop D.J., Ala-Kokko L. (1995) The human *COL11A2* gene structure indicates that the gene has not evolved with the genes for the major fibrillar collagens. *J. Biol. Chem.* **270**: 22 873–22 881.

Walker C.W., Boon J.D.G., Maarsh A.G. (1992) First non-vertebrate member of the *myc* gene family is seasonally expressed in an invertebrate testis. *Oncogene* **7**: 2007–2012.

Wallis M., Lioupis A., Wallis O.C. (1998) Duplicate growth hormone genes in sheep and goat. *J. Molec. Endocrinol.* **21**: 1–5.

Wang W., Wu W., Desai T., Ward D.C., Kaufman S.J. (1995) Localization of the α7 integrin gene (*ITGA7*) on human chromosome 12q13: clustering of integrin and Hox genes implies parallel evolution of these gene families. *Genomics* **26**: 563–570.

Watkins D.I. (1995) The evolution of major histocompatibility class I genes in primates. *Crit. Rev. Immunol.* **15**: 1–29.

Watkins D.I., Chen Z.W., Garger T.L., Hughes A.L., Letvin N.L. (1991) Segmental exchange between MHC class I genes in a higher primate: recombination in the gorilla between the ancestor of a human non-functional gene and an *A* locus gene. *Immunogenetics* **34**: 185–191.

Webb G.C., Baker R.T., Fagan K., Board P.G. (1990) Localization of the human UbB polyubiquitin gene to chromosome band 17p11.1–17p12. *Am. J. Hum. Genet.* **46**: 308–315.

Wedekind C., Seebeck T., Bettens F., Paepke A.J. (1995) MHC-dependent mate preferences in humans. *Proc. Roy. Soc. Lond. B Biol. Sci.* **260**: 245–262.

Wes P.D., Chevesich J., Jeromin A., Rosenberg C., Stetten G. (1995) *TRPC1*, a human homolog of a *Drosophila* store-operated channel. *Proc. Natl. Acad. Sci. USA* **92**: 9652–9656.

Westin G., Zabielski J., Hammarstrom K., Monstein H.-J., Bark C., Petterson U. (1984) Clustered genes for human U2 RNA. *Proc. Natl. Acad. Sci. USA* **81**: 3811–3815.

Wiborg O., Pedersen M.S., Wind A., Berglund L.E., Marcker K.A., Vuust J. (1985) The human ubiquitin multigene family: some genes contain multiple directly repeated ubiquitin coding sequences. *EMBO J.* **4**: 755–759.

Wilkie T.M., Gilbert D.J., Olsen A.S., Chen X.-N., Amatruda T.T., Korenberg J.R., Trask B.J., de Jong P., Reed R.R., Simon M.I., Jenkins N.A., Copeland N.G. (1992) Evolution of the mammalian G protein α subunit multigene family. *Nature Genet.* **1**: 85–91.

Wistow G. (1993) Lens crystallins: gene recruitment and evolutionary dynamism. *Trends Biochem. Sci.* **18**: 301–306.

Wolffe A.P. (1994) Structural and functional properties of the evolutionarily ancient Y-box family of nucleic acid binding proteins. *BioEssays* **16**: 244–251.

Wolffe A.P., Tafuri S., Ranjan M., Familari M. (1992) The Y-box factors: a family of nucleic acid binding proteins conserved from *Escherichia coli* to man. *New Biologist* **4**: 290–298.

Worton R.G., Sutherland J., Sylvester J.E., Willard H.F., Bodrug S., Dubé I., Duff C., Kean V., Ray P.N., Schmickel R.D. (1988) Human ribosomal RNA genes: orientation of the tandem array and conservation of the 5′ end. *Science* **239**: 64–68.

Wright H.T. (1993) Introns and higher-order structure in the evolution of serpins. *J. Mol. Evol.* **36**: 136–143.

Wu C.I., Li W.H., Shen J.J., Scarpulla R.C., Limbach K.J., Wu R. (1986) Evolution of cytochrome *c* genes and pseudogenes. *J. Mol. Evol.* **23**: 61–75.

Xu G., O'Connell P., Viskochil D., Cawthon R., Robertson M., Culver M., Dunn D., Stevens J., Gesteland R., White E., Weiss R. (1990) The neurofibromatosis type 1 gene encodes a protein related to GAP. *Cell* **62**: 599–608.

Yamada Y., Avvedimento V.E., Mudryj M., Ohkubo H., Vogeli G., Irani M., Pastan I., de Crombrugghe B. (1980) The collagen gene: evidence for its evolutionary assembly by amplification of a DNA segment containing an exon of 54 bp. *Cell* **22**: 887–892.

Yeager M., Hughes A.L. (1999) Evolution of the mammalian MHC: natural selection, recombination and convergent evolution. *Immunol. Rev.* **167**: 45–58.

Yokoyama S., Yokoyama R., Rotwein P. (1987) Molecular characterization of cDNA clones encoding the human alcohol dehydrogenase β1 and the evolutionary relationship to the other class I subunits α and γ. *Jpn. J. Genet.* **62**: 241–256.

Yokoyama S., Starmer W.T. (1996) Evolution of the G-protein-coupled receptor superfamily. In *Human Genome Evolution*. (eds M Jackson, T Strachan, G Dover). BIOS Scientific Publishers, Oxford. pp 93–119.

Yoshida Y., Noshiro M., Aoyama Y., Kawamoto T., Horiuchi T., Gotoh O. (1997) Structural and evolutionary studies on sterol 14-demethylase P450 (CYP51), the most conserved P450 monooxygenase: II. Evolutionary analysis of protein and gene structures. *J. Biochem.* **122**: 1122–1128.

Yoshida A., Rzhetsky A., Hsu L.C., Chang C. (1998) Human aldehyde dehydrogenase gene family. *Eur. J. Biochem.* **251**: 549–557.

Zangenberg G., Huang M., Arnheim N., Erlich H.A. (1995) New *HLA-DPB1* alleles generated by interallelic gene conversion detected by analysis of sperm. *Nature Genet.* **10**: 407–414.

Zhang J., Rosenberg H.F., Nei M. (1998) Positive Darwinian selection after gene duplication in primate ribonuclease genes. *Proc. Natl. Acad. Sci. USA* **95**: 3708–3713.

Zinyk D.L., McGonnigal B.G., Dearold C.R. (1993) *Drosophila awd^{k-pn}*, a homologue of the metastasis suppressor gene *nm23*, suppresses the *Tum-I* haematopoietic oncogene. *Nature Genet.* **4**: 195–196.

5

Promoters and transcription factors

The eukaryotic transcriptional apparatus is much more complex than that of prokaryotes and this complexity is bound up with the fact that many eukaryotic genes are silenced by being tightly packed in chromatin (Zuckerkandl, 1997). Transcriptional repression was a necessary pre-condition for the development of large genomes since increasing gene number required gene expression to be restricted both spatially and temporally. That nucleosomal packing and transcriptional repression have, in eukaryotes, long been associated with each other is evidenced by the discovery that the core histone fold is not confined to the histones but has also turned up in a number of transcriptional coactivators and repressors (Ouzounis and Kyrpides, 1996a). This ancient association is also evident in the High Mobility Group (HMG) superfamily (Section 5.2) which may be divided into two distinct subfamilies comprising transcription factors and chromatin structure regulatory proteins, respectively.

The evolution of cellular processes probably involved several distinct stages: metabolism and translation preceded transcription whereas transcription must have preceded transcriptional repression by chromosomal packing at the dawn of the eukaryotic era. Genome expansion was potentiated by chromosomal packing whilst gene regulation became ever more complex as new transcription factors emerged (Ouzounis and Kyrpides, 1996b). The last universal ancestor probably possessed basic molecular components of metabolism and translation while having a prokaryote-like genome organization together with a transcriptional system reminiscent of the archaea.

5.1 Promoters and enhancers

The basic mechanism of transcriptional activation has been conserved across the spectrum of eukaryotes from yeast to human (Schena, 1989). It is therefore not surprising that two sequence elements, the TATAAA and CCAAT boxes which serve to coordinate the basal transcriptional machinery, are found in all eukaryotes. Indeed, these motifs occur in the promoters of a wide variety of RNA polymerase II-transcribed genes. Similarly, the initiator element, a functional analogue of the TATA box, appears to be ubiquitous in eukaryotes (Liston and Johnson, 1999). Other sequence motifs also occur, in different combinations and

Human Gene Evolution, David N. Cooper.

permutations, thereby giving eukaryotic promoters their modular character (Mushegian and Koonin, 1996; reviewed by Nussinov, 1990).

Transcriptional activators bind to promoter sequences in different combinations and permutations, some factors being coordinated by *cis*-acting DNA sequence motifs, others by protein-protein interactions on the promoter. The effect of any given activator is therefore likely to depend on the other activator and repressor proteins present which may bind either competitively or cooperatively. Such a process of 'combinatorial control' has, by maximizing flexibility, potentiated the rapid evolution of quite elaborate gene regulatory networks (Ptashnel, 1997). Those seeking a reference source to *cis*-acting DNA sequence motifs may consult the Transcription Regulatory Regions Database (**http://www.bionet.nsc.ru/trrd**) or the Eukaryotic Promoter Database (**http://www.epd.isb-sib.ch**) both of which contain information on regulatory regions of eukaryotic genes and include data on transcription factor binding sites, promoters, enhancers and silencers.

If we take the simple case of two genes recently duplicated at the genomic DNA level, then we might reasonably expect that their promoter regions would initially be very similar. As a consequence, the expression profiles of these genes might also be expected to be very similar if not identical. However, gene duplication generates redundancy allowing the promoters to be freed from the constraints of selection thereby enabling them to acquire mutations. Such mutations can lead to the inactivation of the gene by abolishing its expression (see Chapter 6) or alternatively can, in rather more subtle ways, serve to change its expression pattern.

In principle, promoters can change either by the slow steady acquisition of single base-pair substitutions in pre-existing sequence motifs or, more dramatically and abruptly, by *promoter shuffling*, the gain or loss of individual regulatory elements exchanged between genes in cassette fashion (Surguchov, 1991). Promoter shuffling may have occurred in cases of homologous genes that contain dissimilar upstream regulatory elements but also in cases of nonhomologous genes containing similar upstream regulatory elements. For promoter shuffling to be viable, the transposed sequence element must be capable of exerting an influence on the transcription of the gene that has just acquired it. That this proposition is a reasonable one is evidenced by the results of studies reported by Kermekchiev *et al.* (1991). These authors tested 27 combinations of different promoters and enhancers and found that the relative *in vitro* efficiency of the enhancers was roughly the same irrespective of the promoter used. Another way in which promoter sequences can change abruptly is through gene conversion as described for the human growth hormone (*GH1*; 17q22-q24; Giordano *et al.*, 1997) and γ-globin (*HBG1, HBG2*; 11p15.5; Chiu *et al.*, 1997) gene promoters.

In subsequent sections, the similarities and differences between the promoters of extant mammalian genes will be explored. In addition, some of the mechanisms by which evolution has recruited specific sequences to a promoter function and fine-tuned their interactions with transcription factors will be described.

5.1.1 Evolutionary conservation of cis-acting elements and 'phylogenetic footprinting'

Sequence conservation in promoter regions is usually held to be an indicator of functional importance and may therefore be used as a rough guide to the location of *cis*-acting regulatory elements that might bind *trans*-acting factors. Such *phylogenetic footprinting* has proven successful in locating *trans*-acting factor binding sites in a considerable number of human genes including the ε-globin (*HBE1*; 11p15.5) gene (Gumucio *et al.*, 1993; Hardison *et al.*, 1997), the γ1-globin (*HBG1*; 11p15.5) gene (Tagle *et al.*, 1988), the β-globin (*HBB*; 11p15.5) gene locus control region (Shelton *et al.*, 1997), the cytochrome *c* oxidase subunit Vb (*COX5B*; 2cen-q13) gene (Bachman *et al.*, 1996), the Duchenne muscular dystrophy (*DMD*; Xp21) gene (Fracasso and Patarnello 1998), the hormone-sensitive lipase (*LIPE*; 19q13) gene (Talmud *et al.*, 1998) and the cystic fibrosis transmembrane conductance regulator (*CFTR*; 7q31.3) gene (Vuillaumier *et al.*, 1997) among others. It was also instrumental in identifying the CArG sequence motif [CC(A/T rich)$_6$GG] important for the regulation of the cardiac α-actin (*ACTC*; 15q14) gene (Taylor *et al.*, 1988).

Phylogenetic footprinting has been used to identify regulatory elements in the promoter regions of the rapidly evolving *SRY* (Yp11.3) genes of 10 mammalian species (Margarit *et al.*, 1998). A total of 10 putative regulatory elements were identified by these means (*Figure 5.1*). Interestingly, differences were apparent between the *SRY* promoters not only in terms of the presence/absence of specific motifs and motif copy number, but also in the spacing between motifs, their relative location and orientation (sense or antisense strand). Conserved elements in the *SRY* gene promoters within each taxonomic group (primates, bovids, rodents) did however tend to occupy orthologous positions.

The study of Vuillaumier *et al.* (1997) identified eleven DNase I-hypersensitive phylogenetic footprints in a 3.5 kb region of the *CFTR* gene (7q31.3) promoter when eight mammals from four different orders were compared. Two of these footprints, which corresponded to the cAMP response element and the PMA-responsive element, were conserved in all species examined, a finding which probably reflects the vital importance of transcriptional control through the cAMP and diacylglyerol pathways respectively. By contrast, a 300 bp purine.pyrimidine stretch, thought to represent a negative element of basal transcription, was found to be present only in the *Cftr* genes of rodents. Studies of the globin genes have also shown that some motifs may be of functional importance in one species but not be conserved between species, for example two Sp1-binding sites in the human *HBB* gene locus control region (Shelton *et al.*, 1997). Such inter-specific differences are explored further in Section 5.1.4.

A variation on this theme is *differential phylogenetic footprinting* which utilizes the promoter sequences of two extant species representing the most closely related lineages between which a difference in developmental expression pattern can be detected. An example of its successful use was in the study of the galago and human γ-globin (*HBG1*) gene promoters (Gumucio *et al.*, 1994; Section 5.1.8). Differential phylogenetic footprinting has also been used to study the promoters of the apolipoprotein AI (*APOA1*; 11q23.3) genes of human and African green monkey in order to assess the functional significance of species-specific

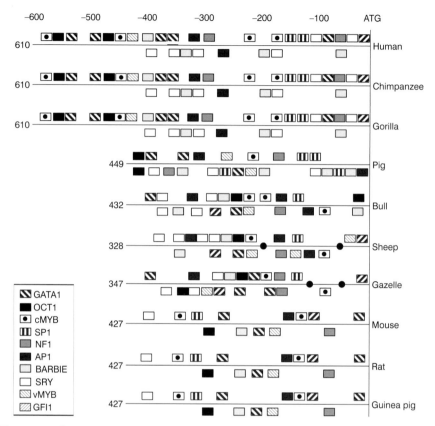

Figure 5.1. Comparison of the promoter sequences from the *SRY* genes of ten mammals (redrawn from Margarit *et al.*, 1998). Putative transcription factor binding sites are denoted by boxes above or below the sequence indicating their position on the sense or antisense strands respectively. Solid circles indicate gap positions present in the sheep and gazelle sequences as compared with the bull sequence. GATA; CMYB; NF1; BARBIE; vMYB; OCT1; SP1; AP1; SRY; GFI1.

differences in expression. Sorci-Thomas and Kearns (1995) identified seven sites within the proximal promoter region of the *APOA1* gene which differed between human and African green monkey. These authors then tested the functional significance of these sequence changes by mutating the human promoter to the simian sequence and testing promoter strength by reporter gene expression assay. Substitutions at three of the sites (−189, −144 and −48 relative to the transcriptional initiation site) were found individually to increase the activity of the wild-type human promoter to ~60–65% of that of the African green monkey. In addition, two double mutations (−144/−48 and −189/−144) restored promoter activity to the same level as found in the monkey. Thus, we may therefore infer that several substitutions in the *APOA1* gene must have occurred during primate evolution which together served to determine the specific level of *APOA1* gene transcription in the different species.

Even if promoter elements are conserved, their relative location may not be. Thus the presence of four perhaps five DNase I hypersensitive sites in the β-globin gene locus control region is conserved between placental mammals, marsupials and monotremes (Hardison *et al.*, 1997) but the spacing between the elements differs between species.

5.1.2 Nonhomologous genes containing similar regulatory elements

Some *cis*-acting elements are common to a wide variety of different genes and may be evolutionarily very ancient. Thus, the TGTGACGTCTTTCAGA cAMP-responsive element in the promoter of the human vasoactive intestinal polypeptide (*VIP*; 6q26-q27) gene is similar not only to elements found in the promoters of nonorthologous murine and avian genes but also to sequences described in yeast and adenoviral promoters and to the *E. coli* consensus sequence recognized by the cAMP receptor protein (Lin and Green, 1989).

Even if we confine ourselves to the same species, completely unrelated genes often also possess similar upstream regulatory elements. Thus, the promoters of the human E-selectin (*SELE*; 1q22-q25) and β-interferon (*IFNB1*; 9p22) genes both contain NFκB, ATF-2 and HMG I(Y) binding sites (Whitley *et al.*, 1994). Other examples of promoter similarity include the serum response elements present in the regulatory regions of the human apolipoprotein E (*APOE*; 19q13.2), c-*fos* (*FOS*; 14q24.3), and β-actin (*ACTB*; 7p12-p15) genes or the AGGCGGCCCTTT motif in the apolipoprotein B (*APOB*; 2p24), apolipoprotein CIII (*APOC3*; 11q23-qter) and α1-antitrypsin (*PI*; 14q32.1) genes (Surguchov, 1991). It is as yet unclear if these elements have evolved by the slow fine tuning of existing sequences or if they have instead been introduced by promoter shuffling.

5.1.3 Paralogous genes containing dissimilar regulatory elements

Evolutionarily related genes in the same organism often differ in their expression profiles. Thus, the various paralogous members of the family of murine cytochrome P450–16 (testosterone 16-α-hydroxylase) genes are evolutionarily related yet are regulated quite differently (Wong *et al.*, 1989). Similarly, the chromosomally unlinked human annexin I (*ANX1*; 9q11-q22), VI (*ANX6*; 5q32-q34), and VII (*ANX7*; 10q21.1-q21.2) genes are evolutionarily related and encode proteins that are structurally and biochemically very similar. However, these genes possess very different sets of promoter and enhancer elements which are presumably responsible for their distinct patterns of tissue expression (Donnelly and Moss, 1998; Shirvan *et al.*, 1994). Can we relate differences in expression profile of paralogous genes to changes in upstream regulatory elements?

A number of studies of paralogous gene promoters have been performed and differences in promoter sequence between members of the same multigene family have indeed been found that may account for differences in transcriptional efficiency. Some paralogous promoters differ with respect to the presence or absence of specific regulatory elements. Others differ in terms of more subtle single base-pair substitutions introduced into specific *cis*-acting motifs which serve to increase or decrease the binding affinity of their cognate transcription factors. Such changes in promoter structure have contributed to post-duplicational gene diversification by providing the divergent gene products with their own specific expression profiles. Steroid hormone receptor responsive elements (Chapter 7, section 7.5.2, *The DNA-binding specifiicity of steroid receptors*) are a classic example of this but a number of other studies of paralogous gene promoters have been reported which serve to illustrate the basic principles.

The murine immunoglobulin variable region heavy chain (V_H) genes may be grouped into 15 families on the basis of coding sequence homology, and the promoter sequences are conserved in a family-specific manner (Buchanan *et al.*, 1997). However, these paralogous V_H gene promoters vary in terms of their transcriptional strength, at least *in vitro*, by as much as 60 to 70-fold. This variation appears to be due to several specific sequence differences between these promoters: (i) the presence/absence of a TATA box, (ii) the presence/absence of initiator elements, (iii) the extent of divergence of the octamer sequence element from its consensus (ATGCAAAT), and (iv) the spacing between the octamer motif and the heptamer motif located between 2 bp and 20 bp upstream (Buchanan *et al.*, 1997).

The human genome contains a cluster of 11 pregnancy-specific glycoprotein (*PSG*) genes on the long arm of chromosome 19q (*PSG1, PSG2, PSG3, PSG4, PSG5, PSG6, PSG7, PSG8, PSG11, PSG12, PSG13*). These genes are highly homologous and this homology extends to their promoter regions. The promoters of six *PSG* genes have been characterized and display differences based upon the minimal promoter required for optimal expression (Chamberlin *et al.*, 1994). The class 1 *PSG* genes require nucleotides −172 to −34 (relative to the transcriptional initiation site) whereas the comparable minimal region of class 2 *PSG* genes lies between −172 and −80. This difference has been attributed to the presence of an imperfect SP1 binding site between −148 and −141 in the class 1 promoters. The perfect SP1 element (CCCCGCCC) present in the class 2 promoters is altered either to CCCTGCCC or CCCCACCC (NB. both represent mutations in a CpG dinucleotide compatible with the mechanism of methylation-mediated deamination) in the class 1 promoters (Chamberlin *et al.*, 1994). The SP1 binding sites of the class 1 promoters bind SP1 with lower affinity and require additional activator elements for optimal expression (Chamberlin *et al.*, 1994).

One of the best characterized differences between paralogous gene promoters is that manifested by one of the human small proline-rich protein genes. The *SPRR1A, SPRR1B, SPRR2A*, and *SPRR2C* genes possess an Ets binding site at position −55 (relative to the transcriptional initiation sites) which is critical for promoter activity. The corresponding site in the *SPRR3* gene has however been lost through a C→T transition, although this loss appears to be compensated for, at least in part, by the presence of an Ets binding site at position −239 (Fischer *et al.*, 1999). This single base-pair substitution has been shown to account for the lower rate of expression of the *SPRR3* gene in cultured keratinocytes as compared with the *SPRR1A* and *SPRR2A* genes. This is despite the fact that the −239 Ets binding site in the *SPRR3* gene has a higher affinity for its cognate transcription factor ESE-1 than the −55 sites in the *SPRR1A* and *SPRR2A* gene promoters. Fischer *et al.* (1999) found that removal of the −239 Ets binding site coupled with the restoration of the −55 binding site by a T→C mutation yielded a promoter activity comparable to that of the *SPRR1A* and *SPRR2A* promoters and threefold higher than that of the wild-type *SPRR3* promoter. The location of the Ets binding site in the *SPRR3* gene therefore appears to be critical for determining the activation potential of ESE-1. In the *SPRR3* promoter, three transcription factor binding sites (AP-1, ATF, and Ets) function cooperatively but these sites are not interdependent as in the *SPRR2A* promoter. Thus, the loss of the proximal Ets binding site in the *SPRR3* gene promoter during the evolution of the gene family may have been responsible for changing the promoter from one which required highly synergistic interactions between its cognate transcription factors to one which functions with less stringent, cooperative interactions.

Although the diversity of glycoproteins encoded by the *HLA-B* locus is greater than that encoded by the *HLA-A* locus, the *HLA-B* locus promoter sequences are more homogeneous than those of the *HLA-A* locus (Vallejo and Pease, 1995). We may speculate that selection for diversity of B glycoproteins may be directly related to the conservation of the promoter sequences and *vice versa* for the A glycoproteins and their gene promoters. Although the human *HLA-A* (6p21.3) and *HLA-B* genes are coordinately expressed, they differ in their responsiveness to interferon. A functional interferon-responsive element (IRE) is present in the promoters of the *HLA-B* genes in the great apes but this IRE has been inactivated by an A → T transversion in the promoters of the *HLA-A* genes (Vallejo *et al.*, 1995). Since this lesion is present in the *HLA-A* genes of orangutan, gorilla, chimpanzee and human, it is likely to have occurred before the divergence of the great apes.

The chorionic gonadotropin β gene (*CGB*; 19q13.32) has evolved by duplication of an ancestral luteinizing hormone β gene and is expressed exclusively in the placenta. It exhibits some 94% homology with the pituitary expressed luteinizing hormone-β (*LHB*; 19q13.32) gene between coding regions and 90% homology between promoters (Hollenberg *et al.*, 1994). Both *LHB* and *CGB* genes possess TATAAA boxes at identical locations. This motif directs the initiation of an *LHB* transcript with a 9 nucleotide 5′ UTR. The *CGB* gene does not however use this box for transcriptional initiation but instead employs a TATA-less promoter (with putative initiator element) that serves to initiate transcription 357 bp upstream of the *LHB* transcriptional initiation site (Jameson *et al.*, 1986). This difference in transcription site usage is evolutionarily conserved, being found in rat, cow and human (Jameson *et al.*, 1986). The experimental exchange of motifs between the two promoters has allowed the identification of three distinct regions between −362 and +104 necessary for placental expression of the *CGB* gene (Hollenberg *et al.*, 1994). It would appear as if the placental expression pattern of this gene has resulted from the use of an alternative transcriptional initiation site in conjunction with the acquisition of multiple regulatory elements that exert their effects in combinatorial fashion.

The paralogous members of the human growth hormone gene family on chromosome 17q23 differ in terms of their promoter sequences: a T → C transition at position −112 of the *CSH1* gene promoter serves to reduce binding of the pituitary-specific transcription factor Pit-1 (Tansey and Catanzaro, 1991). By contrast, the paralogous *GH1*, *GH2* and *CSH2* genes possess a T at this location within the Pit-1 binding site. Pit-1 binding is thought to represent a negative control mechanism by virtue of its interference with the binding of SP1 to an adjacent site in the promoter. The T→C transition in the *CSH1* gene may thus abrogate this negative control mechanism since it facilitates SP1 binding and promoter gene activation by SP1.

Several other examples of promoter differences in paralogous genes are known. Thus the human γB (*CRYGB*; 2q33-q35) and γC (*CRYGC*; 2q33-q35) crystallin genes possess similarly located TATA boxes but the *CRYGC* gene lacks the CCAAT box present in the *CRYGB* gene (Graw *et al.*, 1993). The expression of the human main-type histone H1 (*H1F1*; 6p21.3) genes is coordinated with DNA replication whereas the regulation of the replacement H1 subtype H1° (*H1F0*; 22q13) gene is more complex. This difference appears to be reflected in the structures of the respective gene promoters with a CCAAT box present in the main-type histone H1 gene promoters being absent in the H1° gene promoter (Doenecke *et al.*, 1994). Similarly, the promoters of the human mammaglobin 1 (*MGB1*) and 2 (*MGB2*) genes, which are closely linked to the related uteroglobin (*UGB*) gene on chromosome 11q13, are homologous for the first 132 bp but then exhibit major differences that are probably responsible for the different expression patterns of these genes (Becker *et al.*, 1998). Finally, the highly homologous human placental (*ALPP*; 2q37) and intestinal (*ALPI*; 2q37) alkaline phosphatase genes possess several nucleotide substitutions and deletions in their 5' flanking regions which could account for their differing tissue specificity (Knoll *et al.*, 1988).

5.1.4 Orthologous genes containing dissimilar regulatory elements

Differences between orthologous gene promoter sequences have also been described which may explain differences in expression of the same gene in different species. In some cases, the orthologous promoters differ with respect to the presence or absence of specific *cis*-acting sequences or alternatively in terms of the number of such motifs. In other cases, single base-pair substitutions have been introduced that alter the affinity of the *cis*-acting sequences for their cognate transcription factors. Orthologous genes may therefore acquire different expression profiles in different species. Thus, the expression of the human G protein-coupled receptor 1 (*GPR1*) gene (15q21.6) which is hippocampus-specific contrasts with that of its rat counterpart which is not expressed in the hippocampus (Marchese *et al.*, 1994). The human regulatory myosin light chain (*MYL5*; 4p16.3) gene is expressed in human adult retina and fetal muscle but not in adult skeletal muscle suggesting that this gene is developmentally regulated (Collins *et al.*, 1992). By contrast, the orthologue of this gene is abundantly expressed in the adult skeletal muscle of the African green monkey, providing a good example of a gene that is differentially expressed between humans and another primate. The tissue-specific expression of many genes is known to be determined by the presence of enhancer elements that bind tissue-specific transcription factors. The evolution of novel tissue specificities can therefore be investigated by comparison of the sequences of gene regulatory elements between different species.

The human apolipoprotein A1 (*APOA1*; 11q23) gene is expressed predominantly in the liver whilst its rabbit counterpart is expressed predominantly in the intestine. The rabbit *Apoa1* gene promoter contains two *cis*-acting elements, E2 (GGAGAAGAGAGGTCA) and E3 (GAAAGTCTCTCTTCTGTT) both similar to enhancer sequences found in murine immunoglobulin genes, that are absent from the human *APOA1* gene promoter (Bochkanov *et al.*, 1990; Higuchi *et al.*, 1988; Pan *et al.*, 1987). These sequences may help to explain the differing tissue specificity of this gene between the two species.

The rat tissue-type plasminogen activator gene differs from that of mouse and human (*PLAT*; 8p12) in that it is induced by gonadotropins via a cAMP-dependent pathway. The rat *Plat* gene promoter possesses a cAMP-responsive element (CRE) but, at the corresponding location in the murine and human gene promoters, a single base-pair change is present that serves to reduce drastically the binding affinity to CRE-binding protein (Holmberg *et al.*, 1995). This difference is sufficient to account for the unresponsiveness of the human and murine gene promoters to forskolin and follicle-stimulating hormone.

The promoters of the murine and human lactoferrin (*LTF*; 3p21.2-p21.3) genes have numerous *cis*-acting regulatory motifs in common. However, in the bovine lactoferrin gene, several of these motifs (GATA-1, Oct-1, COUP, AP-2, and glucocorticoid and acute phase response elements) are absent, providing a possible explanation of the relatively weak expression of bovine lactoferrin in comparison to human and mouse (Seyfert *et al.*, 1994).

Both tumor necrosis factor α and lipopolysaccharide stimulate the expression of P-selectin in murine endothelial cells but this does not occur in human endothelial cells. This difference is thought to be explicable in terms of structural differences in the promoters of the P-selectin genes between the two species (Pan *et al.*, 1998). The promoter of the human P-selectin (*SELP*; 1q23-q25) gene possesses a unique NFκB binding site which is thought to be specific for p50 or p52 homodimers. Its murine counterpart contains two tandem NFκB binding sites and a variant activating transcription factor/cAMP response element which is similar to that found in the E-selectin gene required for tumor necrosis factor α- and lipopolysaccharide-inducible expression.

The promoters of the human and murine *XRCC5* DNA repair genes (2q35) differ with respect to the number of copies of a 21 bp near-perfect palindromic element (Kpb); the mouse *Xrcc5* gene possesses a single copy whereas the human *XRCC5* gene possesses seven copies (Ludwig *et al.*, 1997). Amplification of this *cis*-acting element occurred in the human lineage and this may help to account for the human *XRCC5* promoter being at least three to five-fold stronger than its murine counterpart (Ludwig *et al.*, 1997).

Calbindin-D_{9k} is a mammalian-specific cytosolic calcium-binding protein which is encoded by a gene (*CALB3*; Xp) that is ubiquitously expressed in the intestine. In the rat, this gene is also expressed in the uterus under the control of estrogen whose effects are mediated by an estrogen response element (ERE) (Darwish *et al.*, 1990). By contrast, uterine expression is not found in human or baboon. In these primates, two nucleotide substitutions are present that have abolished the binding affinity of the ERE to the estrogen receptor and probably account for the loss of uterine expression (Jeung *et al.*, 1994; 1995).

The chorionic gonadotropin α-chain gene is expressed in the pituitary of all mammals but is expressed in the placenta only in primates and horses. Since horses and primates are only distantly related, and species such as the cow and rodents which are more closely related to horses than primates do not express the gene in placenta, it would appear as if the placental expression of the gene has been independently derived, perhaps by the *convergent evolution* (Chapter 4, section 4.3) of placenta-specific enhancer elements. Support for this postulate has come from a comparison of human and equine chorionic gonadotropin α-chain (*CGA*; 6q21.1-q23) gene promoter sequences (Steger *et al.*, 1991). The human gene contains a trophoblast-specific element (TSE) and two copies of a cAMP response element (CRE) all of which are required for full placental expression. Both the CRE and TSE are bound by the leucine zipper-containing protein, CREB. By contrast, cAMP regulation of the equine *Cga* gene is not mediated by CREB but instead by aACT, a GATA-related protein that binds to the promoter at a site distinct from the CREB-binding site. It would thus appear as if the conversion of the *CGA* gene from being pituitary-specific to being also placentally expressed was a consequence of independent evolutionary processes in primates and horses.

In the New World monkey, *Cebus apella*, the γ2-globin (*HBG2*) gene is expressed at a 20-fold higher level than the closely linked γ1-globin (*HBG1*) gene (Johnson *et al.*, 1996). The most obvious difference between the promoters of the two genes and the one most likely to account for the difference in expression, is a CCAAC motif instead of the canonical CCAAT found in the *HBG1* promoter (Chiu *et al.*, 1996). Intriguingly, the *HBG1* gene has been inactivated by deletion in the Atelidae whereas in the Pitheciini, the CCAAT box has been changed to CCGAT (Chiu *et al.*, 1996). In the New World monkey *Aotus azarae*, a single hybrid *HBG* gene has been created as a result of an unequal crossing over event (Chiu *et al.*, 1996); this gene possesses the promoter and 5′UTR of the *HBG1* gene coupled to the coding region of the *HBG2* gene. Why the *HBG1* gene has been inactivated several times independently in platyrrhines is unclear; expression from both *HBG1* and *HBG2* genes has been preserved in catarrhines (Chapter 4, section 4.2.1, *Globin genes*).

The human prolactin (*PRL*; 6p22) gene possesses two promoters, the proximal one reponsible for directing expression in the anterior pituitary, the distal one for expression in the decidualized endometrium and the mammary gland. Sequences homologous to both promoters are present in the rat *Prl* gene promoter but the distal sequence is nonfunctional (Shaw-Bruha *et al.*, 1998). Whether the human *PRL* gene has gained a functional promoter in the last 100 Myrs since the divergence of our common ancestors, or whether the distal promoter in the rat gene has ceased to function during this time, is unclear.

Other examples of inter-specific differences in orthologous promoter regions include the thymidine kinase 1 (*TK1*; 17q25.2-q25.3) gene promoter which contains functional CCAAT elements in humans, chickens and Chinese hamsters but not in mice and rats (Arcot *et al.*, 1991) and the osteonectin (*SPARC*; 5q31-q32) gene promoter whose GGA-box sequences contribute to cell-type specific expression in human but not in the bovine (Hafner *et al.*, 1995). The mammalian *SPARC* gene also differs from its *Xenopus* counterpart in that the latter contains a TATA box but lacks a GGA-box (Damjanovski *et al.*, 1998).

Various species-specific sequence differences have also been reported in nuclear factor binding sites in the promoter regions of the *D7Rp2e* genes in *Mus domesticus* and *M. pahari* that alter both the pattern of binding site occupancy and the ability of the bound factors to repress transcription (Singh and Berger, 1998; Singh *et al.*, 1998).

Various examples of the recruitment of metabolic enzymes to a novel lens-specific role have been documented among the taxon-specific crystallins (see Chapter 4, section 4.2.1, *Crystallin genes*). Recruitment of these genes to a new lens function has been achieved by acquiring the potential for lens expression by the development of novel promoter elements. Two strategies appear to have been employed viz. the modification of (i) distinct regions of the same non-functional intronic sequence to perform a role in lens-specific expression and (ii) pre-existing promoters previously utilized for nonlens tissue expression (see Chapter 4, section 4.2.1, *Crystallin genes*).

Not surprisingly, some differences between orthologous promoters are apparently neutral and do *not* obviously affect promoter function. One example of this is the 54 bp insert in the promoter of the human liver arginase (*ARG1*; 6q22.3-q23.1) gene which is absent from the same gene in macaques (Goodman *et al.*, 1994).

Not all regulatory sequences occur upstream of the transcriptional initiation site. Some occur in introns (Chapter 3, section 3.1) but such regulatory elements are not necessarily conserved between orthologous genes. Thus, the 83 bp intron 1 of the human *CD68* (17p13) gene contains a macrophage-specific enhancer but the equivalent intron of the orthologous murine (macrosialin; *Ms*) gene does not despite ~80% nucleotide sequence homology (Greaves *et al.*, 1998).

5.1.5 Bidirectional promoters

There is now a growing list of divergently transcribed gene pairs arranged in head-to-head fashion separated by a bidirectional promoter. Such an organization has often but not always arisen through a process of gene duplication followed by inversion. In true cases of a bidirectional promoter, the gene promoters overlap and contain common elements that can allow the coordinate regulation of expression of both genes. Bidirectional promoters are often associated with CpG islands which therefore serve as useful markers for their location (Brenner *et al.*, 1997; Lavia *et al.*, 1987).

Examples of bidirectional promoters are thought to include the human histone *H2A* and *H2B* (1q21-q23) genes (Hentschel and Birnstiel, 1981), type IV collagen (*COL4A1* and *COL4A2*, 13q34, *Figure 5.2*; *COL4A3* and *COL4A4*, 2q36-q37; *COL4A5* and *COL4A6*, Xq22) genes (Schmidt *et al.*, 1993; Oohashi *et al.*, 1995), and the *TAP1* and *LMP2* (6p21.3) genes (Wright *et al.*, 1995). These gene pairs encode proteins that are involved in the same biological processes whether as components of multi-chain proteins (collagens) or protein complexes (histones) or as proteins with associated functions (both TAP1 and LMP2 have a role in antigen processing). This is no coincidence since evolutionarily related pairs of genes that have arisen by a process of gene duplication/inversion would be predicted to have similar functions. As a consequence of their mode of creation, they will often possess the potential for coordinate regulation as long as the promoter region(s) are still intact and the genes remain closely linked. In principle, the newly created functional redundancy of the promoter elements (Section 5.1.11) can be reduced by element removal and subsequent element sharing while still retaining the potential for coordinate regulation (Section 5.1.14). Alternatively,

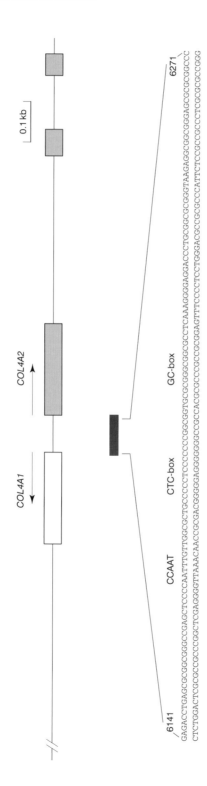

Figure 5.2. Bidirectional promoter separating the human collagen α1(IV) and α2(IV) genes (*COL4A1* and *COL4A2*) (redrawn from Schmidt *et al.*, 1993).

retention of the functional redundancy could allow diversification of promoter elements leading to independent regulation.

There are examples of divergently transcribed gene pairs that are not obviously related evolutionarily and which are unlikely to have arisen by gene duplication and inversion. Thus the human phosphoribosylaminoimidazole carboxylase (*PAICS*) and phosphoribosylpyrophosphate amidotransferase (*PPAT*) genes, which encode enzymes of the *de novo* purine biosynthesis pathway, are only 229 bp apart on chromosome 4q12 but encode products that are not structurally related (Brayton *et al.*, 1994). Similarly, the human isocitrate dehydrogenase 3 (*IDH3G*) and signal sequence receptor 8 (*SSR4*) genes on chromosome Xq28 are separated by a 133 bp sequence of CpG island character, share overlapping promoter elements but encode proteins which have no obvious common function or evolutionary history (Brenner *et al.*, 1997). Finally, some genes may be transcribed in opposite directions but do not necessarily share promoter elements nor encode proteins which share any obvious function, for example the human minichromosome maintenance 4 (*MCM4*) and DNA-activated protein kinase (*PRKDC*) genes which are separated by ~700 bp on chromosome 8q11 (Connelly *et al.*, 1998).

5.1.6 5′ and 3′ untranslated regions of genes

The 5′ and 3′ untranslated regions (UTRs) of vertebrate genes are often evolutionarily conserved (Duret *et al.*, 1993; Lipman 1997) but to a lesser extent than their corresponding coding sequences. Thus, whereas the average degree of nucleotide sequence identity shared between the coding regions of human and murine genes is ~85%, the 5′ and 3′ UTRs exhibit sequence identities of 67% and 69% respectively (Makalowski *et al.*, 1996). In practical terms, this degree of sequence divergence is sufficient to allow the discrimination of different mammals simply by reference to their UTR sequences (e.g. Soteriou *et al.*, 1995).

Within the 5′ and 3′ UTRs, *highly conserved regions* (HCRs) can be found. These were defined by Duret *et al.* (1993) as sequences of at least 100 bp that exhibit >70% homology between species, and which diverged more than 300 Myrs ago. Since such sequences would be expected, in the absence of selective pressure, to share only ~30% similarity, evolutionary conservation implies function. Comparison of mammalian and avian genes reveals that ~17% and ~30% of genes contain HCRs in their 5′ UTRs and 3′ UTRs respectively (Duret *et al.*, 1993). When mammalian and fish genes are compared, the proportion of genes whose 3′ UTRs possess an HCR falls to 5% (Duret *et al.*, 1993). Since HCRs occur relatively infrequently within introns, the evolutionary constraints would appear to operate at the level of the mature mRNA (Duret *et al.*, 1993). The HCRs in the 5′ UTRs of the creatine kinase B, c-*jun* and actin genes span the CCAAT and TATA boxes and are therefore likely to play a role in transcriptional regulation (Duret *et al.*, 1993). Other HCRs in the 5′ UTRs of the transforming growth factor β3 and ferritin heavy chain genes span elements known to be involved in translation (Duret *et al.*, 1993). Some HCRs in 3′ UTRs are thought to play a role in mRNA degradation (e.g. c-*fos* and transferrin receptor genes) whereas others may be important for mRNA transport and translation (Duret *et al.*, 1993). HCRs in 3′ UTRs appear to be preferentially associated with widely expressed genes especially those encoding DNA-binding proteins and cytoskeletal proteins (Duret *et al.*, 1993).

The human surfeit 2 and 4 (*SURF2*, *SURF4*; 9q34.1) genes differ from their murine counterparts in that whilst the mouse genes overlap by 133 bp in their 3′ UTRs, the human genes are separated by 302 bp (Duhig *et al.*, 1998). The human *SURF2* gene contains two alternative polyadenylation sites resulting in short 3′ UTRs of 17 and 25 bp whereas the mouse *Surf2* gene contains a 359 bp 3′ UTR. The much shorter human *SURF2* 3′ UTR probably accounts for the absence of overlap with the human *SURF4* gene.

One example of the emergence of a functional difference between the 3′ UTRs of two paralogous human genes is provided by the α-globin (*HBA1*; 16p13.3) and ζ-globin (*HBZ*; 16p13.3) genes (Russell *et al.*, 1998). The *HBA1* and *HBZ* genes are coexpressed in the embryonic yolk sac. A switch to exclusive expression of the *HBA1* gene in the fetus and adult involves the developmental silencing of the *HBZ* gene. Silencing is achieved both by transcriptional control but also through a post-transcriptional mechanism that serves to reduce the relative stability of *HBZ* mRNA. The *HBA1* and *HBZ* genes both assemble an mRNP stability-determining complex on their 3′ UTRs but these complexes form with different affinities on the two genes. The diminished efficiency of complex assembly on the *HBZ* 3′ UTR results from a C→G transversion in a polypyrimidine tract that is common to both genes. This substitution is associated with a shortened poly(A) tail on the *HBZ* mRNA that may mediate accelerated *HBZ* mRNA decay.

Another example of a functional difference between the 3′ UTRs of evolutionarily related human genes is provided by the human alcohol dehydrogenase (*ADH2*; 4q22) gene which differs from the other paralogous *ADH* family members by virtue of a T→C transition within a canonical polyadenylation site 3′ to the gene. This substitution appears to be at least in part responsible for the use of alternative polyadenylation sites leading to the formation of multiple *ADH2* mRNAs (Trezise *et al.*, 1989).

Alterations in the 5′ UTR may also influence the expression pathway. For example, the efficiency of mRNA translation of the renal ornithine decarboxylase (*Odc*) gene is significantly lower in *Mus pahari* as compared to *Mus domesticus* (Johannes and Berger, 1992). This is thought to be due to the acquisition of several single nucleotide substitutions and a 12 bp deletion in the 5′ UTR of the *Odc* gene in *M. pahari*. These sequence changes are predicted to alter the secondary structure of the mRNA molecule and this may influence translation efficiency.

5.1.7 Inter-specific differences in promoter selection

Three distinct mRNA species (L1, M and L2 respectively) are generated from the human aldolase A (*ALDOA*; 16q22-q24) gene via the differential incorporation of three exons encoding the 5′ UTR (Mukai *et al.*, 1991; *Figure 5.3*). The production of the three mRNAs is controlled by three different promoters (*Figure 5.3*) which are utilized singly, doubly or all together depending upon the expressing tissue. The DNA sequences corresponding to these promoters are present in the rat gene but the L1 promoter is not utilized (Mukai *et al.*, 1991). Thus the L1 promoter has either been acquired in the human lineage or, (perhaps more likely) lost in the rat lineage since the divergence of primates and rodents.

The vast majority of studies that have mapped transcriptional initiation sites have been performed on cultured cells and it is by no means clear that those sites

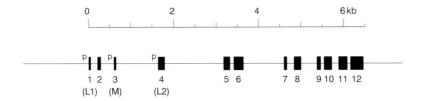

Figure 5.3. Exon/intron distribution in the human aldolase A (*ALDOA*) gene. Exons 1, 3, and 4 correspond to leader exons L1, M2, and L2. The positions of the three alternative promoters are denoted by p (redrawn from Mukai *et al.*, 1991).

identified by *in vitro* studies are actually always utilized *in vivo*. White *et al.* (1998) identified the *in vivo* transcriptional initiation sites of the cystic fibrosis trans-membrane conductance regulator (*CFTR*; 7q31.3) gene in both human and mouse. Tissue-specific variation in the position of the transcriptional initiation sites was noted in both species but the sites were not conserved between equivalent tissues. This finding suggests that the precise mechanism of transcriptional initiation for a given gene may not be absolutely conserved between species.

5.1.8 Developmental changes in gene expression

In higher primates (both platyrrhine and catarrhine), the γ-globin (*HBG1*, *HBG2*; 11p15.5) genes are expressed during fetal life whereas in nonprimates and prosimians, the genes are expressed in the embryo. The conversion of the *HBG1* and *HBG2* genes to a fetal pattern of expression must therefore have occurred after the divergence of simians from prosimians some 55 Myrs ago. Implicit in this conversion is the activation of the γ-globin gene in fetal life (a stage at which it was previously repressed) and repression of the γ-globin gene in embryonic life at which stage it was previously active. Prosimians are characterized by the possession of only one γ-globin gene whereas higher primates possess two copies. The promoter of one of the duplicated γ-globin genes may thus have been able to escape the influence of natural selection and in so doing accumulate mutations that served to alter the timing of its developmental expression. Any beneficial changes thus acquired could then have been readily transferred to the other γ-globin gene by gene conversion.

A burst of sequence change occurred after the divergence of simians from prosimians but before the divergence of Old World monkeys from New World monkeys (Fitch *et al.*, 1990, 1991). Most of these changes were then conserved during the subsequent evolution of the simian γ-globin genes. Studies of the γ-globin gene promoter in transgenic mice have shown that all the sequence changes necessary for changing to the fetal pattern of gene expression are located within a 4 kb fragment containing the γ-globin gene (TomHon *et al.*, 1997). However, within 260 bp of the transcriptional initiation site, there are 19 nucleotide changes specific to simians, 16 of which are located in or near highly conserved sequence motifs (Chiu *et al.*, 1997; Fitch *et al.*, 1990; 1991). Further, comparison of the human and galago (a prosimian) sequences reveals 57 nucleotide differences over the same region with the majority again being located near highly conserved sequence motifs. It may be some time before the precise sequence changes responsible for the change to fetal expression are unequivocally determined.

One way of approaching this question has been through the use of 'differential phylogenetic footprinting' (Section 5.1.1) and gel retardation analysis. By these means, Gumucio *et al.* (1994) identified several proteins (G1, G2, G3, and G4) that bound the galago sequence but did not bind the corresponding human sequence. Phylogenetic reconstruction and gel retardation analysis were used to demonstrate that the promoter sequence of the embryonically expressed γ-globin gene of the primate common ancestor would have bound proteins G1 and G2 some 4–6-fold more strongly than the promoter of the fetally expressed simian ancestor (Gumucio *et al.*, 1994). The binding strength of these proteins correlated with repression of promoter activity *in vitro*, suggesting that the loss of the binding sites for these proteins in the ancestral simian γ-globin gene could have potentiated the conversion of this gene to a fetal onset of expression (Gumucio *et al.*, 1994).

5.1.9 Promoter polymorphisms

Promoter polymorphisms affecting the expression of the downstream gene are probably not infrequent. However, as yet, relatively few examples of promoter polymorphisms in human genes have been properly characterized by means of functional (e.g. reporter gene) studies. Examples of such polymorphisms include plasminogen activator inhibitor type 1 (*PAI1*, 7q21.3-q22.1; Dawson *et al.*, 1993), tumor necrosis factor α (*TNF*; 6p21.3; Wilson *et al.*, 1997), apolipoprotein AI (*APOA1*, 11q23.3; Angotti *et al.*, 1994), lipoprotein(a) (*LPA*, 6q27; Suzuki *et al.*, 1997; Wade *et al.*, 1993), lipoprotein lipase (*LPL*; 8p22; Hall *et al.*, 1997), interleukin 6 (*IL6*, 7p21-p15; Fishman *et al.*, 1998), factor VII (*F7*; 13q34; Pollak *et al.*, 1996), hormone-sensitive lipase (*LIPE*; 19q13; Talmud *et al.*, 1998) and monoamine oxidase A (*MAOA*, Xp11.23; Sabol *et al.*, 1998). The presence of polymorphisms in gene promoter regions is not unusual *per se* since all gene regions harbor polymorphisms. Indeed, such variants are quite consistent with, and explicable in terms of, a neutralist model. However, it is possible that those polymorphisms in the promoter region which specifically affect gene expression confer, or have conferred, a selective advantage.

The *PAI1* promoter polymorphism constitutes the insertion or deletion of a single G residue at position −675 (Dawson *et al.*, 1993). The *ins* allele contains an interleukin 1-responsive element which is not present in the *del* allele suggesting that individuals homozygous for the *del* allele may exhibit an altered PAI-1 response during the acute phase reaction (Dawson *et al.*, 1993). Similarly, a common insertion polymorphism (G) at position −1607 in the human matrix metalloproteinase-1 (*MMP1*; 11q22–23) gene promoter creates a binding site for members of the Ets family of transcription factors which results in the increased transcription of the gene (Rutter *et al.*, 1998).

A highly unusual 76 bp length polymorphism (f = ~ 0.20/0.80) in the human antithrombin III (*AT3*; 1p31.3-qter) gene promoter results from the alternative presence of two apparently distinct sequences of 108 bp (L) and 32 bp (S) at the same position, ~345 bp upstream of the ATG translational initiation codon (Bock and Levitan, 1983). How did this unusual polymorphism evolve and become established in the general population? Winter *et al.* (1995) noted that the sequence flanking the L allele contained numerous homologous motifs suggestive of an ancient duplication event. Some residual homology was also noted between the L and S alleles (*Figure 5.4*) but the L allele did not obviously contain duplicated sequence. The most parsimonious explanation was therefore the emergence of the S allele from the L allele by partial deletion followed by sequence divergence. The deletion event(s) could have occurred by homologous recombination mediated by the homologies between the L-specific sequence and the region immediately downstream. Interestingly, however, no difference in expression could be discerned between the two alternative alleles (Winter *et al.*, 1995).

A repeat length polymorphism in the human solute carrier family member *SLC6A4* (17q11.1-q12; Delbrück *et al.*, 1997) gene also appears to be present in the gorilla although not in the chimpanzee. It is unclear whether this is an example of an ancient trans-species polymorphism (Chapter 1, section 1.2.2) or whether it has arisen independently in the two lineages.

```
     -330
L      CTGATTTAGTTAACGAGAAACAAAAAATCCTGCAGACAAGTTTC   TCCTCAGTCAGGTA
                 T      A   AAC                   CAAGTTT    TCTT   GT AG
S                TGGGTATGAAC                      CAAGTTTGTTTCCTTGGTTAG
             -304
```

Figure 5.4. Alignment of the alternative L- and S-specific sequences in the promoter region of the human *AT3* gene indicating regions of homology (redrawn from Winter *et al.*, 1995). For explanation, see text.

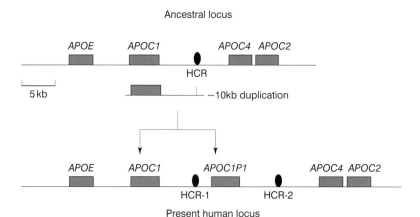

Figure 5.5. Evolution of the human apolipoprotein E2/CI/CIV/CII loci (adapted from Allan *et al.*, 1995). The relative locations of the human *APOE, APOC1, APOC2,* and *APOC4* genes and the *APOC1P1* pseudogene are shown by closed boxes. The locations of the two hepatic control regions (HCR) are denoted by ovals.

5.1.10 Promoter duplication

Promoters may occasionally be duplicated in the absence of the duplication of their associated genes. Thus the *BRCA1* (17q21) and 1A1.3b (*M17S2*; 17q21.1) gene promoters are duplicated in tandem fashion (Barker *et al.*, 1996) but the functional significance of this observation is unknown.

Two copies of a motif known as the hepatic control region (HCR), located, respectively, 15 kb and 5 kb downstream of the human *APOE* and *APOC1* genes on chromosome 19q13.2, arose by a regional duplication event which encompassed the *APOC1* gene ~40 Myrs ago (Raisonnier *et al.*, 1991; *Figure 5.5*). Whilst the duplicated *APOC1* gene has become a pseudogene (*APOC1P1*), the duplicated HCR (HCR2) has retained 85% homology with HCR1 and appears to be able in its new location to direct the transcription (albeit infrequent transcription) of a sequence resembling an exon which is spliced to *APOC4* 5' exons even in the absence of conventional promoter elements (Allan *et al.*, 1995).

5.1.11 Functional redundancy of promoter elements

The human 7SK RNA gene (*RN7SK*; chromosome 6) contains a proximal sequence element (PSE) between -49 and -65 and a distal sequence element (DSE) between -243 and -210 which display sequence similarity and functional homology (Boyd *et al.*, 1995). The PSE can retain function after extensive mutation but only if the DSE is intact. How this apparent functional redundancy has been maintained is unclear.

The promoter of the human ε-globin (*HBE1*; 11p15.5) gene displays functional redundancy of a different type; it contains eight YY1 binding sites, five binding sites for a putative stage selector protein and seven binding sites for a hitherto unidentified protein (Gumucio *et al.*, 1993). Other probable examples of functional redundancy include the eight SP1 sites present in the human muscle phosphofructokinase (*PFKM*; 12q13.3) gene promoter (Johnson and McLachlan, 1994). Without detailed functional studies, however, it is unclear if the multiple *cis*-acting regulatory elements are truly redundant or if each additional copy leads to an incremental increase in transcription factor binding potential and hence promoter strength.

5.1.12 Recruitment of repetitive sequences as promoter and silencer elements

Alu sequences. A number of examples of the recruitment of repetitive sequence elements, mostly *Alu* sequences, as gene promoter and silencer elements have been studied (reviewed by Britten, 1996; 1997; Robins and Samuelson, 1992; Table 5.1). Over evolutionary time, the insertion of *Alu* repetitive sequence elements in the vicinity of genes has served to introduce different motifs that have altered the expression level or tissue specificity of the associated gene either immediately or after subsequent fine tuning by natural selection. One such motif is a retinoic acid response element (RARE) in the human *KRT18* (12q13) gene: three hexamer half sites, related to the consensus AGGTCA, arranged as direct repeats with a spacing of 2 bp (Vansant and Reynolds, 1995). These sites are

Table 5.1. Repetitive sequence insertions in mammalian genes that have altered the expression of the gene

Gene	Insertion	Comments	Reference
KRT18	Alu	Alu sequence contains retinoic acid receptor binding site	Thorey et al. (1993); Vansant and Reynolds (1995)
CD8A	Alu	Alu sequence carries two Lyf-1, one bHLH and one GATA-3 binding sites	Hambor et al. (1993)
FCER1G	Alu	Alu sequence carries positive and negative elements	Brini et al. (1993)
PTH	Alu	Alu sequence carries negative calcium response element	McHaffie and Ralston (1995)
BRCA1	Alu	Alu sequence serves as estrogen receptor-dependent enhancer	Norris et al. (1995)
MPO	Alu	Alu sequence contains composite SP1-thyroid hormone-retinoic acid response element	Piedrafita et al. (1996)
WT1	Alu	Silencer in intron 3, 12 kb from promoter	Hewitt et al. (1995)
F9	LINE	Located at position − 800	Kurachi et al. (1997)
LPA	LINE	Located ~20 kb upstream of transcriptional initiation site; contains enhancer element	Yang et al. (1998)
AMY1A, AMY1B, AMY1C	Endogenous retrovirus	Androgen response element	Ting et al. (1992)
ZNF80	Endogenous retrovirus	ERV9 element	Di Cristofano et al. (1995a); (1995b)
PLA2L	Endogenous retrovirus	HERV-H element	Kowalski et al. (1996); (1997)
PTN	Endogenous retrovirus	HERV element	Schulte and Wellstein (1998)
HLA-DRB6	Endogenous retrovirus	MMTV-like sequence	Mayer et al. (1993)
Sex-linked C4 protein, murine	Endogenous retrovirus	ERV element contains androgen-responsive sites	Stavenhagen and Robins (1988); Adler et al. (1992, 1993); Robins et al. (1994)

Sequences are human unless otherwise specified.

capable of binding the retinoic acid receptor and functioning as RAREs in transiently transfected cells (Vansant and Reynolds, 1995). We may thus surmise that the random insertion of thousands of *Alu* sequences in the primate genome could have altered the expression of numerous genes over evolutionary time.

An *Alu* sequence in the last intron of the human *CD8A* (2p12) gene operates so as to modulate the activity of an adjacent T lymphocyte-specific enhancer (Hambor *et al.*, 1993). The *Alu* sequence appears to contain four functional transcription factor binding sites [Lyf-1 (2), bHLH (1), GATA-3 (1)]. Hambor *et al.* (1993) noted seven (non-CpG) nucleotide differences by comparing this *Alu* sequence with its probable source gene. Two of these differences were in the GATA-3 binding site and both were shown by site-directed mutagenesis to be necessary for its function (Hambor *et al.*, 1993). This was therefore proposed to be a possible example of positively selected change in an inserted *Alu* sequence. However, since the *Alu* sequence appears to be capable of modulating the activity of the enhancer through the formation of a cruciform (stem-loop) structure with another downstream *Alu* sequence (Hanke *et al.*, 1995), it is unclear what if any role the putative binding sites might have in modulating enhancer activity.

An *Alu* sequence in the promoter region of the gene (*FCER1G*; 1q23) encoding the gamma chain of the high affinity IgE receptor contains positive and negative *cis*-acting elements which contribute to the hematopoietic cell specificity of expression of this gene (Brini *et al.*, 1993). An *Alu* sequence in the myeloperoxidase (*MPO*; 17q23.1) gene promoter contains four repeats related to the consensus recognition sequence for nuclear hormone receptors (AGGTCA) (Piedrafita *et al.*, 1996). This sequence acts as a composite SP1-thyroid hormone-retinoic acid response element interacting with SP1 as well as the retinoic acid and thyroid hormone receptors (Piedrafita *et al.*, 1996). An *Alu* sequence may also have been recruited to perform a regulatory function in the human θ1-globin gene (*HBQ1*; 16p13.3; Kim *et al.*, 1989). Finally, the estrogen responsiveness of the human breast cancer (*BRCA1*; 17q21) gene appears to have been conferred by an *Alu* repeat located within the promoter region of the gene (Norris *et al.*, 1995). *Alu* sequences have thus introduced a variety of different DNA sequence motifs capable of binding a range of *trans*-acting factors that have altered the expression level or tissue specificity of the associated genes.

Not all *Alu* sequences inserted into gene regions function as promoters or enhancers of transcription. Indeed, an *Alu* sequence in the third intron of the Wilms' tumor (*WT1*; 11p13) gene, 12 kb downstream of the promoter, acts as a transcriptional silencer, repressing transcription of the *WT1* gene in cells of non-renal origin (Hewitt *et al.*, 1995). Since this silencer can function in an orientation- and distance-independent fashion, Hewitt *et al.* (1995) suggested that it may have acquired silencer function rather than having simply possessed a silencer function intrinsic to the *Alu* sequence. Another example of a silencer is the RRE repetitive sequence element 1 kb upstream of the murine erythropoietin receptor gene (Youssoufian and Lodish, 1993). This sequence, one of ~10^5 copies in the mouse genome, may exert its *cis*-mediated repressor effect on the *EpoR* gene by read-through transcription. Finally, a 27 bp sequence that is important in the negative regulation of a murine immunoglobulin κ light chain gene appears to have been derived from a B1 repetitive element (Saksela and Baltimore, 1993).

A motif within an *Alu* sequence, 3.6 kb 5′ to the transcriptional initiation site of the human parathyroid hormone (*PTH*; 11p15.1-p15.3) gene, contributes toward a negative calcium response element (McHaffie and Ralston, 1995; Okazaki *et al.*, 1991). This element possesses a 12 bp palindromic core (*TGA-GACAGGGTCTCA*) and since it is common in the human genome courtesy of the wide distribution of *Alu* sequences, it may have provided the means for the expression of other genes to be down-regulated by extracellular calcium.

Wu *et al.* (1990) have proposed that an *Alu* repeat 2.2 kb upstream of the human ε-globin (*HBE1*; 11p15.5) gene abolishes down-regulation of the gene mediated by a silencer element 4.5 kb upstream. These authors suggested that the down-regulation was caused by transcriptional interference and that the *Alu* repeat somehow blocks transcription from the upstream element specifically in embryonic erythroid cells where it is transcriptionally active.

Upon insertion, *Alu* sequences may also provide alternative sites for transcriptional initiation, as found in the human apolipoprotein B mRNA-editing enzyme (*APOBEC1*) gene (12p13.1; Fujino *et al.*, 1998). The human *APOBEC1* gene is expressed exclusively in the small intestine whereas in rodents, the gene is more widely expressed. Whilst the human gene contains two *Alu* sequences and two major transcriptional initiation sites, its murine counterpart lacks *Alu* repeats and contains a single transcriptional initiation site (*Figure 5.6*). Insertion of the *Alu* sequences must therefore have occurred during the last 100 Myrs since the divergence of the primate and rodent lineages. In the human gene, the first *Alu* repeat contains the first of the transcriptional initiation sites and lies upstream of a region exhibiting strong homology to the murine intestinal promoter. The second *Alu* repeat contains the second transcriptional initiation site and is located in the first intron. Comparison with the murine gene suggests that *Alu* sequence insertion may have split the human intestinal promoter leading to utilization of the downstream *Alu* sequence as an alternative site of transcriptional initiation. In this case, it would appear as if promoter function has simply adapted to the presence of the *Alu* repeat rather than being qualitatively altered by it.

Endogenous retroviral elements. Not all inserted sequences implicated in influencing gene expression are *Alu* repeats. One example of the recruitment of

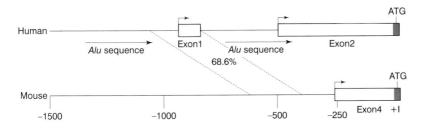

Figure 5.6. Comparison of human and mouse *APOBEC1* gene promoters (from Fujino *et al.*, 1998). Transcriptional initiation sites are denoted by vertical arrows. *Alu* sequences are denoted by horizontal arrows. The murine exon 4 is homologous to a portion of human exon 2. The region of transcriptional initiation between nucleotides -848 and -1034 of the human gene is ~70% homologous to the region of the murine gene promoter responsible for intestinal expression.

non-*Alu* repetitive sequences within a mammalian promoter is provided by the human amylase genes. These genes are located in a 230 kb region of chromosome 1p21 (*Figure 5.7*); two genes (*AMY2A* and *AMY2B*) are expressed in the pancreas, three (*AMY1A, AMY1B, AMY1C*) are expressed in the salivary gland and one (*AMYP1*) represents a truncated pseudogene. A complete γ-actin processed pseudogene is located immediately upstream of *AMY2B*, the consequence of an insertion event ~40 Myrs ago, after the divergence of the New World monkeys but before the divergence of the Old World monkeys from the human-ape lineage (Emi *et al.*, 1988; Samuelson *et al.*, 1996, 1988, 1990). Whether this pseudogene plays a role in regulating amylase gene expression is unclear but the observation that New World monkeys do not possess the pseudogene or express salivary amylase suggests that the pseudogene insertion could have been involved in the switch in tissue specificity of amylase gene expression. In the other four amylase genes, the actin pseudogene is interrupted by an endogenous retroviral sequence, the result of a second insertion which occurred in the human-ape lineage after the divergence of Old World monkeys (Samuelson *et al.*, 1996) and prior to the more recent gene duplications (*Figure 5.7*). The results of gene expression studies in transgenic mice were initially consistent with the view that the retroviral sequence was essential for salivary gland-specific expression of the *AMY1C* gene (Ting *et al.*, 1992). The *AMY2A* gene, which contains only a residual long terminal repeat (LTR) left by excision of the retroviral sequence, is expressed in the pancreas. Excision of the retroviral sequence from *AMY2A* thus appeared to be associated with reversion to a pancreas-specific pattern of expression (*Figure 5.8*). Transcription of the pancreatic amylase genes was found to be initiated at exon *a* whereas transcription of the salivary gland amylase genes appeared to be initiated from an untranslated exon within the actin pseudogene (*Figure 5.8*). It was therefore considered possible that insertion of the retroviral element activated a cryptic promoter within the actin pseudogene that specified the transcriptional initiation site for the expression of the salivary amylase genes. However, more recent studies in various Old World monkey species lacking the retroviral sequence have indicated that salivary amylase gene expression predated the retroviral insertion which cannot therefore be regarded as essential for amylase expression in the salivary gland (Samuelson *et al.*, 1996).

There are now a number of other examples of endogenous retroviral elements which have been recruited to promoter function. These elements have been aptly

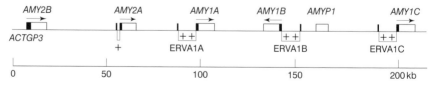

Figure 5.7. Distribution of the amylase genes on human chromosome 1p21. The positions of the five amylase genes (*AMY2B, AMY2A, AMY1A, AMY1B* and *AMY1C*) and the pseudogene *AMYP1* are indicated by boxes. ERVA denotes the endogenous retroviral element present in three copies upstream of each salivary amylase gene. The positions of the retroviral LTRs are marked +. *ACTGP3* is a γ-actin pseudogene located upstream of the *AMY2B* gene. Sequences related to this pseudogene are represented by solid boxes (redrawn from Meisler and Ting ,1993).

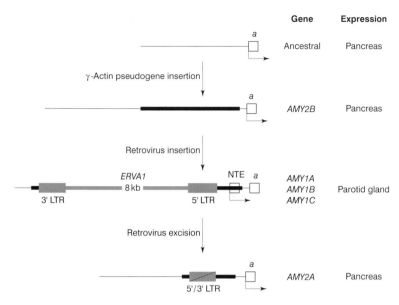

Figure 5.8. Evolution of the human amylase genes. The insertions of the γ-actin pseudogene (solid bar) and the retrovirus (*ERVA1*) occurred around 40 Myrs ago. Exon *a* and the untranslated exon (NTE) are represented by open boxes. An arrow denotes the transcriptional start site (redrawn from Meisler and Ting, 1993).

described as 'perpetually mobile footprints of ancient infections' (Sverdlov, 1998) and this mobility has sometimes been put to constructive use by evolution. For example, the hematopoietic cell-specific expression of the human zinc finger gene *ZNF80* (3q13.3) is driven by the LTR of ERV9, a member of a low copy number family of endogenous retroviral elements (Di Cristofano *et al.*, 1995a, 1995b). Since the ERV9 insertion was not found in the African green monkey, rhesus macaque or orangutan, the integration of this element must have occurred after the divergence of the orangutan from the other great apes but before the divergence of the gorilla.

The LTR of a HERV-H-related sequence within an intron of a human phospholipase A2-like (*Pla2l*; 8q24) gene has been shown to be important for *Pla2l* gene expression (Feuchter-Murthy *et al.*, 1993; Kowalski *et al.*, 1996; 1997). Since the retroviral LTR is also present in the orthologous genes of chimpanzee and gorilla but not in orangutan and lower primates, we may infer that it was integrated into the ancestral primate genome about 15–20 Myrs ago. Upon further analysis, it has become clear that the teratocarcinoma cell-specific *Pla2l* transcript is actually a fusion transcript between two once distinct genes, the HERV–4-associating 1 (*HHLA1*; 8q24) gene and the otoconin 90 (*OC90*; 8q24) gene (Kowalski *et al.*, 1999). Presumably the LTR acts not only as a strong promoter but also as an inducer of transcriptional fusion, at least in teratocarcinoma cells where HERV–H LTRs are known to be active transcriptionally.

A 6.3 kb endogenous retroviral element of the HERV family has been inserted into the human pleiotrophin (*PTN*; 7q33) gene between the exons specifying the 5′ untranslated region and those encoding the protein product (Schulte and

Wellstein, 1998). This HERV element has been shown to drive the expression of the *PTN* gene in trophoblasts and in choriocarcinoma cell lines (Schulte *et al.*, 1996). The HERV sequence is also present in the *Ptn* genes of chimpanzee and gorilla but not in that of the rhesus macaque indicative of a genomic insertion event which occurred after the divergence of the great apes from the Old World monkeys some 23 Myrs ago. The expression of the human *HLA-DRB6* gene is driven by the LTR of a Mouse Mammary Tumor Virus (MMTV)-like sequence which substituted for the original promoter upon insertion (Mayer *et al.*, 1993). Since this MMTV-LTR is also present in the macaque, the insertion event must have occurred >23 Myrs ago. Finally, Feuchter *et al.* (1992) have employed a systematic screening strategy to demonstrate that the expression of a number of other human cellular genes may be influenced by endogenous retroviral elements; these include the cell division cycle 4-like (*CDC4L*) gene.

Retroviral insertion may affect gene expression even if the element is inserted outwith the promoter region. One example is the LTR of an RTV_L-H element which is present in the 3' UTR of a human placentally expressed gene (termed 'PLT' by the authors, Goodchild *et al.*, 1992); the *Plt* mRNA undergoes alternative splicing at its 3' end with polyadenylation occurring within the LTR in one of the transcripts. This sequence is present in the *Plt* genes of the great apes and Old World monkeys and therefore must have been inserted prior to the divergence of these groups.

An LTR of an ERV9 element has been found at the 5' boundary of the human β-globin (*HBB*; 11p15) gene locus control region (LCR; see Chapter 1, section 1.2.8) just upstream of the DNAse I-hypersensitive site HS5 (Long *et al.*, 1998). This LTR is composed of 14 tandem repeats containing recurring GATA, CACCC, and CCAAT motifs that are potentially capable of binding GATA-binding factor, BKLF/TEF2 and C/EBP transcription factors respectively. The orthologous sequence in gorilla has only five repeats whilst the repeat number is polymorphic in humans. Reporter gene studies have demonstrated that this LTR possesses both enhancer and promoter activity in erythroid cells (Long *et al.*, 1998). Moreover, in erythroid cells, the LTR activates transcription of the downstream retroviral R and H5 regions and of genomic regions still further downstream (Long *et al.*, 1998). The HS5 LTR may therefore play a role in regulating the transcription of the human β-globin LCR (which is preferentially transcribed in erythroid cells) which may in turn serve to open up the chromatin structure of the β-globin gene domain.

Retroviral insertions have also influenced gene expression in other mammalian species. Thus the promoter of the rat oncomodulin gene (human counterpart *OCM*; 7p13-p11) contains a long terminal repeat of an intracisternal A particle (IAP, a family of endogenous retroviral elements) which has been recruited to perform a gene regulatory function (Banville and Boie, 1989). Since the mouse lacks the integrated IAP element, the IAP insertion must have occurred after the divergence of the two rodent species 40 Myrs ago (Banville *et al.*, 1992).

LINE elements. LINE elements may also have served as mobile regulatory sequences altering the expression of target genes. They have a tendency to acquire promoter sequences from non-LINE sources, with different sequence lineages acquiring different promoters (Adey *et al.*, 1994a). These regulatory sequences may then confer a novel specificity of expression upon genes with which they become associated. Thus, a LINE element at position -800 in the human factor IX (*F9*; Xq27.1-q27.2) gene promoter has been implicated in conferring high level liver-specific expression on the *F9* gene (Kurachi *et al.*, 1997). An enhancer some 20 kb 5′ to the apolipoprotein(a) gene (*LPA*; 6q27) which contains binding sites for Ets and Sp1 transcription factors, also resides within a LINE element (Yang *et al.*, 1998).

The polyadenylation signal of the murine thymidylate synthase (*Tyms*) gene has been derived from an inserted LINE element (Harendez and Johnson, 1990). Similarly, the polyadenylation sites of several human genes including the β-tubulin (*TUBB*; 6p21.3) gene have been derived from MIR elements, mammalian wide interspersed repeats (Murnane and Morales, 1995).

As we have seen (Chapter 1, section 1.4.3), only a proportion of LINE elements are transpositionally active, the remainder having been inactivated by truncation and rearrangement. Adey *et al.* (1994b) performed phylogenetic analysis to deduce the sequence of an ancestral murine transpositionally active LINE element. This element was then 'resurrected' by chemical synthesis and shown *in vitro* to possess promoter activity.

Minisatellites and microsatellites. Several minisatellites are known to have become recruited as gene regulatory elements. These include minisatellites 1 kb 3′ to the polyadenylation signal of the human *HRAS* proto-oncogene (11p15.5; 28 bp repeat unit; Green and Krontiris 1993), 600 bp 5′ of the transcriptional initiation site of the human insulin (*INS*; 11p15.5; 14 bp repeat unit; Catigani-Kennedy *et al.*, 1995) gene and 4.1 kb upstream of the human insulin-like growth factor II (*IGF2*; 11p15; Paquette *et al.*, 1998) gene, and the minisatellite in the D_H/J_H intron of the human immunoglobulin heavy chain (*IGHD/IGHJ*; 14q32.33) gene cluster (Treppicchio and Krontiris, 1992). The *HRAS* minisatellite binds members of the *rel*/NFκB family (Treppicchio and Krontiris, 1993) whilst the *INS* minisatellite binds the transcription factor Pur-1 (Kennedy *et al.*, 1995). The *IGHD/IGHJ* minisatellite binds a mycHLH protein closely related to USF/MLTF (Treppicchio and Krontiris, 1992). This element may influence *IGHD/IGHJ* gene expression since sequestration of the transcription factor by the minisatellite inhibits transcriptional activation through a *bone fide* USF enhancer element. Since the *HRAS*, *INS* and *IGHD/IGHJ* minisatellites are absent from the analogous positions in orthologous non-primate genes, it would appear that evolution has recruited these elements during primate evolution. Such minisatellites may have provided the raw material for promoter and enhancer sequences which have then been optimized by selection. Alternatively, if transcriptional effects emanating from these sequences are comparatively minor, selection would probably have been unable either to improve or remove them and they would have remained as transcriptional control elements with minor effect.

Microsatellites can also serve as regulatory elements and indeed some are conserved at orthologous positions in the genomes of different species (Meyer *et al.*, 1995; Moore *et al.*, 1991; Stallings, 1994, 1995; Stallings *et al.*, 1991). One example is the $(TCAT)_n$ repeat element in the first intron of the human tyrosine hydroxy-lase (*TH*; 11p15.5) gene (Meloni *et al.*, 1998). This repeat is similar to the consensus thyroid response element (TRE) present in the human and rat *TH* genes and gel shift assays have provided evidence for the formation of specific complexes between the tetranucleotide repeat and proteins found in HeLa cell extracts. A $(CCTTT)_n$ pentanucleotide repeat polymorphism (51–72 copies) found within the promoter region of the human inducible nitric oxide synthase 2A (*NOS2A*; 17q11-q12) gene is also polymorphic in chimpanzees and gorillas but is monomorphic in orangutans and macaques (Xu *et al.*, 1997); its influence, if any, on promoter function is not known.

5.1.13 mRNA editing

The existence of mRNA editing represents something of an evolutionary puzzle since its selective advantage is not immediately obvious (Covello and Gray, 1993). The best understood example in humans is that involving the apolipoprotein B mRNA. ApoB-100, produced in the liver, is essential for the production of very low density lipoprotein whereas ApoB-48 is required for fat absorption in the small intestine. Both proteins are encoded by the same (*APOB*; 2p23-p24) gene. ApoB-48 mRNA is generated as a result of the introduction of an in-frame translational termination codon at the mRNA level by the deamination of cytosine to uracil (C6666T) in the first base of the CAA codon encoding Gln2153. A conserved 29 nucleotide element flanking the edited base (6662–6690) is found in mammals; this includes a regulator region, a spacer and a mooring sequence which is required for the mRNA editing process. By contrast to the situation found in mammals, chicken *Apob* mRNA is not edited although the various tissue-specific factors that serve to mediate the modification in mammals are present (Teng and Davidson, 1992). The absence of mRNA editing in chicken is thought to be due to the presence of several single base-pair substitutions in the mooring region of the chicken *Apob* mRNA since the experimental introduction of A6671T, G6674T and C6680T substitutions into the chicken gene served to confer mRNA editing ability upon chicken cells (Nakamuta *et al.*, 1999).

The editing of the *APOB* mRNA is performed by a multi-protein complex ('editosome') whose catalytic component has been termed apobec-1. The *APOBEC1* gene, localized to chromosome 12p13.1 in human, is not highly conserved when compared with its homologues in other mammals, consistent with its recent rapid evolution (Chan *et al.*, 1997; Fujino *et al.*, 1998). Apobec-1 shows substantial sequence homology to cytidine/cytidylate deaminases (Chan *et al.*, 1997) and it would thus appear that a protein with a nucleoside as substrate has evolved from a protein with a nucleotide as a substrate. It is unlikely, however, that apobec-1 evolved simply by gene duplication and divergence since mRNA editing requires multiple factors which would have had to have coevolved in order to function as a cohesive complex. Even although the separation of apobec-1 and the cytidine/cytidylate deaminases is ancient (Chan *et al.*, 1997), apobec-1 is only found in mammals suggesting that the ancestral apobec-1 protein might have had a function other than mRNA editing.

mRNA editing is not however unique to *APOB* mRNA. Different types of mRNA editing have also been found in several other human mRNAs viz. a T→C transition in the Wilms' tumor (*WT1*; 11p13) mRNA (Sharma *et al.*, 1994), a T→A transversion in the α-galactosidase (*GLA*; Xq21.3-q22) mRNA (Novo *et al.*, 1995) and a C→T transition in the neurofibromatosis type 1 (*NF1*; 17q11.2) mRNA (Skuse *et al.*, 1996).

5.1.14 Coordinate regulation

The clustering of certain genes may be important for their coordinate regulation by common control elements. Possible examples of this include the genes encoding the spermatid-specific nucleoprotamines (*PRM1* and *PRM2*; 16p13.2), the platelet membrane glycoproteins (*ITGA2B* and *ITGA3*; 17q21-q22), the albumin family (*ALB*, *AFP*, *AFM*, *GC*; 4q11-q13), the pregnancy-specific glycoproteins (*PSG1*, *PSG2*, *PSG3*, *PSG4*, *PSG5*, *PSG6*, *PSG7*, *PSG8*, *PSG11*, *PSG12*, *PSG13*; 19q13.2) and the fibrinogens α, β, and γ (*FGA*, *FGB* and *FGG*; 4q31) (Chapter 8.5). In some cases, a head-to-head arrangement may potentiate the use of bidirectional promoters (Section 5.1.5).

Coordinate regulation does not however require close linkage as evidenced by the case of the α- (*HBA1*) and β-globin (*HBB*) genes on chromosomes 16p13.3 and 11p15.5 respectively. Another example is that of the human ribosomal protein genes which encode the 80–90 ribosomal proteins that together constitute the ribosome. The expression of the different ribosomal protein genes, which can account for 7–9% of total cellular RNA, is coordinately regulated in response to the cell's varying requirements for protein synthesis. However, the ribosomal protein genes are highly dispersed in the human genome (Feo *et al.*, 1992; Kenmochi *et al.*, 1998; *Table 5.2*) indicating that their coordinate regulation must be brought about by the action of *trans*-acting factors rather than the influence of shared regulatory sequences.

5.1.15 Changes in expression of developmentally significant gene products

> If we possessed a thorough knowledge of all parts of the seed of any animal (e.g. man), we could from that alone, by reasons entirely mathematical and certain, deduce the whole conformation and figure of each of its members, and, conversely if we knew several peculiarities of this conformation, we would from those deduce the nature of its seed.
> René Descartes *Oeuvres* iv, 494

The study of the molecular genetics of vertebrate development is still very much in its infancy. However, the identification of developmental control genes important in morphogenesis is proceeding apace as a result of studies of (i) model organisms including mouse, zebrafish, *Drosophila* and *Caenorhabditis elegans* (Postlethwait and Talbot, 1997) and (ii) human dysmorphic syndromes and congenital malformations (Epstein, 1995; Kondo *et al.*, 1998; Semenza, 1998).

The spatial and temporal distribution patterns of expression of HOX genes have played an important role in the evolutionary emergence of novel body plans among the metazoa (Belting *et al.*, 1998; Gellon and McGinnis *et al.*, 1998;

Table 5.2. Chromosomal locations of chromosomally assigned human ribosomal protein genes

Ribosomal protein gene	Gene symbol	Chromosomal location
Ribosomal protein L3	RPL3	22
Ribosomal protein L4	RPL4	15
Ribosomal protein L5	RPL5	1
Ribosomal protein L6	RPL6	12q23-q24
Ribosomal protein L7	RPL7	8
Ribosomal protein L8	RPL8	8
Ribosomal protein L9	RPL9	4p13
Ribosomal protein L10	RPL10	X
Ribosomal protein L11	RPL11	1
Ribosomal protein L13	RPL13	16q24
Ribosomal protein L15	RPL15	3
Ribosomal protein L17	RPL17	18
Ribosomal protein L19	RPL19	17p11-q12
Ribosomal protein L22	RPL22	3q26
Ribosomal protein L23A	RPL23A	17q11
Ribosomal protein L24	RPL24	3
Ribosomal protein L27	RPL27	17
Ribosomal protein L27A	RPL27A	11
Ribosomal protein L28	RPL28	19q13
Ribosomal protein L29	RPL29	3q29-qter
Ribosomal protein L30	RPL30	8
Ribosomal protein L31	RPL31	2
Ribosomal protein L32	RPL32	3q13.3-q21
Ribosomal protein L35A	RPL35A	3q29-qter
Ribosomal protein L36A	RPL36A	14
Ribosomal protein L37	RPL37	5
Ribosomal protein L38	RPL38	17
Ribosomal protein L41	RPL41	12
Ribosomal protein S2	RPS2	16p13.3
Ribosomal protein S3	RPS3	11q13
Ribosomal protein S3A	RPS3A	4
Ribosomal protein S4	RPS4X/Y	Xq13.1/Yp11.3
Ribosomal protein S5	RPS5	19q13.4
Ribosomal protein S6	RPS6	9p21
Ribosomal protein S7	RPS7	2p25
Ribosomal protein S8	RPS8	1p32-p34
Ribosomal protein S9	RPS9	19p13.4
Ribosomal protein S10	RPS10	6
Ribosomal protein S11	RPS11	19q13
Ribosomal protein S12	RPS12	6
Ribosomal protein S13	RPS13	11
Ribosomal protein S14	RPS14	5q31-q33
Ribosomal protein S15A	RPS15A	16
Ribosomal protein S17	RPS17	11p13-pter
Ribosomal protein S18	RPS18	6p21.3
Ribosomal protein S24	RPS24	10q22-q23
Ribosomal protein S25	RPS25	11q23.3

Chapter 4.2.1, *Homeobox genes*). The molecular basis of such a role has long been a puzzle but new studies are now providing a glimpse of how subtle genetic changes can have fairly dramatic effects on morphology. One of the best characterized examples is provided by a promoter mutation which appears to have morphological consequences for the evolution of the mammalian body plan: a 4 bp deletion in element C of the early enhancer region of the *Hoxc8* gene of baleen whales (Shashikant *et al.*, 1998). This lesion (TTAATTG→TT-G) is specific to baleen whales (five species tested) and is not found in the highly conserved *Hoxc8* gene promoters of humans (*HOXC8*; 12q12-q13), rodents, artiodactyls or the toothed whales including sperm whales (Shashikant *et al.*, 1998). In mice, the early enhancer region of the *Hoxc8* gene promoter is required to initiate expression of the gene in the posterior region of the day 8.5 mouse embryo and to establish spatial domains of expression in the neural tube and mesoderm. After day 9.0, the late enhancer maintains anterior *Hoxc8* gene expression and down-regulates posterior expression. The baleen whale *Hoxc8* early enhancer region containing the 4 bp deletion has been assayed in transgenic mouse embryos where it was found to direct the expression of the reporter gene to more posterior regions (4–5 somite levels posterior as compared with human or murine *Hoxc8* enhancers) of the neural tube but failed to direct expression to the posterior mesoderm (Shashikant *et al.*, 1998). Similar results were obtained when site-directed mutagenesis was used to introduce the same lesion into the murine *Hoxc8* early enhancer region. Thus the 4 bp deletion in the *Hoxc8* gene promoter of baleen whales may have played a role in modifying the developmental program of these cetaceans. One wider implication of this work is that there may be additional mutations yet to discover in the *cis*-acting regulatory sequences of other Hox genes and these lesions could have contributed to the evolution of body plan diversity during mammalian evolution.

It is anticipated that mutational changes in a number of other genes encoding transcription factors that play a role in embryonic development will be characterized in the coming years thereby shedding new light on the molecular basis of morphogenesis. Possible candidate genes would include the Pax gene family (Chapter 4.1.6; Balczarek *et al.*, 1997; Noll 1993), the Sox gene family (Wegner, 1999), the engrailed (*EN1*, 2q13-q21; *EN2*, 7q36) and *Wingless* (*WNT1*; 12q12-q13) genes (Joyner 1996), the brachyury (*T*; 6q27) gene (Yasuo and Satoh, 1998) and the snail family of transcription factors (Sefton *et al.*, 1998) as well as the heat shock protein 90 (HSP90) family of signal transduction chaperonins (Rutherford and Lindquist, 1998). Sequence differences between orthologous developmental regulatory genes should provide insights into the process of molecular evolution underlying morphological change (Budd, 1999; Eizinger *et al.*, 1999).

5.2 Transcription factors

You geneticists may know something about the hereditary mechanisms that distinguish a red-eyed from a white-eyed fruit fly but you haven't the slightest inkling about the hereditary mechanism that distinguishes fruit flies from elephants.
W.J. Osterhout (1925)

5.2.1 Transcription factor families

Many mammalian transcription factors belong to families whose members bind to very similar or identical DNA sequence motifs. Thus, there are at least eight cAMP-reponsive transcription factors of the CREB/ATF family that bind to the octanucleotide TGACGTCA (Hai *et al.*, 1989). The evolutionary conservation of DNA sequence recognition can be fairly dramatic as for example in the case of the various members of the Brn-3 class of POU domain transcription factors found in both mammals and nematodes (Gruber *et al.*, 1997). On the other hand, some transcription factor families possess members that exhibit a considerable degree of divergence in terms of their DNA binding specificities. For example, at least four members of the nuclear factor 1 (NF1) family recognize sequences containing the trinucleotide TGG (Gil *et al.*, 1988) and at least eight members of the mammalian *Ets* family bind to an 11 bp purine-rich motif containing a conserved GGA core (Wang *et al.*, 1992). The DNA binding specificity of different *Ets* family members is determined by the nucleotides at the 3' end of the *Ets*-binding site (Wang *et al.*, 1992).

The evolutionary subdivisions of transcription factor families as revealed by phylogenetic analysis may be paralleled by functional subdivisions (Elsen *et al.*, 1995). If function is conserved within (although not between) subfamilies, then the functions of novel transcription factors may be to some extent predictable by comparison with other members of the same subfamily. One example of this is provided by the high mobility group (HMG) protein superfamily of DNA-binding proteins. These proteins possess one or more copies of an 80 amino acid domain termed the HMG box and have an evolutionary history that dates back 1000 Myrs (Laudet *et al.*, 1993). The HMG superfamily may be divided into two sub-families (i) the TCF/SOX family which comprises transcription factor proteins that contain a single sequence-specific HMG box and (ii) the UBF/HMG family of chromatin structure regulatory proteins which possess multiple HMG boxes that exhibit little if any sequence specificity. Representatives of the first family in the human genome include the chromosomally dispersed SRY-related HMG-box (SOX) genes (*SOX1*, 13q34; *SOX2*, 3q26-q27; *SOX3*, Xq26-q27; *SOX4*, 6p23; *SOX5*, 12p12; *SOX9*, 17q23; *SOX10*, 22q13; *SOX11*, 2p25; *SOX20*, 17p13; *SOX22*, 20p13; Wegner 1999), the lymphoid enhancer-binding factor 1 (*LEF1*; 4q23-q25) and the hepatocyte nuclear factor 1α (*TCF1*; 12q24) genes. Human representatives of the second family include the high mobility group (non-histone chromosomal) protein genes, *HMG1* (13q12), *HMG2* (4q31), *HMG4* (Xq28), *HMG14* (21q22) and *HMG17* (1p35-p36) and upstream binding transcription factor (*UBTF*; 17q21). The diversity of function exhibited by members of the HMG superfamily is a result of the action of a number of different evolutionary processes including gene duplication, intragenic duplication, exon shuffling and single base-pair substitution mediated divergence (Laudet *et al.*, 1993).

Duplications or amplifications of transcription factor genes are sometimes very ancient as is evidenced by the limited homology still evident between nuclear factor 1 (NF1) and the protein kinase family (Mannermaa and Oikarinen, 1989). On the other hand, some duplications have occurred during mammalian evolution [e.g. the transcription factors *TCF1* (12q24) and *LEF1* (4q23-q25); Gastrop *et al.*,

1992] and the *hedgehog* gene family, *sonic* (*SHH*; 7q36) and *Indian* (*IHH*; 2q33-q35); Zardoya *et al.*, 1996) and some as recently as 55 Myrs ago in the common ancestor of the simians e.g. the ZNF91 family (*ZNF91*; 19p12-p13.1; Bellefroid *et al.*, 1995).

5.2.2 Functional conservation of orthologous transcription factors

The general transcription factor, TBP (TFIID), binds to the upstream TATA box element and is essential for transcription in all eukaryotes. TBP is highly conserved throughout the eukaryotes and this conservation even extends to the archaebacteria (Hoffman *et al.*, 1990; Rowlands *et al.*, 1994). Functional conservation at the protein level is evidenced by the ability of yeast TBP to replace human TBP in *in vitro* transcription reactions and *vice versa* (Buratowski *et al.*, 1988). Not surprisingly, therefore, yeast and human TBP have virtually identical sequence recognition characteristics, at least *in vitro* (Wobbe and Struhl, 1990). However, *in vivo*, human TBP cannot properly substitute for yeast TBP and the yeast cells in question grow extremely poorly (Cormack *et al.*, 1991; Gill and Tjian, 1991). This difference between the *in vitro* and *in vivo* situations is salutary and probably reflects subtle differences in the interactions of TBP with activator proteins or other components of the transcriptional initiation complex.

Interactions between *cis*-acting DNA sequence motifs and the *trans*-acting transcription factors binding them may thus be conserved over fairly long periods of evolutionary time. Another example is that of the rat pituitary-specific transcription factor Pit-1 which is able to bind to and activate the growth hormone gene promoter of the rainbow trout (Argenton *et al.*, 1993). Similarly, mammalian *ets* genes are functionally homologous to the *pointed* gene of *Drosophila* (Albagli *et al.*, 1996). Functional conservation is also evident with other factors such as c-*jun*, the serum response factor SRF, the CCAAT box-binding factor CP1 and the glucocorticoid and estrogen receptors; the mammalian proteins can substitute for their yeast counterparts in order to activate gene transcription in yeast (Guarente and Bermingham-McDonogh, 1992). This functional conservation owes its existence to the fact that once specificity has been established between a given transcriptional activator and its cognate binding sites in the regulatory regions of different genes, both the structure of the DNA-binding domain of the activator and the sequence of the recognition site are evolutionarily constrained. Evolutionary change between both paralogous and orthologous transcription factors has nevertheless occurred in a number of different ways and some of these are illustrated in the following sections.

5.2.3 Functional redundancy of paralogous transcription factors

Gene duplication and amplification initially generate functional redundancy as in the case of the c-Ets-1 (*ETS1*; 11q23) and c-Ets-2 (*ETS2*; 21q22) genes in mammals where the former appears to be dispensable and indeed replaceable by the latter (Albagli *et al.*, 1994). Functional redundancy is also apparent in the MyoD family of transcription factors, for example between *MYOD1* (11p15.1) and *MYF5* (12q21), and between *MYOD1* (11p15.1) and *MYF6* (12q21) (Atchley *et al.*, 1994). It has also been noted that the transgenic inactivation of several genes known to be important in mammalian development does not automatically lead to a deleterious phenotype, implying a degree of functional redundancy. We may surmise that developmental gene redundancy will have endured if its maintenance has served to confer a selective advantage (Cooke *et al.*, 1997). On the other hand, functional redundancy could also have originated merely as a consequence of certain genes being either especially prone to duplication (Iwabe *et al.*, 1996) or manifesting an increased probability of survival after duplication (Gibson and Spring, 1998).

Functional redundancy does not however always endure and the relaxation of selective pressure consequent to gene duplication/amplification potentiates the diversification of paralogous transcription factors leading eventually to the emergence of families of related transcription factors with different DNA target sequence specificities and hence distinct functions.

5.2.4 Paralogous transcription factors

Perhaps the best characterized example of the divergence of paralogous transcription factors is provided by the nuclear receptor family (Chapter 4, section 4.2.3, *Nuclear receptor genes*). In this family, dimerized receptors typified by the estrogen receptor group, bind to two 6 bp half sites of the sequence TGACCT whereas those of the glucocorticoid receptor group recognise the related sequence TGTTCT (Chapter 7, section 7.5.2). The amino acids involved in discriminating between these motifs are located in the 'P-box' of the DNA recognition helix. The P-box of

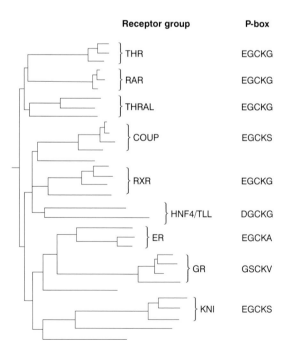

Figure 5.9. P-box sequences of different receptor groups within the nuclear receptor family (redrawn from Zilliacus *et al.*, 1994).

the glucocorticoid receptor group appears to have evolved from a progenitor resembling the present-day estrogen receptor (Amero *et al.*, 1992; Laudet *et al.*, 1992; Martinez *et al.*, 1991; see *Figure 5.9*). Thus, mutations in the P-box have altered the DNA-binding specificity of a receptor with high affinity for TGACCT sites and low affinity for TGTTCT sequences, to proteins with the opposite speci-ficity. Residues Glu439 and Ser440 are critical for conferring binding specificity on the receptor (Zilliacus *et al.*, 1994).

Another example of a functional change in the DNA-binding domain within a family of paralogous transcription factors is provided by *Pax6* (*PAX6*; 11p13) which possesses an Asn at amino acid residue 47 in the third α-helix of the paired domain (Balczarek *et al.*, 1997); this amino acid recognizes the nucleotide T. All other known *Pax* genes encode proteins with a His at this position; this amino acid shows higher affinity toward the nucleotide G.

5.2.5 Orthologous transcription factors

Orthologous transcription factors have also diverged over evolutionary time. Divergence has occurred by, for example, incorporation of novel motifs or the amplification of existing motifs. Thus, the human transcription factor MOK2 (*MOK2*; 19q13.2-q13.3) contains 10 zinc-finger motifs in comparison to seven in its murine homologue (Ernoult-Lange *et al.*, 1995). Similarly, the human ery-throid-specific transcription factor Eryth 1 contains different numbers of repeat motifs as compared with its chicken counterpart (Trainor *et al.*, 1990). Finally, gene sequences encoding TBP, the general transcription factor, exhibit a considerable degree of sequence simplicity as a direct result of simple repeat amplification, per-haps by replication slippage (Hancock, 1993). The incorporation of new repeats and the consequent enlargement of TBP may have permitted novel interactions with domains of other proteins leading to the acquisition of new functions.

5.2.6 Alternative splicing of transcription factor genes

Alternative splicing provides the means to generate transcription factor diversity in the absence of gene duplication. Thus, alternative splicing of the *PAX8* (2q12-q14) gene results in the alternative presence or absence of a single Ser residue in the recognition helix of the paired domain which is critical for DNA binding (Kozmik *et al.*, 1997). The two forms of Pax8 differ in their binding specificity.

5.2.7 Promoter shuffling in transcription factor genes

Promoter modularity arising from the shuffling of component motifs (Section 5.1) often occurs in the promoters of paralogous transcription factor genes. This provides the means for changes in the expression of single genes to lead to changes in the expression of many downstream target genes, a process which has contributed significantly to the evolution of complex gene expression networks.

5.2.8 Transcription factor-binding site interactions

DNA sequence elements that play a role in gene regulation have evolved so as to provide appropriate binding sites for their cognate transcription factors. In many cases, element function (measured in terms of influence on transcription) is directly proportional to the probability that a specific site is bound by the transcription factor protein. It follows that the same function can be achieved with a strong binding site and a small amount of protein as with a weak binding site and a large amount of protein. The extent of selective pressure on the DNA sequence is therefore likely to be determined by a combination of the cellular abundance of the protein, the functional activity of the protein and the initial binding strength of the DNA-protein interaction (Berg, 1992). Transcription factor binding may however also be influenced by the number of binding sites available on the promoter (Section 5.1.11). Thus, the pituitary-specific growth hormone (*GH1*; 17q22-q24) gene of humans and rats contains two binding sites for the transcription factor Pit-1 whereas four such sequences occur in the promoter of the homologous *Gh1* gene of *Oncorhynchus mykiss*, the rainbow trout (Argenton *et al.*, 1993). Rat Pit-1 has been shown to be capable of binding to three regions of the trout gene promoter thereby driving expression of a downstream reporter gene (Argenton *et al.*, 1993).

5.2.9 Exon shuffling in the evolution of transcription factors

Transcription factors are sometimes encoded by genes that are evolutionarily unrelated yet share the same type of DNA-binding domain. Such genes may have arisen by exon shuffling (Chapter 3, section 3.6), the process by which functional domains encoded by one or more exons have been dispersed to a variety of different proteins. That some DNA-binding domains are encoded by several exons implies that these exons must have been shuffled together as a single block. Matsuo *et al.* (1994) have suggested that the presence of short unconserved introns with different types of splice junction within the mammalian Oct-2 (*POU2F2*; chromosome 19) gene may have served to prevent recombination between the exons comprising the conserved POU domain without inhibiting the shuffling of this domain in its entirety between different transcription factor-encoding genes during evolution.

The divergence of the mammalian T-box family of transcription factors, which began before the separation of the vertebrate, arthropod and nematode lineages, has occurred both by the insertion or deletion of specific introns, or by *intron sliding* (Chapter 3, section 3.4) leading to variations in exon length (Wattler *et al.*, 1998).

References

Adey N.B., Schichman S.A., Graham D.K., Peterson S.N., Edgell M.H., Hutchison C.A. (1994a) Rodent L1 evolution has been driven by a single dominant lineage that has repeatedly acquired new transcriptional regulatory sequences. *Mol. Biol. Evol.* **11**: 778–789.

Adey N.B., Tollefsbol T.O., Sparks A.B., Edgell M.H., Hutchison C.A. (1994b) Molecular resurrection of an extinct ancestral promoter for mouse L1. *Proc. Natl. Acad. Sci. USA* **91**: 1569–1573.

Adler A.J., Danielsen M., Robins D.M. (1992) Androgen-specific gene activation via a consensus glucocorticoid response element is determined by interaction with nonreceptor factors. *Proc. Natl. Acad. Sci. USA* **89**: 11 660–11 663.

Adler A.J., Scheller A., Robins D.M. (1993) The stringency and magnitude of androgen-specific gene activation are combinatorial functions of receptor and nonreceptor binding site sequences. *Mol. Cell. Biol.* **13**: 6326–6335.

Albagli O., Soudant N., Ferreira E., Dhordain P., Dewitte F., Begue A., Flourens A., Stehelin D., Leprince D. (1994) A model for gene evolution of the *ets-1/ets-2* transcription factors based on structural and functional homologies. *Oncogene* **9**: 3259–3271.

Albagli O., Klaes A., Ferreira E., Leprince D., Klambt C. (1996) Function of *ets* genes is conserved between vertebrates and *Drosophila*. *Mechanisms Devel.* **59**: 29–40.

Allan C.M., Walker D., Taylor J.M. (1995) Evolutionary duplication of a hepatic control region in the human apolipoprotein E gene locus. *J. Biol. Chem.* **270**: 26278–26281.

Amero S.A., Kretsinger R.H., Moncrief N.D., Yamamoto K.R., Pearson W.R. (1992) The origin of nuclear receptor proteins: a single precursor distinct from other transcription factors. *Molec. Endocrinol.* **6**: 3–7.

Angotti E., Mele E., Costanzo F. (1994) A polymorphism (G→A transition) in the -78 position of the apolipoprotein A-I promoter increases transcriptional efficiency. *J. Biol. Chem.* **269**: 17 371–17 374.

Arcot S.S., Traina-Dorge C.L., Deininger P.L. (1991) The rat thymidine kinase gene 59 region: evolution of a promoter. *DNA Sequence* **2**: 129–131.

Argenton F., Vianello S., Bernardini S., Jacquemin P., Martial J., Belayew A., Colombo L., Bortolussi M. (1993) The transcriptional regulation of the growth hormone gene is conserved in vertebrate evolution. *Biochem. Biophys. Res. Commun.* **192**: 1360–1366.

Atchley WR, Fitch WM, Bronner-Fraser M. (1994) Molecular evolution of the MyoD family of transcription factors. *Proc. Natl. Acad. Sci. USA* **91**: 11522–11526.

Bachman N.J., Yang T.L., Dasen J.S., Ernst R.E., Lomax M.I. (1996) Phylogenetic footprinting of the human cytochrome *c* oxidase subunit VB promoter. *Arch. Biochem. Biophys.* **333**: 152–162.

Balczarek K.A., Lai Z.-C., Kumar S. (1997) Evolution and functional diversification of the paired box (*Pax*) DNA-binding domains. *Mol. Biol. Evol.* **14**: 829–842.

Banville D., Boie Y. (1989) Retroviral long terminal repeat is the promoter of the gene encoding the tumor-associated calcium-binding protein oncomodulin in the rat. *J. Mol. Biol.* **207**: 481–490.

Banville D., Rotaru M., Boie Y. (1992) The intracisternal A particle derived solo LTR promoter of the rat oncomodulin gene is not present in the mouse gene. *Genetics* **86**: 85–97.

Barker D.F., Liu X., Almeida E.R.A. (1996) The *BRCA1* and 1A1.3B promoters are parallel elements of a genomic duplication at 17q21. *Genomics* **38**: 215–222.

Becker R.M., Darrow C., Zimonjic D.B., Popescu N.C., Watson M.A., Fleming T.P. (1998) Identification of mammaglobin B, a novel member of the uteroglobin gene family. *Genomics* **54**: 70–78.

Bellefroid E.J., Marine J.-C., Matera A.G., Bouguignon C., Desai T., Healy K.C., Bray-Ward P., Martial J.A., Ihle J.N., Ward D.C. (1995) Emergence of the *ZNF91* Kruppel-associated box-containing zinc finger family in the last common ancestor of Anthropoidea. *Proc. Acad. Sci. USA* **92**: 10 757–10 761.

Belting H.G., Shashikant C.S., Ruddle F.H. (1998) Modification of expression and *cis*-regulation of Hoxc8 in the evolution of diverged axial morphology. *Proc. Natl. Acad. Sci. USA* **95**: 2355–2360.

Berg O.G. (1992) The evolutionary selection of DNA base pairs in gene-regulatory binding sites. *Proc. Natl. Acad. Sci. USA* **89**: 7501–7505.

Bochkanov S.S., Surguchov A.P., Smirnov V.N. (1990) Rabbit *apoA-I* gene: organization of the upstream region. *Doklady Akademii Nauk SSSR* **313**: 482–484.

Bock S.C., Levitan D.J. (1983) Characterization of an unusual DNA length polymorphism 59 to the human antithrombin III gene. *Nucleic Acids Res.* **11**: 8569–8582.

Boyd D.C., Turner P.C., Watkins N.J., Gerster T., Murphy S. (1995) Functional redundancy of promoter elements ensures efficient transcription of the human 7SK gene *in vivo*. *J. Mol. Biol.* **253**: 677–690.

Brayton K.A., Chen Z., Zhou G., Nagy P.L., Gavalas A., Trent J.M., Deaven L.L., Dixon J.E., Zalkin H. (1994) Two genes for *de novo* purine nucleotide synthesis on human chromosome 4 are closely linked and divergently transcribed. *J. Biol. Chem.* **269**: 5313–5321.

Brenner V., Nyakatura G., Rosenthal A., Platzer M. (1997) Genomic organization of two novel genes on human Xq28: compact head-to-head arrangement of IDH-γ and TRAP-δ is conserved in rat and mouse. *Genomics* **44**: 8–14.

Brini A.T., Llee G.M., Kinet J.-P. (1993) Involvement of *Alu* sequences in the cell-specific regulation of transcription of the g chain of Fc and T cell receptors. *J. Biol. Chem.* **268**: 1355–1361.

Britten R.J. (1996) DNA sequence insertion and evolutionary variation in gene regulation. *Proc. Natl. Acad. Sci. USA* **93**: 9374–9377.

Britten R.J. (1997) Mobile elements inserted in the distant past have taken on important functions. *Gene* 205: 177–182.

Buchanan K.L., Smith E.A., Dou S., Corcoran L.M., Webb C.F. (1997) Family-specific differences in transcription efficiency of Ig heavy chain promoters. *J. Immunol.* **159**: 1247–1254.

Budd G.E. (1999) Does evolution in body patterning genes drive morphological change – or vice versa? *Bioessays* 21: 326–332.

Buratowski S., Hahn S., Sharp P.A., Guarente L. (1988) Function of a yeast TATA element-binding protein in a mammalian transcription system. *Nature* 334: 37–42.

Catignani-Kennedy G., German M.S., Rutter W.J. (1995) The minisatellite in the diabetes susceptibility locus IDDM2 regulates insulin transcription. *Nature Genet.* **9**: 293–298.

Chamberlin M.E., Lei K.-J., Chou J.Y. (1994) Subtle differences in human pregnancy-specific glycoprotein gene promoters allow for differential expression. *J. Biol. Chem.* **269**: 17 152–17 159.

Chan L., Chang B.H.-J., Nakamuta M., Li W.-H., Smith L.C. (1997) Apobec-1 and apolipoprotein B mRNA editing. *Biochim. Biophys. Acta* **1345**: 11–26.

Chiu C.-H., Schneider H., Schneider M., Sampaio I., Meireles C.M., Sightom J., Gumucio D., Goodman M. (1996) Reduction of two functional γ-globin genes to one: an evolutionary trend in New World monkeys (infraorder Platyrrhini). *Proc. Natl. Acad. Sci. USA* **93**: 6510–6515.

Chiu C.H., Schneider H., Slightom J.L., Gumucio D.L., Goodman M. (1997) Dynamics of regulatory evolution in primate β-globin gene clusters: *cis*-mediated acquistion of simian gamma fetal expression patterns. *Gene* 205: 47–57.

Collins C., Schappert K., Hayden M.R. (1992) The genomic organization of a novel regulatory myosin light chain gene (*MYL5*) that maps to chromosome 4p16.3 and shows different patterns of expression between primates. *Hum. Molec. Genet.* **1**: 727–733.

Connelly M.A., Zhang H., Kieleczawa J., Anderson C.W. (1998) The promoters for human DNA-PK(cs) (PRKDC) and MCM4: divergently transcribed genes located at chromosome 8 band q11. *Genomics* 47: 71–83.

Cooke J., Nowak M.A., Boerlijst M., Maynard-Smith J. (1997) Evolutionary origins and maintenance of redundant gene expression during metazoan development. *Trends Genet.* 13: 360–364.

Cormack B.P., Strubin M., Ponticelli A.S., Struhl K. (1991) Functional differences between yeast and human TFIID are localized to the highly conserved region. *Cell* **65**: 341–348.

Covello P.S., Gray M.W. (1993) On the evolution of RNA editing. *Trends Genet.* 9: 265–268.

Damjanovski S., Huynh M.H., Motamed K., Sage E.H., Ringuette M. (1998) Regulation of *SPARC* expression during early *Xenopus* development: evolutionary divergence and conservation of DNA regulatory elements between amphibians and mammals. *Devel. Genes Evol.* 207: 453–461.

Darwish H.M., Krisinger J., Furlow J.D., Smith C., Murdoch F.E., DeLuca H.F. (1990) An estrogen-responsive element mediates the transcriptional regulation of calbindin D-9k in rat uterus. *J. Biol. Chem.* **266**: 551–558.

Dawson S.J., Wiman B., Hamsten A. (1993) The two allele sequences of a common polymorphism in the promoter of the plasminogen activator inhibitor-1 (PAI-1) gene respond differently to interleukin-1 in HepG2 cells. *J. Biol. Chem.* **268**: 10 739–10 745.

Delbrück S.J.W., Wendel B., Grunewald I., Sander T., Morris-Rosendahl D., Crocq M.A., Berrettini W.H., Hoehe M.R. (1997) A novel allelic variant of the human serotonin transporter gene regulatory polymorphism. *Cytogenet. Cell Genet.* **79**: 214–220.

Di Cristofano A., Strazullo M., Longo L., La Mantia G. (1995a) Characterization and genomic mapping of the *ZNF80* locus: expression of this zinc-finger gene is driven by a solitary LTR of ERV9 endogenous retroviral family. *Nucleic Acids Res.* **23**: 2823–2830.

Di Cristofano A., Strazullo M., Parisi T., La Mantia G. (1995b) Mobilization of an ERV9 human endogenous retroviral element during primate evolution. *Virology* **213**: 271–275.

Doenecke D., Albig W., Bouterfa H., Drabent B. (1994) Organization and expression of H1 histone and H1 replacement histone genes. *J. Cell. Biochem.* 54: 423–431.

Donnelly S.R., Moss S.E. (1998) Functional analysis of the human annexin I and VI gene promoters. *Biochem. J.* **332**: 681–687.

Duhig T., Ruhrberg C., Mor O., Fried M. (1998) The human surfeit locus. *Genomics* 52: 72–78.

Duret L., Dorkeld F., Gautier C. (1993) Strong conservation of non-coding sequences during vertebrates evolution: potential involvement in post-transcriptional regulation of gene expression. *Nucleic Acids Res.* **21**: 2315–2322.

Eizinger A., Jungblut B., Sommer R.J. (1999) Evolutionary change in the functional specificity of genes. *Trends Genet.* **15**: 197–202.

Elsen J.A., Sweder K.S., Hanawalt P.C. (1995) Evolution of the SNF2 family of proteins: subfamilies with distinct sequences and functions. *Nucleic Acids Res.* **23**: 2715–2723.

Emi M., Horii A., Tomita N., Nishide T., Ogawa M., Mori T., Matsubara K. (1988) Overlapping two genes in human DNA: a salivary amylase gene overlaps with a gamma-actin pseudogene that carries an integrated human endogenous retroviral DNA. *Gene* **62**: 229–235.

Epstein C.J. (1995) The new dysmorphology: application of insights from basic developmental biology to the understanding of human birth defects. *Proc. Natl. Acad. Sci. USA* **92**: 8566–8573.

Ernoult-Lange M., Arranz V., Le Coniat M., Berger R., Kress M. (1995) Human and mouse Kruppel-like (MOK2) orthologue genes encode two different zinc finger proteins. *J. Mol. Evol.* **41**: 784–794.

Feo S., Davies B., Fried M. (1992) The mapping of seven intron-containing ribosomal protein genes shows that they are unlinked in the human genome. *Genomics* **13**: 201–207.

Feuchter A.E., Freeman J.D., Mager D.L. (1992) Strategy for detecting cellular transcripts promoted by human endogenous long terminal repeats: identification of a novel gene (*CDC4L*) with homology to yeast CDC4. *Genomics* **13**: 1237–1246.

Feuchter-Murthy A.E., Freeman J.D., Mager D.L. (1993) Splicing of a human endogenous retrovirus to a novel phospholipase A2 related gene. *Nucleic Acids Res.* **21**: 135–143.

Fischer D.F., Sark M.W.J., Lehtola M.M., Gibbs S., van de Putte P., Backendorf C. (1999) Structure and evolution of the human *SPRR3* gene: implications for function and regulation. *Genomics* **55**: 88–99.

Fishman D., Faulds G., Jeffery R., Mohamed-Ali V., Yudkin J.S., Humphries S., Woo P. (1998) The effect of novel polymorphisms in the interleukin-6 (IL-6) gene on IL-6 transcription and plasma IL-6 levels, and an association with systemic-onset juvenile chronic arthritis. *J. Clin. Invest.* **102**: 1369–1376.

Fitch D.H.A., Mainone C., Goodman M., Slightom J.L. (1990) Molecular history of gene conversions in the primate γ-globin genes. *J. Biol. Chem.* **265**: 781–793.

Fitch D.H.A., Bailey W.J., Tagle D.A., Goodman M., Sieu L., Slightom J.L. (1991) Duplication of the γ-globin gene mediated by L1 long interspersed repetitive elements in an early ancestor of simian primates. *Proc. Natl. Acad. Sci. USA* **88**: 7396–7400.

Fracasso C., Patarnello T. (1998) Evolution of the dystrophin muscular promoter and 59 flanking region in primates. *J. Mol. Evol.* **46**: 168–179.

Fujino T., Navaratnam N., Scott J. (1998) Human apolipoprotein B editing deaminase gene (*APOBEC1*). *Genomics* **47**: 266–275.

Gastrop J., Hoevenagel R., Young J.R., Clevers H.C. (1992) A common ancestor of the mammalian transcription factors TCF-1 and TCF-1 alpha/LEF-1 expressed in chicken T cells. *Eur. J. Immunol.* **22**: 1327–1330.

Gellon G., McGinnis W. (1998) Shaping animal body plans in development and evolution by modulation of Hox expression patterns. *Bioessays* **20**: 116–125.

Gibson T.J., Spring J. (1998) Genetic redundancy in vertebrates: polyploidy and persistence of genes encoding multidomain proteins. *Trends Genet.* **14**: 46–49.

Gil G., Smith J.R., Goldstein J.L., Slaughter C.A., Orth K., Brown M.S., Osborne T.F. (1988) Multiple genes encode nuclear factor 1-like proteins that bind to the promoter for 3-hydroxy-3-methylglutaryl-coenzyme A reductase. *Proc. Natl. Acad. Sci. USA* **85**: 8963–8967.

Gill G., Tjian R. (1991) A highly conserved domain of TFIID displays species specificity *in vivo*. *Cell* **65**: 333–340.

Giordano M., Marchetti C., Chiorboli E., Bona G., Richiardi P.M. (1997) Evidence for gene conversion in the generation of extensive polymorphism in the promoter of the growth hormone gene. *Hum. Genet.* **100**: 249–255.

Goodchild N.L., Wilkinson D.A., Mager D.L. (1992) A human endogenous long terminal repeat provides a polyadenylation signal to a novel, alternatively spliced transcript in normal placenta. *Gene* **121**: 287–294.

Goodman B.K., Klein D., Tabor D.E., Vockley J.G., Cederbaum S.D., Grody W.W. (1994) Functional and molecular analysis of liver arginase promoter sequences from man and *Macaca fascicularis*. *Somat. Cell Molec. Genet.* **20**: 313–325.

Graw J., Liebstein A., Pietrowski D., Schmitt-John T., Werner T. (1993) Genomic sequences of murine γB- and γC-crystallin-encoding genes: promoter analysis and complete evolutionary pattern of mouse, rat and human γ-crystallins. *Gene* **136**: 145–156.

Greaves D.R., Quinn C.M., Seldin M.F., Gordon S. (1998) Functional comparison of the murine macrosialin and human CD68 promoters in macrophage and nonmacrophage cell lines. *Genomics* **54**: 165–168.

Green M., Krontiris T.G. (1993) Allelic variation of reporter gene activation by the *HRAS1* minisatellite. *Genomics* **17**: 429–434.

Gruber C.A., Rhee J.M., Gleiberman A., Turner E.E. (1997) POU domain factors of the Brn-3 class recognize functional DNA elements which are distinctive, symmetrical and highly conserved in evolution. *Mol. Cell. Biol.* **17**: 2391–2400.

Guarente L., Bermingham-McDonogh O. (1992) Conservation and evolution of transcriptional mechanisms in eukaryotes. *Trends Genet.* **8**: 27–31.

Gumucio D.L., Shelton D.A., Bailey W.J., Slightom J.L., Goodman M. (1993) Phylogenetic footprinting reveals unexpected complexity in *trans* factor binding upstream from the ε-globin gene. *Proc. Natl. Acad. Sci. USA* **90**: 6018–6022.

Gumucio D.L., Shelton D.A., Blanchard-McQuate K., Gray T., Tarle S., Heilstedt-Williamson H., Slightom J.L., Collins F., Goodman M. (1994) Differential phylogenetic footprinting as a means to identify base changes responsible for recruitment of the anthropoid γ gene to a fetal expression pattern. *J. Biol. Chem.* **269**: 15 371–15 380.

Hafner M., Zimmermann K., Pottgiesser J., Krieg T., Nischt R. (1995) A purine-rich sequence in the human BM-40 gene promoter region is a prerequisite for maximum transcription. *Matrix Biol.* **14**: 733–741.

Hai T.W., Liu F., Coukos W.J., Green M.R. (1989) Transcription factor ATF cDNA clones: an extensive family of leucine zipper proteins able to selectively form DNA-binding heterodimers. *Genes Dev.* **3**: 2083–2091.

Hall S., Chu G., Miller G., Cruickshank K., Cooper J.A., Humphries S.E., Talmud P.J. (1997) A common mutation in the lipoprotein lipase gene promoter, -93T/G, is associated with lower plasma triglyceride levels and increased promoter activity *in vitro*. *Arterioscl. Thromb. Vasc. Biol.* **17**: 1969–1976.

Hambor J.E., Mennone J., Coon M.E., Hanke J.H., Kevathas P. (1993) Identification and characterization of an *Alu*-containing, T-cell-specific enhancer located in the last intron of the human CD8α gene. *Mol. Cell. Biol.* **13**: 7056–7070.

Hancock J.M. (1993) Evolution of sequence repetition and gene duplications in the TATA-binding protein TBP (TFIID). *Nucleic Acids Res.* **21**: 2823–2830.

Hanke J.H., Hambor J.E., Kavathas P. (1995) Repetitive *Alu* elements form a cruciform structure that regulates the function of the human CD8α T cell-specific enhancer. *J. Mol. Biol.* **246**: 63–73.

Hardison R., Slightom J.L., Gumucio D.L., Goodman M., Stojanovic N., Miller W. (1997) Locus control regions of mammalian β-globin gene clusters: combining phylogenetic analyses and experimental results to gain functional insights. *Gene* **205**: 73–94.

Harendez C., Johnson L. (1990) Polyadenylation signal of the mouse thymidylate synthase gene was created by insertion of an L1 repetitive element downstream of the open reading frame. *Proc. Natl. Acad. Sci. USA* **87**: 2531–2535.

Hentschel C.C., Birnstiel M.L. (1981) The organization and expression of histone gene families. *Cell* **25**: 301–313.

Hewitt S.M., Fraizer G.C., Saunders G.F. (1995) Transcriptional silencer of the Wilms' tumor gene *WT1* contains an *Alu* repeat. *J. Biol. Chem.* **270**: 17908–17912.

Higuchi K., Law S.W., Hoeg J.M., Schmacher U.K., Meglin N., Brewer H.B. (1988) Tissue-specific expression of apolipoprotein A-I is regulated by the 5′ flanking region of the human *apoA-I* gene. *J. Biol. Chem.* **263**: 18530–18536.

Hoffman A., Sinn E., Yamamoto T., Wang J., Roy A., Horikoshi M., Roeder R.G. (1990) Highly conserved core domain and unique N terminus with presumptive regulatory motifs in a human TATA factor (TFIID). *Nature* **346**: 387–390.

Hollenberg A.N., Pestell R.G., Albanese C., Boers M.-E., Jameson J.L. (1994) Multiple promoter elements in the human chorionic gonadotropin β subunit genes distinguish their expression from the luteinizing hormone β gene. *Molec. Cell. Endocrinol.* **106**: 111–119.

Holmberg M., Leonardsson G., Ny T. (1995) The species-specific differences in the cAMP regulation of the tissue-type plasminogen activator gene between rat, mouse and human is caused by a one-nucleotide substitution in the cAMP-responsive element of the promoters. *Eur. J. Biochem.* **231**: 466–474.

Iwabe N., Kuma K., Miyata T. (1996) Evolution of gene families and relationship with organismal evolution: rapid divergence of tissue-specific genes in the early evolution of chordates. *Mol. Biol. Evol.* **13**: 483–493.

Jameson J.L., Lindell C.M., Habener J.F. (1986) Evolution of different transcriptional start sites in the human luteinizing hormone and chorionic gonadotropin β-subunit genes. *DNA* **5**: 227–234.

Jeung E.-B., Leung P.C.K., Krisinger J. (1994) The human calbindin-D$_{9k}$ gene: complete structure and implications on steroid hormone regulation. *J. Mol. Biol.* **235**: 1231–1238.

Jeung E.-B., Fan N.C., Leung P.C.K., Herr J.C., Freemerman A., Krisinger J. (1995) The baboon expresses the calbindin-D9K gene in intestine but not in uterus and placenta: implication for conservation of the gene in primates. *Molec. Reprod. Devel.* **40**: 400–407.

Johannes G., Berger F.G. (1992) Alterations in mRNA translation as a mechanism for the modification of enzyme synthesis during evolution. The ornithine decarboxylase model. *J. Biol. Chem.* **267**: 10 108–10 115.

Johnson J.L., McLachlan A. (1994) Novel clustering of Sp1 transcription factor binding sites at the transcription initiation site of the human muscle phosphofructokinase P1 promoter. *Nucleic Acids Res.* **22**: 5085–5092.

Johnson R.M., Buck S., Chiu C.-H., Schneider H., Samaio I., Gage D.A., Shen T.-L., Schneider M.P.C., Muniz J.A., Gumucio D.L., Goodman M. (1996) Fetal globin expression in New World monkeys. *J. Biol. Chem.* **271**: 14 684–14 691.

Joyner A.L. (1996) Engrailed, Wnt and Pax genes regulate midbrain-hindbrain development. *Trends Genet.* **12**: 15–20.

Kenmochi N., Kawaguchi T., Rozen S., Davis E., Goodman N., Hudson T.J., Tanaka T., Page D.C. (1998) A map of 75 human ribosomal protein genes. *Genome Res.* **8**: 509–523.

Kennedy G.C., German M.S., Rutter W.J. (1995) The minisatellite in the diabetes susceptibility locus IDDM2 regulates insulin transcription. *Nature Genet.* **9**: 293–298.

Kermekchiev M., Pettersson M., Matthias P., Schaffner W. (1991) Every enhancer works with every promoter for all the combinations tested: could new regulatory pathways evolve by enhancer shuffling? *Gene Expression* **1**: 71–81.

Kim J., Yu C., Bailey A., Hardison R., Shen C. (1989) Unique sequence organization and erythroid cell-specific nuclear factor-binding of mammalian θ1-globin promoters. *Nucleic Acids Res.* **17**: 5687–5701.

Knoll B.J., Rothblum K.N., Longley M. (1988) Nucleotide sequence of the human placental alkaline phosphatase gene. *J. Biol. Chem.* **263**: 12 020–12 027.

Kondo T., Herault Y., Zakany J., Duboule D. (1998) genetic control of murine limb morphogenesis: relationships with human syndromes and evolutionary relevance. *Mol. Cell. Endocrinol.* **140**: 3–8.

Kowalski P.E., Freeman J.D., Nelson D.T., Mager D.L. (1996) Structure and expression of an endogenous retrovirus-controlled human gene with duplicated phospholipase A2-like domains. *Am. J. Hum. Genet.* **59**: A153.

Kowalski P.E., Freeman J.D., Nelson D.T., Mager D.L. (1997) Genomic structure and evolution of a novel gene (*PLA2L*) with duplicated phospholipase A2-like domains. *Genomics* **39**: 38–46.

Kowalski P.E., Freeman J.D., Mager D.L. (1999) Intergenic splicing between a HERV-H endogenous retrovirus and two adjacent human genes. *Genomics* **57**: 371–379.

Kozmik Z., Czerny T., Busslinger M. (1997) Alternatively spliced insertions in the paired domain restrict the DNA sequence specificity of Pax6 and Pax8. *EMBO J.* **16**: 6793–6803.

Kurachi S., Deyashiki Y., Takeshita J., Kurachi K. (1997) Comprehensive studies on the human factor IX gene regulation: critical novel role of LINE 1 sequence in age-dependent and liver-specific expression. *Thromb. Haemost. Suppl.* **78**: 460.

Laudet V., Hänni C., Coll J., Catzeflis F., Stéhelin D. (1992) Evolution of the nuclear receptor gene superfamily. *EMBO J.* **11**: 1003–1013.

Laudet V., Stehelin D., Clevers H. (1993) Ancestry and diversity of the HMG box superfamily. *Nucleic Acids Res.* **21**: 2493–2501.

Lavia P., Macleod D., Bird A. (1987) Coincident start sites for divergent transcripts at a randomly selected CpG-rich island of mouse. *EMBO J.* **6**: 2773–2779.

Lin Y.-S., Green M.R. (1989) Similarities between prokaryotic and eukaryotic cyclic AMP-responsive elements. *Nature* **340**: 656–659.

Lipman D.J. (1997) Making (anti)sense of non-coding sequence conservation. *Nucleic Acids Res.* **25**: 3580–3583.

Liston D.R., Johnson P.J. (1999) Analysis of a ubiquitous promoter element in a primitive eukaryote: early evolution of the initiator element. *Mol. Cell. Biol.* **19**: 2380–2388.

Long Q., Bengra C., Li C., Kutlar F., Tuan D. (1998) A long terminal repeat of the human endogenous retrovirus ERV-9 is located in the 59 boundary area of the human β-globin locus control region. *Genomics* **54**: 542–555.

Ludwig D.L., Chen F., Peterson S.R., Nussenzweig A., Li G.C., Chen D.J. (1997) Ku80 gene expression is Sp1-dependent and sensitive to CpG methylation within a novel *cis* element. *Gene* **199**: 181–194.

McHaffie G.S., Ralston S.H. (1995) Origin of a negative calcium response element in an *Alu*-repeat: implications for regulation of gene expression by extracellular calcium. *Bone* **17**: 11–14.

Makalowski W., Zhang J., Boguski M.S. (1996) Comparative analysis of 1196 orthologous mouse and human full-length mRNA and protein sequences. *Genome Res.* **6**: 846–857.

Mannermaa R.M., Oikarinen J. (1989) Homology of nuclear factor I with the protein kinase family. *Biochem. Biophys. Res. Commun.* **162**: 427–434.

Marchese A., Cheng R., Lee M.C., Porter C.A., Heiber M., Goodman M., George S.R., O'Dowd B.F. (1994) Mapping studies of two G protein-coupled receptor genes: an amino acid difference may confer a functional variation between a human and rodent receptor. *Biochem. Biophys. Res. Commun.* **205**: 1952–1958.

Margarit E., Guillén A., Rebordosa C., Vidal-Taboada J., Sanchez M., Ballesta F., Oliva R. (1998) Identification of conserved potentially regulatory sequences of the *SRY* gene from 10 different species of mammals. *Biochem. Biophys. Res. Commun.* **245**: 370–377.

Matrinez E., Givel F., Wahli W. (1991) A common ancestor DNA motif for invertebrate and vertebrate hormone response elements. *EMBO J.* **10**: 263–268.

Matsuo K., Clay O., Kunzler P., Georgiev O., Urbanek P., Schaffner W. (1994) Short introns interrupting the Oct-2 POU domain may prevent recombination between POU family genes without interfering with potential POU domain 'shuffling' in evolution. *Biol. Chem. Hoppe-Seyler* **375**: 675–683.

Mayer W.E., O'hUigin C., Klein J. (1993) Resolution of the *HLA-DR6* puzzle: a case of grafting a *de novo*-generated exon on an existing gene. *Proc. Natl. Acad. Sci. USA* **90**: 10 720–10 724.

Meisler M.H., Ting C.-N. (1993) The remarkable evolutionary history of the human amylase genes. *Crit. Rev. Oral Biol. Med.* **4**: 503–509.

Meyer E., Wiegand P., Rand S.P., Kuhlmann D., Brack M., Brinkmann B. (1995) Microsatellite polymorphisms reveal phylogenetic relationships in primates. *J. Mol. Evol.* **41**: 10–14.

Moore S., Sargeant L., King T., Mattik J., Georges M., Hetzel D. (1991) The conservation of dinucleotide microsatellites among mammalian genomes allows the use of heterologous PCR primer pairs in closely related species. *Genomics* **10**: 654–660.

Mukai T., Arai Y., Yatsuki H., Hori K. (1991) An additional promoter functions in the human aldolase A gene, but not in rat. *Eur. J. Biochem.* **195**: 781–787.

Murnane J.P., Morales J.F. (1995) Use of a mammalian interspersed repetitive (MIR) element in the coding and processing sequences of mammalian genes. *Nucleic Acids Res.* **23**: 2837–2839.

Mushegian A.R., Koonin E.V. (1996) Sequence analysis of eukaryotic developmental proteins: ancient and novel domains. *Genetics* **144**: 817–828.

Nakamuta M., Tsai A., Chan L., Davidson N.O., Teng B.-B. (1999) Sequence elements required for apolipoprotein B mRNA editing enhancement activity from chicken enterocytes. *Biochem. Biophys. Res. Commun.* **254**: 744–750.

Noll M. (1993) Evolution and role of Pax genes. *Curr. Opin. Genet. Devel.* **3**: 595–605.

Norris J., Fan D., Aleman C., Marks J.R., Futreal P.A., Wiseman R.W., Iglehart J.D., Deininger P.L., McDonnell D.P. (1995) Identification of a new subclass of *Alu* DNA repeats which can function as estrogen receptor-dependent transcriptional enhancers. *J. Biol. Chem.* **270**: 22 777–22 782.

Novo F.J., Kruszewski A., MacDermot K.D., Goldspink G., Gorecki D.C. (1995) Editing of human α-galactosidase RNA resulting in a pyrimidine to purine conversion. *Nucleic Acids Res.* **23**: 2636–2640.

Nussinov R. (1990) Sequence signals in eukaryotic upstream regions. *Crit. Rev. Biochem. Molec. Biol.* **25**: 185–224.

Oohashi T., Ueki Y., Sugimoto M., Ninomiya Y. (1995) Isolation and structure of the *COL4A6* gene encoding the human α6(IV) collagen chain and comparison with other type IV collagen genes. *J. Biol. Chem.* **270**: 26 863–26 867.

Okazaki T., Zajac J.D., Igarashi T., Ogata E., Kronenberg H.M. (1991) Negative regulatory elements in the human parathyroid hormone gene. *J. Biol. Chem.* **266**: 21 903–21 910.

Ouzounis C.A., Kyrpides N.C. (1996a) Parallel origins of the nucleosome core and eukaryotic transcription from Archaea. *J. Mol. Evol.* **42**: 234–239.

Ouzounis C.A., Kyrpides N.C. (1996b) The emergence of major cellular processes in evolution. *FEBS Letts* **390**: 119–123.

Pan T.C., Hao Q.L., Yamin T.T., Dai P.H., Chen B.S., Chen S.L., Kroon P.L., Chao Y.S. (1987) Rabbit apolipoprotein A-I mRNA and gene. *Eur. J. Biochem.* **170**: 99–104.

Pan J., Xia L., McEver R.P. (1998) Comparison of promoters for the murine and human P-selectin genes suggests species-specific and conserved mechanisms for transcriptional regulation in endothelial cells. *J. Biol. Chem.* **273**: 10058–10067.

Paquette J., Giannoukakis N., Polychronakos C., Vafiadis P., Deal C. (1998) The *INS* 5′ variable number of tandem repeats is associated with *IGF2* expression in humans. *J. Biol. Chem.* **273**: 14 158–14 164.

Piedrafita F.J., Molander R.B., Vansant G., Orlova E.A., Pfahl M., Reynolds W.F. (1996) An *Alu* element in the myeloperoxidase promoter contains a composite SP1-thyroid hormone-retinoic acid response element. *J. Biol. Chem.* **271**: 14412–14420.

Pollak E.S., Hung H.-L., Godin W., Overton G.C., High K.A. (1996) Functional characterization of the human factor VII 5′-flanking region. *J. Biol. Chem.* **271**: 1738–1747.

Postlethwait J.H., Talbot W.S. (1997) Zebrafish genomics: from mutants to genes. *Trends Genet.* **13**: 183–189.

Ptashne M. (1997) Control of gene transcription: an outline. *Nature Medicine* **3**: 1069–1072.

Raisonnier A. (1991) Duplication of the apolipoprotein C-I gene occurred about forty million years ago. *J. Mol. Evol.* **32**: 211–219.

Robins D.M., Samuelson L.C. (1992) Retroposons and the evolution of mammalian gene expression. *Genetica* **86**: 191–201.

Robins D.M., Scheller A., Adler A.J. (1994) Specific steroid response from a nonspecific DNA element. *J. Steroid Biochem. Mol. Biol.* **49**: 251–255.

Rowlands T, Baumann P, Jackson SP. (1994) The TATA-binding protein: a general transcription factor in eukaryotes and archaebacteria. *Science* **264**: 1326–1328.

Russell J.E., Morales J., Makeyev A.V., Liebhaber S.A. (1998) Sequence divergence in the 3′ untranslated regions of human ζ- and α-globin mRNAs mediates a difference in their stabilities and contributes to efficient α-ζ gene developmental switching. *Mol. Cell. Biol.* **18**: 2173–2183.

Rutherford S.L., Lindquist S. (1998) Hsp90 as a capacitor for morphological evolution. *Nature* **396**: 336–341.

Rutter J.L., Mitchell T.I., Butticè G., Meyers J., Gusella J.F., Ozelius L.J., Brinckerhoff C.E. (1998) A single nucleotide polymorphism in the matrix metalloproteinase-1 promoter creates an Ets binding site and augments transcription. *Cancer Res.* **58**: 5321–5325.

Sabol S.Z., Hu S., Hamer D. (1998) A functional polymorphism in the monoamine oxidase A gene promoter. *Hum. Genet.* **103**: 273–279.

Saksela K., Baltimore D. (1993) Negative regulation of immunoglobulin kappa light-chain gene transcription by a short sequence homologous to the murine B1 repetitive element. *Mol. Cell. Biol.* **13**: 3698–3705.

Samuelson L.C., Wiebauer K., Gumucio D.L., Meisler M.H. (1988) Expression of the human amylase genes: recent origin of a salivary amylase promoter from an actin pseudogene. *Nucleic Acids Res.* **16**: 8261–8276.

Samuelson L.C., Wiebauer K., Snow C.M., Meisler M.H. (1990) Retroviral and pseudogene insertion sites reveal the lineage of human salivary and pancreatic amylase genes from a single gene during primate evolution. *Molec. Cell. Biol.* **10**: 2513–2520.

Samuelson L.C., Phillips R.S., Swanberg L.J. (1996) Amylase gene structure in primates: retroposon insertions and promoter evolution. *Mol. Biol. Evol.* **13**: 767–779.

Schena M. (1989) The evolutionary conservation of eukaryotic gene transcription. *Experientia* **45**: 972–983.

Schmidt C., Fischer G., Kadner H., Genersch E., Kuhn K., Poschl E. (1993) Differential effects of DNA-binding proteins on bidirectional transcription from the common promoter region of human collagen type IV genes *COL4A1* and *COL4A2*. *Biochim. Biophys. Acta* **1174**: 1–10.

Schulte A.M., Wellstein A. (1998) Structure and phylogenetic analysis of an endogenous retrovirus inserted into the human growth factor gene pleiotrophin. *J. Virol.* **72**: 6065–6072.

Schulte A.M., Lai S., Kurtz A., Czubayko F., Riegel A.T., Wellstein A. (1996) Human trophoblast and choriocarcinoma expression of the growth factor pleiotrophin attributable to germ-line insertion of an endogenous retrovirus. *Proc. Natl. Acad. Sci. USA* **93**: 14 759–14 764.

Sefton M., Sanchez S., Nieto M.A. (1998) Conserved and divergent roles for members of the Snail family of transcription factors in the chick and mouse embryo. *Development* **125**: 3111–3121.

Semenza G.L. (1998) *Transcription Factors and Human Disease*. Oxford University Press, New York.

Seyfert H.-M., Tuckoricz A., Interthal H., Koczan D., Hobom G. (1994) Structure of the bovine lactoferrin-encoding gene and its promoter. *Gene* **143**: 265–269.

Sharma P., Bowman M., Madden S., Rauscher F.I., Sukumar S. (1994) RNA editing in the Wilms' tumour susceptibility gene, *WT1*. *Genes Dev.* **8**: 720–731.

Shashikant C.S., Kim C.B., Borbély M.A., Wang W.C.H., Ruddle F.H. (1998) Comparative studies on mammalian *Hoxc8* early enhancer sequence reveal a baleen whale-specific deletion of a *cis*-acting element. *Proc. Natl. Acad. Sci. USA* **95**: 15 446–15 451.

Shaw-Bruha C.M., Pennington K.L., Shull J.D. (1998) Identification in the rat prolactin gene of sequences homologous to the distal promoter of the human prolactin gene. *Biochim. Biophys. Acta* **1442**: 304–313.

Shelton D.A., Stegman L., Hardison R., Miller W., Bock J.H., Slightom J.L., Goodman M., Gumucio D.L. (1997) Phylogenetic footprinting of hypersensitive site 3 of the β-globin locus control region. *Blood* **89**: 3457–3469.

Shirvan A., Srivastava M., Wang M.G., Cultraro C., Magendzo K., McBridge O.W. (1994) Divergent structure of the human synexin (annexin VII) gene and assignment to chromosome 10. *Biochemistry* **33**: 6888–6901.

Singh N., Berger F.G. (1998) Evolution of a mammalian promoter through changes in patterns of transcription factor binding. *J. Mol. Evol.* **46**: 639–648.

Singh N., Barbour K.W., Berger F.G. (1998) Evolution of transcriptional regulatory elements within the promoter of a mammalian gene. *Mol. Biol. Evol.* **15**: 312–325.

Skuse G., Cappione A., Sowden M., Metheny L., Smith H. (1996) The neurofibromatosis type 1 messenger RNA undergoes base-modification RNA editing. *Nucleic Acids Res.* **24**: 478–486.

Sorci-Thomasa M., Kearns M.W. (1995) Species-specific polymorphism in the promoter of the apolipoprotein A-I gene: restoration of human transcriptional efficiency by substitution at positions -189, -144 and -48bp. *Biochim. Biophys. Acta* **1256**: 387–395.

Soteriou B., Fisher R.A., Khan I.M., Kessling A.M., Archard L.C., Buluwela L. (1995) Conserved gene sequences for species identification: PCR analysis of the 3′UTR of the SON gene distinguishes human and other mammalian DNAs. *Forensic Sci. Int.* **73**: 171–181.

Stallings R.L. (1994) Distribution of trinucleotide microsatellites in different categories of mammalian genomic sequence: implication for human genetic diseases. *Genomics* **21**: 116–121.

Stallings R.L. (1995) Conservation and evolution of $(CT)_n/(GA)_n$ microsatellite sequences at orthologous positions in diverse mammalian genomes. *Genomics* **25**: 107–113.

Stallings R.L., Ford A.F., Nelson D., Torney D.C., Hildebrand C.E., Moyzis R.K. (1991) Evolution and distribution of $(GT)_n$ repetitive sequences in mammalian genomes. *Genomics* **10**: 807–815.

Stavenhagen J.B., Robins D.M. (1988) An ancient provirus has imposed androgen regulation on the adjacent mouse sex-limited protein gene. *Cell* **55**: 247–254.

Steger D.J., Altschmied J., Buscher M., Mellon P.L. (1991) Evolution of placenta-specific gene expression: comparison of the equine and human gonadotropin α-subunit genes. *Molec. Endocrinol.* **5**: 243–255.

Surguchov A. (1991) Migration of promoter elements between genes: a role in transcriptional regulation and evolution. *Biomed. Sci.* **2**: 22–28.

Suzuki K., Kuriyama M., Saito T., Ichinose A. (1997) Plasma lipoprotein(a) levels and expression of the apolipoprotein(a) gene are dependent on the nucleotide polymorphisms in its 59 flanking region. *J. Clin. Invest.* **99**: 1361–1366.

Sverdlov E.D. (1998) Perpetually mobile footprints of ancient infections in human genome. *FEBS Letts.* **428**: 1–6.

Tagle D.A., Koop B.F., Goodman M., Slightom J.L., Hess D.L., Jones R.T. (1988) Embryonic ε and γ globin genes of a prosimian primate (*Galago crassicaudatus*). *J. Mol. Biol.* **203**: 439–455.

Talmud P.J., Palmen J., Walker M. (1998) Identification of genetic variation in the human hormone-sensitive lipase gene and 59 sequences: homology of 5′ sequences with mouse promoter and identification of potential regulatory elements. *Biochem. Biophys. Res. Commun.* **252**: 661–668.

Tansey W.P., Catanzaro D.F. (1991) Sp1 and thyroid hormone receptor differentially activate expression of human growth hormone and chorionic somatomammotropin genes. *J. Biol. Chem.* **266**: 9805–9813.

Taylor A., Erba H.P., Muscat G.E.O., Kedes L. (1988) Nucleotide sequence and expression of the human skeletal α-actin gene: evolution of functional regulatory domains. *Genomics* **3**: 323–336.

Teng B., Davidson N.O. (1992) Evolution of intestinal apolipoprotein B mRNA editing. Chicken apolipoprotein B mRNA is not edited but chicken enterocytes contain *in vitro* editing enhancement factor(s). *J. Biol. Chem.* **267**: 21 265–21 272.

Thorey I.S., Cecena G., Reynolds W., Oshima R.G. (1993) *Alu* sequence involvement in transcriptional insulation of the keratin 18 gene in transgenic mice. *Molec. Cell. Biol.* **13**: 6742–6751.

Ting C.-N., Rosenberg M.P., Snow C.M., Samuelson L.C., Meisler M.H. (1992) Endogenous retroviral sequences are required for tissue-specific expression of a human salivary amylase gene. *Genes Devel.* **6**: 1457–1465.

TomHon C., Zhu W., Millinoff D., Hayasaka K., Slightom J.L., Goodman M., Gumucio D.L. (1997) Evolution of a fetal expression pattern via *cis* changes near the γ-globin gene. *J. Biol. Chem.* **272**: 14 062–14 066.

Trainor C.D., Evans T., Felsenfeld G., Boguski M.S. (1990) Structure and evolution of a human erythroid transcription factor. *Nature* **343**: 92–96.

Treppicchio W.L., Krontiris T.G. (1992) Members of the *rel*/NF-κB family of transcriptional regulatory factors bind the *HRAS1* minisatellite DNA sequence. *Nucleic Acids Res.* **20**: 2427–2434.

Treppicchio W.L., Krontiris T.G. (1993) *IGH* minisatellite suppression of USF-binding site- and Em-mediated transcriptional activation of the adenovirus major late promoter. *Nucleic Acids Res.* **21**: 977–985.

Trezise A.E., Godfrey E.A., Holmes R.S., Beacham I.R. (1989) Cloning and sequencing of cDNA encoding baboon liver alcohol dehydrogenase: evidence for a common ancestral lineage with the human alcohol dehydrogenase β subunit and for class I *ADH* gene duplications predating primate radiation. *Proc. Natl. Acad. Sci. USA* **86**: 5454–5458.

Vallejo A.N., Allen K.S., Pease L.R. (1995) A common mutation in the hominoid class I *A*-locus IFN-responsive element results in the loss of enhancer activity. *Int. Immunol.* **7**: 853–859.

Vallejo A.N., Pease L.R. (1995) Structure of the MHC A and B locus promoters in hominoids. *J. Immunol.* **154**: 3912–3921.

Vansant G., Reynolds W.F. (1995) The consensus sequence of a major *Alu* subfamily contains a functional retinoic acid response element. *Proc. Natl. Acad. Sci. USA* **92**: 8229–8233.

Vuillaumier S., Dixmeras I., Messai H., Lapoumeroulie C., Lallemand D., Gekas J., Chehab F.F., Perret C., Elion J., Denamur E. (1997) Cross-species characterization of the promoter region of the cystic fibrosis conductance regulator gene reveals multiple levels of regulation. *Biochem. J.* **327**: 651–662.

Wade D.P., Clarke J.G., Lindahl G.E., Liu A.C., Zysow B.R., Meer K., Schwartz K., Lawn R.M. (1993) 5′ control regions of the apolipoprotein(a) gene and members of the related plasminogen gene family. *Proc. Natl. Acad. Sci. USA* **90**: 1369–1373.

Wang C.-Y., Petryniak B., Ho I.-C., Thompson C.B., Leiden J.M. (1992) Evolutionarily conserved Ets family members display distinct DNA binding specificities. *J. Exp. Med.* **175**: 1391–1399.

Wattler S., Russ A., Evans M., Nehls M. (1998) A combined analysis of genomic and primary protein structure defines the phylogenetic relationship of new members of the T-box family. *Genomics* **48**: 24–33.

Wegner M. (1999) From head to toes: the multiple facets of Sox proteins. *Nucleic Acids Res.* **27**: 1409–1420.

White N.L., Higgins C.F., Trezise A.E.O. (1998) Tissue-specific *in vivo* transcription start sites of the human and murine cystic fibrosis genes. *Hum. Molec. Genet.* **7**: 363–369.

Whitley M.Z., Thanos D., Read M.A., Maniatis T., Collins T. (1994) A striking similarity in the organization of the E-selectin and beta-interferon gene promoters. *Mol. Cell. Biol.* **14**: 6464–6475.

Wilson A.G., Symons J.A., McDowell T.L., McDevitt H.O., Duff G.W. (1997) Effects of a polymorphism in the human tumor necrosis factor alpha promoter on transcriptional activation. *Proc. Natl. Acad. Sci. USA* **94**: 3195–3199.

Winter P.C., Scopes D.A., Berg L.-P., Millar D.S., Kakkar V.V., Mayne E.E., Krawczak M., Cooper D.N. (1995) Functional analysis of an unusual length polymorphism in the human antithrombin III (*AT3*) gene promoter. *Blood Coag. Fibrinol.* **6**: 659–664.

Wobbe C.R., Struhl K. (1990) Yeast and human TATA-binding proteins have nearly identical DNA sequence requirements for transcription *in vitro*. *Mol. Cell. Biol.* **10**: 3859–3867.

Wong G., Itakura T., Kawajiri K., Skow L., Negishi M. (1989) Gene family of male-specific testosterone 16α-hydroxylase (C-P450$_{16}$) in mice. *J. Biol. Chem.* **264**: 2920–2927.

Wright K.L., White L.C., Kelly A., Beck S., Trowsdale J., Ting J.P.-Y. (1995) Coordinate regulation of the human *TAP1* and *LMP2* genes from a shared bidirectional promoter. *J. Exp. Med.* **181**: 1459–1471.

Wu J., Grindlay G.J., Bushel P., Mendelsohn L., Allan M. (1990) Negative regulation of the human ϵ-globin gene by transcriptional interference: role of an *Alu* repetitive element. *Mol. Cell. Biol.* **10**: 1209–1216.

Xu W., Liu L., Emson P.C., Harrington C.R., Charles I.G. (1997) Evolution of a homopurine-homopyrimidine pentanucleotide repeat sequence upstream of the human inducible nitric oxide synthase gene. *Gene* **204**: 165–170.

Yang Z., Boffelli D., Boonmark N., Schwartz K., Lawn R. (1998) Apolipoprotein(a) gene enhancer resides within a LINE element. *J. Biol. Chem.* **273**: 891–897.

Yasuo H., Satoh N. (1998) Conservation of the developmental role of *Brachyury* in notochord formation in a urochordate, the ascidian *Balocynthia roretzi*. *Devel. Biol.* **200**: 158–170.

Youssoufian H., Lodish H.F. (1993) Transcriptional inhibition of the murine erythropoietin receptor gene by an upstream repetitive element. *Mol. Cell. Biol.* **13**: 98–104.

Zardoya R., Abouheif E., Meyer A. (1996) Evolution and orthology of *hedgehog* genes. *Trends Genet.* **12**: 496–497.

Zilliacus J., Carlstedt-Duke J., Gustafsson J.-A., Wright A.P.H. (1994) Evolution of distinct DNA-binding specificities within the nuclear receptor family of transcription factors. *Proc. Natl. Acad. Sci. USA* **91**: 4175–4179.

Zuckerkandl E. (1997) Junk DNA and sectorial gene repression. *Gene* **205**: 323–343.

Pseudogenes and their formation

6.1 Pseudogene formation

Pseudogenes may be regarded as 'floating hulks,' gene-derived DNA sequences that are no longer capable of being expressed as protein products. Some are the remnants of once active genes that have acquired inactivating mutations which either preclude their transcription or at the very least prevent mRNA translation. Others are retrotransposed copies of expressed mRNAs which, since they almost always lack promoter sequences, are incapable of being expressed and therefore also tend to accumulate deleterious mutations. Pseudogenes ought not, however, to be regarded entirely as evolutionary *cul-de-sacs*. Although it is unlikely that reactivating (reverse) mutations will occur so as to restore their function, pseudogenes may nevertheless influence the evolution of other functionally significant sequences by for example mediating recombination events (Beck *et al.*, 1996; Takahashi *et al.*, 1982) or acting as sequence donors in gene conversion (Section 6.1.6).

6.1.1 The generation of pseudogenes by duplication

A considerable number of pseudogenes have been described in the human genome that have retained the exon-intron structure of their functional source genes (*Figure 6.1*). A selection of the known pseudogenes of this type is presented in *Table 6.1*. One assumes that such sequences arose by simple duplication of functional gene sequences but became inactivated since their intrinsic redundancy prevented selection from maintaining their potential to be expressed (Wilde, 1985). Many of the examples listed in *Table 6.1* have therefore accumulated nonsense mutations, frameshift deletions and insertions, or single base-pair substitutions within splice sites, any one of which would have been sufficient to render the expression of the sequences impossible. Perhaps the archetypal example is that of the human pseudogene (*CYP21P*) for cytochrome P450c21 which is closely linked to its cognate gene (*CYP21*; 6p21.3) as a result of a duplication event which occurred before the separation of apes from Old World monkeys >23

Human Gene Evolution, David N. Cooper.
© 1999 BIOS Scientific Publishers Ltd, Oxford.

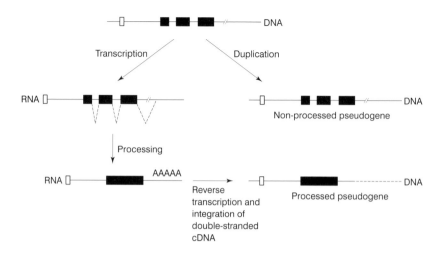

Figure 6.1. Schematic diagram depicting the different mechanisms of pseudogene formation.

Table 6.1. Examples of human pseudogenes (ψ) generated by duplication

Pseudogene	Complete/partial	Comments	Reference
Cytochrome P450c21	Complete	Linked to *CYP21* (6p21.3) gene	Harada *et al*. (1987)
Amylase	Partial (lacks exons 1–3)	Linked to *AMY* gene cluster (1p21)	Groot *et al*. (1990)
Apolipoprotein C1	Complete	Linked to *APOC1* (19q13.2) gene	Raisonnier (1991)
Tyrosinase	Partial (exons 4 and 5 and non-coding)	11p11.2-cen. Unlinked to *TYR* gene (11q14-q21). Not transcribed.	Giebel *et al*. (1991)
Dopamine D5 receptor (2)	Complete (intronless ψ derived from intronless gene)	ψ on chromosomes 1 and 2. Unlinked to *DRD5* gene (4p15-p16).	Marchese *et al*. (1995) (1995)
Olfactory receptor	Complete	Linked to olfactory receptor gene cluster on chromosome 17. Transcribed.	Crowe *et al*. (1996)
Histone H2B	Complete	Linked to other histone genes (1q21-q23).	Albig *et al*. (1997)
Keratin K14	Complete	Unclear if linked to *KRT14* gene at 17q12-q21	Savtchenko *et al*. (1988)
Fibroblast growth factor (16?)	Partial (exon 2, intron 2, exon 3 and 3′ UTR)	Transcribed. Dispersed. Unlinked to *FGF7* gene (15q15-q21).	Kelley *et al*. (1992)
XG blood group	Partial (exons 1, 2A, 2B, 3)	Yq11.21. *XG* gene maps to pseudo-autosomal boundary region, Xp22-pter Transcribed.	Weller *et al*. (1995)

Pseudogene	Complete/partial	Comments	Reference
Interleukin-8 receptor	Complete	2q34-q35. Linked to *IL8RA* and *ILR8B* genes.	Ahuja *et al.* (1992)
Kallmann syndrome gene	Complete	Yq11. Unlinked to *KAL1* gene (Xp22.3).	del Castillo *et al.* (1992)
α2-Macroglobulin	Complete	12p12-p13. Linked to *A2M* gene.	Devriendt *et al.* (1989)
von Willebrand factor	Partial (exons 23–28)	Chromosome 22. Unlinked to *VWF* gene (12p13)	Marchetti *et al.* (1991)
Mannose-binding protein	Complete	10q22. Probably linked to *MBP* gene (10q11-q21)	Guo *et al.* (1998)
Serum amyloid A (SAA3)	Complete	Not transcribed. Linked to *SAA* gene cluster (11p15).	Kluve-Beckerman *et al.* (1991)
Carbonic anhydrase V	Partial (exons 3–7, introns 3–6)	16p11.2-p12. Unlinked to *CA5* gene (16q24.3).	Nagao *et al.* (1995)
Calcitonin	Partial (exons 2 and 3)	Linked to *CALCA* gene (11p15).	Lips *et al.* (1989)
Protein S	Partial (exons 2–15 present)	Linked to *PROS1* gene (3p11-q11). Not transcribed.	Ploos van Amstel *et al.* (1990)
β-Glucocerebrosidase	Complete	Linked to *GBA* gene (1q21). Transcribed.	Horowitz *et al.* (1989)
Metaxin	Complete	Linked to *MTX* gene (1q21). Not transcribed.	Long *et al.* (1996)
β-Glucuronidase (2)	Complete and partial	5p13 and 5q13. Unlinked to *GUSB* gene (7q11).	Speleman *et al.* (1996)
α-Globin	Complete	Linked to *HBA1* (16p13.3-pter) gene. Transcribed.	Whitelaw and Proudfoot (1983)
ζ-Globin	Complete	Linked to *HBA1* (16p13.3-pter) gene. Not transcribed.	Proudfoot *et al.* (1984)
Immunoglobulin Cε2	Partial	Linked to immunoglobulin Cε1 *IGHE* gene (14q32).	Hisajima *et al.* (1983)
Immunoglobulin γ	Complete	Linked to *IGHG* gene cluster (14q32).	Takahashi *et al.* (1982)
Immunoglobulin Vκ (3)	Complete	Chromosomes 1, 15, 22. Unlinked to *IGKV* gene cluster (2p12)	Bentley and Rabbitts (1980); Lotscher *et al.* (1986)
Immunoglobulin V$_H$	Complete	Linked to immunoglobulin V$_H$ (*IGHV*) gene cluster (14q32.3)	Cook and Tomlinson (1995)
HLA class I (9)	Complete	Linked to *HLA-A*, *HLA-E*, *HLA-G* and *HLA-F* genes (6p21.3)	Hughes (1995); Gruen *et al.* (1996) Geraghty *et al.* (1992)
HLA class II	Complete	Linked to *HLA-DQB1* genes (6p21.3)	Figueroa *et al.* (1994)

Table 6.1 (continued)

Pseudogene	Complete/partial	Comments	Reference
Prostaglandin EP4 receptor (2)	Complete	Unclear if linked to *PTGER4* gene (5p13)	Foord *et al.* (1996)
Adrenoleukodystrophy	Partial (exons 7–10)	Unclear if linked to *ALD* gene at Xq28.	Braun *et al.* (1996)
β-Globin (1984)	Complete	Linked to *HBB* gene (11p15)	Chang and Slightom
Aldehyde dehydrogenase (*ALDH8*)	Complete	Transcribed	Hsu and Chang (1996); Hsu *et al.* (1997)
Iduronate-2-sulfatase	Partial (exons 2 and 3, introns 2 and 7)	Linked to *IDS* gene (Xq28). Not transcribed.	Timms *et al.* (1995); Rathmann *et al.* (1995)

Numbers in brackets denote numbers of pseudogenes in cases where more than one has been found.

Myrs ago (Horiuchi *et al.*, 1993). This pseudogene contains three inactivating mutations: an 8 bp deletion in exon 3, the insertion of a T in exon 7 and a nonsense mutation in exon 8 (Kawaguchi *et al.*, 1992). The 8 bp deletion is present in both humans and chimpanzees whereas the other two mutations are human-specific (Kawaguchi *et al.*, 1992) consistent with the 8 bp deletion being the original inactivating mutation in the human-chimpanzee lineage. In gorilla and orangutan, the extra *Cyp21* gene copies are inactivated by various other mutations (Kawaguchi *et al.*, 1992).

From inspection of *Table 6.1*, it is apparent that such pseudogenes are essentially of two types: either complete copies of the source gene or partial copies of that gene. Complete copies tend to be linked to the source gene (examples given in *Figure 6.2*) whereas partial copies are often (although not always) unlinked to

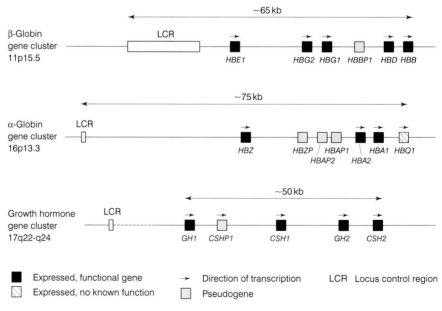

Figure 6.2. Human β-globin, α-globin, and growth hormone gene clusters illustrating relative locations of nonprocessed pseudogenes.

the gene from which they originated. This may reflect mechanistic differences in pseudogene generation.

Since many of the pseudogenes listed in *Table 6.1* are duplicated copies of entire genes that include their original promoters, it is not surprising that some are capable of being transcribed. However, some of the partial, truncated pseudogenes also appear to be transcribed. We may nevertheless assume that the nonsense mutations and frameshift insertions and deletions acquired by these sequences serve to preclude the translation of any mRNAs synthesized.

A particularly good example of the generation of multiple partial pseudogenes is that of the numerous sequences related to the neurofibromatosis type 1 (*NF1*; 17q11.2) gene identified on human chromosomes 2, 12, 14, 15, 18, 20, 21 and 22 (Gasparini *et al.*, 1993; Hulsebos *et al.*, 1996; Kehrer-Sawatzki *et al.*, 1997; Legius *et al.*, 1992; Marchuk *et al.*, 1992; Purandare *et al.*, 1995; Suzuki *et al.*, 1994). These NF1-homologous sequences contain numerous nucleotide substitutions, small deletions and insertions (some inactivating) but still exhibit >90% homology to the corresponding sequences of the *NF1* gene. They appear to have arisen by partial duplication of the *NF1* gene between 22 Myrs and 33 Myrs ago with subsequent rounds of duplication generating new copies which were then transposed to the pericentromeric regions of the other chromosomes (Régnier *et al.*, 1997).

In cases where the parental source gene is intronless [e.g. some argininosuccinate synthetase pseudogenes (Nomiyama *et al.*, 1986) and the dopamine D5 receptor pseudogene (Marchese *et al.*, 1995)], the pseudogenes may have been secondarily created by the genomic duplication of pre-existing processed pseudogenes (see Section 6.1.2 below).

The occasional human pseudogene manifests a presence/absence polymorphism, for example the T-cell receptor β v6.10 pseudogene which has been found in the genome of most but not all individuals tested (Li *et al.*, 1993).

6.1.2 The generation of pseudogenes by retrotransposition

The second main category of pseudogenes probably accounts for the majority of inactivated gene sequences found in the human genome. A selection of pseudogenes of this type is presented in *Table 6.2*. They are termed *processed* pseudogenes since they are thought to have originated by retrotransposition of a correctly processed mRNA intermediate lacking intervening sequences (Vanin, 1984; Weiner, 1986; *Figure 6.1*). The mRNA origin is evidenced by the lack of introns and the frequent presence of poly(A) tracts at their 3′ ends (Vanin, 1984). Flanking direct repeats, commonly of 9–17 bp (Vanin, 1984), are usually present and have probably been acquired during the process of integration. Many processed pseudogenes correspond to the entire length of the coding region of their source genes but others are truncated (e.g. *CFTR* exon 9 pseudogene fragments; Rozmahel *et al.*, 1997). The vast majority of processed pseudogenes are located on different chromosomes from their functional source genes (*Table 6.2*). This is not very surprising in view of the fact that retrotransposition necessarily requires a mobile mRNA intermediate.

Many processed pseudogenes have, like their duplication-derived counterparts, acquired multiple genetic lesions that preclude their expression. This is not however an invariant finding as evidenced by the human metallothionein processed pseudogene which has no such lesions and retains an open reading frame (Karin and Richards, 1982). This sequence is nevertheless probably not transcribed since, owing to its mRNA origin, it necessarily lacks the promoter elements necessary for transcription to occur. In general, retrotransposed pseudogenes are rarely associated with active promoter elements and are not therefore usually expressed. Again there are always exceptions, for example the human glutamine synthetase pseudogene which contains a functional promoter/enhancer in its 5′ flanking sequence that allows it to be transcribed although not expressed (owing to the presence of numerous frameshift mutations) (Chakrabarti *et al.*, 1995). Those processed 'pseudogenes' which possess both open reading frames and some inherent or acquired promoter activity such that they can be expressed, are really not pseudogenes at all but rather examples of gene creation by retrotransposition, a topic covered in Chapter 9, section 9.6.

To have been inherited, processed pseudogenes must have originated in the germline. It therefore follows that their functional source genes must have been expressed in the germline. Consistent with this assumption, many processed pseudogenes correspond either to ubiquitously expressed 'housekeeping' genes (e.g. snRNA genes) or to genes that are known to be expressed in germ cells (e.g. β-tubulin; Lewis and Cowan, 1990) (*Table 6.2*). However, some processed pseudogenes are derived from transcripts of tissue-specific genes such as those encoding the immunoglobulins (*Table 6.2*). These pseudogenes may have originated from ectopic transcripts, the consequence of the 'leaky' transcription which appears to be a property of every gene in every cell. In a very few cases, processed pseudogenes have been shown to be derived from antisense transcripts (Rozmahel *et al.*, 1997; Zhou *et al.*, 1992).

Table 6.2. Examples of human pseudogenes generated by retrotransposition

Pseudogene	Chromosomal location/comments	Reference
Dihydrofolate reductase (2)	Unlinked to functional source gene *DHFR* (5q11-q13)	Chen *et al.* (1982) Shimada *et al.* (1983) Anagnou *et al.* (1985)
ADP-ribosyltransferase, NAD⁺ (2)	Chromosomes 13 and 14. Unlinked to functional source gene *ADPRT* (1q42). Present in gorilla (98% homologous). Originated ~27 Myrs ago.	Lyn *et al.* (1995)
Ribosomal protein L7	Located within intron 1 of c-*fms* (*CSF1R*) gene (5q33)	Sapi *et al.* (1994)
High mobility group protein	Flanked by 15 bp direct repeat. Originated only ~1 Myrs ago.	Stros and Dixon (1993)
Topoisomerase 1	Chromosome 1. Truncated. Unlinked to functional source gene *TOP1* (20q12-q13)	Zhou *et al.* (1992)
Glutamine synthetase	Flanked by 9 bp direct repeat. Transcribed.	Chakrabarti *et al.* (1995)
FAU proto-oncogene	Chromosome 18. Integrated within sequence homologous to promoter of islet amyloid polypeptide (*IAPP*) gene. 317 amino acid open reading frame. Not transcribed. Unlinked to *FAU* gene (11q13)	Kas *et al.* (1995)

Pseudogene	Chromosomal location/comments	Reference
28S ribosomal RNA	Flanked by 16 bp direct repeats	Wang *et al*. (1997)
Ribosomal protein L23A	Chromosome 13. Unlinked to functional source gene *RPL23A* (17q11)	de Fatima Bonaldo *et al*. (1996)
Ribosomal protein L38	Located in promoter region of type-1 angiotensin II receptor gene (*AGTR1*; 3q21-q25)	Espinosa *et al*. (1997)
Ferritin L (5)	Unlinked to functional source gene *FTL* (19q13).	Santoro *et al*. (1986)
Ferritin H (18)	Highly dispersed (1p, 1q, 2q, 3q, 4, 6p, 13q, 14, 17p, X). Unlinked to functional source gene *FTH1* (11q12-q13).	Dugast *et al*. (1990); Zheng *et al*. (1995; 1997)
Glyceraldehyde-3-phosphate dehydrogenase (5)	Dispersed including Xp21-p11. Not linked to functional source gene *GAPD* (12p13).	Arcari *et al*. (1989)
Cytochrome *c* oxidase subunit VIb (3)	Unlinked to functional source gene *COX6B* (19q13)	Taanman *et al*. (1991)
Cytochrome b_5 (2)	Partial. Transcribed. 14q31-q32 and 20p11. Unlinked to functional source gene *CYB5* (18q23).	Giordano *et al*. (1993)
Prothymosin α (5)	Unlinked to functional source gene *PTMA* (chromosome 2). Not transcribed.	Manrow *et al*. (1992)
Argininosuccinate synthetase (14)	Highly dispersed. Originated during last 40 Myrs. Unlinked to functional source gene *ASS* (9q34).	Nomiyama *et al*. (1986); Freytag *et al*. (1984)
Immunoglobulin J	Chromosome 8q13-q21. Unlinked to functional source gene *IGJ* (4q21). Originated ~40–50 Myrs ago.	Max *et al*. (1994)
Immunoglobulin ε	Present in Old World monkeys and gorilla but appears to have been lost in chimpanzee.	Ueda *et al*. (1985)
Immunoglobulin λ	Unlinked to functional source immunoglobulin λ genes *IGLC1* (22q11).	Hollis *et al*. (1982)
Calmodulin (5)	Unlinked to functional source gene *CALM1* (14q24-q31). Originated ~21–49 Myrs ago.	Koller *et al*. (1991); Rhyner *et al*. (1994)
Phosphoglycerate kinase		McCarrey (1990)
Metallothionein		Karin and Richards (1982)
β-Actin (3)	Chromosomes 6, 15 and 18. Unlinked to functional source gene *ACTB* (7p22)	Moos and Gallwitz (1982; 1983); Ueyama *et al*. (1996)
γ-Actin (3)	Chromosomes 3q23, 20p13, 6p21. Unlinked to functional source gene *ACTG1* (17q25)	Ueyama *et al*. (1996)
β-Tubulin	8q21–8pter. Unlinked to functional source gene *TUBB* (6p21-pter).	Wilde *et al*. (1982a); (1982b); Floyd-Smith *et al*. (1986)
Laminin receptor	Flanked by 18 bp direct repeats. Not transcribed. Originated 3.5–5 Myrs ago.	Richardson *et al*. (1998)
Enolase 1	Chromosome 1q41-q42. Unlinked to functional source gene *ENO1* (1p36-pter)	Ribaudo *et al*. (1996)

Table 6.2 (continued)

Pseudogene	Chromosomal location/comments	Reference
Uracil-DNA glycosylase (2)	Located on chromosomes 14 and 16. Unlinked to functional source gene *UNG* (12q23-q24)	Lund *et al.* (1996)
Tropomyosin, non-muscle		MacLeod and Talbot (1983)
Zinc finger protein ZNF75	12q13. Unlinked to functional source gene *ZNF75* (Xq26)	Villa *et al.* (1996)
Adenylate kinase 3	Located within intron 10 of *NF1* gene at 17q11.2	Xu *et al.* (1992)
Nucleotide excision repair protein RAD52	Chromosome 2. Unlinked to functional source gene *RAD52* (12p12-q13)	Johnson and Campbell 1996)
Serotonin 5-hydroxytryptamine 1D receptor	Transcribed. Capable of being translated into 28 amino acid. polypeptide. Unlinked to functional source gene *HTR1D* (1p34-p36)	Nguyen *et al.* (1993); Bard *et al.* (1995)
Fatty acid-binding protein 3	13q13-q14. Unlinked to functional source gene *FABP3* (1p32-p33).	Prinsen *et al.* (1997)
Cysteine and glycine-rich protein 2	3q21. Unlinked to functional source gene *CSRP2* (12q21).	Weiskirchen *et al.* (1997)
c-Ki-*ras*1 proto-oncogene	6p12-p11. Unlinked to functional source gene *KRAS2* (12p12).	McGrath *et al.* (1983)
c-*elk*-1 proto-oncogene (2)	14q32 within *IGHV* locus. Unlinked to functional source gene *ELK1* (Xp11.2). Insertion occurred 30–60 Myrs ago prior to *IGHV* locus duplication.	Harindranath *et al.* (1998)
Interleukin-6 signal transducer	17p11. Unlinked to functional source gene *IL6ST* (5q11).	Rodriguez *et al.* (1995)
SHC transforming protein	Xq12-q13. Unlinked to functional source gene *SHC1* (1q21).	Harun *et al.* (1997)
Mitochondrial-3-glycerol phosphate dehydrogenase 2	Chromosome 17. Unlinked to functional source gene *GPD2* (2q24)	Brown *et al.* (1996)
Mitochondrial elongation factor Tu	17q11. Unlinked to functional source gene *TUFM* (16p11)	Ling *et al.* (1997)
Mitochondrial NADH: ubiquinone oxidoreductase 24-kDa subunit	19q13.3-qter. Unlinked to functional source gene *NDUFV2* (18p11.2-p11.31). Partially processed.	de Coo *et al.* (1995)
NADH: ubiquinone oxido reductase B13 subunit	11p15. Unlinked to functional source gene *NDUFA5* (7q32)	Russell *et al.* (1997)
Solute carrier family 9 (NHE3)	Chromosome 10. Unlinked to functional source gene *SLC9A3* (5p15)	Kokke *et al.* (1996)
B-*raf*	Xq13. Unlinked to functional source gene *BRAF* (7q34)	Eychene *et al.* (1992)
GTP-binding protein Gαq	2q14-q21. Unlinked to functional source gene *GNAQ* (9q21)	Dong *et al.* (1995)
αE-catenin	5q22 [Unlinked to functional source gene *CTNNA1* (5q11)	Nollet *et al.* (1995)

Pseudogene	Chromosomal location/comments	Reference
Acyl-CoA binding protein	Chromosome 6. Unlinked to functional source gene *DBI* (2q12-q21).	Gersuk *et al.* (1995)
Methylthioadenosine phosphorylase	3q28. Unlinked to functional source gene *MTAP* (9p21).	Tran *et al.* (1997)
Serine hydroxymethyltransferase	1p32–33. Unlinked to functional source gene *SHMT1* (17p11-p12).	Byrne *et al.* (1996); Devor and Dill-Devor (1997)
Hexokinase II	X chromosome. Unlinked to functional source gene *HK2* (2p12). Originated 14–16 Myrs ago. Integrated into LINE element.	Ardehali *et al.* (1995)
Dihydrolipoamide succinyltransferase	1p31. Unlinked to functional source gene *DLST* (14q24)	Nakano *et al.* (1994)
S-adenosylmethionine decarboxylase	Xq28. Unlinked to functional source gene *AMD1* (6q21-q22).	Maric *et al.* (1995)
Dual specificity phosphatase 8	10q11.2. Unlinked to functional source gene *DUSP8* (11p15).	Nesbit *et al.* (1997)
G protein-coupled receptor kinase GRK6	Chromosome 13. Unlinked to functional source gene *GPRK6* (5q35)	Gagnon and Benovic 1997)
Estrogen-related receptor ERRα	13q12. Unlinked to functional source gene *ESRRA* (11q12-q13).	Sladek *et al.* (1997)
Phosphatidylinositol glycan class F	Chromosome 5q35. Unlinked to functional source gene *PIGF* (2p16-p21)	Ohishi *et al.* (1995)
Nucleophosmin (7)	4 full-length, 3 truncated. Unlinked to functional source gene *NPM1* (5q35)	Liu and Chan (1993)
RNA helicase, DDX10	9q21-q22. Unlinked to functional source gene *DDX10* (11q22-q23)	Savitsky *et al.* (1996)
Transcriptional elongation factor TFIIS	Unlinked to functional source gene *TCEA1* (3p21-p22)	Park *et al.* (1994)
U1 snRNA		Dennison *et al.* (1981); Manser and Gesteland (1981)
U2 snRNA		Van Arsdell and Weiner (1984)
U3 snRNA		Bernstein *et al.* (1983)
U4 snRNA		Hammarström *et al.* (1982)
U6 snRNA		Hayashi (1981)
U13 snRNA		Baserga *et al.* (1991)

Numbers in brackets denote numbers of pseudogenes in cases where more than one has been found.

Some genes possess only a single processed pseudogene copy whereas others (e.g. the U1 and U6 small nuclear RNA (snRNA) genes) can possess hundreds. In some cases, the processed pseudogenes greatly outnumber their functional counterparts. One example is that of the human ribosomal protein multigene family (Chapter 5, section 5.1.14) whose members are composed predominantly of multiple processed pseudogenes (Davies *et al.*, 1989). Other examples are given in *Table 6.2*. The proportion of processed pseudogenes to functionally active genes in any one gene family may reflect the level of transcription of the source gene in the germline. It is also likely to be influenced by the private sequence characteristics of the source gene which will determine both the efficiency of priming of reverse transcriptase and of their eventual integration of reverse transcripts. Thus, for example, the plethora of U3 snRNA pseudogenes may be explicable in terms of the inherent ability of U3 snRNA to act as a self-priming template for reverse transcriptase (Bernstein *et al.*, 1983). Both human and murine cells possess an endogenous reverse transcriptase activity (Maestre *et al.*, 1995; Tchenio *et al.*, 1993) although it is unclear whether this is derived from retroviral infection (Carlton *et al.*, 1995), endogenous sources of reverse transcriptase such as retroviral elements (Chapter 1, section 1.4.4) or, perhaps more likely, LINE elements (Dhellin *et al.*, 1997; Jurka, 1997; Chapter 1, section 1.4.3).

Some human processed pseudogenes have been inserted into the introns of other genes, for example a ribosomal protein L7 pseudogene in intron 1 of the c-*fms* (*CSF1R*; 5q33) proto-oncogene (Sapi *et al.*, 1994), a phosphoglycerate mutase pseudogene in intron 1 of the Menkes disease (*ATP7A*; Xq13.3) gene (Dierick *et al.*, 1997), an L5 ribosomal protein pseudogene in an intron of the small nuclear ribonucleoprotein N (*SNRPN*; 15q12) gene (Buiting *et al.*, 1996), an L21 ribosomal protein pseudogene in intron 13 of the breast cancer (*BRCA1*; 17q21) gene (Smith *et al.*, 1996) and an adenylate kinase 3 pseudogene in intron 10 of the neurofibromatosis type 1 (*NF1*; 17q11.2) gene (Xu *et al.*, 1992). If integrated upstream of a gene, the pseudogene sequence may affect expression of that gene. For instance, a processed ribosomal protein S25 pseudogene, which has become integrated into the promoter region of a rat class I alcohol dehydrogenase gene, inactivated a glucocorticoid response element and contributed a novel suppressor element (Cortese *et al.*, 1994). Similarly, the human L38 ribosomal protein processed pseudogene has become integrated into the promoter of the type-1 angiotensin II receptor (*AGTR1*; 3q21-q25) gene although the effect, if any, on the expression of the gene is unclear (Espinosa *et al.*, 1997). In general, the sites into which processed pseudogenes integrate appear to be AT-rich and various models have been proposed which attempt to describe the process of integration (Vanin, 1984; Wilde, 1985). Co-retrotransposition sometimes occurs with other sequences e.g. endogenous retroviral elements (Lyn *et al.*, 1995), LINE elements (Rozmahel *et al.*, 1997) and possibly *Alu* sequences (Koller *et al.*, 1991).

Some pseudogenes may be duplicated copies of pre-existing processed pseudogenes, for example the tandemly arranged proliferating cell nuclear antigen pseudogenes on chromosome 4q34 (Taniguchi *et al.*, 1996), the L7a ribosomal protein pseudogenes associated with the closely linked lymphotactin α (*LTNA*) and β (*LTNB*) genes on chromosome 1q23 (Yoshida *et al.*, 1996) or the *elk-1* pseudogene associated with the immunoglobulin heavy chain locus (*IGHV*) on 14q32 (Harindrath *et al.*, 1998).

A highly unusual presence/absence polymorphism has been noted for a human dihydrofolate reductase processed pseudogene (Anagnou *et al.*, 1984). This pseudogene is present in only a proportion of individuals and this proportion varies between racial groups. However, it is unclear whether the polymorphism is a consequence of the recent acquisition of the pseudogene or whether it instead results from its occasional loss.

6.1.3 The generation of pseudogenes by other means

Although the vast majority of known pseudogenes are either inactivated genomic copies (partial or complete) of functional genes or processed pseudogenes, there are a few unusual examples which do not fit easily into either of these two groupings. One such case is that of the hybrid β-δ-globin pseudogene found in lemurs (Jeffreys *et al.*, 1982). The first and second exons plus intron 1 appear to be derived from the ψβ-globin pseudogene whilst intron 2, exon 3, and the 3′ UTR originated from the δ-globin gene. The hybrid pseudogene probably originated from an unequal crossover which fused sequences derived from the ψβ-globin pseudogene and the δ-globin gene.

The cadherin pseudogene on human chromosome 5q13 described by Selig *et al.* (1995) is also unusual. It contains one exon derived from the source gene encoding cadherin 12 (*CDH12*; 5p13-p14) plus flanking introns, an open reading frame of 794 amino acids, and is transcribed at a 10-fold higher rate than the source gene. Selig *et al.* (1995) proposed that this pseudogene had been only partially processed in so far as the intron sequences had not been completely removed from the mRNA before reverse transcription and integration had taken place. Other examples of *semi-processed* human pseudogenes include one at 1q24-q25 derived from the MADS box transcription factor 2 (*MEF2A*; 5q26) gene (Suzuki *et al.*, 1996) and another at 19q13.3-qter derived from the mitochondrial NADH: ubiquinone oxidoreductase (*NDUFV2*; 18p11.2-p11.31) gene (de Coo *et al.*, 1995) and at least 13 pseudogenes which have independently originated from an identical mis-spliced transcript of the protein geranylgeranyltransferase type Iβ (*PGGT1B*) gene (Dhawan *et al.*, 1998).

6.1.4 Origin and age of pseudogenes

Some indication of the relative age of a pseudogene can be obtained by the extent of homology between the pseudogene and its parent or source gene. Thus, in the human α-globin gene cluster on chromosome 16, the ψζ-globin pseudogene is >99.5% homologous to its parental ζ-globin (*HBZ*; 16p13.3-pter) gene whereas the rather older ψα-globin pseudogene is only 75–80% homologous to its parental α-globin (*HBA1*; 16p13.3-pter) gene.

The approximate age of a pseudogene can be estimated by determining whether or not an orthologous sequence is present at an identical location in other related species of known phylogeny. Thus, the XG blood group pseudogene is present in the great apes but not in Old World monkeys, New World monkeys or prosimians (Weller *et al.*, 1995) whilst multiple copies of the keratinocyte growth factor pseudogene are present in chimpanzee, gorilla but not in gibbons or Old World monkeys (Kelley *et al.*, 1992). Similarly, the ψβ1-globin pseudogene has been found in prosimians and New World monkeys as well as the anthropoid apes and human and is therefore thought to have originated very early in primate evolution (Harris *et al.*, 1984). Rouquier *et al.* (1998) noted distinct human- and gorilla-specific inactivating mutations in the olfactory receptor pseudogene (termed 912–93) that is located on human chromosome 11q11-q12 and which exhibits synteny in the hominoid primates (Rouquier *et al.*, 1998). Finally, the serine hydroxymethyltransferase processed pseudogene, derived from the *SHMT1* gene on chromosome 17p11.2, arose during the evolution of the primates but different lesions have occurred in different phylogenetic branches of the family making this pseudogene a potentially useful marker in molecular studies (Devor *et al.*, 1998).

Attempts have also been made to estimate the age of pseudogenes (time since duplication and/or inactivation) by comparing the proportion of silent and replacement changes in the pseudogene sequence to that exhibited by the active parent gene. Such estimates assume that following inactivation, pseudogenes will accumulate mutations at the same rate as silent changes in active genes. However, what is often ignored is that there is still some selection against silent changes in functional genes. Moreover, such estimates are extremely sensitive to the confounding effects of gene conversion (see Section 6.1.6 and Chapter 9, section 9.6) whose homogenizing influence serves continually to restart the molecular clock leading to the underestimation of the age of the pseudogene.

6.1.5 Patterns of mutation in pseudogenes

Pseudogenes represent extremely useful tools for the study of mutation because, since they lack a biological function and are not subject to selective constraints, all mutations that occur in pseudogenes are selectively neutral and will become fixed in the population with equal probability.

Pseudogenes exhibit a very high rate of nucleotide substitution with the CpG dinucleotide being a hotspot for mutation (Bulmer *et al.*, 1986; Gojobori *et al.*, 1982; Li *et al.*, 1981; Li *et al.*, 1984; Miyata and Hayashida, 1981). By contrast, deletions occur about once every 40 nucleotide substitutions whilst insertions occur about once every 100 nucleotide substitutions (Ophir and Graur, 1997). The age of the pseudogene, however, is not always linearly related to the numbers of deletions and insertions present (Ophir and Graur, 1997), and this may be due at least in part to gene conversion (see Section 6.1.6).

The rate of DNA loss through deletion from processed pseudogenes appears to be considerably higher in rodents than in humans (Graur *et al.*, 1989; Ophir and Graur 1997). Interpretation of such interspecies comparisons can however be confounded by possible differences in the frequency of gene conversion between species.

The mutation rate within processed pseudogenes appears to be higher than in regions flanking the site of insertion (Casane *et al.*, 1997). This is potentially explicable if one considers that those sites which are inherently the most mutable have been maintained by selection within the retrotransposed sequence up until pseudogene formation whereas such sites have been removed in the flanking regions which have been unconstrained by the effects of natural selection for much longer periods of evolutionary time.

6.1.6 Pseudogenes and gene conversion

Gene conversion is the modification of one of two alleles by the other. It involves the nonreciprocal correction of an 'acceptor' gene or DNA sequence by a 'donor' sequence which remains physically unchanged. In most known cases, gene conversion has occurred between highly homologous and closely linked gene sequences (Cooper and Krawczak, 1993). Examples of gene conversion that have occurred during primate evolution are discussed in Chapter 9, section 9.5. Gene conversion has also served as a mutational mechanism causing human inherited disease; probable examples involve the genes and pseudogenes for steroid 21-hydroxylase (*CYP21*; 6p21.3; Tusié-Luna and White, 1995), polycystic kidney disease (*PKD1*; 16p13), neutrophil cytosolic factor p47-*phox* (*NCF1*; 7q11.2; Görlach *et al.*, 1997), immunoglobulin l-like polypeptide 1 (*IGLL1*; 22q11; Minegishi *et al.*, 1998), glucocerebrosidase (*GBA*; 1q21; Eyal *et al.*, 1990), von Willebrand factor (*VWF*; 12p13; Eikenboom *et al.*, 1994), and phosphomanno-mutase (*PMM2*; 16p13; Schollen *et al.*, 1998). These gene/pseudogene pairs are all closely linked with the exception of the *VWF* gene (12p13) and its pseudogene (22q11-q13), and the *PMM2* gene (16p13) and its pseudogene (18p). Together, these two exceptions would seem to establish a precedent for the occurrence of gene conversion between unlinked loci in the human genome.

Could pseudogenes (both the highly dispersed processed variety and/or the closely linked duplication-derived pseudogenes) also have templated advantageous changes in their single copy functional source genes over evolutionary time? Certainly the converse is true since sequence changes in pseudogenes can be templated by their functional homologues (DeBry, 1998). If pseudogene-templated changes in functional genes were not deleterious, they could eventually have become fixed in populations or even species. Thus, in principle, pseudogenes, whether processed or non-processed, could act as a reservoir of sequence variation which could at some stage be transferred back to the functional gene. In this way, different mutational combinations might be put together within the pseudogene, all the time being shielded from selective pressure, and then functionally tested after simultaneous transfer to the expressed gene copy. For any one gene, the contribution of pseudogene-mediated gene conversion to the process of evolutionary change might be expected to depend on:

(i) *The number of homologous pseudogenes (processed or unprocessed) in the genome that are homologous to the gene in question.*
As we have seen in Section 6.1.2, this can vary by at least two orders of magnitude.

(ii) *Whether the pseudogenes are linked or unlinked to the functional gene.*
Gene conversion does occur between unlinked homologous sequences (Fitzgerald *et al.*, 1996; Murti *et al.*, 1994). Gene conversion between multi-copy, dispersed retrotransposons such as *Alu* sequences (Kass *et al.*, 1995) and LINE elements (Burton *et al.*, 1991) has also been reported. However, inter-chromosomal gene conversion events are predicted to be much less frequent than intra-chromosomal events (Liao *et al.*, 1997).

(iii) *The length of sequence involved in the gene conversion event.* Were this to be too large, inactivating mutations present in the pseudogene would be more likely to be transferred to the functional gene thereby inactivating it rather than altering it. In practice, however, gene conversion events are often quite localized (Kim *et al.*, 1993; Pamilo and Bianchi, 1993; Zhou and Lee, 1996), one example being the *HLA-DQB1* gene (6p21.3) in which gene conversion events have been confined to exon 2 without extending into the adjacent

introns (Bergstrom *et al.*, 1998). Some gene conversion events can nevertheless involve rather longer stretches of DNA, for example over the entire rhesus monkey γ1 and γ2 globin genes (Slightom *et al.*, 1988) and over the entire
hominoid immunoglobulin Cα genes (Kawamura *et al.*, 1992). The interpre-
tational problem in such studies lies in the fact that it is difficult to distinguish a single large gene conversion event from multiple overlapping short
gene conversions.

(iv) *The age of the pseudogenes.* The degree of sequence homology to the functional
gene will decay with time and this could serve to reduce the frequency of gene
conversion as well as the length of sequence involved. Further, the number of
inactivating mutations acquired by the pseudogene will increase with time
and this would have the effect of reducing the likelihood of a functionally
productive gene conversion event.

At the present time, definitive evidence for pseudogene-mediated gene conversion is fairly sparse. Since by its very nature, gene conversion tends to cover its
own traces, such evidence may prove hard to obtain. Further, although the donor
sequence might be a pseudogene in an extant genome, this does not mean that it
was necessarily a pseudogene at the time of the gene conversion event. The
demonstration that the sequence in question has been accumulating mutations
over a considerable period of time would provide evidence in favor of an ancient
inactivating event. Again, however, gene conversion could confound estimates of
the length of evolutionary time elapsed. The above notwithstanding, one possible
example of pseudogene-mediated gene conversion involves a human
immunoglobulin V_H pseudogene (V4–55P) which may have served as a donor
sequence in the conversion of two functional V_H (*IGHV*; 14q32) genes, V4–4b and
V4–28 (Haino *et al.*, 1994). Similarly, a human pseudogene in the human growth
hormone gene cluster may have templated sequence changes in a functional
source gene: the growth hormone 1 (*GH1*; 17q22-q24) gene promoter region contains five single base polymorphic alleles (at –57, –1, +3, +16, and +26) which are
identical to the bases present at the homologous locations in the promoter of the
closely linked and evolutionarily related pseudogene (*CSHL1*) but are different to
those present in the more distal *CSH1*, *GH2* and *CSH2* genes (Giordano *et al.*,
1997). A gorilla *HLA-A* gene has been shown to be similar to an *HLA-AR* pseudogene only in exon 2 with the remainder of the gene being closely related to other
primate *A* locus genes (Watkins *et al.*, 1991). Finally, immunoglobulin V gene
diversity in chickens is known to be increased by gene conversion during B-cell
development; the germline pool of donor sequence information for somatic gene
conversion is found in the families of V pseudogenes located 5′ to the single functional V gene at each locus (McCormack *et al.*, 1993).

6.1.7 Pseudogene reactivation

Pseudogenes may be by definition inactive but this does not mean that they can
never regain their activity. Indeed, Marshall *et al.* (1994) calculated that the resurrection of pseudogenes is probabilistically feasible within about 6 Myrs of formation but that it is unlikely after more than 10 Myrs have elapsed owing to the
accumulation of multiple inactivating mutations.

One putative example of the natural resurrection of a pseudogene is that of a bovine seminal ribonuclease gene which appears to have become reactivated after the divergence of the kudu but before the divergence of the ox, between ~5 Myrs and 10 Myrs ago (Trabesinger-Ruef et al., 1996). Since the seminal ribonuclease genes are highly homologous to the pancreatic ribonuclease gene family (the seminal proteins emerged about 35 Myrs ago by a process of duplication and divergence from a pancreatic protein gene), the repair of the seminal ribonuclease gene may have been effected by gene conversion, involving the transfer of genetic information from a pancreatic ribonuclease gene.

An example of the natural reactivation of a pseudogene in humans is that of a truncated γE crystallin gene (*CRYGEP1*) found in association with Coppock-like cataract (Brakenhoff et al., 1994). As we have seen in Chapter 4, section 4.2.1, *Crystallin genes*, the γ-crystallin gene cluster at 2q33-q35 comprises four genes and two pseudogenes, γE and γF, both of which contain in-frame stop codons. In the abovementioned patient, a number of sequence changes occurred within and around the TATA box in the promoter region of the γE pseudogene which appear to have resulted in a 10-fold increase in promoter activity. The predicted product of the γE pseudogene is a 6 kDa N-terminal γ-crystallin fragment which could be responsible for the increased opacity of the lens in the patient.

In an evolutionary context, the reactivation of the human ψζ1-globin pseudogene (*HBZP*; 16p13) by removal of its sole inactivating mutation appears to have occurred by gene conversion (Hill et al., 1985). This change, which has probably been templated by the neighboring ζ-globin (*HBZ*; 16p13) gene, is a common polymorphism in a number of different populations (Hill et al., 1985).

The above examples of the reactivation of pseudogenes were naturally occurring. Reactivation of an inactivated pseudogene can however also be effected artificially by correction of the inactivating mutation(s) and restoration of the sequence originally present prior to inactivation. One example of this is the artificial reactivation by *in vitro* mutagenesis of the human T-cell receptor γ variable V10 pseudogene which was originally inactivated by a single base-pair substitution in a donor splice site (Zhang et al., 1996). Such experimental studies serve to demonstrate that the corrected lesion was the sole impediment to expression and therefore likely to be the original inactivating mutation.

6.2 Gene loss/inactivation in primates

Genes can be created but they may also be destroyed. As we have seen, one form of gene inactivation involves pseudogene formation. This process involves the creation of a copy of a specific gene (whether cDNA or genomic) followed by its inactivation by the sudden acquisition, and subsequent steady accumulation of, deleterious mutations. Owing to the redundancy of genetic information available upon gene duplication, there is often no selective pressure to retain the integrity of the new sequence because the sequence from which it was derived still exists and functions normally.

By contrast, the loss or inactivation of a single copy essential gene is likely to have a dramatic phenotypic effect. Many of the lesions responsible will be evident by their clinical manifestations (assuming that they are not embryonic lethal) and we should expect them to be rapidly removed from the population by natural selection. On the other hand, the inactivation of a nonessential gene could in principle have relatively few deleterious consequences and might even have a beneficial effect under the appropriate circumstances. One possible example of this is the C4b-binding protein β-chain gene (*C4BPB*; 1q32) gene which occurs as a single copy functional gene in the human genome but has been inactivated in the mouse (Rodriguez de Cordoba *et al.*, 1994). That the loss of this gene has been fixed evolutionarily in the murine lineage suggests that it is at least neutral with respect to fitness. It's loss may even have conferred a selective advantage (antithrombotic effect?) in which case, it could have become fixed within a relatively short period of evolutionary time. Several similar examples of the inactivation of single copy genes from primate genomes have also been documented and these *single copy pseudogenes* will now be described in some detail.

6.2.1 Urate oxidase gene

Urate oxidase is a copper binding enzyme found in most vertebrates which catalyses the conversion of uric acid to allantoin. Although found in Old World monkeys and in the majority of New World monkeys, urate oxidase activity has been lost in the hominoid line. The human urate oxidase (*UOX*; 1p22) 'gene' was isolated by screening a genomic library with a porcine cDNA probe (Wu *et al.*, 1989). It was found to contain two nonsense (CGA→TGA) mutations at codons 33 and 187 (Wu *et al.*, 1989; Yeldandi *et al.*, 1990). These lesions were also found to be present in chimpanzee and gorilla but only the codon 33 mutation was detected in orangutan (Yeldandi *et al.*, 1991). This implies that the codon 33 mutation was the original inactivating mutation and that it must have occurred before the divergence of orangutan from the chimpanzee/human line, between 7 Myrs and 13 Myrs ago. Urate oxidase activity is also absent in the gibbon but this has been shown to be due to a 13 bp deletion in exon 2 of the *Uox* gene (Wu *et al.*, 1992). The loss of urate oxidase activity in primates has therefore been due to the occurrence of at least two distinct gene lesions in different lineages. That this gene could have been lost in primates as a result of at least two independent mutational events is consistent with its loss being advantageous. Although somewhat far-fetched perhaps, it has been suggested that since uric acid is a potent antioxidant, an increased uric acid concentration might have contributed to a lowering of the somatic mutation rate with a consequent increase in hominoid lifespan. The loss of urate oxidase may however be responsible for at least some cases of renal stones and gouty arthritis in humans.

6.2.2 α-1,3-Galactosyltransferase gene

α-1,3-Galactosyltransferase (α-1,3 GT) is responsible for the synthesis of the α-galactosyl epitope present in the cell surface receptors of most mammals including prosimians and New World monkeys. However, the catarrhines (Old World monkeys, apes and humans) appear to lack this epitope and produce large amounts of antibodies against it. Two inactivating single base-pair deletions (del

C822 and del G904) were found in the human α-1,3 GT (*GGTA1*) 'gene', located at chromosome 9q34 (Larsen *et al.*, 1990). Since chimpanzees possess both deletions whereas orangutan and gorilla only have del904, it was reasoned that del904 was the original inactivating mutation (Galili and Swanson, 1991). Old World monkeys were found not to possess either lesion and no other inactivating mutation was obvious. Clearly, as with urate oxidase, at least two distinct inactivating mutations have occurred in apes and Old World monkeys. Selective pressure to suppress α-galactosyl transferase expression may have been mediated by a pathogen which infected primates via cell surface receptors containing the α-galactosyl epitope. Alternatively, the pathogen might have expressed α-galactosyl epitopes and an effective host immune response would have required the suppression of autologous α-galactosyl epitope synthesis.

6.2.3 Elastase I gene

The human pancreatic elastase I (*ELA1*; 12q13) 'gene' is transcriptionally silent even though the coding sequence and splice junctions appear to be intact. Studies of the corresponding (transcriptionally active) rat gene have demonstrated that some 205 bp immediately upstream of the transcriptional start site are required for pancreatic expression. This regulatory region comprises an enhancer between −205 and −72 that contains all the information necessary for tissue-specific expression, and a minimal promoter element between −71 and +8. The enhancer contains three functional elements (A, B, and C), two of which (A and B) bind pancreas-specific transcription factors (PTF1 and PDX1) that mediate the cell type specificity of the enhancer. The 5′ flanking region of the human *ELA1* 'gene' was studied by Rose and MacDonald (1997). Its nucleotide sequence differs by 28% from its rat counterpart whilst its enhancer/promoter strength (as measured by reporter gene assays) was 2.6% that of the homologous rat sequence. Comparison of the rat and human gene 5′ flanking regions revealed a total of 13 nucleotide differences within the A, B, and C elements. A combination of the three human enhancer elements, 'repaired' by reference to the rat sequence, together with the rat promoter, partially restored the activity of the human enhancer (Rose and MacDonald, 1997). These authors went on to demonstrate that two mutations in the A element and four mutations in the B element served to abolish the binding of the cognate transcription factors. The degree to which these lesions individually exert a deleterious effect on enhancer function will not be apparent without further functional studies. In addition, without phylogenetic studies of the *ELA1* genes from other mammals/primates, it is unclear in which order these mutations occurred during evolution. This notwithstanding, we may still conclude that the silencing of the human *ELA1* 'gene' has come about through the mutational inactivation of the upstream enhancer rather than through the alteration of the coding region.

6.2.4 L-gulono-γ-lactone oxidase gene

Primates (including humans) and guinea pigs have at least one thing in common: they are, unlike other mammals, unable to synthesize L-ascorbic acid from

D-glucose owing to a deficiency of the enzyme L-gulono-γ-lactone oxidase. Using a rat cDNA as a probe, Nishikimi *et al.* (1994) isolated the once active human L-gulono-γ-lactone oxidase (*GULOP*) 'gene' which is located at chromosome 8p21.1. A considerable number of inactivating mutations were found including the deletion of exon VIII, two 1 bp deletions, one 1 bp insertion and two nonsense mutations. The presence of two non-consensus dinucleotides at donor splice sites (GC and GG instead of GT) also suggested the introduction of potentially inactivating mutations at splice sites. Since both Old World and New World monkeys are deficient in L-gulono-γ-lactone oxidase whilst prosimians possess this enzyme, its loss must have preceded the divergence of New World from Old World monkeys (~45 Myrs ago) but occurred after the divergence of the prosimians from the simian line (50–65 Myrs ago).

6.2.5 ADP-ribosyltransferase 1 gene

ADP-ribosyltransferase 1 (RT6) is a glycosyl phosphatidylinositol-anchored cell membrane protein expressed in peripheral T cells and intra-epithelial lymphocytes. In the rat, expression of RT6 has been used to distinguish subsets of T cells and a defect in RT6 expression has been associated with susceptibility to autoimmune type I diabetes. Haag *et al.* (1994) have shown that the human (*ART2P*; 11p15) and chimpanzee 'genes' contain three nonsense mutations, at codons 47, 141, and 193 respectively, which serve to inactivate them. No further copies of the *ART2P* 'gene' were detectable in the human genome consistent with its presumed status of single copy pseudogene. It is possible that inactivation of the *ART2P* gene conferred a selective advantage if, for example, it resulted in the loss of a membrane receptor for a pathogenic virus. It is as yet unclear if loss of the RT6 protein confers increased susceptibility to disease.

6.2.6 Haptoglobin gene

Haptoglobin, a hemoglobin-binding protein, is encoded by three closely linked genes in chimpanzee (*HP, HPR,* and *HPP*) but only two (*HP, HPR;* 16q22.1) in human. The loss of the *HPP* 'gene' in human occurred after the separation of the human and chimpanzee lineages by an unequal homologous crossover that deleted most of the the *HPP* gene (McEvoy and Maeda, 1988).

6.2.7 Fertilin-α gene

In most mammals including macaques and baboons, α-fertilin contributes a subunit to a heterodimeric membrane glycoprotein on the sperm surface. The macaque and baboon possess two α-fertilin genes which appear to lack introns (indicative perhaps of a retrotranspositional origin). Only a single copy homologue (*FTNA*) is present in the human genome and, although this sequence is transcribed in the testis, it represents a nonfunctional pseudogene since it contains numerous mutations which render translation of the transcript impossible (Jury *et al.*, 1997). What is puzzling is that although the 5′ half of the human *FTNA* 'gene' is highly homologous to the macaque *Ftna1* gene, the 3′ half differs markedly from both *Ftna1* and *Ftna2* genes. Jury *et al.* (1997) attempted to explain the creation of a solitary nonfunctional human gene in terms of a recombination event between the two ancestral primate *Ftna* genes.

6.2.8 T-Cell receptor γ V10 variable region gene

The 'gene' encoding the V10 variable region of the T-cell receptor γ (*TCRG*; 7p15) has a single base-pair substitution in a donor splice site which serves to inactivate the gene in humans (Zhang *et al.*, 1996). The counterparts of this gene in chimpanzee and gorilla contain functional splice sites and therefore this lesion must have occurred in the human lineage. Restoration of the defective splice site by *in vitro* mutagenesis has been shown to generate a correctly spliced product.

6.2.9 Cytidine monophospho-N-acetylneuraminic acid hydroxylase gene

Humans differ from the great apes in that they lack the enzyme CMP-*N*-acetylneuraminic acid hydroxylase which, as the name suggests, hydroxylates *N*-acetylneuraminic acid. This cell surface sialic acid molecule is thought to be involved in cell-cell recognition and in cell-pathogen interactions. The loss of enzyme activity in human is caused by a 92 bp deletion in the coding region of the CMP-*N*-acetylneuraminic acid hydroxylase (*CMAH*; 6p22-p23) gene (Chou *et al.*, 1998; Irie *et al.*, 1998). This deletion results in a frameshift leading to premature termination of translation. It is specific to the human lineage since it is not found in any of the great apes (Chou *et al.*, 1998) but it is not of recent origin since it has been found to be present in Caucasians, African Americans, Japanese, Kung bushmen and Khwe pigmies (Chou *et al.*, 1998). Chou *et al.* (1998) speculated that the loss of *N*-acetylneuraminic acid hydroxylation may have had implications for susceptibility to infectious disease since some pathogens utilize sialic acids as specific binding sites on mammalian cells.

6.2.10 Flavin-containing monooxygenase 2 gene

The flavin-containing monooxygenase 2 gene is one of five FMO genes found in mammals which encode a series of NADPH-dependent flavoenzymes that are capable of catalyzing the oxidation of numerous drugs and xenobiotics. FMO2 is expressed predominantly in the lung where it constitutes the major form of FMO. The human FMO2 gene (*FMO2*; 1q) differs from those of other mammals in that it possesses a CAG→TAG transition converting Gln472 to a stop codon (Dolphin *et al.*, 1998). This lesion predicts a truncated polypeptide lacking 64 amino acids from its C-terminus. The *FMO2* mRNA transcript is both abundant and stable. *In vitro* expression studies have demonstrated that the truncated protein product is translated, correctly targeted to the endoplasmic reticulum but has lost its catalytic activity. This mutation is not present in the *Fmo2* genes of gorilla and chimpanzee and must therefore have arisen in the human lineage in the last 5–7 Myrs. Interestingly, about 4% of individuals in African populations possess a CAG (Gln) triplet at codon 472. It is possible that this CAG allele represents a remnant of the original gene sequence in which case the lesion may have occurred during early human history. The alternative is that it is a reverse mutation brought about perhaps by gene conversion templated by a related FMO gene. The biochemical and toxicological significance of the loss of FMO2 activity in human lung is unclear and so it is premature to speculate as to whether its loss conferred a selective advantage upon bearers or whether the inactivating lesion could have become fixed by genetic drift alone.

6.2.11 Overview

Several genes are thus known to have become inactivated during primate/human evolution. Inactivation does not however always imply the deletional removal of the gene. Rather, the once functional gene sequences often remain in the human genome to be detected as the inactive orthologues of their still active counterparts in the genomes of lower vertebrates. A variety of subtle lesions are usually responsible for the inactivation e.g. micro-deletions or insertions, or single base-pair substitutions which create in-frame stop codons, alter the invariant bases at splice junctions or adversely affect the binding of transcription factors to the promoter region. The greater the time that has elapsed since the initial inactivation event, the greater the number of mutations that have accumulated, and therefore the harder it is to discern the identity of the initial inactivation event. We may only speculate as to the reasons for the loss of these genes. In some cases, gene loss may have been neutral with respect to fitness and the null allele would have become fixed through genetic drift. In the case of the flavin-containing monooxygenase 2 (*FMO2*) gene, Dolphin *et al.* (1998) calculated, assuming a generation time of 15 years and an effective population size of 10 000, that it would have taken some 600 000 years for the Gln472Term mutation to approach fixation. In the case of other genes lost from the human or primate genomes, some selective advantage, for example resistance to pathogens may have accrued to bearers of the inactivated genes, resulting in the more rapid spread of the mutant allele.

References

Ahuja S.K., Ozcelik T., Milatovitch A., Francke U., Murphy P.M. (1992) Molecular evolution of the human interleukin-8 receptor gene cluster. *Nature Genet.* **2**: 31–36.

Albig W., Meergans T., Doenecke D. (1997) Characterization of the H1.5 gene completes the set of human H1 subtype genes. *Gene* **184**: 141–148.

Anagnou N.P., O'Brien S.J., Shimada T., Nash W.G., Chen M.Y., Nienhuis A.W. (1984) Chromosomal organization of the human dihydrofolate reductase genes: dispersion, selective amplification and a novel form of polymorphism. *Proc. Natl. Acad. Sci. USA* **81**: 5170–5174.

Anagnou N.P., Antonarakis S.E., O'Brien S.J., Nienhuis A.W. (1985) Chromosomal localization and racial distribution of the polymorphic hDHFR-psi-1 pseudogene. *Clin. Res.* **33**: 328A.

Arcari P., Martinelli R., Salvatore F. (1989) Human glyceraldehyde-3-phosphate dehydrogenase pseudogenes: molecular evolution and a possible mechanism for amplification. *Biochem. Genet.* **27**: 439–450.

Ardehali H., Printz R.L., Koch S., Phillips J.A., Granner D.K. (1995) Isolation, characterization and chromosomal localization of a human pseudogene for hexokinase II. *Gene* **164**: 357–361.

Bard J.A., Nawoschik S.P., O'Dowd B.F., George S.R., Brancek T.A., Weinshank R.L. (1995) The human 5-hydroxytryptamine$_{1D}$ receptor pseudogene is transcribed. *Gene* **153**: 295–296.

Baserga S.J., Yang X.W., Steitz J.A. (1991) Three pseudogenes for human U13 snRNA belong to class III. *Gene* **107**: 347–348.

Beck S., Abdulla S., Alderton R.P., Glynne R.J., Gut I.G., Hosking L.K., Jackson A., Kelly A., Newell W.R., Sanseau P., Radley E., Thorpe K.L., Trowsdale J. (1996) Evolutionary dynamics of non-coding sequences within the class II region of the human MHC. *J. Mol. Biol.* **255**: 1–13.

Bentley D.L., Rabbitts T.H. (1980) Human immunoglobulin variable region genes – DNA sequences of two V_k genes and a pseudogene. *Nature* **288**: 730–733.

Bergstrom T.F., Josefsson A., Erlich H.A., Gyllensten U. (1998) Recent origin of HLA-DRB1 alleles and implication for human evolution. *Nature Genet.* **18**: 237–242.

Bernstein L.B., Mount S.M., Weiner A.M. (1983) Pseudogenes for small nuclear RNA U3 appear to arise by integration of self primed reverse transcripts of the RNA into new chromosomal sites. *Cell* **32**: 461–467.

Brakenhoff R.H., Henskens H.A.M., van Rossum M.W.P.C., Lubsen N.H., Schoenmakers J.G.G. (1994) Activation of the γE-crystallin pseudogene in the human hereditary Coppock-like cataract. *Hum. Molec. Genet.* **3**: 279–283.

Braun A., Kammerer S., Ambach H., Roscher A.A. (1996) Characterization of a partial pseudogene homologous to the adrenoleukodystrophy gene and application to mutation detection. *Hum. Mutation* **7**: 105–108.

Brown L.J., Stoffel M., Moran S.M., Fernald A.A., Lehn D.A., LeBeau M.M., MacDonald M.J. (1996) Structural organization and mapping of the human mitochondrial glycerol phosphate dehydrogenase-encoding enzyme gene and pseudogene. *Gene* **172**: 309–312.

Buiting K., Kaya-Westerloh S., Horsthemke B. (1996) A pseudogene for the human ribosomal protein L5 (RPL5P1) maps within an intron of the *SNRPN* transcription unit on human chromosome 15. *Cytogenet. Cell Genet.* **75**: 224–226.

Bulmer M. (1986) Neighbouring base effects on substitution rates in pseudogenes. *Mol. Biol. Evol.* **3**: 322–329.

Burton F.H., Loeb D.D., Edgell M.H., Hutchison C.A. (1991) L1 gene conversion or same site transposition. *Mol. Biol. Evol.* **8**: 609–619.

Byrne P.C., Shipley J.M., Chave K.J., Sanders P.G., Snell K. (1996) Characterisation of a human serine hydroxymethyltransferase pseudogene and its localisation to 1p32.3–33. *Hum. Genet.* **97**: 340–344.

Carlton M.B.I., Colledge W.H., Evans M.J. (1995) Generation of a pseudogene during retroviral infection. *Mamm. Genome* **6**: 90–95.

Casane D., Boissinot S., Chang B.H., Shimmin L.C., Li W. (1997) Mutation pattern variation among regions of the primate genome. *J. Mol. Evol.* **45**: 216–226.

Chakrabarti R., McCracken J.B., Chakrabarti D., Souba W.W. (1995) Detection of a functional promoter/enhancer in an intron-less human gene encoding a glutamine synthetase-like enzyme. *Gene* **153**: 163–169.

Chang L.-Y.E., Slightom J.L. (1984) Isolation and nucleotide sequence analysis of the β-type globin pseudogene from human, gorilla and chimpanzee. *J. Mol. Biol.* **180**: 767–784.

Chen M.J., Shimada T., Moulton A.D., Harrison M., Nienhuis A.W. (1982) Intronless human dihydrofolate reductase genes are derived from processed RNA molecules. *Proc. Natl. Acad. Sci. USA* **79**: 7435–7439.

Chou H.-H., Takematsu H., Diaz S., Iber J., Nickerson E., Wright K.L., Muchmore E.A., Nelson D.L., Warren S.T., Varki A. (1998) A mutation in human CMP-sialic acid hydroxylase occurred after the *Homo-Pan* divergence. *Proc. Natl. Acad. Sci. USA* **95**: 11 751–11 756.

Cook G.P., Tomlinson I.M. (1995) The human immunoglobulin V_H repertoire. *Immunology Today* **16**: 237–242.

Cooper D.N., Krawczak M. (1993) *Human Gene Mutation.* BIOS Scientific Publishers, Oxford.

Cortese J.F., Majewska J.L., Crabb D.W., Edenberg H.J., Yang V.W. (1994) Characterization of the 5′-flanking sequence of rat class I alcohol dehydrogenase gene. *J. Biol. Chem.* **269**: 21 898–21 906.

Crowe M.L., Perry B.N., Connerton I.F. (1996) Olfactory receptor-encoding genes and pseudogenes are expressed in humans. *Gene* **169**: 247–249.

Davies B., Feo S., Heard E., Fried M. (1989) A strategy to detect and isolate an intron-containing gene in the presence of multiple processed pseudogenes. *Proc. Natl. Acad. Sci. USA* **86**: 6691–6695.

de Coo R., Buddiger P., Smeets H., van Kessel A.G., Morgan-Hughes J., Weghuis D.O., Overhauser J., van Oost B. (1995) Molecular cloning and characterization of the active human mitochondrial NADH: ubiquinone oxidoreductase 24-kDa gene (*NDUFV2*) and its pseudogene. *Genomics* **26**: 461–466.

de Fatima Bonaldo M., Jelenc P., Su L., Lawton L., Yu M.T., Warburton D., Soares M.B. (1996) Identification and characterization of three genes and two pseudogenes on chromosome 13. *Hum. Genet.* **97**: 441–452.

DeBry R.W. (1998) Comparative analysis of evolution in a rodent histone H2a pseudogene. *J. Mol. Evol.* **46**: 355–360.

del Castillo I., Cohen-Salmon M., Blanchard S., Lutfalla G., Petit C. (1992) Structure of the X-linked Kallmann syndrome gene and its homologous pseudogene on the Y chromosome. *Nature Genet.* **2**: 305–310.

Denison R.A., Van Arsdell S.W., Bernstein L.B., Weiner A.W. (1981) Abundant pseudogenes for small nuclear RNAs are dispersed in the human genome. *Proc. Natl. Acad. Sci. USA* **78**: 810–815.

Devor E.J., Dill-Devor R.M. (1997) Nucleotide sequence, chromosome localization, and evolutionary conservation of a serine hydroxymethyltransferase-processed pseudogene. *Hum. Hered.* **47**: 125–130.

Devor E.J., Dill-Devor R.M., Magee H.J., Waziri R. (1998) Serine hydroxymethyltransferase pseudogene, SHMT-ps1: a unique genetic marker of the order primates. *J. Exp. Zool.* **282**: 150–156.

Devriendt K., Zhang J., van Leuven F., van den Berghe H., Cassiman J.J., Marynen P. (1989) A cluster of α2-macroglobulin-related genes (α2M) on human chromosome 12p: cloning of the pregnancy-zone protein gene and an α2M pseudogene. *Gene* **81**: 325–334.

Dhawan P., Yang E., Kumar A., Mehta K.D. (1998) Genetic complexity of the human geranylgeranyltransferase I beta-subunit gene: a multigene family of pseudogenes derived from mis-spliced transcripts. *Gene* **210**: 9–15.

Dhellin O., Maestre J., Heidmann T. (1997) Functional differences between the human LINE retrotransposon and retroviral reverse transcriptases for *in vivo* mRNA reverse transcription. *EMBO J.* **16**: 6590–6602.

Dierick H.A., Mercer J.F., Glover T.W. (1997) A phosphoglycerate mutase brain isoform (PGAM1) pseudogene is localized within the human Menkes disease gene (*ATP7A*). *Gene* **198**: 37–41.

Dolphin C.T., Beckett D.J., Janmohamed A., Cullingford T.E., Smith R.L., Shephard E.A., Phillips I.R. (1998) The flavin-containing monooxygenase 2 gene (*FMO2*) of humans, but not of other primates, encodes a truncated, nonfunctional protein. *J. Biol. Chem.* **273**: 30 599–30 607.

Dong Q., Shenker A., Way J., Haddad B.R., Lin K., Hughes M.R., McBride O.W., Spiegel A.M., Battey J. (1995) Molecular cloning of human G alpha q cDNA and chromosomal localization of the G alpha q gene (*GNAQ*) and a processed pseudogene. *Genomics* **30**: 470–475.

Dugast I.J., Papadopoulos P., Zappone E., Jones C., Theriault K., Handelman G.J., Benarous R., Drysdale J.W. (1990) Identification of two human ferritin H genes on the short arm of chromosome 6. *Genomics* **6**: 204–211.

Eikenboom J.C., Vink T., Briët E., Sixma J.J., Reitsma P.H. (1994) Multiple substitutions in the von Willebrand factor gene that mimic the pseudogene sequence. *Proc. Natl. Acad. Sci. USA* **91**: 2221–2224.

Espinosa L., Martin M., Nicolas A., Fabre M., Navarro E. (1997) Primary sequence of the human, lysine-rich, ribosomal protein RPL38 and detection of an unusual RPL38 processed pseudogene in the promoter region of the type-1 angiotensin II receptor gene. *Biochim. Biophys. Acta* **1354**: 58–64.

Eyal N., Wilder S., Horowitz M. (1990) Prevalent and rare mutations among Gaucher patients. *Gene* **96**: 277–283.

Eychene A., Barnier J.V., Apiou F., Dutrillaux B., Calothy G. (1992) Chromosomal assignment of two human B-raf(Rmil) proto-oncogene loci: B-raf-1 encoding the p94Braf/Rmil and B-raf-2, a processed pseudogene. *Oncogene* **7**: 1657–1660.

Figueroa F., O'hUigin C., Tichy H., Klein J. (1994) The origin of the primate *Mhc-DRB* genes and allelic lineages as deduced from the study of prosimians. *J. Immunol.* **152**: 4455–4464.

Fitzgerald J., Dahl H.-H.M., Jakobsen I.B., Easteal S. (1996) Evolution of mammalian X-linked and autosomal *Pgk* and *Pdh E1a* subunit genes. *Mol. Biol. Evol.* **13**: 1023–1031.

Floyd-Smith G., De Martinville B., Francke U. (1986) An expressed β-tubulin gene, *TUBB*, is located on the short arm of human chromosome 6 and two related sequences are dispersed on chromosomes 8 and 13. *Exp. Cell Res.* **163**: 539–548.

Foord S.M., Marks B., Stolz M., Bufflier E., Fraser N.J., Lee M.G. (1996) The structure of the prostaglandin EP4 receptor gene and related pseudogenes. *Genomics* 35: 182–188.

Freytag S.O., Bock H.-G., Beaudet A.L., O'Brien W.E. (1984) Molecular structures of human argininosuccinate synthetase pseudogenes. Evolutionary and mechanistic implications. *J. Biol. Chem.* 259: 3160–3167.

Gagnon A.W., Benovic J.L. (1997) Identification and chromosomal localization of a processed pseudogene of human GRK6. *Gene* 184: 13–19.

Galili U., Swanson K. (1991) Gene sequences suggest inactivation of α-1,3-galactosyltransferase in catarrhines after the divergence of apes from monkeys. *Proc. Natl. Acad. Sci. USA* 88: 7401–7404.

Gasparini P., Grifa A., Origone P., Coviello D., Antonacci R., Rocchi M. (1993) Detection of a neurofibromatosis type 1 (*NF1*) homologous sequence by PCR: implications for the diagnosis and screening of genetic diseases. *Mol. Cell. Probes* 7: 415–418.

Geraghty D.E., Koller B.H., Pel J., Hansen J.A. (1992) Examination of four HLA class I pseudogenes. *J. Immunol.* 149: 1947–1956.

Gersuk V.H., Rose T.M., Todaro G.J. (1995) Molecular cloning and chromosomal localization of a pseudogene related to the human acyl-CoA binding protein/diazepam binding inhibitor. *Genomics* 25: 469–476.

Giebel L.B., Strunk K.M., Spritz R.A. (1991) Organization and nucleotide sequences of the human tyrosinase gene and a truncated tyrosinase-related segment. *Genomics* 9: 435–445.

Giordano S.J., Yoo M., Ward D.C., Bhatt M., Overhauser J., Steggles A.W. (1993) The human cytochrome b$_5$ gene and two of its pseudogenes are located on chromosomes 18q23, 14q31–32.1 and 20p11.2, respectively. *Hum. Genet.* 92: 615–618.

Giordano M., Marchetti C., Chiorboli E., Bona G., Richiardi P.M. (1997) Evidence for gene conversion in the generation of extensive polymorphism in the promoter of the growth hormone gene. *Hum. Genet.* 100: 249–255.

Gojobori T., Li W.-H., Graur D. (1982) Patterns of nucleotide substitution in pseudogenes and functional genes. *J. Mol. Evol.* 18: 360–369.

Görlach A., Lee P.L., Roesler J., Hopkins P.J., Christensen B., Gree E.D., Chanock S.J., Curnutte J.T. (1997) A p47-*phox* pseudogene carries the most common mutation causing p47-*phox*-deficient chronic granulomatous disease. *J. Clin. Invest.* 100: 1907–1918.

Graur D., Shuali Y., Li W.-H. (1989) Deletions in processed pseudogenes accumulate faster in rodents than in humans. *J. Molec. Evol.* 28: 279–285.

Groot P.C., Mager W.H., Henriquez N.V., Pronk J.C., Arwert F., Planta R.J., Eriksson A.W., Frants R.R. (1990) Evolution of the α-amylase multigene family through unequal, homologous, and inter- and intrachromosomal crossovers. *Genomics* 8: 97–105.

Gruen J.R., Nalabolu S.R., Chu T.W., Bowlus C., Fan W.F., Goei V.L., Wei H., Sivakamasundari R., Liu Y., Xu Y., Xu H.X., Parimoo S., Nallur G., Ajioka R., Shukla H., Bray-Ward P., Pan J., Weissman S.M. (1996) A transcription map of the major histocompatibility complex (MHC) class I region. *Genomics* 36: 70–85.

Guo N., Mogues T., Weremowicz S., Morton C.C., Sastry K.N. (1998) The human ortholog of rhesus mannose-binding protein-A gene is an expressed pseudogene that localizes to chromosome 10. *Mamm. Genome* 9: 246–248.

Haag F., Koch-Nolte F., Kühl M., Lorenzen S., Thiele H.-G. (1994) Premature stop codons inactivate the RT6 genes of the human and chimpanzee species. *J. Mol. Biol.* 243: 537–546.

Haino M., Hayashida H., Miyata T., Shin E.K., Matsuda F., Nagaoka H., Matsumura R., Takaishi S., Fukita Y., Fujikura J., Honjo T. (1994) Comparison and evolution of human immunoglobulin V$_H$ segments located in the 3′ 0.8-megabase region. *J. Biol. Chem.* 269: 2619–2626.

Hammarström K., Westin G., Pettersson U. (1982) A pseudogene for human U4 RNA with a remarkable structure. *EMBO J.* 1: 737–743.

Harada F., Kimura A., Iwanaga T., Shimozawa K., Yata J., Sasazuki T. (1987) Gene conversion-like events cause 21-hydroxylase deficiency in congenital adrenal hyperplasia. *Proc. Natl. Acad. Sci. USA* 84: 8091–8094.

Harindranath N., Mills F.C., Mitchell M., Meindl A., Max E.E. (1998) The human *elk-1* gene family: the functional gene and two processed pseudogenes embedded in the IgH locus. *Gene* 221: 215–224.

Harris S., Barrie P.A., Weiss M.L., Jeffreys A.J. (1984) The primate ψβ1 gene: an ancient β-globin pseudogene. *J. Mol. Biol.* **180**: 785–801.

Harun R.B., Smith K.K., Leek J.P., Markham A.F., Norris A., Morrison J.F. (1997) Characterization of human SHC p66 cDNA and its processed pseudogene mapping to Xq12-q13.1. *Genomics* **42**: 349–352.

Hayashi K. (1981) Organization of sequences related to U6 RNA in the human genome. *Nucleic Acids Res.* **9**: 3379–3387.

Hill A.V.S., Nicholls R.D., Thein S.L., Higgs D.R. (1985) Recombination within the human embryonic ζ-globin locus: a common ζ-ζ chromosome produced by gene conversion of the psi-ζ gene. *Cell* **37**: 809–819.

Hisajima H., Nishida Y., Nakai S., Takahashi N., Ueda S., Honjo T. (1983) Structure of the human immunoglobulin Cε2 gene, a truncated pseudogene: implications for its evolutionary origin. *Proc. Natl. Acad. Sci. USA* **80**: 2995–2999.

Hollis G.F., Hieter P.A., McBride O.W., Swan D., Leder P. (1982) Processed genes: a dispersed human immunoglobulin gene bearing evidence of RNA-type processing. *Nature* **296**: 321–325.

Horiuch Y., Kawaguchi H., Figueroa F., O'hUigin C., Klein J. (1993) Dating the primigenial *C4-CYP21* duplication in primates. *Genetics* **134**: 331–339.

Hsu L.C., Chang W.C. (1996) Sequencing and expression of the human *ALDH8* gene encoding a new member of the aldehyde dehydrogenase family. *Gene* **174**: 319–322.

Hsu L.C., Chang W.C., Yoshida A. (1997) Human aldehyde dehydrogenase genes, *ALDH7* and *ALDH8*: genomic organization and gene structure comparison. *Gene* **189**: 89–94.

Hughes A.L. (1995) Origin and evolution of *HLA* class I pseudogenes. *Mol. Biol. Evol.* **12**: 247–258.

Hulsebos T.J.M., Bijleveld E.H., Riegman P.H.J., Smink L.J., Dunham I. (1996) Identification and characterization of *NF1*-related loci on human chromosomes 22, 14 and 2. *Hum. Genet.* **98**: 7–11.

Horowitz M., Wilder S., Horowitz Z., Reiner O., Gelbart T., Beutler E. (1989) The human glucocerebrosidase gene and pseudogene: structure and evolution. *Genomics* **4**: 87–96.

Irie A., Koyama S., Kozutsumi Y., Kawasaki T., Suzuki A. (1998) The molecular basis for the absence of *N*-glycolylneuraminic acid in humans. *J. Biol. Chem.* **273**: 15 866–15 871.

Jeffreys A.J., Barrie P.A., Harris S., Fawcett D.M., Nugent Z.J., Boyd A.C. (1982) Isolation and sequence analysis of a hybrid δ-globin pseudogene from the brown lemur. *J. Mol. Biol.* **156**: 487–495.

Johnson B.L., Campbell C. (1996) Identification of a human RAD52 pseudogene located on chromosome 2. *Gene* **169**: 229–232.

Jurka J. (1997) Sequence patterns indicate an enzymatic involvement in integration of mammalian retroposons. *Proc. Natl. Acad. Sci. USA* **94**: 1872–1877.

Jury J.A., Frayne J., Hall L. (1997) The human fertilin α gene is non-functional: implications for its proposed role in fertilization. *Biochem. J.* **321**: 577–581.

Karin M., Richards R.I. (1982) Human metallothionein genes – primary structure of the metallothionein-II gene and a related processed gene. *Nature* **299**: 797–802.

Kas K., Stickens D., Merregaert J. (1995) Characterization of a processed pseudogene of human *FAU1* on chromosome 18. *Gene* **160**: 273–276.

Kass D.H., Batzer M.A., Deininger P.L. (1995) Gene conversion as a secondary mechanism of short interspersed element (SINE) evolution. *Mol. Cell. Biol.* **15**: 19–25.

Kawaguchi H., O'hUigin C., Klein J. (1992) Evolutionary origin of mutations in the primate cytochrome P450c21 gene. *Am. J. Hum. Genet.* **50**: 766–780.

Kawamura S., Saitou N., Ueda S. (1992) Concerted evolution of the primate immunoglobulin alpha-gene through gene conversion. *J. Biol. Chem.* **267**: 7359–7367.

Kehrer-Sawatzki H., Schwickardt T., Assum G., Rocchi M., Krone W. (1997) A third neurofibromatosis type 1 (NF1) pseudogene at chromosome 15q11.2. *Hum. Genet.* **100**: 595–600.

Kelley M.J., Pech M., Seuanez H.N., Rubin J.S., O'Brien S.J., Aaronson S.A. (1992) Emergence of the keratinocyte growth factor multigene family during the great ape radiation. *Proc. Natl. Acad. Sci. USA* **89**: 9287–9291.

Kim H.S., Lyons K.M., Saitoh E., Azen E.A., Smithies O., Maeda N. (1993) The structure and evolution of the human salivary proline-rich protein gene family. *Mamm. Genome* **4**: 3–14.

Kluve-Beckerman B., Drumm M.L., Benson M.D. (1991) Nonexpression of the human serum amyloid A3 (SAA3) gene. *DNA Cell Biol.* **10**: 651–661.

Kokke F.T.M., Elsawy T., Bengtsson U., Wasmuth J.J., Jabs E.W., Tse C.-M., Donowitz M., Brant S.R. (1996) A NHE3-related pseudogene is on human chromosome 10; the functional gene maps to 5p15.3. *Mamm. Genome* **7**: 235–236.

Koller M., Baumer A., Strehler E.E. (1991) Characterization of two novel human retropseudogenes related to the calmodulin-encoding gene, *CaMII*. *Gene* **97**: 245–251.

Larsen R.D., Rivera-Marrero C.A., Ernst L.K., Cummings R.D., Lowe J.B. (1990) Frameshift and nonsense mutations in a genomic sequence homologous to a murine UDP-Gal: beta-D-Gal(1,4)-D-GlcNAc alpha (1,3)-galactosyltransferase cDNA. *J. Biol. Chem.* **265**: 7055–7061.

Legius E., Marchuk D.A., Hall B.K., Andersen L.B., Wallace M., Collins F.S., Glover T.W. (1992) *NF1*-related locus on chromosome 15. *Genomics* **13**: 1316–1318.

Lewis S.A., Cowan N.J. (1990) Tubulin genes: structure, expression and regulation. In: *Microtubule Proteins*. (ed. J Avila). CRC Press Inc, Boca Raton. pp 37–66.

Li W.-H., Gojobori T., Nei M. (1981) Pseudogenes as a paradigm of neutral evolution. *Nature* **292**: 237–239.

Li W.-H., Wu C.-I., Luo C.-C. (1984) Nonrandomness of point mutation as reflected in nucleotide substitutions in pseudogenes and its evolutionary implications. *J. Mol. Evol.* **21**: 58–71.

Li Y., Wong A., Szabo P., Posnett D.N. (1993) Human Tcrb-V6.10 is a pseudogene with *Alu* repetitive sequences in the promoter region. *Immunogenetics* **37**: 347–355.

Liao D., Paveliz T., Kidd J.R., Kidd K.K., Weiner A.M. (1997) Concerted evolution of the tandemly repeated genes encoding human U2 snRNA (the RNU2 locus) involves rapid intrachromosomal homogenization and rare interchromosomal gene conversion. *EMBO J.* **16**: 588–598.

Ling M., Merante F., Chen H.S., Duff C., Duncan A.M., Robinson B.H. (1997) The human mitochondrial elongation factor Tu (EF-Tu) gene: cDNA sequence, genomic localization, genomic structure, and identification of a pseudogene. *Gene* **197**: 325–336.

Lips C.J.M., Geerdink R.A., Nieuwenhuis M.G., van der Sluys Veer J. (1989) Evolutionary pathways of the calcitonin genes. *Henry Ford Hosp. Med. J.* **37**: 201–203.

Liu Q.R., Chan P.K. (1993) Characterization of seven processed pseudogenes of nucleophosmin/B23 in the human genome. *DNA Cell Biol.* **12**: 149–156.

Long G.L., Winfield S., Adolph K.W., Ginns E.I., Bornstein P. (1996) Structure and organization of the human metaxin gene (*MTX*) and pseudogene. *Genomics* **33**: 177–184.

Lotscher E., Grzeschik K.-H., Bauer H.G., Pohlenz H.-D., Straubinger B., Zachau H.G. (1986) Dispersed human immunoglobulin κ light-chain genes. *Nature* **320**: 456–458.

Lund H., Eftedal I., Haug T., Krokan H.E. (1996) Pseudogenes for the human uracil-DNA glycosylase on chromosomes 14 and 16. *Biochem. Biophys. Res. Commun.* **224**: 265–270.

Lyn D., Istock N.L., Smulson M. (1995) Conservation of sequences between human and gorilla lineages: ADP-ribosyltransferase (NAD[+]) pseudogene 1 and neighboring retroposons. *Gene* **155**: 241–245.

McCarrey J.R. (1990) Molecular evolution of the human Pgk-2 retroposon. *Nucleic Acids Res.* **18**: 949–955.

McCormack W.T., Hurley E.A., Thompson C.B. (1993) Germ line maintenance of the pseudogene donor pool for somatic immunoglobulin gene conversion in chickens. *Mol. Cell. Biol.* **13**: 821–830.

McEvoy S.M., Maeda N. (1988) Complex events in the evolution of the haptoglobin gene cluster in primates. *J. Biol. Chem.* **263**: 15 740–15 747.

McGrath J.P., Capon D.J., Smith D.H., Chen E.Y., Seeburg P.H., Goeddel D.V., Levinson A.D. (1983) Structure and organization of the human Ki-*ras* proto-oncogene and a related processed pseudogene. *Nature* **304**: 501–506.

MacLeod A.R., Talbot K. (1983) A processed gene defining a gene family encoding a human non-muscle tropomyosin. *J. Mol. Biol.* **167**: 523–534.

Maestre J., Tchenio T., Dhellin O., Heidmann T. (1995) mRNA retroposition in human cells: processed pseudogene formation. *EMBO J.* **14**: 6333–6338.

Manser T., Gesteland R.F. (1981) Characterization of small nuclear RNA U1 gene candidates and pseudogenes from the human genome. *J. Mol. Genet.* **1**: 117–128.

Marchese A., Beischlag T.V., Nguyen T., Niznik H.B., Weinshank R.L., George S.R., O'Dowd B.F. (1995) Two gene duplication events in the human and primate dopamine D5 receptor gene family. *Gene* **154**: 153–158.

Marchetti G., Patracchini P., Volinia S., Aiello V., Schiavoni M., Ciavarella N., Calzolari E., Schwienbacher C., Bernardi F. (1991) Characterization of the pseudogenic and genic homologous regions of von Willebrand factor. *Brit. J. Haematol.* **78**: 71–79.

Maric S.C., Crozat A., Louhimo J., Knuutila S., Janne O.A. (1995) The human S-adenosylmethionine decarboxylase gene: nucleotide sequence of a pseudogene and chromosomal localization of the active gene (*AMD1*) and the pseudogene (*AMD2*). *Cytogenet. Cell Genet.* **70**: 195–199.

Marshall C.R., Raff E.C., Raff R.A. (1994) Dollo's law and the death and resurrection of genes. *Proc. Natl. Acad. Sci. USA* **91**: 12 283–12 287.

Max E.E., Jahan N., Yi H., McBride W.O. (1994) A processed J chain pseudogene on human chromosome 8 that is shared by several primate species. *Molec. Immunol.* **31**: 1029–1036.

Minegishi Y., Coustan-Smith E., Wang Y.H., Cooper M.D., Campana D., Conley M.E. (1998) Mutations in the human λ5/14.1 gene results in B cell deficiency and agammaglobulinemia. *J. Exp. Med.* **187**: 71–77.

Miyata T., Hayashida H. (1981) Extraordinarily high evolutionary rate of pseudogenes: evidence for the presence of selective pressure against changes between synonymous codons. *Proc. Natl. Acad. Sci. USA* **78**: 5739–5743.

Moos M., Gallwitz D. (1982) Structure of a human β-actin-related pseudogene which lacks intervening sequences. *Nucleic Acids Res.* **10**: 7843–7849.

Moos M., Gallwitz D. (1983) Structure of two human β-actin-related processed genes one of which is located next to a simple repetitive sequence. *EMBO J.* **2**: 757–764.

Murti J.R., Bumbulis M., Schimenti J.C. (1994) Gene conversion between unlinked sequences in the germline of mice. *Genetics* **137**: 837–843.

Nagao Y., Batanian J.R., Clemente M.F., Sly W.S. (1995) Genomic organization of the human gene (CA5) and pseudogene for mitochondrial carbonic anhydrase V and their localization to chromosomes 16q and 16p. *Genomics* **28**: 477–484.

Nakano K., Takase C., Sakamoto T., Nakagawa S., Inazawa J., Ohta S., Matuda S. (1994) Isolation, characterization and structural organization of the gene and pseudogene for the dihydrolipoamide succinyltransferase component of the human 2-oxoglutarate dehydrogenase complex. *Eur. J. Biochem.* **224**: 179–189.

Nesbit M.A., Hodges M.D., Campbell L., de Meulemeester T.M., Alders M., Rodrigues N.R., Talbot K., Theodosiou A.M., Mannens M.A., Nakamura Y., Little P.F., Davies K.E. (1997) Genomic organization and chromosomal localization of a member of the MAP kinase phosphatase gene family to human chromosome 11p15.5 and a pseudogene to 10q11.2. *Genomics* **42**: 284–294.

Nguyen T., Marchese A., Kennedy J.L., Petronis A., Peroutka S.J., Wu P.H., O'Dowd B.F. (1993) An *Alu* sequence interupts a human 5-hydroxytryptamine$_{1D}$ receptor pseudogene. *Gene* **124**: 295–301.

Nishikimi M., Fukuyama R., Minoshima S., Shimizu N., Yagi K. (1994) Cloning and chromosomal mapping of the human nonfunctional gene for L-gulono-γ-lactone oxidase, the enzyme for L-ascorbic acid biosynthesis missing in man. *J. Biol. Chem.* **269**: 13685–13688.

Nollet F., van Hengel J., Berx G., Molemans F., van Roy F. (1995) Isolation and characterization of a human pseudogene (*CTNNAP1*) for α E-catenin (*CTNNA1*): assignment of the pseudogene to 5q22 and the α E-catenin gene to 5q31. *Genomics* **26**: 410–413.

Nomiyama H., Obaru K., Jinno Y., Matsuda I., Shimada K., Miyata T. (1986) Amplification of human argininosuccinate synthetase pseudogenes. *J. Mol. Biol.* **192**: 221–233.

Ohishi K., Inoue N., Endo Y., Fujita T., Takeda J., Kinoshita T. (1995) Structure and chromosomal localization of the GPI-anchor synthesis gene *PIGF* and its pseudogene psi PIGF. *Genomics* **29**: 804–807.

Ophir R., Graur D. (1997) Patterns and rates of indel evolution in processed pseudogenes from humans and murids. *Gene* **205**: 191–202.

Pamilo P., Bianchi N.O. (1993) Evolution of the Zfx and Zfy genes: rates and interdependence between the genes. *Mol. Biol. Evol.* **10**: 271–281.

Park H., Baek K., Jeon C., Agarwal K., Yoo O. (1994) Characterization of the gene encoding the human transcriptional elongation factor TFIIS. *Gene* **139**: 263–267.

Ploos van Amstel H.K., Reitsma P.H., van der Logt C.P.E., Bertina R.M. (1990) Intron-exon organization of the active human protein S gene PSα and its pseudogene PSβ: duplication and silencing during primate evolution. *Biochemistry* **29**: 7853–7861.

Prinsen C.F., Weghuis D.O., Kessel A.G., Veerkamp J.H. (1997) Identification of a human heart FABP pseudogene located on chromosome 13. *Gene* **193**: 245–251.

Proudfoot N.J., Rutherford T.R., Partington G.A. (1984) Transcriptional analysis of human zeta globin genes. *EMBO J.* **3**: 1533–1542.

Purandare S.M., Breidenbach H.H., Li Y., Zhu X.L., Sawada S., Neil S.M., Brothman A., White R., Cawthon R., Viskochil D. (1995) Identification of neurofibromatosis 1 (*NF1*) homologous loci by direct sequencing, fluorescence *in situ* hybridization and PCR amplification of somatic cell hybrids. *Genomics* 30: 476–485.

Raissier A. (1991) Duplication of the apolipoprotein C-I gene occurred about forty million years ago. *J. Mol. Evol.* 32: 211–219.

Rathmann M., Bunge S., Steglich C., Schwinger E., Gal A. (1995) Evidence for an iduronate-sulfatase pseudogene near the functional Hunter syndrome gene in Xq27.3-q28. *Hum. Genet.* 95: 34–38.

Régnier V., Meddeb M., Lecointre G., Richard F., Duverger A., Nguyen V.C., Dutrillaux B., Bernheim A., Danglot G. (1997) Emergence and scattering of multiple neurofibromatosis (*NF1*)-related sequences during hominoid evolution suggest a process of pericentromeric interchromosomal transposition. *Hum. Molec. Genet.* 6: 9–16.

Rhyner J.A., Ottiger M., Wicki R., Greenwood T.M., Strehler E.E. (1994) Structure of the human *CALM1* calmodulin gene and identification of two *CALM1*-related pseudogenes *CALM1P1* and *CALM1P2*. *Eur. J. Biochem.* 225: 71–82.

Ribaudo M.R., Di Leonardo A., Rubino P., Giallongo A., Feo S. (1996) Assignment of enolase processed pseudogene (*ENO1P*) to human chromosome 1 bands 1q41→q42. *Cytogenet. Cell Genet.* 74: 201–202.

Richardson M.P., Braybrook C., Tham M., Moore G.E., Stanier P. (1998) Molecular cloning and characterization of a highly conserved human 67-kDa laminin receptor pseudogene mapping to Xq21.3. *Gene* 206: 145–150.

Rodriguez C., Grosgeorge J., Nguyen V.C., Gaudray P., Theillet C. (1995) Human gp130 transducer chain gene (*IL6ST*) is localized to chromosome band 5q11 and possesses a pseudogene on chromosome band 17p11. *Cytogenet. Cell Genet.* 70: 64–67.

Rodriguez de Cordoba S., Perez-Blas M., Ramos-Ruiz R., Sanchez-Corral P., Pardo-Manuel de Villena F., Rey-Campos J. (1994) The gene coding for the β-chain of C4b-binding protein (*C4BPB*) has become a pseudogene in the mouse. *Genomics* 21: 501–509.

Rose S.D., MacDonald R.J. (1997) Evolutionary silencing of the human elastase I gene (*ELA1*). *Hum. Molec. Genet.* 6: 897–903.

Rouquier S., Friedman C., Delettre C., van den Engh G., Blancher A., Crouau-Roy B., Trask B.J., Giorgi D. (1998) A gene recently inactivated in human defines a new olfactory receptor family in mammals. *Hum. Molec. Genet.* 7: 1337–1345.

Rozmahel R., Heng H.H.Q., Duncan A.M.V., Shi X.-M., Rommens J.M., Tsui L.-C. (1997) Amplification of *CFTR* exon 9 sequences to multiple locations in the human genome. *Genomics* 45: 554–561.

Russell M.W., du Manoir S., Collins F.S., Brody L.C. (1997) Cloning of the human NADH: ubiquinone oxidoreductase subunit B13: localization to chromosome 7q32 and identification of a pseudogene on 11p15. *Mamm. Genome* 8: 60–61.

Santoro C., Marone M., Ferrone M., Costanzo F., Colombo M., Minganti C., Cortese R., Silengo L. (1986) Cloning of the gene coding for human L-apoferritin. *Nucleic Acids Res.* 14: 2863–2876.

Sapi E., Flick M.B., Kacinski B.M. (1994) The first intron of human c-*fms* proto-oncogene contains a processed pseudogene (RPL7P) for ribosomal protein L7. *Genomics* 22: 641–645.

Savitsky K., Ziv Y., Bar-Shira A., Gilad S., Tagle D.A., Smith S., Uziel T., Sfez S., Nahmias J., Sartiel A., Eddy R.L., Shows T.B., Collins F.S., Shiloh Y., Rotman G. (1996) A human gene (*DDX10*) encoding a putative DEAD-box RNA helicase at 11q22-q23. *Genomics* 33: 199–206.

Savtchenko E.S., Freedberg J.M., Choi I.Y., Blumenberg M. (1988) Inactivation of human keratin genes: the spectrum of mutations in the sequence of an acidic keratin pseudogene. *Mol. Biol. Evol.* 5: 97–108.

Schollen E., Pardon E., Heykants L., Renard J., Doggett N.A., Callen D.F., Cassiman J.J., Matthijs G. (1998) Comparative analysis of the phosphomannomutase genes *PMM1*, *PMM2* and *PMM2psi*: the sequence variation in the processed pseudogene is a reflection of the mutations found in the functional gene. *Hum. Molec. Genet.* 7: 157–164.

Selig S., Bruno S., Scharf J.M., Wang C.H., Vitale E., Gilliam T.C., Kunkel L.M. (1995) Expressed cadherin pseudogenes are localized to the critical region of the spinal muscular atrophy gene. *Proc. Natl. Acad. Sci. USA* 92: 3702–3706.

Shimada T., Chen M.-J., Nienhuis A.W. (1983) Genomic fluidity: multiple DNA insertions at a single chromosomal site. *Clin. Res.* 31: 479A.

Sladek R., Beatty J., Squire J., Copeland N.G., Gilbert D.J., Jenkins N.A., Giguere V. (1997) Chromosomal mapping of the human and murine orphan receptors ERRα (*ESRRA*) and ERRβ (*ESRRB*) and identification of a novel human ERRα-related pseudogene. *Genomics* **45**: 320–326.

Slightom J.L., Koop B.F., Xu P.L., Goodman M. (1988) Rhesus fetal globin genes. Concerted gene evolution in the descent of higher primates. *J. Biol. Chem.* **263**: 12 427–12 438.

Smith T.M., Lee M.K., Szabo C.I., Jerome N., McEuen M., Taylor M., Hood L., King M.C. (1996) Complete genomic sequence and analysis of 117 kb of human DNA containing the gene *BRCA1*. *Genome Res.* **6**: 1029–1049.

Speleman F., Vervoort R., van Roy N., Liebaers I., Sly W.S., Lissens W. (1996) Localization by fluorescence *in situ* hybridization of the human functional β-glucuronidase gene (*GUSB*) to 7q11.21→q11.22 and two pseudogenes to 5p13 and 5q13. *Cytogenet. Cell Genet.* **72**: 53–55.

Stros M., Dixon G.H. (1993) A retropseudogene for non-histone chromosomal protein HMG-1. *Biochim. Biophys. Acta* **1172**: 231–235.

Suzuki H., Ozawa N., Taga C., Kano T., Hattori M., Sakaki Y. (1994) Genomic analysis of a *NF1*-related pseudogene on human chromosome 21. *Gene* **147**: 277–280.

Suzuki E., Lowry J., Sonoda G., Testa J.R., Walsh K. (1996) Structures and chromosome locations of the human *MEF2A* gene and a pseudogene *MEF2AP*. *Cytogenet. Cell Genet.* **73**: 244–249.

Taanman J.-W., Schrage C., Reuvekamp P., Bijl J., Hartog M., de Vries H., Agsteribbe E. (1991) Identification of three human pseudogenes for subunit VIb of cytochrome *c* oxidase: a molecular record of gene evolution. *Gene* **102**: 237–244.

Takahashi N., Ueda S., Obata M., Nikaido T., Nakai S., Honjo T. (1982) Structure of human immunoglobulin gamma genes: implications for evolution of a gene family. *Cell* **29**: 671–679.

Taniguchi Y., Katsumata Y., Koido S., Suemizu H., Yoshimura S., Moriuchi T., Okumura K., Kagotani K., Taguchi H., Imanishi T., Gojobori T., Inoko H. (1996) Cloning, sequencing, and chromosomal localization of two tandemly arranged human pseudogenes for the proliferating cell nuclear antigen (PCNA). *Mamm. Genome* **7**: 906–908.

Tchenio T., Segal-Bendirdjian B., Heidmann T. (1993) Generation of processed pseudogenes in murine cells. *EMBO J.* **12**: 1487–1494.

Timms K.M., Lu F., Shen Y., Pierson C.A., Muzny D.M., Gu Y., Nelson D.L., Gibbs R.A. (1995) 130 kb of DNA sequence reveals two new genes and a regional duplication distal to the human iduronate-2-sulfate sulfatase locus. *Genome Res.* **5**: 71–78.

Trabesinger-Ruef N., Jermann T., Zankel T., Durrant B., Frank G., Benner S.A. (1996) Pseudogenes in ribonuclease evolution: a source of new biomacromolecular function? *FEBS Letts.* **382**: 319–322.

Tran P.T., Hori H., Hori Y., Okumura K., Kagotani K., Taguchi H., Carson D.A., Nobori T. (1997) Molecular cloning of the human methylthioadenosine phosphorylase processed pseudogene and localization to 3q28. *Gene* **186**: 263–269.

Tusie-Luna M.-T., White P.C. (1995) Gene conversions and unequal crossovers between *CYP21* (steroid 21-hydroxylase gene) and *CYP21P* involve different mechanisms. *Proc. Natl. Acad. Sci. USA* **92**: 10 796–10 800.

Ueda S., Takenaka O., Honjo T. (1985) A truncated immunoglobulin ε pseudogene is found in gorilla and man but not in chimpanzee. *Proc. Natl. Acad. Sci. USA* **82**: 3712–3715.

Ueyama H., Inazawa J., Nishino H., Ohkubo I., Miwa T. (1996) FISH localization of human cytoplasmic actin genes *ACTB* to 7q22 and *ACTG1* to 17q25 and characterization of related pseudogenes. *Cytogenet. Cell Genet.* **74**: 221–224.

Van Arsdell S.W., Weiner A.M. (1984) Pseudogenes for human U2 small nuclear RNA do not have a fixed site of 39 truncation. *Nucleic Acids Res.* **12**: 1463–1471.

Vanin E.F. (1984) Processed pseudogenes. Characteristics and evolution. *Biochim. Biophys. Acta* **782**: 231–241.

Villa A., Strina D., Frattini A., Faranda S., Macchi P., Finelli P., Bozzi F., Susani L., Archidiacono N., Rocchi M., Vezzoni P. (1996) The *ZNF75* zinc finger gene subfamily: isolation and mapping of the four members in humans and great apes. *Genomics* **35**: 312–320.

Wang S., Pirtle I.L., Pirtle R.M. (1997) A human 28S ribosomal RNA retropseudogene. *Gene* **196**: 105–111.

Watkins D.I., Chen Z.W., Garger T.L., Hughes A.L., Letvin N.L. (1991) Segmental exchange between MHC class I genes in a higher primate: recombination in the gorilla between the ancestor of a human non-functional gene and an *A* locus gene. *Immunogenetics* **34**: 185–191.

Watnick T.J., Gandolph M.A., Weber H., Neumann H.P.H., Germino G.G. (1998) Gene conversion is a likely cause of mutation in *PKD1*. *Hum. Molec. Genet.* **7**: 1239–1243.

Weiner A.M., Deininger P.L., Efstratiadis A. (1986) Nonviral retroposons: genes, pseudogenes, and transposable elements generated by the reverse flow of genetic information. *Annu. Rev. Biochem.* **55**: 631–636.

Weiskirchen R., Erdel M., Utermann G., Bister K. (1997) Cloning, structural analysis, and chromosomal localization of the human *CSRP2* gene encoding the LIM domain protein CRP2. *Genomics* **44**: 83–93.

Weller P.A., Critcher R., Goodfellow P.N., German J., Ellis N.A. (1995) The human Y chromosome homologue of *XG*: transcription of a naturally truncated gene. *Hum. Molec. Genet.* **4**: 859–868.

Whitelaw E., Proudfoot N. (1983) Transcriptional activity of the human pseudogene ψα globin compared with α globin, its functional counterpart. *Nucleic Acids Res.* **11**: 7717–7724.

Wilde C.D. (1985) Pseudogenes. *CRC Crit. Rev. Biochem.* **19**: 323–352.

Wilde C.D., Crowther C.E., Cripe T.P., Lee M.G.-S., Cowan N.J. (1982a) Evidence that a human β-tubulin pseudogene is derived from its corresponding mRNA. *Nature* **297**: 83–87.

Wilde C.D., Crowther C.E., Cowan N.J. (1982b) Diverse mechanisms in the generation of human β-tubulin pseudogenes. *Science* **217**: 549–552.

Wu X., Lee C.C., Muzny D.M., Caskey C.T. (1989) Urate oxidase: primary structure and evolutionary implications. *Proc. Natl. Acad. Sci. USA* **86**: 9412–9416.

Wu X.W., Muzny D.M., Lee C.C., Caskey C.T. (1992) Two independent mutational events in the loss of urate oxidase during hominoid evolution. *J. Mol. Evol.* **34**: 78–84.

Xu G., O'Connell P., Stevens J., White R. (1992) Characterization of human adenylate kinase 3 (*AK3*) cDNA and mapping of the *AK3* pseudogene to an intron of the *NF1* gene. *Genomics* **13**: 537–542.

Yeldandi A.V., Wang X., Alvares K., Kumar S., Rao M.S., Reddy J.K. (1990) Human urate oxidase gene: cloning and partial sequence analysis reveals a stop codon within the fifth exon. *Biochem. Biophys. Res. Commun.* **171**: 641–646.

Yeldandi A.V., Yeldandi V., Kumar S., Murthy C.V.N., Wang X., Alvares K., Rao M.S., Reddy J.K. (1991) Molecular evolution of the urate oxidase-encoding gene in hominoid primates: nonsense mutations. *Gene* **109**: 281–284.

Yoshida T., Imai T., Takagi S., Nishimura M., Ishikawa I., Yaoi T., Yoshie O. (1996) Structure and expression of two highly related genes encoding SCM-1/human lymphotactin. *FEBS Letts.* **395**: 82–88.

Zhang X.M., Cathala G., Soua Z., Lefranc M.P., Huck S. (1996) The human T-cell receptor γ variable pseudogene V10 is a distinctive marker of human speciation. *Immunogenetics* **43**: 196–203.

Zheng H.-D., Bhavsar D., Drysdale J. (1995) An unusual human ferritin H sequence from chromosome 4. *DNA Sequence* **5**: 173–175.

Zheng H.-D., Bhavsar D., Dugast I., Zappone E., Drysdale J. (1997) Conserved mutations in human ferritin H pseudogenes: a second functional sequence or an evolutionary quirk? *Biochim. Biophys. Acta* **1351**: 150–156.

Zhou Y.H., Li W.H. (1996) Gene conversion and natural selection in the evolution of X-linked color vision genes in higher primates. *Mol. Biol. Evol.* **13**: 780–783.

Zhou B.-S., Beidler D.R., Cheng Y.-C. (1992) Identification of antisense RNA transcripts from a human DNA topoisomerase I pseudogene. *Cancer Res.* **52**: 4280–4285.

Zimonjic D.B., Kelley M.J., Rubin J.S., Aaronson S.A., Poescu N.C. (1997) Fluorescence *in situ* hybridization analysis of keratinocyte growth factor gene amplification and dispersion in evolution of great apes and humans. *Proc. Natl. Acad. Sci. USA* **94**: 11 461–11 465.

PART 3
MUTATIONAL MECHANISMS IN EVOLUTION

Single base-pair substitutions

7.1 Single base-pair substitutions in evolution

7.1.1 The selectionist and neutralist perspectives

> Nowadays a certain number of believers in evolution do not regard natural selection as a cause of it. We must therefore carefully distinguish between two quite different doctrines which Darwin popularised, the doctrine of evolution, and that of natural selection. It is quite possible to hold the first and not the second.
>
> J.B.S. Haldane *The Causes of Evolution* (1932)

The vast majority of mutations that occur are either neutral with respect to *fitness* (defined as the individual's ability to survive and reproduce) or are disadvantageous. If they are disadvantageous, they will tend to be removed from the population since their bearers will be less likely to survive and/or reproduce (*negative* or *purifying selection*). Occasionally, a new mutation confers a selective advantage and increases the fitness of individuals bearing it so that it will eventually reach *fixation* (the point at which the allele frequency in the population becomes 100%). This is termed *positive selection*. Selection is thus nothing more than the differential and nonrandom reproduction of genotypes resulting from the superior or inferior fitness of their associated phenotypes. In vertebrates, direct evidence at the molecular level for the occurrence of selection, whether positive or negative, has however often been hard to obtain and there are as yet relatively few good practical examples (reviewed in Chapter 2, section 2.3.7 and Sections 7.1.2 and 7.5.2).

Gene evolution does not however invariably require selection since changes in allele frequency can also occur by chance owing to random sampling of gametes (*genetic drift*). Whereas selection implies directed change, genetic drift may be viewed as a stochastic process of undirected change. Genetic drift can cause rapid changes in small populations but its effect will be fairly minimal in large ones. This is one of the central conclusions of the *neutral theory of molecular evolution* (Kimura, 1983). The importance of this theory to our understanding of the evolutionary process cannot be understated. Its major points may be summarized as follows:

(i)   Most polymorphic variation within a species and most amino acid substitutions between species are likely to be neutral with respect to selection. This is not to say that variant alleles have no biological function or phenotypic effect, merely that the alleles are indistinguishable (neutral) in terms of their function.

(ii)   The amino acid substitution rate can vary between genes and this is likely to be related to the nature and extent of structural and functional constraints upon the different protein products.

(iii)   Those parts of a protein with the fewest functional constraints will evolve most rapidly. Conversely, those regions of a protein which are functionally important will be those that are most highly conserved evolutionarily.

(iv)   Genes whose protein products have a stable function manifest a *molecular clock* i.e. the amino acid substitution rate is similar for a given gene in different evolutionary lineages.

(v)   A high level of genetic polymorphism will exist and this is predicted to increase with population size since the probability of fixation of a specific allele is lower in large populations.

(vi)   In coding sequences, nucleotides in the third codon position evolve faster than the first two positions. The proportion of substitutions in the third position should be higher for proteins that are evolving more slowly.

(vii)   Any change in the function of a gene sequence will serve to alter the proportion of amino acid substitutions that are neutral. Genes that are no longer capable of expression (pseudogenes) will accumulate mutations rapidly (*Figure 7.1*) and are destined eventually to lose recognizable homology to the original sequence.

(viii) Sequences or portions of sequences that are evolving faster than the neutral rate for the species are likely to have been subject to positive selection.

(ix)   Genes manifesting a high rate of variation should also evolve at a higher rate.

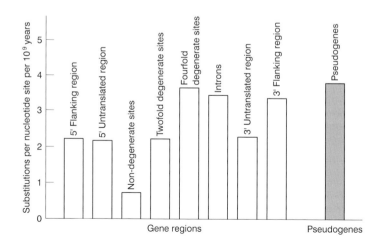

Figure 7.1. Average rates of substitution in different parts of genes and in pseudogenes (Li and Graur, 1991).

7.1.2 Synonymous and nonsynonymous substitutions

> The preservation of favourable variations and the rejection of injurious varia-
> tions, I call Natural Selection, or Survival of the Fittest. Variations neither
> useful nor injurious would not be affected by natural selection and would be
> left a fluctuating element.
> Charles Darwin (1859) *The Origin of Species*

Rates of nucleotide substitution vary between different gene regions, tending to
be higher in introns and flanking regions than in coding sequences (Li and Graur,
1981; *Figure 7.1*). Single base-pair changes in noncoding regions do not usually
affect gene expression unless they occur in a promoter or regulatory region or
alternatively impair mRNA splicing efficiency. Such changes are not usually subject to the effects of negative selection and may therefore accumulate in the
genome. Using anonymous DNA segments, Cooper *et al.* (1985) estimated the
unique DNA sequence heterozygosity in the human genome to be 0.0037, indicating that ~1/250 bases vary polymorphically in human populations and that
most of these variants may be expected to occur in noncoding regions. More
recently, and employing rather larger sample sizes, other workers have estimated
that polymorphisms in the human genome occur with a frequency of between
1/200 and 1/1000 bp (Collins *et al.*, 1997; Li and Sadler, 1991; Nickerson *et al.*,
1998; Wang *et al.*, 1998; see Chapter 1, section 1.2.2).

   In coding regions, single base-pair substitutions that are silent are termed *synonymous* in that they do not change the amino acid sequence of the gene product.
Most of these changes occur at the third base positions of codons. Such mutations
are likely to be neutral with respect to fitness, assuming that they do not alter gene
expression or mRNA splicing, stability or transport. *Nonsynonymous* mutations,
on the other hand, alter the amino acid sequence of a polypeptide and can be
either *conservative* or *nonconservative*. Conservative substitutions (e.g. Asp→Glu,
Leu→Val) result in the replacement of one amino acid by another that is chemically similar to it and may thus alter protein structure and function only mini-
mally (Creighton, 1993). Nonconservative substitutions (e.g. Arg→Gly,
Arg→Pro), however, result in the replacement of one amino acid by another that
possesses a chemically dissimilar side chain thereby altering either its charge or
its polarity. Nonconservative substitutions are likely to have a more deleterious
effect than conservative substitutions and will therefore tend to be removed from
the population with a higher probability (Creighton, 1993). The rates of nonsynonymous and synonymous substitutions per site have been obtained for a variety
of eukaryotic genes and are of the order of $0-2 \times 10^{-9}$ and $2-12 \times 10^{-9}$, respectively
(MacIntyre, 1994; Li, 1997). The study of 1880 human-rodent orthologue pairs
yielded average estimates of 0.52×10^{-9} and 2.92×10^{-9} for nonsynonymous and
synonymous substitutions, respectively (Makalowski and Boguski, 1998). In general, eukaryotic pseudogenes evolve at approximately the same rate as the synonymous sites in functional genes (Li *et al.*, 1981), a rate that can therefore be
considered to represent the *neutral rate of substitution*.

   Owing to the nonrandom nature of the genetic code, different amino acids are
encoded with different degrees of degeneracy. Base positions in codons can thus
belong to any one of three classes:

(i)   *Nondegenerate sites* (~65% of base positions in human codons): all possible substitutions are nonsynonymous. The base substitution rate at these sites is very low (*Figure 7.1*) on account of the strong selection pressure to avoid amino acid exchanges.

(ii) *Two-fold degenerate sites* (~19% of base positions in human codons): base positions at which only one of the three possible substitutions is synonymous.

(iii) *Four-fold degenerate sites* (~16% of base positions in human codons): base positions in which all three possible substitutions are synonymous. The base substitution rate at these sites is higher than those noted for non-degenerate sites and two-fold degenerate sites (*Figure 7.1*). The comparison of human and rodent genes has allowed the rates of transitional and transversional silent substitutions in four-fold degenerate sites to be estimated (1.71×10^{-9} and 1.22×10^{-9} per site per year, respectively) since the human-rodent divergence (Collins and Jukes, 1994).

7.1.3 Positive and negative selection in protein evolution

> It may be said that natural selection is daily and hourly scrutinizing, throughout the world, the slightest variations; rejecting those that are bad, preserving and adding up all that are good; silently and invisibly working, whenever and wherever opportunity offers, at the improvement of each organic being in relation to its organic and inorganic conditions of life.
> Charles Darwin (1859) *The Origin of Species*

Protein coding genes exhibit considerable variation in their rates of synonymous (k_s) and nonsynonymous (k_a) substitutions (reviewed by Li, 1997). The k_s/k_a ratio serves as a rough guide to the degree of evolutionary conservation and therefore the functional constraints placed upon a protein by natural selection (Li *et al.*, 1985). Thus, some human gene sequences are very highly conserved (k_s/k_a ratio high), for example ubiquitins (Vrana and Wheeler, 1996), histones H3 and H4 (Thatcher and Gorovsky, 1994), ribosomal proteins (De Falco *et al.*, 1993), *H19* (Hurst and Smith, 1999), calmodulin (Thomas and Wilson, 1991), and the G protein α-subunits (Yokoyama and Starmer, 1992) indicating that these proteins have evolved under negative or purifying selection. By contrast, other human/primate genes have evolved very rapidly (k_s/k_a less than unity), for example the sex determining locus, *SRY* (Yp11.3; Whitfield *et al.*, 1993), and those genes encoding apolipoprotein C-I (*APOC1*; 19q13.2; Pastorcic *et al.*, 1992), protamines P1 and P2 (*PRM1, PRM2*; 16p13.3; Retief and Dixon 1993), myelin proteolipid protein (*PLP*; Xq22; Kurihara *et al.*, 1997), pregnancy-specific glycoprotein 1 and carcinoembryonic antigen (*PSG1, CEA*; 19q13.2; Streydio *et al.*, 1990; Teglund *et al.*, 1994), eosinophil cationic protein (*RNASE3*; 14q24-q31; Zhang *et al.*, 1998), the rhesus blood group genes (*RHD*, 1pp34-p36.2; *RHAG*, 6p11-p21.1; Kitano *et al.*, 1998) and β-microseminoprotein (*MSMB*, 10q11.2; Nolet *et al.*, 1991). Such examples of a significantly higher rate of nonsynonymous nucleotide substitution than synonymous substitution provide strong evidence for the action of positive selection.

Lysozyme, originally a bacteriolytic enzyme whose origin preceded the emergence of the vertebrates, has been independently recruited as a digestive enzyme by two groups of mammals, the ruminant artiodactyls and the leaf-eating colobine monkeys (Prager, 1996; Stewart *et al.*, 1987). Both groups are able to ferment plant material in the foregut and possess stomachs that contain high levels of lysozyme. Advanced ruminants such as cows, sheep and deer possess ~10 lysozyme genes as a result of gene amplification events which occurred after the divergence from the pig lineage. Sequence comparison of the lysozyme genes from human (*LYZ*; chromosome 12) and other primates has indicated a k_s/k_a ratio significantly less than unity in both the colobine and hominoid lineages (Messier and Stewart, 1997), indicative of the action of positive selection for amino acid replacements. Interestingly, several of the amino acid replacements noted in the colobine lineage also occurred in parallel or convergently in the ruminant artiodactyls (Messier and Stewart, 1997; Stewart *et al.*, 1987; Swanson *et al.*, 1991). Comparison of the nucleotide sequences of the coding regions of the lysozyme genes of advanced ruminants and pigs revealed no difference in the rate of synonymous substitution consistent with the view that it was a change in selective pressure rather than the mutation rate that was responsible for changes in the rate of stomach lysozyme evolution (Yu and Irwin, 1996). The reasons for positive selection for lysozyme amino acid replacements in hominoids are at present unclear but Messier and Stewart (1997) suggested that it might have been associated with the increased neutrophil expression of lysozyme in hominoids as compared with other catarrhines.

Since synonymous substitutions are likely to be neutral with respect to selection, they have been employed in numerous attempts to calibrate *molecular clocks* (Easteal *et al.*, 1995 Fitch and Ayala, 1994; Li, 1997). However, substitution rates can vary quite widely between orthologous gene sequences in different taxonomic groups or lineages, as well as between different genes in the same species (Britten, 1986; Easteal, 1988; Easteal and Collet, 1994; Gibbs and Dugaiczyk, 1994; Li *et al.*, 1990; Li *et al.*, 1996; Ohta and Ina, 1995). The speed of the molecular clock appears to vary according to the lineage. In order to estimate the relative rates of nucleotide substitutions in two lineages leading to extant species A and B, a *relative rate test* is employed. This involves the use of a third (reference) species, C, which is known to have branched off prior to the divergence of A and B. Pairwise comparisons of orthologues in A and C, and in B and C, are used to calculate the k value, the number of substitutions per 100 sites. The k_{AC} and k_{BC} values then provide a measure of the relative rates of mutation in the lineages leading to species A and B, respectively. Such calculations have suggested that the substitution rates in the lineages leading to mouse and rat are approximately equal (~7.9 × 10^{-9} (3.9–11.8); Li and Graur, 1991; O'hUigin *et al.*, 1992; Wolfe and Sharpe, 1993) whereas comparable estimates for humans and Old World monkeys (~2.2 × 10^{-9} (1.8–2.8)) and humans and chimpanzees [~1.3 × 10^{-9} (0.9–1.9)] are considerably lower. This is held to be indicative of a slowdown in the substitution rate in primates which appears to be at its greatest in hominoids (Bailey *et al.*, 1991; Ellsworth *et al.*, 1993; Gu and Li, 1992; Koop *et al.*, 1986; Li and Graur, 1991; Li and Tanimura, 1987; Li *et al.*, 1996; Seino *et al.*, 1992). Despite these data, the *hominoid slowdown* is not a universal phenomenon in primate evolution since it is not apparent with some gene sequences (Easteal, 1991; Kawamura *et al.*, 1991; Shaw *et al.*, 1989).

Some genes/proteins show a rather different, discontinuous pattern of evolution. Thus, for example, in both primates and artiodactyls, the growth hormone (*GH1*; 17q23) gene exhibits a pattern of near stasis punctuated by bursts of rapid evolution during which the rate of evolutionary change has increased at least 25-fold (Ohta, 1993; Wallis, 1994, 1996). Interestingly, during mammalian evolution, only the coding region of the growth hormone gene corresponding to the mature protein has exhibited rapid change whereas other regions of the gene (signal peptide, nonsynonymous substitutions and 5' and 3' untranslated regions) have remained relatively unchanged. This would be consistent with the rapid bursts of evolution being of adaptive significance and Wallis (1997) suggested that this may have involved what he termed *function switching*. Briefly, if primate GH were to have had biological functions other than growth promotion viz. lactogenic (prolactin-like) activity, GH could then also have played a role in maintaining the nutritional balance between mother and young. Acquisition of its lactogenic function could have involved changes in GH structure away from that best adapted to growth regulation to one that represented an evolutionary compromise which optimized the protein for its dual function. If the secondary function of GH had then changed (e.g. the adoption of a pattern of seasonal breeding following migration or climatic change, in which gestation and suckling no longer overlapped), selection for dual function would no longer have operated and the selection pressure on GH structure would have been related to its growth promoting function alone. If several amino acid substitutions had already occurred, then the process of functional reversal involving simply the reversal of each amino acid change would have been very unlikely. Rather, GH would have adopted a new structural form, slightly different from the original, but nevertheless adapted to its primary function. Switching back to dual function might then have led to a new cycle of structural change, driven by selection but without any overall change in function. This 'pushme-pullyou' mechanism of *function switching* can in principle be resolved by gene duplication with subsequent divergence of the paralogues to perform different functions. Indeed, the gene duplications giving rise to the GH cluster occurred after the rapid burst of evolution in primates. Once duplication had occurred, the rapid evolution of the pituitary expressed *GH1* gene would probably have ceased although the relatively high evolutionary rate would still have been maintained by the placentally expressed genes. Whilst this idea is very appealing, it should be pointed out that the evolution of the placental lactogens early on in the mammalian radiation might reasonably be expected to have removed any selective pressure on the lactogenic properties of GH.

Another example of the intermittent acceleration and deceleration of the nucleotide substitution rate is provided by the cytochrome *c* oxidase subunit IV (*COX4*; 16q22-qter) gene at different stages of primate evolution (Wu *et al.*, 1997). In passing, it is interesting to note that conceptually, the idea of function switching had its origin in the days of the early evolutionists:

> It is an error to imagine that evolution signifies a constant tendency to increased perfection. That process undoubtedly involves a constant remodeling of the organism in adaptation to new conditions; but it depends on the nature of those conditions whether the direction of the modifications effected shall be upward or downward.
>
> T.H. Huxley (1888) *The Stuggle for Existence in Human Society*

Intriguingly, Miyata *et al.* (1994) have claimed that the rate of evolutionary change of tissue-specific genes varies with the site of expression. Thus, brain-specific genes were reported to evolve at a slower rate than immune system-specific genes. However, since the identity of the compared genes was not revealed and the sample size small, it is hard to comment on the validity or otherwise of the authors' conclusions.

7.1.4 Mutation rates and their evolution

> The fundamental importance of mutation for any account of evolution is clear. It enables us to escape from the impasse of the pure line. Selection within a pure line will only be ineffective until a mutation arises.
> J.B.S. Haldane *The Causes of Evolution* (1932)

Although very much dependent upon the type of mutation being considered and the identity of the gene in question, studies of disease pathology have suggested that the mutation rate in males is higher than that in females (reviewed by Cooper and Krawczak, 1993). This has been held to reflect the rather higher number of cell divisions required from zygote to mature germ cell in the male as compared to the female. In an evolutionary context, Miyata *et al.* (1987) developed a method for estimating the male-to-female ratio of mutation rates (α_m) from rates of nucleotide substitution in sex-linked and autosomal sequences. They predicted that if α_m were very large, the rate of synonymous substitution in X-linked genes would be ~2/3 of that in autosomal genes on the basis that the X chromosome is twice as likely to be present in a female as in a male. This prediction was borne out by their analysis of human and rodent gene sequences. Further, the rate at which the Y chromosome accumulated substitutions was found to be twice that of the autosomes (the Y chromosome mutates at relative rate α_m as mutation always occurs in a male). The comparison of rodent *Ube1* genes and pseudogenes allowed α_m to be estimated to be of the order of 2.0 (Chang and Li, 1995). Similarly, comparison of *SMCX/SMCY* (Xp11.21-p11.22/Yq) genes from mouse, horse and human yielded an estimate for α_m of 3.0 (Agulnik *et al.*, 1997). By contrast, comparison of intronic sequences of the homologous *ZFX* (Xp21.3-p22.3) and *ZFY* (Yp11.32) loci in humans, orangutans, baboons and squirrel monkeys suggested that α_m may be as high as 6.0 in primates (Shimmin *et al.*, 1993). The actual value of α_m may however be very much dependent upon the species compared. Thus, η-globin pseudogene sequence data were used to derive estimates of α_m between 3 and 6 in higher primates but only ~2 in mice and rats (Li *et al.*, 1996). If studies of many different systems indicate that the mutation rate is consistently higher in males than in females, then it may indeed be possible to view evolution as being 'male-driven' (Hurst and Ellegren, 1998).

That some sequences are hypermutable in the human genome is clear from studies of pathological lesions responsible for human genetic disease (Cooper and Krawczak, 1993; Krawczak *et al.*, 1998). Hotspots for somatic mutation are also apparent from studies of mammalian immunoglobulin genes (consensus sequences, G C/T A/T and TAA; Rogozin and Kolchanov, 1992). Such hotspots can be interpreted in terms of nearest neighbor effects on nucleotide substitution rates (Krawczak *et al.*, 1998; Section 7.5.1). In an evolutionary context, these effects are also apparent from sequence comparison studies of human genes and pseudogenes (Blake and Hess, 1992; Hess *et al.*, 1994).

The mutation rate in higher organisms has long been assumed to be a compromise between keeping the frequency of deleterious substitutions low at the same time as not completely abolishing the potential for generating adaptive variation. One prediction of this *trade-off theory* is that a lower equilibrium mutation rate should evolve if the deleterious effect of mutation is increased. To examine the validity of this postulate, McVean and Hurst (1997) compared rates of synonymous nucleotide substitution in 33 X-linked genes and 238 autosomal genes in mouse and rat. Since the X chromosome is hemizygous in male mammals, deleterious recessive mutations arising on it might reasonably be expected to have a greater effect on fitness than those arising on the autosomes. If, however, synonymous substitutions were completely neutral with respect to fitness, then they should accumulate at a rate equal to the mutation rate and could be used directly to estimate mutation rates. McVean and Hurst (1997) found that the X-linked genes exhibited significantly lower rates of synonymous substitution than their autosomal counterparts and this was held to be explicable in terms of the X-chromosomal mutation rate being reduced by natural selection.

Mutational pressure may have influenced the evolution of the eukaryotic genetic code in that the code's organization at least appears to provide at least some protection against the deleterious consequences of single base-pair substitutions (Goldman, 1993; Haig and Hurst, 1991; Jukes, 1993; Kuhn and Waser, 1994; Osawa and Jukes, 1988; see Section 7.2). However, the converse argument is more persuasive: that the design of the genetic code and the functional similarity (or dissimilarity) of amino acids to one another may have affected the relative mutabilities of individual codons during evolution. Codon usage varies quite widely between different vertebrates (Nakamura *et al.*, 1998; CUTG Database at **http://www.dna.affrc.go.jp/~nakamura/CUTG.html**), a finding which may be related to the influence of the codon frequencies of highly expressed genes on translation efficiency via tRNA pools (Britten, 1993). Codon usage may also vary between different genes in the same species. In humans, such a finding has been suggested to be related to chromosomal location (D'Onofrio *et al.*, 1991; see Section 7.4).

7.1.5 The deleterious mutation rate in humans

> Man's yesterday may ne'er be like his tomorrow;
> Nought may endure but mutability.
> Percy Bysshe Shelley (1816) *Mutability*

It has been suggested that humans may experience a high deleterious mutation rate (Crow, 1997; Kondrashov and Crow, 1993). If a significant proportion of these mutations were even mildly deleterious, such lesions would tend to accumulate in populations with small effective sizes or in which selection had been relaxed. Eyre-Walker and Keightley (1999) estimated the human deleterious mutation rate per diploid genome per generation, U, by comparing the expected and observed rates of nonsynonymous substitution in 46 orthologous gene pairs from human and chimpanzee. Under conservative assumptions of 60 000 genes in the human genome, an average length of protein coding sequence of 1.52 kb, a human-chimpanzee divergence time of 6 Myrs ago, and an average generation time of 25 years in the human lineage, Eyre-Walker and Keightley (1999) estimated the rate of nonsynonymous substitutions, M, to be 4.2 ± 0.5 mutations per diploid genome per generation, and the deleterious mutation rate, U, to be 1.6 ± 0.8 mutations per diploid genome per generation. Estimates of U for chimpanzee (1.7 ± 0.8) and gorilla (1.2 ± 0.6) were found to be similar to that obtained for humans. Eyre-Walker and Keightley (1999) concluded that the human deleterious mutation rate is close to the upper limit tolerable by a species with a low reproductive rate. This implies that in hominids, synergistic epistasis may have occurred between deleterious mutations. Further, the level of selective constraint (U/M) in human protein coding sequences (0.38 ± 0.17) was judged to be atypically low as were estimates from chimpanzee (0.53 ± 0.16) and gorilla (0.38 ± 0.17). A large number of slightly deleterious mutations may therefore have become fixed in the hominids and the most likely explanation for this, is small long-term effective population size.

7.2 Mutations in pathology and evolution; two sides of the same coin

Genotypes never have votes. Phenotypes sometimes do.
A.L. Mackay

In the early days of genetics, many thinkers saw spontaneous variation purely and simply in terms of its role as the evolutionary fuel for speciation. The first to draw parallels with disease was probably the British geneticist William Bateson who, at the turn of the century, maintained that disease presented 'a discontinuity closely comparable with that of many variations'. Indeed, he speculated

that the problem of species [may well] be solved by the study of pathology; for the likeness between variation and disease goes far to support the view which Virchow has forcibly expressed, that 'every deviation from the type of the parent animal must have its foundation on a pathological accident'
Bateson (1894)

Couched in modern terms, single base-pair substitutions in human gene pathology and evolution may be viewed as two sides of the same coin. This appealing supposition has recently been corroborated by an in-depth comparison of mutations causing inherited disease with mutations in noncoding DNA that have become fixed during the evolutionary divergence of human and other mammalian genomes (Krawczak and Cooper, 1996a). Mutations in noncoding DNA may date back some millions of years and their survival has been independent of natural selection. By contrast, disease-associated substitutions are of fairly recent origin by comparison with the evolutionary timescale. This difference notwithstanding, under the assumption that the underlying molecular mechanisms of mutation have not changed substantially during mammalian evolution, some resemblance between the two mutational spectra was to be expected. Consistent with this explanation, the relative base substitution rates observed in the context of human genetic disease (Cooper and Krawczak, 1993; Krawczak and Cooper, 1996a) were found to be remarkably similar to those derived by Hess *et al.* (1994) in an extensive analysis of human gene/pseudogene alignments. This pattern of similarity was still apparent after allowance had been made for the DNA sequence environment at the site of mutation by the use of nearest neighbor-dependent mutation rates (Krawczak and Cooper, 1996a); the only notable difference was a slight under-representation among pseudogene mutations of C→T and G→A substitutions. This was held to be explicable in terms of either a lower level of germline methylation or, perhaps more likely, a deficiency of the 5mC-containing CG dinucleotides ('CG suppression'; Bird, 1980) in non-coding DNA sequences.

The claim that relative mutation rates exhibit a strong positive correlation between pathological mutations in coding sequences and evolutionarily fixed mutations in non-coding sequences (Krawczak and Cooper, 1996a) might appear surprising in the light of other experimental results. For example, a study of murine *aprt* gene sequences has claimed that silent mutations accumulate more slowly in transcribed sequences, possibly due to preferential DNA repair (Boulikas, 1992; Turker *et al.*, 1993). Further, Hanawalt (1990) has shown that *in vitro*-induced pyrimidine dimers and interstrand DNA crosslinks are repaired with a substantially higher efficiency in active genes than in noncoding regions. Although this type of lesion is specific to the action of particular exogenous chemical mutagens and irradiation, the idea of a system which is generally more effective at removing endogenous mutations from coding DNA as opposed to noncoding DNA is appealing. This is because efficient DNA repair should only have conferred a substantial selective advantage in coding regions. By contrast, the results of Krawczak and Cooper (1996a) suggest that the relative contribution (via variable efficiency) of different DNA repair pathways to the generation of mutations is unlikely to differ substantially between intragenic and intergenic sequences.

To what extent is the likelihood of generation of a mutation related to its phenotypic consequences? When codon substitutions causing genetic disease were categorised according to whether they were neutral or whether they changed the hydrophobicity or polarity of the encoded amino acid residue, it emerged that neutral changes were characterized by larger likelihoods of generation via mutation than nonneutral substitutions (Krawczak and Cooper, 1996a). This disparity suggested that selection has operated on the cellular DNA repair machinery in such a way as to optimize the removal of the latter type of mutation. If nonneutral changes were more likely to result in a disadvantageous (disease) phenotype than neutral substitutions, then any repair bias operating against these changes at the DNA level would have had a selective advantage.

The hypothesis of a mutational repair bias was further supported by the finding that the likelihood of generation of an amino acid substitution in humans is negatively correlated with its likelihood of coming to clinical attention (Krawczak and Cooper, 1996a). The extent of the correlation was, however, found to differ dramatically between different types of substitution. Thus, a significant decrease in mutation generation likelihood was only associated with an increased likelihood of clinical observation for substitutions which affected hydrophilic and polar residues. Various explanations may be considered to account for these findings. First, substitutions of hydrophilic and polar amino acid residues could have resulted in more severe and/or variable consequences for the phenotype than other types of substitution during most of the time period that the DNA repair mechanisms were evolving. Second, it has been observed that, during evolution, the effect of an amino acid change in the hydrophobic protein core is often compensated for by another change in the immediate vicinity (Schirmer, 1979). Thus, any selective pressure acting upon the organism to avoid substitutions at hydrophobic core residues may have been balanced by the requirement for evolutionary fixation of a second compensatory mutation. Finally, the majority of adaptive events shaping the eukaryotic DNA repair process will have occurred in organisms other than human. It therefore follows that, were current clinical observation likelihoods to be specific to human, this could have obscured any correlation between mutation generation likelihoods and the phenotypic consequences of substitutions at hydrophobic or nonpolar residues.

If mutations in human genes are biased against amino acid replacements with a high present day probability of resulting in a disease phenotype, the question arises as to whether such a bias might also be reflected in the evolutionary history of protein sequences. To explore this possibility, Krawczak and Cooper (1996a) calculated a quantity termed the *relative evolutionary acceptability* for each possible amino acid substitution from the data of Collins and Jukes (1994) who had reported the numbers and types of amino acid mismatches deduced from the alignment of 337 pairs of human-rodent cDNA sequences. Krawczak and Cooper (1996a) found that clinical observation likelihoods were negatively correlated with evolutionary acceptability values. This implied that the more likely a given amino acid substitution was to result in a disease phenotype (at least in contemporary humans), the less often has it been tolerated during the evolution of mammalian protein sequences.

In summary, it is evident that the evolutionary requirement to avoid a deleterious phenotype has left its footprints in the mechanisms of mutation generation. In this context, the most promising target for selection would appear to be the intracellular DNA repair mechanism. Although the effect of a given amino acid replacement upon protein structure is known to be heavily dependent upon its precise location within the tertiary structure of the molecule (Alber, 1989; Pakula and Sauer, 1989; Wacey *et al.*, 1994), some basic rules which relate local causes and consequences may nevertheless be perceived (De Filippis *et al.*, 1994). If the efficiency of mutation removal were directed by the immediate DNA sequence context of a lesion, it may be that this has facilitated the avoidance during evolution of hazardous amino acid replacements by consideration of the genetic code.

7.3 The importance of evolutionary conservation in the study of pathological mutations at the protein level using human factor IX as a model

The nature, frequency and location of gene lesions causing human genetic disease are highly specific and, as outlined in Chapter 1, section 1.5, are determined in part by the local DNA sequence environment. Once a given mutation has arisen, however, the likelihood that it will come to clinical attention is a complex function of the nature of the resulting amino acid substitution, its precise location and immediate environment within the protein molecule, and its effects upon protein structure and function. Although in-depth investigations of the *in vivo* effects of missense mutations upon specific human proteins are generally rare, two such studies have nevertheless been performed for human factor IX (Bottema *et al.*, 1991; Wacey *et al.*, 1994), the liver-expressed zymogen of a vitamin K-dependent serine protease that activates factor X in the presence of factor VIIIa. These studies will be discussed in some detail.

The vast majority of known lesions in the *F9* (Xq26-q27) gene causing hemophilia B are missense mutations (Giannelli *et al.*, 1996), causing ~59% of severe (<1% FIX: C) and moderate (1–5% FIX: C) hemophilia, and perhaps as much as 97% of mild (>5% FIX: C) hemophilia (Sommer *et al.*, 1992). On the basis of 95 independent missense mutations, Bottema *et al.* (1991) concluded that substitutions of 'generic' factor IX residues (conserved in factor IX of other mammals and in related human serine proteases) almost invariably cause hemophilia B. Mutations at factor IX-specific residues (conserved only in mammalian factor IX) and nonconserved residues were, by contrast, found to be some six to 33-fold less likely to result in a disease phenotype. Even though the study of Bottema *et al.* (1991) provided new insights into the identity of amino acid residues of structural or functional importance to factor IX, the authors did not employ models of the tertiary structure of the protein or its constituent domains. The significance of the location of specific amino acid residues within the structure of the factor IX molecule to the consequences of mutation could therefore not be assessed. Moreover, neither the variable propensity of different regions of the *F9* gene to mutate nor the nature of the resulting amino acid exchanges were considered.

In many ways, factor IX represents an ideal system in which to assess the influence of positional determinants upon the disease-associated mutational spectrum of a single protein. Firstly, the number of known *F9* missense mutations is among the highest of all human genes (*Human Gene Mutation Database*; **http://www.uwcm.ac.uk/uwcm/mg/hgmd0.html**). Secondly, the amino acid sequences of numerous other vertebrate factor IX proteins and evolutionarily related serine proteases are available for direct comparison (Sarkar *et al.*, 1990; Bottema *et al.*, 1991). Finally, the structure of factor IX has been determined by X-ray crystallography (Brandstetter *et al.*, 1995) and the three-dimensional structures of a number of homologous serine proteases are also known. Wacey *et al.* (1994) constructed by comparative methods (Swindells and Thornton, 1991), a multidomain model of the quaternary structure of activated factor IX (FIXa) and used this model to study the expression pathway of *F9* gene lesions from genotype to clinical phenotype: a total of 277 different single base-pair substitutions in the *F9* gene, comprising 241 missense mutations and 36 nonsense mutations, were analysed. Comparison of the relative nearest neighbor-dependent single base-pair substitution rates in the *F9* gene with estimates derived from a wide range of other human genes revealed similar profiles (with CpG dinucleotides representing hotspots for mutation), suggesting that similar mutational mechanisms were operating at the DNA level.

Wacey *et al.* (1994) classified *F9* missense mutations as either *conservative* or *nonconservative* on the basis of the chemical difference between the wild-type and the mutant amino acid residue. *Chemical difference* is a measure, originally devised by Grantham (1974), that combined the three interdependent properties of chemical composition, polarity and molecular volume of an amino acid residue. When the magnitude of biochemical change upon substitution was measured in this way and related to the clinical severity of the resulting disease phenotype, nonconservative substitutions (i.e. those characterized by large chemical differences) were found to result in severe rather than mild or moderate hemophilia B approximately 1.7 times more often than conservative substitutions. Conversely, conservative substitutions were 3- to 4-fold more likely to be associated with a moderate or mild phenotype than their nonconservative counterparts. The possibility was considered that conservative amino acid substitutions might be more likely to come to clinical attention in the tightly packed core of the protein as opposed to the surface of the molecule where they mght be more readily tolerated. However, since mutations from all chemical difference classes appeared to be scattered over all domains of the protein and since no one chemical difference class was found to be associated solely with surface or buried residues, there appeared to be no relationship between the magnitude of the amino acid exchange and the location of the affected residue (Wacey *et al.*, 1994).

The extent of evolutionary sequence conservation exhibited by amino acid residues in factor IX was also found to correlate with disease severity. Whilst 71% of mutations at 'highly conserved' residues (residues conserved in all mammalian factor IX proteins and in three other serine proteases) caused severe rather than mild or moderate hemophilia B, this was the case for only 50% of mutations at less conserved residues. Furthermore, Wacey *et al.* (1994) estimated that missense mutations at non-conserved residues were 15–20 times less likely than mutations at conserved residues to result in a disease phenotype at all. Although this implies that many missense mutations at evolutionarily unconserved residues are tolerated by the molecule and do not come to clinical attention, the relative importance of such residues was considered to be greater than previously claimed by Bottema *et al.* (1991). Several explanations were suggested for this discrepancy. Firstly, the sample of mutations used by Wacey *et al.* (1994) was three times larger than that of Bottema *et al.* (1991). Secondly, Wacey *et al.* (1994) allowed for determinants neglected by Bottema *et al.* (1991) viz. the actual *F9* gene coding sequence and the redundancy of the genetic code. Finally, the two studies were not directly comparable since Wacey *et al.* (1994) confined their estimation of clinical observation likelihoods to severe cases of hemophilia B in order to cope with the problem of identical-by-descent mutations which are likely to be more prevalent in cases of mild or moderate disease.

Amino acid residues which are sequence conserved both in mammalian factor IX proteins and four different human serine proteases are likely to be critical for functions common to all serine proteases. These residues are located predominantly in the interior of factor IX, within α-helices or β-turns, and Wacey *et al.* (1994) found that substitutions tended to cluster in the Gla domain, the EGF domains and the serine protease domain (around the reactive site and oxyanion hole). All but one of the Cys residues known to be involved in disulfide bonding were affected by mutation as were the reactive site residues, residues involved in carboxylase recognition and activation peptide cleavage site, and residues which contribute to factor X binding. By contrast, mutations at factor IX residues which are sequence conserved in mammals and one or two other serine proteases were found to cluster in the serine protease domain and also at domain boundaries within the protein structure. Presumably, docking of the constituent domains of factor IX and the other homologous serine proteases during protein folding requires amino acid conservation at these boundaries. Finally, factor IX residues which are sequence conserved between mammals but not between other serine proteases may be exclusively important for the structure and function of factor IX. When the 3D structure of the factor IX protein was considered, spatially clustered groups of mutations at such residues became apparent on the surface of the EGF domains, in regions implicated in the binding of factors Va and VIIIa, and of the serine protease domain.

Regions which exhibit similar functions in different homologous proteins may not only be sequence conserved but may also exhibit structural conservation (Greer, 1990). Although *sequence conserved regions* (SeqCRs) of mammalian serine proteases are invariably structurally conserved, *structurally conserved regions* (SCRs) may differ markedly with respect to their amino acid sequences. SCRs were defined by Greer (1990) as those portions of known protein structures that overlap very well when superimposed. In serine proteases, SCRs usually comprise secondary structure elements, the active site and other essential structural framework residues of the molecule.

The locations of structurally conserved regions in human factor IXa were determined by Wacey *et al.* (1994) employing their homology model of the factor IX protein. No clear relationship was noted between the severity of the hemophilia B phenotype and the level of structural conservation of a mutated factor IX amino acid residue, although substitutions at structurally conserved residues were estimated to have an approximately two-fold higher likelihood of resulting in a disease state than mutations at nonconserved residues. Interestingly, mutated sites which were not sequence-conserved were nearly all structurally conserved. The only exception involved two missense mutations at Gly59. However, Gly59 lies immediately adjacent to a type β hairpin SCR and would be predicted to be critical in defining this structural element (Swindells and Thornton, 1991). Some SCRs, although not sequence-conserved, may thus serve as structural supports through their backbone interactions and should therefore be regarded as 'scaffolding' residues rather than 'spacers' (Bottema *et al.*, 1991).

The topological properties of a mutated factor IX amino acid residue are also important for determining clinical severity. Mutations at residues with their side chains pointing away from the solvent ('buried residues') were found to cause severe hemophilia 1.5 times more often than mutations at residues with solvent-accessible side chains (Wacey *et al.*, 1994). Finally, the likelihood of a mutation resulting in a severe disease phenotype was higher for substitutions in hydrophobic as opposed to polar regions, probably because of the critical importance of these residues for correct protein folding (Kragelund *et al.*, 1999). This is consistent with the conclusions of other workers that amino acid substitutions occurring in the protein core give rise to a 'continuum of increasingly non-native properties' affecting the stability and/or the folding dynamics of the protein (Alber, 1989; Lim *et al.*, 1992; Pakula and Sauer, 1989).

7.4 Equilibrium of synonymous codon substitutions

Single base-pair substitutions in coding regions that do not change the encoded amino acid sequence can be assumed to be comparatively free of selectional constraints (Creighton, 1993). Although it cannot be entirely excluded that these synonymous (silent) changes might influence gene expression via effects on mRNA translation (e.g. via alterations in RNA secondary structure or local imbalances in the tRNA reservoir of a cell), the probability of survival and ultimate population fixation should in general be much higher for silent mutations than for missense mutations (Nei, 1987). This view is consistent with the fact that the vast majority of evolutionarily stable base substitutions in coding regions of human genes have taken place at the wobble positions of degenerate codons (Wilbur, 1985).

There are 19 groups of triplets which encode the same amino acid such that the constituent triplets of each group can be replaced by each other via single base-pair substitution (see *Table 7.1*; Krawczak and Cooper, 1996b). If mutations that cause an amino acid exchange are ignored, the mutation dynamics within each group of codons can be modelled by a simple system of linear equations involving the relative rates of different single base-pair substitutions. With evolutionary time, this system will approach an equilibrium state and the equilibrium codon frequencies within each group can be determined by solving this system of equations. As can be inferred from *Table 7.1*, the actual frequencies within degenerate codons are still some distance from equilibrium in humans.

When a similar analysis is performed for other vertebrate species with known codon usage (Wada *et al.*, 1991), it turns out that humans are not the closest to their own equilibrium. In 17/19 cases, *Xenopus laevis* ranks first whereas humans and rodents form a second group of species, all ranking equally low (Krawczak

Table 7.1. Euclidean distance between the vectors of current and equilibrium frequencies within degenerate codons of human genes (from Krawczak and Cooper, 1996b)

Encoded amino acid	Codon group	Euclidean distance
Glu	GAA GAG	0.083
Lys	AAA AAG	0.088
Asp	GAT GAC	0.164
Asn	AAT AAC	0.173
Tyr	TAT TAC	0.196
His	CAT CAC	0.222
Pro	CCT CCC CCA CCG	0.240
Thr	ACT ACC ACA ACG	0.275
Gln	CAA CAG	0.294
Ser	TCT TCC TCA TCG	0.295
Cys	TGT TGC	0.311
Gly	GGT GGC GGA GGG	0.314
Ala	GCT GCC GCA GCG	0.318
Phe	TTT TTC	0.367
Arg	CGT CGC CGA CGG AGA AGG	0.371
Ser	AGT AGC	0.416
Leu	TTA TTG CTT CTC CTA CTG	0.431
Val	GTT GTC GTA GTG	0.444
Ile	ATT ATC ATA	0.540

and Cooper, 1996b). This class is followed by chicken, dog, cow, pig, rabbit and sheep in order of distance from equilibrium. There is thus some correlation between distance from equilibrium and generation time and, with the exception of humans, the ranking of species reflects the total synonymous substitution rates estimated by Bulmer *et al.* (1991). These data are therefore consistent with a model of DNA sequence evolution which, after species divergence, allows ancestral gene sequences to approach equilibrium codon usage faster in one species than in another.

If current codon usage in different species were indeed the result of the divergent evolution of common ancestral sequences progressing at different absolute (albeit equal relative) substitution rates, then one species should always be closer to equilibrium than the other in all codon groups. This, however, is not the case. For example, hamster is closer to equilibrium than dog for the tyrosine (TAT, TAC) and histidine (CAT, CAC) encoding triplets, but not for glutamine (CAA, CAG). Furthermore, the codon frequencies for lysine (AAA, AAG), aspartic acid (GAT, GAC) and glutamic acid (GAA, GAG) in *Xenopus* are on opposite sites of the equilibrium when compared to all other vertebrates (Krawczak and Cooper, 1996b).

One may therefore surmise that codon usage has not evolved in a strictly uniform way. Although it cannot be excluded that relative mutation rates differ between species thereby resulting in different equilibria, the fact that *Xenopus* and rodents are very close to an equilibrium which is itself based upon human genetic disease data, argues strongly against this objection. It is thus more likely that species divergence has been accompanied by substantial changes in codon usage, allowing some species to manifest sudden changes with respect to their distance from equilibrium. This would also be consistent with the finding that differences in synonymous codon usage between vertebrates is not explicable merely by differential absolute DNA repair efficiency (Eyre-Walker, 1994).

Ikemura and Wada (1991) were able to demonstrate through an analysis of approximately 2000 human gene sequences that codon usage differs dramatically between different genomic regions. The major proportion of GC-rich genes was observed in T-bands whereas AT-rich genes were located mainly in G-bands. Further, the average $G+C$ percentage at the third position of codons was found to be related to the quinacrine dullness and the mitotic chiasma density of a particular chromosome. Since species divergence has almost always been characterized by gross structural chromosomal rearrangements, and since different genes are known to evolve at different absolute rates even in one and the same species (Bulmer *et al.*, 1991), the piece-wise reconstitution of new genomes from their common ancestors is likely to have altered codon frequencies substantially.

7.5 Single base-pair substitutions in gene regions

> The smallest changes that add to or subtract from a part in the smallest measurable degree may also arise by mutation. We identify these smaller mutational changes as the most probable variants that make a theory of evolution possible both because they do transcend the original types, and because they are inherited.
>
> T. H. Morgan (1925) *Evolution and Genetics*

7.5.1 Neighboring-nucleotide effects on the rate of germline single base-pair substitutions in human genes

In terms of their relative frequency of occurrence, the most important category of single base-pair substitution causing human genetic disease is represented by C→T and G→A transitions within CpG dinucleotides; some 23% of all pathological single base-pair substitutions found within the coding regions of human genes are of this type (Krawczak et al., 1998). Allowing for the confounding effects of codon usage and differential clinical observation likelihoods through consideration of relative mutabilities, this proportion translates into a mean transition rate for either CG→TG or CG→CA that is five times higher than the base mutation rate.

Krawczak et al. (1998) demonstrated that the proportion of pathological CG→(TG,CA) transitions is significantly higher for autosomal genes (25.0%) than for X-linked genes (17.7%). These proportions are a direct reflection of the significantly lower frequency of CpG in the coding sequences of X-linked genes (2.9%) as compared to autosomal genes (3.7%). Krawczak et al. (1998) speculated that the lower CpG frequency in X-chromosomal genes may be a consequence of a generally increased level of DNA methylation resulting from the evolutionary recruitment of this post-synthetic modification to play a role in X-inactivation (Hornstra and Yang, 1994; Jamieson et al., 1996).

For CpG dinucleotides to be hypermutable in the context either of genetic disease or evolution, they must be methylated in the germline (El-Maarri et al., 1998). Since it cannot be excluded that the efficiency of both DNA methyltransferase action (Smith, 1994; Smith and Baker, 1997) and G: T mismatch repair (Sibghat-Ullah and Day, 1993) may be influenced by sequence motifs flanking the CpG dinucleotide, the question arises as to whether some CpGs may be intrinsically more mutable than others by virtue of their DNA sequence context. Significant differences in the relative mutation rate of CpG dinucleotides depending upon their flanking nucleotides were indeed noted by Krawczak et al. (1998). These results were consistent with those of Ollila et al. (1996) who noted a preference for 5′ pyrimidines and 3′ purines flanking mutated CpG dinucleotides.

Comparison of the sequences flanking the human and chimpanzee β-globin genes has shown that the CpG dinucleotide is hypermutable over evolutionary time and subject to high frequency C→T or G→A transitions (Savatier et al., 1985). Indeed, some 40% of the CpG dinucleotides present in either the human or chimpanzee sequences were found to be affected by nucleotide sequence changes. Similar conclusions have been drawn by Perrin-Pecontal et al. (1992). Comparison of the CpG mutation rates exhibited by globin gene and pseudogene sequences from human, chimpanzee and macaque yielded an estimate of the rate of 5mC deamination of $\sim 1 \times 10^{-16}$ (Cooper and Krawczak, 1989). The absence of any significant difference between deamination rate estimates derived from gene and pseudogene sequence data suggested that the action of selection has had a negligible effect on the transition rate at CpG dinucleotides in primate β-globin genes. Indeed, this constancy in the CpG deamination rate is consistent with a neutralist view of gene evolution. Moreover, the successful use of evolutionary comparisons of DNA sequences to derive consistent values of the CpG deamination rate has demonstrated the feasibility of using the CpG deamination rate as a 'molecular clock', at least over relatively short periods of evolutionary time.

Although the nonrandomness of mutation is at its most dramatic for CpG mutations, non-CpG mutations also appear to be distributed nonrandomly. A subtle neighboring nucleotide effect reminiscent of misalignment models of mutagenesis was noted by Krawczak et al. (1998) in their study of pathological mutations; a substantial proportion of the observed single base-pair substitutions exhibited identity of the newly introduced base to one of the bases immediately flanking the site of mutation. Since this effect occurred only at a distance of one base-pair and in the absence of surrounding repetitive sequences, it was potentially explicable in terms of misalignment mutagenesis involving highly localized DNA slippage, misincorporation and realignment events between template and primer at the replication fork (Kunkel, 1990, 1992). Intriguingly, this next-neighbor mutational bias only occurred at specific codon positions and exhibited polarity. Since such a phenomenon is unlikely to be explicable in terms of the primary mutational event, it is probably associated with the DNA repair process. Implicit in this assumption, however, is that the DNA repair machinery is able to recognize the reading frame and to utilize this information as a cue in effecting DNA repair. Consistent with such a relationship, the observed correction bias would operate in such a way as to remove newly introduced termination codons. The ability of a DNA repair mechanism to take the reading frame into account, thereby minimizing the effects of mutation, would have had positive selective value owing to the relatively deleterious nature of in-frame termination codons. Evidence for reading frame sensitivity in the DNA repair process first came from the observation that relative single base-pair susbtitution rates are biased toward the avoidance of those replacements that (i) change the chemical characteristics of the encoded amino acid residue substantially and (ii) have a high likelihood of coming to clinical attention (Krawczak and Cooper, 1996a; Section 7.2). Krawczak et al. (1998) concluded that, by consideration of the genetic code, selection has optimized the DNA repair mechanism in such a way as to avoid the most hazardous of amino acid replacements, a category which would certainly include nonsense mutations.

Although the same substitution types appear to be subject to next neighbor-effects on the coding and noncoding DNA strands, the quantitative differences in non-CpG single base-pair substitution rates observed by Krawczak et al. (1998) confirmed that the two DNA strands are not fully equivalent in terms of their rates and patterns of mutation (Wu and Maeda, 1987). Nucleotide substitutions that have accumulated in noncoding sequences during evolution are also asymmetric between DNA strands (Francino and Ochman, 1997; Maeda et al., 1988). There are several possible (and nonmutually exclusive) reasons for this strand asymmetry. Firstly, the four nuclear DNA polymerases, each associated with its own distinctive mutational spectrum, may be differentially involved in the synthesis of the leading and lagging strands during DNA replication (Bambara et al., 1997; Kunkel, 1992). Secondly, since the transcriptional elongation complex is asymmetric (Kainz and Roberts, 1992), mutation rates may differ between transcribed and non-transcribed strands on account of either unequal exposure to DNA damage or differential repair. Not only may the transiently single-stranded nontranscribed DNA strand be particularly vulnerable to mutation (e.g. by methylation-mediated deamination; Beletskii and Bhagwat, 1996) but transcription-coupled repair (Bhatia *et al.*, 1996; Drapkin *et al.*, 1994; Hanawalt, 1994), a process that corrects lesions specifically on the transcribed DNA strand, could also account for different mutation rates between transcribed and nontranscribed strands. Both these mechanisms would predict a higher mutation rate for the nontranscribed as opposed to the transcribed DNA strand.

Finally, in their dataset of pathological substitutions, Krawczak *et al.* (1998) found that thermodynamic stability of DNA triplets was positively correlated with the average relative rate at which the central nucleotide of a triplet underwent substitution. At first sight, this finding might appear to be counterintuitive since it implies that higher rather than lower DNA duplex stability would render a gene region more prone to single base-pair substitution. However, the consistent absence of flanking repeat elements noted for the mutations analyzed here suggests that extensive strand slippage (which would require the DNA to be single-stranded) is unlikely to play an important role in the generation of single base-pair substitutions. Nevertheless, a high degree of thermodynamic stability could in principle impair DNA replication in various ways. First, the likelihood that DNA helicases would be incapable of unwinding the two DNA strands correctly or efficiently may be expected to be higher in more stable regions (Chen *et al.* 1992). Second, temporary reannealing of the two native DNA strands during replication might be favored and could be more enduring in such regions. In both cases, DNA polymerase activity would be seriously impeded by localized double-stranded DNA structures, which could result in either a cessation of polymerization or the skipping of one or more nucleotides, leaving a gap in the nascent DNA strand. Miscorrection during the post-replicative repair of such nicks would then introduce a single base-pair substitution. Alternatively, the observed correlation could reflect the increased stability of at least some slippage-mediated misalignments during replication of the native and nascent DNA strands, allowing enough time for misincorporation of a noncomplementary nucleotide. In this case, however, the thermodynamic stabilities of the misaligned structures must be comparable to those of the wild-type triplets, an assumption for which there is currently no evidence.

7.5.2 Single base-pair substitutions in evolution which have altered the function of specific amino acid residues

> Natural selection is a mechanism for generating an exceedingly high degree of improbability.
> R.A. Fisher

Some single base-pair substitutions occurring during gene evolution have introduced nonsense mutations into protein coding regions thereby either prematurely truncating the protein product (Chapter 8, section 8.7) or abolishing the expression of that product altogether (Chapter 6, section 6.2). However, other single base-pair substitutions have introduced missense mutations that have served to alter the function of specific amino acid residues and these are the topic of this Section.

Nucleotide substitutions that have occurred and become fixed through the action of genetic drift are readily apparent in any comparative analysis of orthologous genes and provide abundant evidence for a neutralist model of evolution. Nucleotide substitutions resulting in human genetic disease (Cooper and Krawczak, 1993) may be taken as representative examples of negative (purifying) selection since many of the lesions that have come to clinical attention are likely to reduce (or would once have reduced) survival and/or reproductive fitness. By contrast, unequivocal examples of the positive Darwinian selection of nucleotide/amino acid substitutions in higher eukaryotes are much rarer. Indeed, there are relatively few examples of single base-pair substitutions that have occurred within the coding regions of mammalian genes during evolution which have been sufficiently well characterized for us to be able to identify the consequent change in protein structure and function that has been subject to positive selection. Several illustrative examples are discussed below. Since these systems have not so far been fully characterized, further studies are eagerly awaited.

The visual pigments. The visual pigments, which comprise an integral membrane protein (opsin) coupled to a light-sensitive chromophore, are spectrally tuned to a particular wavelength of maximal absorption, λmax. The visual pigments in rods (photoreceptor cells that function in dim light) are rhodopsins which have a λmax of 495 nm. In cones, the photoreceptor cells that mediate colour vision, there are three types of visual pigment which in humans have λmax values of 420 nm (blue/short wavelength-sensitive), 530 nm (green/middle wavelength-sensitive) and 560 nm (red/long wavelength-sensitive). The human genes encoding these opsins are: rhodopsin (*RHO*; 3q21-q24), the blue cone pigment (*BCP*; 7q31-q35), the green cone pigment (*GCP*; Xq28) and red cone pigment (*RCP*; Xq28). These genes have evolved by a process of duplication and divergence from a common ancestor ~500 Myrs ago (Yokoyama, 1997) and encode proteins that harbour specific amino acid changes that are directly responsible for the shifts in λmax values between the different visual pigments.

To understand the molecular basis of spectral tuning of visual pigments, it is necessary to correlate the sequences of the visual pigments with their λmax values. Such an analysis was first performed by Yokoyama and Yokoyama (1990) who compared the red and green visual pigments from human and the Mexican cavefish, *Astyanax fasciatus*. The red pigments in humans and cavefish were found to have evolved from the green pigments *independently* by three specific amino acid changes (Ala180Ser, Phe277Tyr and Ala285Thr, i.e. AFA SYT; *Figure 7.2*). A similar study in primates (Neitz *et al.*, 1991) also concluded that the spectral difference between red and green pigments could be accounted for by the difference between AFA (green) and SYT (red). These three critical amino acid residues are located near the chromophore and their experimental substitution has been shown to alter λmax values not only in human red and green pigments (Asenjo *et al.*, 1994) but also in bovine rhodopsin (Chan *et al.*, 1992).

Most mammals have dichromatic vision, possessing blue visual pigments together with either red or green pigments (Jacobs, 1993). Phylogenetic analysis of a range of vertebrates has suggested that the vertebrate ancestor of the opsin gene was a green pigment gene encoding AFA at the three critical sites and that the common ancestor of tetrapods acquired a red pigment by two amino acid substitutions (Phe277Tyr and Ala285Thr; *Figure 7.2*). The SYT of the red pigments

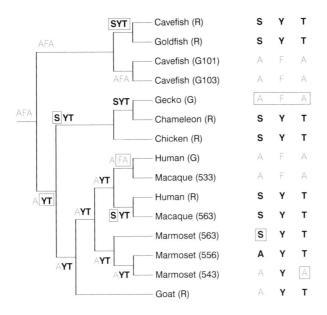

Figure 7.2. Phylogenetic relationships between vertebrate opsins (after Yokoyama, 1997) showing amino acids inferred at sites 180, 277, and 285. The bold and outlined letters refer to red pigment and green pigment characteristic amino acids respectively. Probable amino acid substitutions at different evolutionary stages are boxed.

of extant tetrapods was acquired by a further amino acid substitution, Ala180Ser (*Figure 7.2*).

Old World monkeys have both red and green pigments and therefore possess trichromatic vision. Phylogenetic analysis of the primate opsins (Nei *et al.*, 1997) has indicated that the common ancestor of hominoids, Old World monkeys and New World monkeys possessed a red visual pigment with AYT at the three critical sites (*Figure 7.2*) but no green pigment. The ancestor of the green pigment gene is thought to have arisen by gene duplication ~35 Myrs ago in the Old World monkey lineage (Nathans *et al.*, 1986) and the green pigment was then derived in two distinct steps (AFA→AYT→AFA; *Figure 7.2*).

New World monkeys have only one X-linked opsin gene but exhibit multiple alleles at this locus (Jacobs *et al.*, 1996; Neitz *et al.*, 1991). Thus, in the squirrel monkey (*Saimiri sciureus*), the visual pigments derived from three alternative alleles have λmax values of 532, 547, and 561 nm and manifest amino acids AFA, AFT, and SYT respectively at the critical sites (Neitz *et al.*, 1991). Whilst all male and homozygous female squirrel monkeys are dichromatic, heterozygous females that happen to have pigments with λmax values of 532 nm and 561 nm possess trichromatic vision. Tovee (1994) suggested that color vision in New World monkeys might be an adaptation to allow a wide variety of color vision types within a single family group. However, an alternative explanation, which does not invoke group selection, would be for the alternative alleles to be maintained in the population by overdominant selection. This postulate appears to be supported by the independent evolution of triallelic systems in several other New World monkey lineages represented by the marmoset (*Callithrix jacchus*), the saki monkey

(*Pithecia irrorata*), the capuchin (*Cebus nigrivittatus*) and the tamarin (*Saguinus mystax*) (Boissinot *et al.*, 1998; Shyue *et al.*, 1995).

Zhou *et al.* (1997) studied the X-linked opsin gene of two nocturnal prosimians, the bushbaby species *Galago senegalensis* and *Otolemur garnettii*. At those amino acid positions known to cause spectral differences, however, the cone pigment possessed identical residues to those of the marmoset protein. This suggests that, in spite of the bushbaby's nocturnal existence, its X-linked opsin gene is under functional constraint. Consistent with this view, Zhou *et al.* (1997) noted two amino acid substitutions which may be important for maximizing dim light sensitivity.

The kringle domains of apolipoprotein(a). The human apolipoprotein(a) (*LPA*; 6q27) gene emerged from the plasminogen (*PLG*; 6q27) gene by a process of gene duplication followed by intragenic amplification of exons within the *LPA* gene (Chapter 8, section 8.6). Apolipoprotein(a) is a component of the cholesteryl ester-rich particle lipoprotein(a) within which it is covalently linked to apolipoprotein B100. Lipoprotein(a) has been postulated to play a role in fibrinolysis by competing with plasminogen for binding to fibrin, thereby interfering with clot lysis. Fibrin binding appears to be potentiated by lysine-binding sites in some of the many kringle domains of apolipoprotein(a). Kringle IV-10 of human apolipoprotein(a) most closely resembles that of plasminogen kringle IV and appears to be critical for the fibrin-binding potential of lipoprotein(a). The lysine-binding sites of apolipoprotein(a) consist of a hydrophobic trough containing three aromatic amino acids (Trp62, Phe64, and Trp72), an anionic centre composed of two aspartic acid residues (Asp55 and Asp57), and a cationic centre comprising two residues, Lys35 and Arg71. Kringles IV-1, IV-2, IV-3, and IV-4 of apolipoprotein(a) do not bind to lysine and in each case, this is associated with the absence of Asp57 within the kringle. Two different substitutions in kringle IV-10 have occurred during primate evolution which are associated with the loss of the lysine-binding properties of the lipoprotein(a) particle; a Trp72→Arg substitution in rhesus monkey (Scanu *et al.*, 1993; Tomlinson et al., 1989) and an Asp57→Asn substitution in chimpanzee (Chenivesse *et al.*, 1998). The physiological consequences of these two substitutions are however as yet unknown.

The DNA-binding specificity of steroid receptors. Gene regulation by steroid hormones is mediated by binding of the hormone ligand to its cognate receptor (Chapter 4, section 4.2.3, *Nuclear receptor genes*). Upon ligand binding, most nuclear receptors then interact as dimers with their response elements, each monomer binding to a half-site sequence. These response elements comprise two 6 bp half-site sequences organized as palindromic repeats with 3–5 bp separating the half sites. Thus the thyroid hormone, retinoic acid and estrogen receptors recognize the half-site sequence TGACCT whereas the androgen, progesterone and glucocorticoid receptors recognize the half-site sequence TGTTCT. More specifically, the consensus estrogen response element (ERE) is AGGTCANNNTGACCT whilst the consensus glucocorticoid response element (GRE) is GGTACANNNT-GTTCT (Zilliacus *et al.*, 1995). These response elements exhibit a 3 bp spacer between half-sites. By contrast, the thyroid hormone receptor recognizes a 4 bp spacer and the retinoic acid receptor a 5 bp spacer. Thus, both the sequence and the spacing between the half-sites determines the specificity.

Most nuclear receptors have two or three amino acids (P box residues) in their DNA-binding domains that serve to specify recognition of the response element within the promoter of the target gene. Some residues (e.g. Val443 in the gluco-corticoid receptor or Glu439 in the estrogen receptor) can contribute to specificity both by forming a positive interaction with a base in the cognate response element and by forming a negative interaction with a base in a noncognate response element (Zilliacus *et al.*, 1994, 1995). In the glucocorticoid receptor, Ser440 inhibits the interaction of the receptor with the ERE but at the cost of also reducing affinity for the GRE (Zilliacus *et al.*, 1994, 1995). Thus, the diversification of steroid receptor specificity was probably achieved during evolution by a relatively small number of single base-pair substitutions either in the P-box encoding residues of steroid receptors or in the response elements of their target genes.

The calcium-dependent and -independent synaptotagmins. Synaptotagmins constitute a large family of proteins involved in membrane trafficking. At least five different synaptotagmin genes have been characterized in the human genome (*SYT1*, 12cen-q21; *SYT2*, 1q; *SYT3*, 19q; *SYT4*, 5q; *SYT5*, 11p). Most synapto-tagmins are capable of binding calcium through their calcium-binding C_2A domains. Synaptotagmins IV and XI are unique in their inability to bind calcium and Von Poser *et al.* (1997) have shown that this inability is caused by the substitution of Ser for Asp at residue 230 in the C_2A domain. This substitution is evolutionarily conserved in synaptotagmin IV and it seems likely that synaptotagmins IV and XI shared a common ancestor in which the substitution originated. Von Poser *et al.* (1997) postulated that evolution selected for the loss of calcium binding in two different synaptotagmins while leaving the remainder of the protein structures intact. This would of course imply that these synaptotagmins also possess calcium-independent properties and that these properties have been retained by selection through evolutionary time.

Olfactory receptor-ligand interactions. Certain residues in the α-helical sixth transmembrane domain of human olfactory receptors have been implicated in interactions with their odorant ligands and these residues appear to have been subject to positive selection (Singer *et al.*, 1996). The highly variable residues 6–22 and 6–25 are thought to constitute a receptor sub-site that binds hydroxyl groups on odorant ligands; substitutions at these critical positions may therefore help to determine the odor specificity of olfactory receptor sub-types. A Val/Ile difference at residue 206 of the orthologous rat and mouse I7 olfactory receptor proteins has been shown to be responsible for a species-specific odorant response; rats preferring octanal to heptanal, mice the reverse (Krautwurst *et al.*, 1998).

7.5.3 Single base-pair substitutions in evolution which have affected mRNA splicing

Single base-pair substitutions affecting mRNA splicing make up between 8% and 15% of all mutations causing human genetic disease (Krawczak *et al.*, 1992). These lesions occur disproportionately at the most evolutionarily conserved positions within the splice site and fall into three main categories: (i) mutations within 5' or 3' splice sites which reduce the amount of correctly processed mature RNA and/or activate alternative ('cryptic') splice sites in the vicinity, (ii) mutations out-with actual splice sites which create cryptic splice sites, and (iii) mutations in the branch-point sequence (Krawczak *et al.*, 1992). The vast majority of the patholog-ical lesions affecting mRNA splicing so far reported have been single base-pair substitutions within splice sites. This is not only because these are comparatively frequent but also because they are both readily detectable and highly likely to result in a severe clinical phenotype. Disease-associated mutations affecting 5' splice sites are approximately twice as frequent as mutations at 3' splice sites (Krawczak *et al.*, 1992). This discrepancy coincides with a much higher level of sequence conservation at 5' splice sites and is likely to reflect the strong require-ment for U1 snRNA binding at 5' splice sites to promote alignment and cleavage. Regarding the phenotypic consequences of pathological mutations affecting mRNA splicing, the exclusion of one or more exons from the end-product (*exon skipping*) is observed more frequently than *cryptic splice site utilization* (Krawczak *et al.*, 1992). Some evidence exists, at least for 5' splice sites, that cryptic splice site usage is favored under conditions where a number of potential sites are present in the vicinity of the mutated splice site and where these potential splice sites exhibit sufficient homology to the consensus sequence (Krawczak *et al.*, 1992). In most such cases, the activating mutation improves the similarity between the cryptic site and the splice site consensus sequence. At 3' splice sites, the amount of mRNA product consequent to the utilization of the cryptic splice site appears to be correlated with the level of similarity to the consensus sequences; at 5' splice sites, the distance to the nearest wild-type splice site may also play a role (Krawczak *et al.*, 1992).

Splicing mutations and gene inactivation. The alteration of mRNA processing as a result of mutations in splice sites also occurs during evolution. One dramatic example is that of the once active human L-gulono-γ-lactone oxidase (*GULOP*) gene (chromosome 8p21.1) which was inactivated at least in part by the introduc-tion of single base-pair substitutions at the invariant bases of splice sites (Nishikimi *et al.*, 1994; see Chapter 6, section 6.2.4). Another similar example is provided by the *CSHL1* gene, long regarded as a pseudogene of the human pla-centally expressed growth hormone/chorionic somatomammotropin gene family which is clustered on chromosome 17q22-q24. The *CSHL1* 'pseudogene' was originally thought to have been inactivated by the introduction of a G→A substi-tution in the first position of the intron 2 donor splice site (Press *et al.*, 1994). However, *in vitro* expression studies have shown that it may only have been par-tially inactivated (Misra-Press *et al.*, 1994). Although five alternatively spliced forms of *CSHL1* mRNA are produced from the 'pseudogene,' the majority of these transcripts lack exon 2 which encodes the signal peptide necessary for nor-mal secretion. Some low abundance *CSHL1* mRNAs are nevertheless produced which possess the leader peptide and these could in principle encode novel hor-mones of physiological significance.

Splice site differences between orthologous genes. In an evolutionary context, gene inactivation is likely to be a relatively rare occurrence and alterations in splice junction sequences are more often to be found associated with more subtle changes in mRNA processing such as alternative splicing. A number of examples have been noted in orthologous genes (Chapter 3, section 3.2). Thus, in the erythroid 5-aminolevulinate synthase (*ALAS1*; 3p21.1) gene, exon 4 is involved in alternative splicing in the human but not in the dog or mouse (Conboy *et al.*, 1992). Conversely, the alternative splicing of the 45 bp exon 3 of the murine *Alas1* gene utilizes a major upstream splice site (85% of mRNAs) and a minor downstream site (15% of mRNAs). This is not found in human owing to an A→G transition which abolishes the consensus sequence of the 3′ splice site thereby preventing the possibility of alternative splicing (Conboy *et al.*, 1992). Finally, the alternative splicing pathway involving the mutually exclusive exons 6A and 6B of the β-tropomyosin (*TPM2*; 9q13) gene differs between chicken, rat and *Xenopus* (Pret and Fiszman, 1996). The chicken β-tropomyosin exon 6A is flanked by stronger splicing signals than its rat counterpart and this has been related to inter-specific differences in both the donor and acceptor splice sites (Pret and Fiszman, 1996). A chicken-specific pyrimidine-rich splicing enhancer present upstream of exon 6A may also play a role.

Splice site differences between paralogous genes. Evidence for splice site mutations having occurred during evolution has also come from the study of paralogous genes. The human glycophorin B (*GYPB*; 4q28.2-q31) gene lacks exon 3 by comparison with the related glycophorin A (*GYPA*; 4q28.2-q31) gene owing to a G→T transversion at the exon 3 donor splice site (Kudo and Fukuda, 1989). Comparison with the *GYPB* gene has identified an A→T substitution in the first base of exon 5 of the *GYPA* gene which leads to the alternative use of an acceptor splice site 9 bp upstream and the incorporation of nine extra bases into the *GYPA* coding sequence (Kudo and Fukuda, 1989).

The human genes encoding interleukin-1α (*IL1A*; 2q13-q21), interleukin-1β (*IL1B*: 2q13-q21) and interleukin-1 receptor antagonist (*IL1RN*; 2q14.2) are evolutionarily related members of the interleukin family which remain closely linked on the long arm of chromosome 2. The first exon of *IL1RN* (encoding a leader peptide) is homologous to the untranslated first exon of *IL1B* but the *IL1RN* gene lacks the exons corresponding to the first three expressed exons of the *IL1A* and *IL1B* genes. Hughes (1994) suggested that the common ancestor of the *IL1B* and *IL1RN* genes was an alternatively spliced gene: one transcript could have included exons 1–7 encoding the ancestral IL1B protein whereas the other transcript may have included exons 1 and 5–7 encoding the IL1RN protein. The duplication of this ancestral gene would have freed one copy from functional constraints so that it could encode IL1RN only. Selection would no longer have been able to conserve the intron-exon junctions involved in the splicing of exons 2–4 of the *IL1RN* gene. Consistent with this view of events, the region of the *IL1B* gene between exons 1 and 5 is more than twice the length of the analogous region of the *IL1RN* gene.

Finally, the pituitary-expressed growth hormone (*GH1*; 17q22–q24) gene differs from the closely linked placentally-expressed growth hormone (*GH2*) gene in its pattern of splice site selection. Whereas 9% of *GH1* mRNA transcripts contain exon 2 spliced to an alternative acceptor site located 45 bp into exon 3, the *GH2* gene does not utilize this alternative splicing pathway. Three single base substitutions located between the two alternative acceptor splice sites have been shown to be both necessary and sufficient to define the *GH1* alternative splicing event (Estes *et al.*, 1990). One of these changes is thought to specify a lariat branchpoint essential for alternative acceptor site usage whereas the other two bases may serve to modulate the frequency with which the site is selected.

References

Agulnik A.I., Bishop C.E., Lerner J.L., Agulnik S.I., Solovyev V.V. (1997) Analysis of mutation rates in the *SMCY/SMCX* genes shows that mammalian evolution is male driven. *Mamm. Genome* **8**: 134–138.

Alber T. (1989) Mutational effects on protein stability. *Ann. Rev. Biochem.* **58**: 765–798.

Bailey W.J., Fitch D.H.A., Tagle D.A., Czelusniak J., Slightom J.L., Goodman M. (1991) Molecular evolution of the ψη-globin gene locus: gibbon phylogeny and the hominoid slowdown. *Mol. Biol. Evol.* **8**: 155–184.

Bambara R.A., Murante R.S., Henricksen L.A. (1997) Enzymes and reactions at the eukaryotic DNA replication fork. *J. Biol. Chem.* **272**: 4647–4650.

Bateson W. (1894) *Materials for the Study of Variation Treated with Especial Regard to Discontinuity in the Origin of Species.* MacMillan, London.

Beletskii A., Bhagwat A.S. (1996) Transcription-induced mutations: increase in C to T mutations in the nontranscribed strand during transcription in *Escherichia coli. Proc. Natl. Acad. Sci. USA* **93**: 13 919–13 924.

Bhatia P.K., Wang Z., Friedberg E.C. (1996) DNA repair and transcription. *Curr. Opin. Genet. Devel.* **6**: 146–150.

Bird A.P. (1980) DNA methylation and the frequency of CpG in animal DNA. *Nucleic Acids Res.* **8**: 1499–1504.

Blake R.D., Hess S.T. (1992) The pattern of substitution mutation in different nearest-neighbor environments of the human genome. *Computers Chem.* **16**: 165–170.

Boissinot S., Tan Y., Shyue S.-K., Schneider H., Sampaio I., Neiswanger K., Hewett-Emmett D., Li W.-H. (1998) Origins and antiquity of X-linked triallelic color vision systems in New World monkeys. *Proc. Natl. Acad. Sci. USA* **95**: 13 749–13 754.

Bottema C.D.K., Ketterling R.P., Setsuko L., Yoon H.-S., Phillips J.A., Sommer S.S. (1991) Missense mutations and evolutionary conservation of amino acids: evidence that many of the amino acids in factor IX function as 'spacer' elements. *Am. J. Hum. Genet.* **49**: 820–838.

Boulikas T. (1992) Evolutionary consequences of nonrandom damage and repair of chromatin domains. *J. Mol. Evol.* **35**: 156–180.

Brandstetter H., Bauer M., Huber R., Lollar P., Bode W. (1995) X-ray structure of clotting factor IXa: active site and molecular structure related to Xase activity and hemophilia B. *Proc. Natl. Acad. Sci. USA* **92**: 9796–9800.

Britten R.J. (1986) Rates of DNA sequence evolution differ between taxonomic groups. *Science* **231**: 1393–1398.

Britten R.J. (1993) Forbidden synonymous substitutions in coding regions. *Mol. Biol. Evol.* **10**: 205–220.

Bulmer M., Wolfe K.H., Sharp P.M. (1991) Synonymous nucleotide substitution rates in mammalian genes: implications for the molecular clock and the relationship of mammalian orders. *Proc. Natl. Acad. Sci. USA* **88**: 5974–5978.

Chan T., Lee M., Sakmar T.P. (1992) Introduction of hydroxyl-bearing amino acids causes bathochromic spectral shifts in rhodopsin: amino acid substitutions responsible for red-green color pigment spectral tuning. *J. Biol. Chem.* **267**: 9478–9480.

Chang B.H.J., Li W.-H. (1995) Estimating the intensity of male-driven evolution in rodents by using X-linked and Y-linked *Ube1* genes and pseudogenes. *J. Mol. Evol.* **40**: 70–77.

Chen Y.Z., Zhuang W., Prohofsky E.W. (1992) Energy flow considerations and thermal fluctuational opening of DNA base pairs at a replication fork: unwinding consistent with observed replication rates. *J. Biomol. Struct. Dyn.* **10**: 415–427.

Chenivesse X., Huby T., Wickins J., Chapman J., Thillet J. (1998) Molecular cloning of the cDNA encoding the carboxy terminal domain of chimpanzee apolipoprotein(a): an Asp57→Asn mutation in kringle IV-10 is associated with poor fibrin binding. *Biochemistry* **37**: 7213–7223.

Collins D.W., Jukes T.H. (1994) Rates of transition and transversion in coding sequences since the human-rodent divergence. *Genomics* **20**: 386–396.

Collins F.S., Guyer M.S., Chakravarti A. (1997) Variations on a theme: cataloguing human DNA sequence variation. *Science* **278**: 1580–1581.

Conboy J.G., Cox T.C., Bottomley S.S., Bawden M.J., May B.K. (1992) Human erythroid 5-aminolevinate synthase. *J. Biol. Chem.* **267**: 18 753–18 758.

Cooper D.N., Smith B.A., Cooke H.J., Niemann S., Schmidtke J. (1985) An estimate of unique DNA sequence heterozygosity in the human genome. *Hum. Genet.* **69**: 201–205.

Cooper D.N., Krawczak M. (1989) Cytosine methylation and the fate of CpG dinucleotides in vertebrate genomes. *Hum. Genet.* **83**: 181–188.

Cooper D.N., Krawczak M. (1990) The mutational spectrum of single base-pair substitutions causing human genetic disease: patterns and predictions. *Hum. Genet.* **85**: 55–74.

Cooper D.N., Krawczak M. (1993) *Human Gene Mutation*. BIOS Scientific Publishers, Oxford.

Creighton T.E. (1993) *Proteins: Structures and Molecular Properties*. 2nd Edn. W.H. Freeman, New York.

Crow J.F. (1997) The high spontaneous mutation rate: is it a health risk? *Proc. Natl. Acad. Sci. USA* **94**: 8380–8386.

D'Onofrio G., Mouchiroud D., Aïssani B., Gautier C., Bernardi G. (1991) Correlations between the compositional properties of human genes, codon usage, and amino acid composition of proteins. *J. Mol. Evol.* **32**: 504–510.

De Falco S., Russo G., Angiolillo A., Pietropaolo C. (1993) Human L7a ribosomal protein: sequence, structural organization, and expression of a functional gene. *Gene* **126**: 227–235.

De Filippis V., Sander C., Vriend G. (1994) Predicting local structural changes that result from point mutations. *Protein Engineering* **7**: 1203–1208.

Drapkin R., Sancar A., Reinberg D. (1994) Where transcription meets repair. *Cell* **77**: 9–12.

Easteal S. (1988) Rate constancy of globin gene evolution in placental mammals. *Proc. Natl. Acad. Sci. USA* **85**: 7622–7626.

Easteal S. (1991) The relative rate of DNA evolution in primates. *Mol. Biol. Evol.* **8**: 115–127.

Easteal S., Collet C. (1994) Consistent variation in amino acid substitution rate, despite uniformity of mutation rate: protein evolution in mammals is not neutral. *Mol. Biol. Evol.* **11**: 643–647.

Easteal S., Collet C., Betty D. (1995) *The Mammalian Molecular Clock*. Springer Verlag, New York.

El-Maarri O., Olek A., Balaban B., Montag M., van der Ven H., Urman B., Olek K., Caglayan S.H., Walter J., Oldenburg J. (1998) Methylation levels at selected CpG sites in the factor VIII and *FGFR3* genes in mature female and male germ cells: implications for male-driven evolution. *Am. J. Hum. Genet.* **63**: 1001–1008.

Ellsworth D.L., Hewett-Emmett D., Li W.-H. (1993) Insulin-like growth factor II intron sequences support the hominoid rate-slowdown hypothesis. *Molec. Phylogenet. Evol.* **2**: 315–321.

Estes P.A., Cooke N.E., Liebhaber S.A. (1990) A difference in the splicing patterns of the closely related normal and variant human growth hormone gene transcripts is determined by a minimal sequence divergence between two potential splice acceptor sites. *J. Biol. Chem.* **265**: 19 863–19 870.

Eyre-Walker A. (1994) DNA mismatch repair and synonymous codon evolution in mammals. *Mol. Biol. Evol.* **11**: 88–98.

Eyre-Walker A., Keightley P.D. (1999) High genomic deleterious mutation rates in hominids. *Nature* **397**: 344–347.

Fitch W.M., Ayala F.J. (1994) Molecular clocks are not as bad as you think. In *Molecular Evolution of Physiological Processes*, (Ed. DM Fambrough). Rockefeller University Press, New York. pp 3–12.

Francino M.P., Ochman H. (1997) Strand asymmetries in DNA evolution. *Trends Genet.* **13**: 240–245.

Giannelli F., Green P.M., Sommer S.S., Poon M.C., Ludwig M., Schwaab R., Reitsma P.H., Goossens M., Yoshioka A., Brownlee G.G. (1996) Haemophilia B (sixth edition): a database of point mutations and short additions and deletions. *Nucleic Acids Res.* **24**: 103–118.

Gibbs P.E.M., Dugaiczyk A. (1994) Reading the molecular clock from the decay of internal symmetry of a gene. *Proc. Natl. Acad. Sci. USA* **91**: 3413–3417.

Goldman N. (1993) Further results on error minimization in the genetic code. *J. Mol. Evol.* **37**: 662–664.

Grantham R. (1974) Amino acid difference formula to help explain protein evolution. *Science* **185**: 862–864.

Greer G. (1990) Comparative modelling methods: application to the family of the mammalian serine proteases. *Proteins Struct. Funct. Genet.* **7**: 317–334.

Gu X., Li W.H. (1992) Higher rates of amino acid substitution in rodents than in humans. *Molec. Phylogenet. Evol.* **1**: 211–214.

Haig D., Hurst L.D. (1991) A quantitative measure of error minimization in the genetic code. *J. Mol. Evol.* **33**: 412–417.

Hanawalt P.C. (1990) Selective DNA repair in active genes. *Acta Biologica Hungarica* **41**: 77–91.

Hanawalt P.C. (1994) Transcription-coupled repair and human disease. *Science* **266**: 1957–1958.

Hess S.T., Blake J.D., Blake R.D. (1994) Wide variations in neighbor-dependent substitution rates. *J. Mol. Biol.* **236**: 1022–1033.

Hornstra I.K., Yang T.P. (1994) High-resolution methylation analysis of the human hypoxanthine phosphoribosyltransferase gene 5′ region on the active and inactive X chromosomes: correlation with binding sites for transcription factors. *Mol. Cell. Biol.* **14**: 1419–1430.

Hughes A.L. (1994) Evolution of the interleukin-1 gene family in mammals. *J. Mol. Evol.* **39**: 6–12.

Hurst L.D., Ellegren H. (1998) Sex biases in the mutation rate. *Trends Genet.* **14**: 446–452.

Hurst L.D., Smith N.G.C. (1999) Molecular evolutionary evidence that *H19* mRNA is functional. *Trends Genet.* **15**: 134–135.

Ikemura T., Wada K. (1991) Evident diversity of codon usage patterns of human genes with respect to chromosome banding patterns and chromosome numbers; relation between nucleotide sequence data and cytogenetic data. *Nucleic Acids Res.* **19**: 4333–4339.

Jacobs G.H. (1993) The distribution and nature of colour vision among the mammals. *Biol. Rev.* **68**: 413–471.

Jacobs G.H., Neitz M., Deegan J.F., Neitz J. (1996) Trichromatic color vision in New World monkeys. *Nature* **382**: 156–158.

Jamieson R.V., Tam P.P.L., Gardiner-Garden M. (1996) X-chromosome activity: impact of imprinting and chromatin structure. *Int. J. Dev. Biol.* **40**: 1065–1080.

Jukes T.H. (1993) The genetic code – function and evolution. *Cell. Molec. Biol. Res.* **39**: 685–688.

Kainz M., Roberts J. (1992) Structure of transcription elongation complexes *in vivo*. *Science* **255**: 838–841.

Kawamura S., Tanabe H., Watanabe Y., Kurosaki K., Saitou N., Ueda S. (1991) Evolutionary rate of immunoglobulin alpha noncoding region is greater in hominoids than in Old World monkeys. *Mol. Biol. Evol.* **8**: 743–752.

Kimura M. (1983) *The Neutral Theory of Molecular Evolution.* Cambridge University Press, Cambridge.

Kitano T., Sumiyama K., Shiroishi T., Saitou N. (1998) Conserved evolution of the Rh50 gene compared to its homologous Rh blood group gene. *Biochem. Biophys. Res. Commun.* **249**: 78–85.

Kondrashov A.S., Crow J.F. (1993) A molecular approach to estimating the human deleterious mutation rate. *Hum. Mutation* **2**: 229–234.

Koop B.F., Goodman M., Xu P., Chan K., Slightom J.L. (1986) Primate η-globin DNA sequences and man's place among the great apes. *Nature* **319**: 234–238.

Kragelund B.B., Poulsen K., Andersen K.V., Baldursson T., Krøll J.B., Neergård T.B., Jepsen J., Roepstorff P., Kristiansen K., Poulsen F.M., Knudsen J. (1999) Conserved residues and their role in the structure, function and stability of acyl-coenzyme A binding protein. *Biochemistry* **38**: 2386–2394.

Krautwurst D., Yau K.-W., Reed R.R. (1998) Identification of ligands for olfactory receptors by functional expression of a receptor library. *Cell* **95**: 917–926.

Krawczak M, Cooper DN. (1996a) Single base-pair substitutions in pathology and evolution: two sides to the same coin. *Hum. Mutation* **8**: 23–31.

Krawczak M., Cooper D.N. (1996b) Mutational processes in pathology and evolution. In: *Human Genome Evolution*, (Eds. MS Jackson, T Strachan, G Dover). BIOS Scientific Publishers, Oxford, pp 1–33.

Krawczak M., Reiss J., Cooper D.N. (1992) The mutational spectrum of single base-pair substitutions in mRNA splice junctions of human genes: causes and consequences. *Hum. Genet.* **90**: 41–54.

Krawczak M., Ball E.V., Cooper D.N. (1998) Neighboring nucleotide effects on the rates of inherited single base-pair substitution in human genes. *Am. J. Hum. Genet.* **63**: 474–488.

Kudo S., Fukuda M. (1989) Structural organization of glycophorin A and B genes: glycophorin B gene evolved by homologous recombination at *Alu* repeat sequences. *Proc. Natl. Acad. Sci. USA* **86**: 4619–4623.

Kuhn H., Waser J. (1994) On the origin of the genetic code. *FEBS Letts.* **352**: 259–264.

Kunkel T.A. (1990) Misalignment-mediated DNA synthesis errors. *Biochemistry* **29**: 8003–8011.

Kunkel T.A. (1992) DNA replication fidelity. *J. Biol. Chem.* **267**: 18 251–18 254.

Kurihara T., Sakuma M., Gojobori T. (1997) Molecular evolution of myelin proteolipid protein. *Biochem. Biophys. Res. Commun.* **237**: 559–561.

Li W.-H. (1997) *Molecular Evolution*. Sinauer Associates, Sunderland.

Li W.-H., Graur D. (1991) *Fundamentals of Molecular Evolution*. Sinauer Associates, Sunderland.

Li W.-H., Sadler L.A. (1991) Low nucleotide diversity in man. *Genetics* **129**: 513–523.

Li W.-H., Tanimura M. (1987) The molecular clock runs more slowly in man than in apes and monkeys. *Nature* **326**: 93–96.

Li W.-H., Gojobori T., Nei M. (1981) Pseudogenes as a paradigm of neutral evolution. *Nature* **292**: 237–239.

Li W.-H., Wu C.-I., Luo C.-C. (1985) A new method for estimating synonymous and nonsynonymous rates of nucleotide substitution considering the relative likelihood of nucleotide and codon changes. *Mol. Biol. Evol.* **2**: 66–69.

Li W.-H., Gouy M., Sharp P.M., O'hUigin C., Yang Y.-W. (1990) Molecular phylogeny of rodentia, lagomorpha, primates, artiodactyla and carnivora and molecular clocks. *Proc. Natl. Acad. Sci. USA* **87**: 6703–6707.

Li W.-H., Ellsworth D.L., Krushkal J., Chang B.H., Hewett-Emmett D. (1996) Rates of nucleotide substitution in primates and rodents and the generation-time effect hypothesis. *Mol. Phylogenet. Evol.* **5**: 182–187.

Lim W.A., Faruggio D.C., Sauer R.T. (1992) Structural and energetic consequences of disruptive mutations in a protein core. *Biochemistry* **31**: 4324–4333.

McVean G.T., Hurst L.D. (1997) Evidence for a selectively favourable reduction in the mutation rate of the X chromosome. *Nature* **386**: 388–392.

MacIntyre R.J. (1994) Molecular evolution: codes, clocks, genes and genomes. *BioEssays* **16**: 699–703.

Maeda N., Wu C.-I., Bliska J., Reneke J. (1988) Molecular evolution of intergenic DNA in higher primates: pattern of DNA changes, molecular clock and evolution of repetitive sequences. *Mol. Biol. Evol.* **5**: 1–20.

Makalowski W., Boguski M.S. (1998) Evolutionary parameters of the transcribed mammalian genome: an analysis of 2820 orthologous rodent and human sequences. *Proc. Natl. Acad. Sci. USA* **95**: 9407–9412.

Messier W., Stewart C.-B. (1997) Episodic adaptive evolution of primate lysozymes. *Nature* **385**: 151–154.

Misra-Press A., Cooke N.E., Liebhaber S.A. (1994) Complex alternative splicing partially inactivates the human chorionic somatomammotropin-like (*hCS-L*) gene. *J. Biol. Chem.* **269**: 23 220–23 229.

Miyata T., Hayashida H., Kuma K., Mitsuyasa K., Yasunaga T. (1987) Male-driven molecular evolution: a model and nucleotide sequence analysis. *Cold Spring Harb. Symp. Quant. Biol.* **52**: 863–867.

Miyata T., Kuma K., Iwabe N., Nikoh N. (1994) A possible link between molecular evolution and tissue evolution demonstrated by tissue specific genes. *Jpn. J. Genet.* **69**: 473–480.

Nakamura Y., Gojobori T., Ikemura T. (1998) Codon usage tabulated from the international DNA sequence databases. *Nucleic Acids Res.* **26**: 334–335.

Nei M. (1987) *Molecular Evolutionary Genetics*. Columbia University Press, New York.

Nei M., Zhang J., Yokoyama S. (1997) Color vision of ancestral organisms of higher primates. *Mol. Biol. Evol.* **14**: 611–618.

Neitz M., Neitz J., Jacobs G.H. (1991) Spectral tuning of pigments underlying red-green color vision. *Science* 252: 971–974.

Nickerson D.A., Taylor S.L., Weiss K.M., Clark A.G., Hutchinson R.G., Stengård J., Salomaa V., Vartiainen E., Boerwinkle E., Sing C.F. (1998) DNA sequence diversity in a 9.7-kb region of the human lipoprotein lipase gene. *Nature Genet.* 19: 233–240.

Nishikimi M., Fukuyama R., Minoshima S., Shimizu N., Yagi K. (1994) Cloning and chromosomal mapping of the human nonfunctional gene for L-gulono-γ-lactone oxidase, the enzyme for L-ascorbic acid biosynthesis missing in man. *J. Biol. Chem.* 269: 13 685–13 688.

Nolet S., St-Louis D., Mbikay M., Chrétien M. (1991) Rapid evolution of prostatic protein PSP_{94} suggested by sequence divergence between rhesus monkey and human cDNAs. *Genomics* 9: 775–777.

Ohta T. (1993) Pattern of nucleotide substitutions in growth hormone-prolactin gene family: a paradigm for evolution by gene duplication. *Genetics* 134: 1271–1276.

Ohta T., Ina Y. (1995) Variation in synonymous substitution rates among mammalian genes and the correlation between synonymous and nonsynonymous divergences. *J. Mol. Evol.* 41: 717–720.

O'hUigin C., Li W.-H. (1992) The molecular clock ticks regularly in murine rodents and hamsters. *J. Mol. Evol.* 35: 377–384.

Olilla J., Lappalainen I., Vihinen M. (1996) Sequence specificity in CpG mutation hotspots. *FEBS Letts.* 396: 119–122.

Osawa S., Jukes T.H. (1988) Evolution of the genetic code as affected by anticodon content. *Trends Genet.* 4: 191–198.

Pakula A.A., Sauer R.T. (1989) Genetic analysis of protein stability and function. *Ann. Rev. Genet.* 23: 289–310.

Pastorcic M., Birnbaum S., Hixson J.E. (1992) Baboon apolipoprotein C-I: cDNA and gene structure and evolution. *Genomics* 13: 368–374.

Perrin-Pecontal P., Gouy M., Nigon V.-M., Trabuchet G. (1992) Evolution of the primate β-globin gene region: nucleotide sequence of the δ-β-globin intergenic region of gorilla and phylogenetic relationships between African apes and man. *J. Mol. Evol.* 34: 17–30.

Prager E.M. (1996) Adaptive evolution of lysozyme: changes in amino acid sequence, regulation of expression and gene number. In: *Model Enzymes in Biochemistry and Biology,* (ed. P Jollès). Birkhäuser Verlag, Basel.

Press A.M., Cooke N.E., Liebhaber S.A. (1994) Complex alternative splicing partially inactivates the human chorionic somatomammotropin-like (hCS-L) gene. *J. Biol. Chem.* 269: 23 220–23 229.

Pret A.M., Fiszman M.Y. (1996) Sequence divergence associated with species-specific splicing of the non-muscle β-tropomyosin alternative exon. *J. Biol. Chem.* 271: 11 511–11 517.

Retief J.D., Dixon G.H. (1993) Evolution of pro-protamine P2 genes in primates. *Eur. J. Biochem.* 214: 609–615.

Rogozin I.B., Kolchanov N.A. (1992) Somatic hypermutagenesis in immunoglobulin genes. II. Influence of neighbouring base sequences on mutagenesis. *Biochim. Biophys. Acta* 1171: 11–18.

Sarkar G., Koeberl D.D., Sommer S.S. (1990) Direct sequencing of the activation peptide and the catalytic domain of the factor IX gene in six species. *Genomics* 6: 133–143.

Savatier P., Trabuchet G., Faure C., Chebloune Y., Gouy M., Verdier G., Nigon V.M. (1985) Evolution of the primate β-globin gene region. High rate of variation in CpG dincleotides and in short repeated sequences between man and chimpanzee. *J. Mol. Biol.* 182: 21–29.

Scanu A.M., Miles L.A., Fless G.M., Pfaffinger D., Eisenbart J., Jackson E., Hoover-Plow J.L., Brunck T., Plow E.F. (1993) Rhesus monkey lipoprotein(a) binds to lysine Sepharose and U937 monocytoid cells less efficiently than human lipoprotein(a). Evidence for the dominant role of kringle 4(37). *J. Clin. Invest.* 91: 283–291.

Schirmer R.H. (1979) *Principles of Protein Structure.* Springer, New York.

Seino S., Bell G.I., Li W.-H. (1992) Sequences of primate insulin genes support the hypothesis of a slower rate of molecular evolution in humans and apes than in monkeys. *Mol. Biol. Evol.* 9: 193–203.

Shaw J.-P., Marks J., Shen C.C., Shen C.-K.J. (1989) Anomalous and selective DNA mutations of the Old World monkey α-globin genes. *Proc. Natl. Acad. Sci. USA* 86: 1312–1316.

Shimmin L.C., Chang B.H.-J., Li W.-H. (1993) Male-driven evolution of DNA sequences. *Nature* 362: 745–747.

Shyue S.-K., Hewett-Emmett D., Sperling H.G., Hunt D.M., Bowmaker J.K., Mollon J.D., Li W.-H. (1995) Adaptive evolution of color vision genes in higher primates. *Science* **269**: 1265–1269.

Sibghat-Ullah S., Day R.S. (1993) DNA-substrate sequence specificity of human G: T mismatch repair activity. *Nucleic Acids Res.* **21**: 1281–1287.

Singer M.S., Weisinger-Lewin Y., Lancet D., Shepherd G.M. (1996) Positive selection moments identify potential functional residues in human olfactory receptors. *Recept. Channels* **4**: 141–147.

Smith S.S. (1994) Biological implications of the mechanism of action of human DNA (cytosine-5) methyltransferase. *Prog. Nucleic Acid Res. Mol. Biol.* **49**: 65–111.

Smith S.S., Baker D.J. (1997) Stalling of human methyltransferase at single-strand conformers from the Huntington's locus. *Biochem. Biophys. Res. Commun.* **234**: 73–78.

Sommer S.S., Bowie E.J.W., Ketterling R.P., Bottema C.D.K. (1992) Missense mutations and the magnitude of functional deficit: the example of factor IX. *Hum. Genet.* **89**: 295–297.

Stewart C.-B., Schilling J.W., Wilson A.C. (1987) Adaptive evolution in the stomach lysozymes of foregut fermenters. *Nature* **330**: 401–404.

Streydio C., Swillens S., Georges M., Szpirer C., Vassart G. (1990) Structure, evolution and chromosomal localization of the human pregnancy-specific β1 glycoprotein gene family. *Genomics* **6**: 579–592.

Swanson K.W., Irwin D.M., Wilson A.C. (1991) Stomach lysozyme gene of the langur monkey: tests for convergence and positive selection. *J. Mol. Evol.* **33**: 418–425.

Swindells M.B., Thornton J.M. (1991) Modelling by homology. *Curr. Op. Struct. Biol.* **1**: 219–223.

Teglund S., Olsen A., Khan W.N., Frngsmyr L., Hammarström S. (1994) The pregnancy-specific glycoprotein (*PSG*) gene cluster on human chromosome 19: fine structure of the 11 *PSG* genes and identification of 6 new genes forming a third subgroup within the carcinoemryonic antigen (*CEA*) family. *Genomics* **23**: 669–684.

Thatcher T.H., Gorovsky M.A. (1994) Phylogenetic analysis of the core histones H2A, H2B, H3, and H4. *Nucleic Acids Res.* **22**: 174–179.

Thomas W.K., Wilson A.C. (1991) Mode and tempo of molecular evolution in the nematode *Caenorhabditis*: cytochrome oxidase II and calmodulin sequences. *Genetics* **128**: 269–279.

Tomlinson J.E., McLean J.W., Lawn R.M. (1989) Rhesus monkey apolipoprotein(a). Sequence, evolution, and sites of synthesis. *J. Biol. Chem.* **264**: 5957–5965.

Tovee M.J. (1994) The molecular genetics and evolution of primate colour vision. *Trends Neurosci.* **17**: 3036.

Turker M.S., Cooper G.E., Bishop P.L. (1993) Region-specific rates of molecular evolution: a fourfold reduction in the rate of accumulation of 'silent' mutations in transcribed versus nontranscribed regions of homologous DNA fragments derived from two closely related mouse species. *J. Mol. Evol.* **36**: 31–40.

von Poser C., Ichtchenko K., Shao X., Rizo J., Südhof T.C. (1997) The evolutionary pressure to inactivate. A subclass of synaptotagmins with an amino acid substitution that abolishes Ca^{2+} binding. *J. Biol. Chem.* **272**: 14 314–14 319.

Vrana P.B., Wheeler W.C. (1996) Molecular evolution and phylogenetic utility of the polyubiquitin locus in mammals and higher vertebrates. *Molec. Phylogenet. Evol.* **6**: 259–269.

Wacey A.I., Krawczak M., Kakkar V.V., Cooper D.N. (1994) Determinants of the factor IX mutational spectrum in haemophilia B: an analysis of missense mutations using a multi-domain molecular model of the activated protein. *Hum. Genet.* **94**: 594–608.

Wada K., Wada Y., Doi H., Ishibashi F., Gojibori T., Ikemura T. (1991) Codon usage tabulated from the GenBank genetic sequence data. *Nucleic Acids Res. Suppl.* **19**: 1981–1986.

Wallis M. (1994) Variable evolutionary rates in the molecular evolution of mammalian growth hormones. *J. Mol. Evol.* **38**: 619–627.

Wallis M. (1996) The molecular evolution of vertebrate growth hormones: a pattern of near-stasis interrupted by sustained bursts of rapid change. *J. Mol. Evol.* **43**: 93–100.

Wallis M. (1997) Function switching as a basis for bursts of rapid change during the evolution of pituitary growth hormone. *J. Mol. Evol.* **44**: 348–350.

Wang D.G., Fan J.-B., Siao C.-J. et al. (1998) Large-scale identification, mapping, and genotyping of single-nucleotide polymorphisms in the human genome. *Science* **280**: 1077–1082.

Wilbur J.W. (1985) Codon equilibrium II: its use in estimating silent substitution rates. *J. Mol. Evol.* **21**: 182–191.

Wolfe K.H., Sharp P.M. (1993) Mammalian gene evolution: nucleotide sequence divergence between mouse and rat. *J. Mol. Evol.* **37**: 441–456.

Wu C.-I., Maeda N. (1987) Inequality in mutation rates of the two strands of DNA. *Nature* **327**: 169–170.

Wu W., Goodman M., Lomax M.I., Grossman L.I. (1997) Molecular evolution of cytochrome *c* oxidase subunit IV: evidence for positive selection in simian primates. *J. Mol. Evol.* **44**: 477–491.

Yokoyama R., Yokoyama S. (1990) Convergent evolution of the red- and green-like visual pigment genes in fish, *Astyanax fasciatus*, and human. *Proc. Natl. Acad. Sci. USA* **87**: 9315–9318.

Yokoyama S. (1995) Amino acid replacements and wavelength absorption of visual pigments in vertebrates. *Mol. Biol. Evol.* **12**: 53–61.

Yokoyama S. (1997) Molecular genetic basis of adaptive selection: examples from colour vision in vertebrates. *Annu. Rev. Genet.* **31**: 315–336.

Yokoyama S., Starmer W.T. (1992) Phylogeny and evolutionary rates of G protein α subunit genes. *J. Mol. Evol.* **35**: 230–238.

Yu M., Irwin D.M. (1996) Evolution of stomach lysozyme: the pig lysozyme gene. *Molec. Phylogenet. Evol.* **5**: 298–308.

Zhang J., Rosenberg H.F., Nei M. (1998) Positive Darwinian selection after gene duplication in primate ribonuclease genes. *Proc. Natl. Acad. Sci. USA* **95**: 3708–3713.

Zhou Y.H., Hewett-Emmett D., Ward J.P., Li W.-H. (1997) Unexpected conservation of the X-linked color vision gene in nocturnal prosimians: evidence from two bush babies. *J. Mol. Evol.* **45**: 610–618.

Zilliacus J., Carlstedt-Duke J., Gustafsson J.-Å., Wright A.P.H. (1994) Evolution of distinct DNA-binding specificities within the nuclear receptor family of transcription factors. *Proc. Natl. Acad. Sci. USA* **91**: 4175–4179.

Zilliacus J., Wright A.P.H., Carlstedt-Duke J., Gustafsson J-Å. (1995) Structural determinants of DNA-binding specificity by steroid receptors. *Mol. Endocrinol.* **9**: 389–400.

8

Contractions and expansions in gene size and number

There are now many pathological examples of the deletion, insertion, duplication and expansion of human genes causing inherited disease. Similar mutations have however also occurred over evolutionary time. Far from being invariably disadvantageous, such mutational changes have often been recruited by the opportunistic evolutionary process and now contribute to both gene and genome architecture. These types of mutation have led to significant changes in gene size and number in different lineages and their contribution to the evolution of extant human genes will now be reviewed.

8.1 Gross gene deletions in evolution

8.1.1 Gross gene deletions during primate evolution

Gross gene deletions may arise through a number of different recombinational mechanisms but probably the most common is likely to be *homologous unequal recombination* (occurring either between related gene sequences or between repetitive elements). Thus, *Alu* sequences flanking deletion breakpoints have been noted in a considerable number of human genetic conditions and may represent favored sites for recombination and hotspots for gene deletions (Cooper and Krawczak, 1993; Chapter 1, section 1.5.4). Chromosomally duplicated regions (*duplicons*) are often common sites for pathological rearrangements, particularly gross deletions, since they have the potential to mediate homologous unequal recombination events (e.g. 15q11–q13; Christian *et al.*, 1999).

Not surprisingly, several instances of gross gene deletion have been noted during primate evolution. One such example is the loss of the γ1-globin gene in New World monkeys (with the notable exception of the capuchin monkey, *Cebus albifrons*) due to a 1.8 kb deletion which has removed most of exon 2, all of intron 2, exon 3, and much of the 3′ flanking region (Meireles *et al.*, 1995). As a result, γ2-globin is the primary fetally expressed globin gene in New World monkeys whereas in Old World monkeys, it is γ1-globin.

Another example of gross gene deletion in primates is the loss of one of the haptoglobin (*HPP*; 16q22.2) genes in humans which occurred, probably via homologous unequal recombination, after the separation of the human and chimpanzee lineages (McEvoy and Maeda, 1988; see Chapter 6, section 6.2.6). Finally, although there are two semenogelin genes (*SEMG1*, *SEMG2*; 20q12-q13.1) in humans and most Old World and New World monkeys, the *Semg2* gene has been deleted from the genome of the cotton-top tamarin (*Saguinus oedipus*) (Lundwall, 1998). There is no evidence for any selective advantage resulting from any of these gene deletions. In all cases cited, a similar paralogous gene was available to substitute for the deleted locus. This genetic redundancy probably ensured that, owing to the absence of purifying selection, such deletions came to be fixed through genetic drift.

8.1.2 Gross deletional polymorphisms

Gross gene deletions in two distinct glutathione S-transferase genes have been found to occur as polymorphic variants in various human populations. In humans, four gene families of glutathione S-transferases encode a series of enzymes responsible for the metabolism of a wide range of xenobiotics (DeJong *et al.*, 1991; Pearson *et al.*, 1993). The gene for the mu-class glutathione S-transferase (*GSTM1*; 1p13.3) is absent from between 10% and 64% of individuals depending upon the population under study (Board *et al.*, 1990; Nelson *et al.*, 1995; Seidegard *et al.*, 1988). Loss of the *GSTM1* gene is due to a 15 kb deletion probably brought about by homologous unequal recombination between two almost identical 4.2 kb repeats that flank the *GSTM1* gene (*Figure 8.1*; Xu *et al.*, 1998). These repeats are likely to have originated with the original duplication that gave rise to the mu-class glutathione S-transferase genes more than 20 Myrs ago (Xu *et al.*, 1998). A similar deletional polymorphism also occurs in the theta-class glutathione S-transferase (*GSTT1*; 22q11.2) gene; the gene is absent in ~38% of the Caucasian population (Brockmoller *et al.*, 1992; Pemble *et al.*, 1994). Since the *GSTM1* and *GSTT1* genes are involved in xenobiotic metabolism, it is quite possible that the presence/absence of these polymorphic alleles are not selectively neutral (Weber, 1997).

Dichromacy (color blindness) is very common (occurring at a polymorphic frequency (5–8%) in Caucasian males) with individuals so affected having either a deletion of the green color pigment (*GCP*) gene or possessing a hybrid red/green color pigment (*RCP/GCP*) gene in its place (Deeb *et al.*, 1995). Further variable gene number polymorphisms, which probably also arose by homologous unequal recombination, are apparent in the human α-amylase gene cluster at chromosome 1p21 (Groot *et al.*, 1989) and the pepsinogen A gene cluster at 11q13 (Taggart *et al.*, 1987).

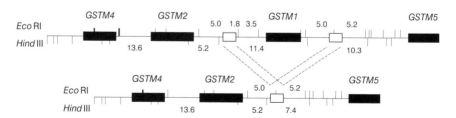

Figure 8.1. Model for homologous recombination between 4.2 kb repeats (open boxes) flanking the human mu-class glutathione S-transferase (*GSTM1*) gene leading to gene deletion (Xu *et al.*, 1998). The *GSTM1*, *GSTM2*, *GSTM4* and *GSTM5* genes are represented by solid bars. Vertical lines denote *Eco*RI and *Hind*III restriction sites. The sizes of the *Eco*RI/*Hind*III fragments are shown.

In the first case, a 'short' haplotype lacks *AMY1A*, *AMY1B* and the pseudogene *AMYP1* whilst a 'long' haplotype contains two extra copies of a duplicated fragment containing *AMY1A*, *AMY1B*, and *AMYP1*. In the second case, the three most common haplotypes were PGA-A (containing the *PGA3*, *PGA4*, and *PGA5* genes), PGA-B (containing the *PGA3* and *PGA4* genes) and PGA-C (containing only the *PGA4* gene). Copy number polymorphisms due to gene deletions have also been reported at the human T-cell receptor β (*TCRB*; 7q35) and γ (*TCRG*; 7p15) loci (Ghanem *et al.*, 1989; Rowen *et al.*, 1996), the ζ-globin (*HBZ*; 16p13.3) gene (Felice *et al.*, 1986), the rhesus blood group D antigen (*RHD*; 1p34–36.2) gene (Colin *et al.*, 1991), the immunoglobulin heavy chain constant region γ4 (*IGHG4*; 14q32) gene (Rabbani *et al.*, 1996) and the complement C4A (*C4A*; 6p21.3) and C4B (*C4B*; 6p21.3) genes (Teisberg *et al.*, 1988). In some gene clusters, it can be difficult to ascertain whether polymorphic alleles have arisen by gene deletion or duplication; some examples of human duplicational polymorphisms are given in Section 8.5.

8.2 Microdeletions in evolution

Is it possible to extrapolate from lessons learned through the study of microdeletions in a pathological context to microdeletions that have occurred during gene evolution? In particular, can we gain insight into the nature of the generative mechanism(s) underlying evolutionarily significant microdeletions and the possible influence of the local DNA sequence environment? Although in principle the answer to this question is likely to be in the affirmative, in practice the DNA sequences that were originally responsible for mediating the microdeletions have often decayed or been lost and it may not always be possible to reconstruct them.

8.2.1 Microdeletions in pathology

Microdeletions (<20 bp) causing human genetic disease were analyzed by Cooper and Krawczak (1993) in an attempt to relate the presence of specific DNA sequence motifs in the vicinity of these lesions to possible mechanisms responsible for their generation. In many cases, slipped mispairing at the replication fork between homologous sequences in close proximity to one another on complementary DNA strands appeared to be the causative mechanism. Slipped mispairing probably occurred either between *direct repeats* or through the formation of secondary structure intermediates potentiated by the presence of *inverted repeats* or *symmetric elements* (Cooper and Krawczak, 1993; Krawczak and Cooper, 1991). A consensus sequence, TGRRKM, common to pathological deletion hotspots has been noted in a number of different human genes (Krawczak and Cooper, 1991). This *deletion hotspot consensus sequence* is similar to the core motifs, TGGGG and TGAGC, found in immunoglobulin switch (Sμ) regions (Gritzmacher, 1989) and to putative arrest sites for DNA polymerase α (Weaver and DePamphilis, 1982). Cooper and Krawczak (1993) also found that a second motif (polypyrimidine runs of at least 5 bp; YYYYY) was over-represented in the vicinity of short human gene deletions whilst Monnat *et al.* (1992) observed a significant association between *HPRT1* (Xq26.1) gene deletion breakpoints and CTY vertebrate topoisomerase I cleavage sites. In principle, such sequence motifs may also have promoted the occurrence of microdeletions during evolution. Indeed, in probably the largest study of its kind to date, a sequence comparison of orthologous and paralogous members of the primate T-cell receptor β (*TCRB*; 7q35) gene family, microdeletion breakpoints appear to be frequently flanked by polypyrimidine runs and sequences which possess marked homology to the deletion hotspot consensus sequence (Funkhouser *et al.*, 1997).

8.2.2 Microdeletions mediated by direct repeats

Micro-deletions occur during gene evolution at a frequency 10% that of nucleotide substitutions (Saitou and Ueda, 1994). Pairwise comparisons of the noncoding regions of human, rabbit and murine β-globin genes have shown that they differ from each other in terms of numerous deletions/insertions and Efstratiadis *et al.* (1980) proposed that the short direct repeat (2–8 bp) sequences immediately flanking these sites could have templated the generation of these lesions by slipped mispairing (*Figure 8.2*). Direct repeats may also have been involved in generating the two inactivating single base-pair deletions (del C822 and del G904) noted in the human α-1,3 galactosyltransferase (*GGTA1*; 9q34) gene (Larsen *et al.*, 1990; see Chapter 6, section 6.2.2). Since chimpanzees possess both these deletions, whereas orangutan and gorilla only have del904, it would

```
5' FLANKING
Human  δ      CCA-------GCATAAAA
Human  β      CCAGGGCTGGGCATAAAA
LARGE INTRON
Human  Aγ     ATGACTTTT----ATTAGAT
Human  Aγ     ATGACTTTTCTTTATTAGAT
Human  Aγ     TGTGTGTGTGTGTGTG-------------------TGTGTGTGTGTG
Human  Gγ     TGTGTGTGTGTGTGTGTGTGTGCGCGCGTGTGTTTGTGTGTGTGTG
Rabbit β      AAGTA---------------CTTTCTCTAATC
Mouse  β      AGTCCTTCTCTCTCTCCTCTCTCTTTCTCTAATC
Rabbit β      TGGTAG-----------------------------AAACAACT
Mouse  β      TTGGCTTTTATGCCAGGGTGACAGGGGAAGAATATATTTTACATAT
Mouse  βmaj   TGACATAGG------------------------------------ATTCT
Mouse  βmin   TGTCATAGAATAATTCTTTTTTATTTTTTTATTTATTATTTTTTTCATAGAATAATTCT
Mouse  βmaj   TGTGTGTGG-----------AGTGTT
Mouse  βmin   GGTGTGTGGATGTGAATTGTGAGTGTT
3' NONCODING
Human  ε      CAGGT----------GTTCCT
Human  β      CCAATTTCTATTAAAGGTTCCT
Human  Aγ     GCAATACAAA-------------------------------------------TAATAAAAT
Human  β      GTCCAACTACTAAACTGGGGGATATTATGAAGGGCCTTGAGCATCTGGATTCTGCCTAATAAAAA
Human  Aγ     ATACAA-------------------------------------TAATAAA
Human  δ      TATTTTCTGAACTTGGGAACAATGAATACTTCAAGGGTATGGCTTCTGCCTAATAAA
Rabbit β      AAAAATTAT------------------------------------GGGGACA
Human  β      AATTTCTATTAAAGGTTAATTTGTTCCCTAAGTCCAACTACTAAACTGGGGGATA
```

Figure 8.2. Examples of deletions flanked by short direct repeats within non-coding sequences of mammalian β-globin genes (redrawn from Efstratiadis *et al.*, 1980). Pairwise alignments of sequences within non-coding regions of mammalian β-globin genes are shown. A deletion is assumed in the upper sequence with respect to the lower sequence. Dashes denote the nucleotides not present in the upper sequence. Short direct repeats are underlined. The two aligned human Aγ large intron sequences are those of two different alleles.

appear that del904 was the original inactivating mutation (Galili and Swanson, 1991). C822 is immediately flanked by imperfect direct repeats (TACAGGC**C**T and TACAAGGCAG, where **C** is nucleotide 822) that could have templated the 1 bp deletion via slipped mispairing. Slipped mispairing may also have been responsible for the G904 deletion since G904 is the 3' most base of a string of five Gs.

A direct repeat may also have templated the single base deletion in the 5' flanking region of the human interferon α10 (*IFNA10*; 9p22) gene relative to the other α-interferon genes (see *Figure 4.27*). Flanking direct AGGT repeats appear to have mediated an AGG deletion exhibited by both orthologous and paralogous members of the primate T-cell receptor β (*TCRB*; 7q35) gene family whilst in the same gene family, overlapping 7 bp direct repeats (CTTTTCTTTTCT) may have served to template a TTTCT deletion (Funkhouser *et al.*, 1997).

8.2.3 Microdeletions mediated by inverted repeats

Inverted repeats may also have mediated the generation of microdeletions during gene evolution. One example is the inactivating 13 bp deletion in exon 2 of the gibbon urate oxidase gene (Wu *et al.*, 1992; see Chapter 6, section 6.2.1); two imperfect inverted repeats (CAAGAAC and GTTCATG) span the breakpoints of this deletion. A 20 bp deletion has been reported from the 5' region of the δ-globin (*Hbd*) gene of the colobus monkey, *Colobus polykomos* (Vincent and Wilson, 1989). This deletion, which spans the transcriptional initiation site used in Old World monkeys and anthropoid apes, is responsible for a five-fold reduction in *Hbd* gene transcription as assessed by *in vitro* transcription assay. Inspection of the putative deleted bases and the flanking DNA sequence reveals the presence of a 13 bp imperfect inverted repeat which could have been responsible for the deletion through formation of a hairpin loop (*Figure 8.3*).

An inverted repeat also appears to have templated the deletion of a GAT codon in the human interferon α2 (*IFNA2*; 9p22) gene relative to the other α-interferon genes (*Figure 4.27*). Finally, a contiguous inverted repeat sequence (ATTC-CCAGTTTCTGGGAAT) may well have templated an 8 bp deletion exhibited by both orthologous and paralogous members of the primate T-cell receptor β (*TCRB*; 7q35) gene family (Funkhouser *et al.*, 1997).

```
                             +1
     C  aagggagggcagag-------------------CTTCTGA
     R        g   a      a  ccaactgttgcttATACTTG
     B        g   g      a  tcaactgttgcttACATTTG
     S        c   g      g  ctaactgttgcttTGACTTG
     H        c   g      a  tcgactgttgcttACACTTT
```

Figure 8.3. Alignment of δ-globin (*Hbd*) gene sequences from colobus monkey (C), rhesus macaque (R), baboon (B), spider monkey (S) and human (H) showing the location of a 20 bp deletion in the colobus monkey (after Vincent and Wilson, 1989). Transcribed sequences are denoted by upper case letters, flanking regions are in lower case letters. +1 denotes the site of trancriptional initiation. Underlined bases represent an imperfect inverted repeat which may contribute to the formation of a hairpin loop.

8.2.4 Microdeletions in vertebrate evolution

The comparison of gene/protein sequences between humans and the great apes also yields examples of *in-frame* microdeletions that must have occurred during primate evolution. Thus, amino acid residue Glu9 of the blue cone pigment protein (*BCP*; 7q31-q35) present in the talapoin monkey *Miopithecus talapoin* (an Old World primate) and in the marmoset *Callithrix jacchus* (a New World primate) is absent from the human protein and appears to have been deleted from the *BCP* gene within the human lineage (Hunt *et al.*, 1995). The functional consequences of the removal of this amino acid residue are however unclear.

Some gene regions harbor a disproportionate number of deletions/insertions inferred from alignment gaps noted in sequence comparisons e.g. exons 6 of the orthologous amelogenin (*AMELX*, Xp22.1-p22.31; *AMELY*, Yp11.2) genes of various vertebrates. These lesions have dramatically reduced the similarity between vertebrate amelogenins in the Pro/Gln-rich region of the protein as compared with that manifested by other regions (Toyosawa *et al.*, 1998).

Some microdeletions occurring during evolution may have been advantageous by virtue of their alteration of a protein product, others through a change in the reading frame bringing about gene inactivation (see Chapter 6, section 6.2). Of course, micro-deletions need not necessarily have conferred any selective advantage; even if merely neutral with respect to fitness, they could have become fixed by genetic drift alone.

8.3 Microinsertions in evolution

Microinsertions that have occurred during evolution have scarcely been studied. However, the underlying generative mechanisms are likely to be broadly similar to those causing human genetic disease. In their study of microinsertions in human genes causing inherited disease, Cooper and Krawczak (1991) concluded that insertional mutation involving the introduction of <10 bp DNA sequence into a gene coding region was not a random process and appeared to be highly dependent upon the local DNA sequence context. Further, mechanistic models which have explanatory value in the context of gene deletions were found to be useful in accounting for the nature and location of gene insertions.

In noncoding DNA, insertions are about half as frequent as deletions and mostly involve single nucleotides (De Jong and Ryden, 1981; Graur *et al.*, 1989; Saitou and Ueda, 1994). The rate of gap formation, regardless of whether caused by insertions or deletions, has been estimated to be $\sim$0.15–0.17 kb^{-1} Myrs^{-1} (Saitou and Ueda, 1994). In practice, studies of the DNA sequence environment of microinsertions that have occurred during evolution are likely to be rather difficult since the original sequence context of the insertion or deletion will often have become obscured by subsequent mutation.

8.3.1 Gene coding region microinsertions

Microinsertions occurring during human gene evolution can be found by sequence comparison of either orthologous or paralogous genes/proteins. Thus, the 54 bp insertion in the promoter of the human liver arginase (*ARG1*; 6q22.3-q23.1) gene is absent in the orthologous gene of *Macaca fascicularis* (Goodman *et al.*, 1994). Similarly, a 37 bp insertion has been introduced into the promoter of the orthologous Duchenne muscular dystrophy (*DMD*; Xp21) gene of the spider monkey, *Ateles geoffroy* (Fracasso and Patarnello, 1998); the inserted sequence is flanked by two TAAA repeats. A 12 bp insertion in the T-cell receptor α-chain (*TCRA*; 14q11.2) gene encodes an Ile-Pro-Ala-Asp tetrapeptide (residues 88–91) that is specific to the primate lineage (Thiel *et al.*, 1995). Interestingly, the region of the T-cell receptor α-chain protein between positions 86 and 91 appears to be a hotspot for insertional events during evolution: the rabbit and rat genes appear to have acquired a 3 bp (single amino acid) insertion, whereas the bovine, ovine, and murine genes manifest a 6 bp (double amino acid) insertion at this position (Thiel *et al.*, 1995). Finally, a 24 bp sequence found in the transmembrane domain region of the human glycophorin E (*GYPE*; 4q28-q31) gene appears to have been derived from the paralogous glycophorin B (*GYPB*; 4q28-q31) gene during primate evolution prior to the divergence of the gorilla from the lineage of the other great apes (Rearden *et al.*, 1993). It is unclear whether this insertion event was mediated by homologous unequal recombination or gene conversion.

8.3.2 Microinsertion polymorphisms

Several microinsertion polymorphisms have been reported in human genes. Thus, a single nucleotide insertion polymorphism has been noted in the promoter region of the insulin promoter factor 1 (*IPF1*; 13q12.1; Yamada *et al.*, 1998) gene. A single nucleotide insertion polymorphism is also present in the ABO blood group (*ABO*; 9q34; Olsson and Chester, 1996) gene which serves to inactivate it. Finally, a 9 bp insertion polymorphism in exon 9 of the cytochrome P450 *CYP2D6* (22q13.1) gene occurs in the Japanese population and is associated with a poor metabolizer phenotype (Yokoi *et al.*, 1996).

8.3.3 Indels

Clearly, in extant proteins, selection must have ensured the retention of essential features of tertiary structure. Indeed, insertions or deletions which altered the reading frame must have been rendered harmless in order for the protein to retain its biological activity. Thus, the insertion of bases inferred in one member of a paralogous protein pair often implies a counterbalancing deletion in the immediate vicinity (or *vice versa*) to restore the reading frame. Such 'indels' tend to involve sequences of between 1 bp and 5 bp in length (Pascarella and Argos, 1992). They are generally found in turn and coil structures and rarely interrupt α-helices and strands (Pascarella and Argos, 1992). One example of a simple indel occurring during evolution is the deletion of an AA doublet and its replacement with a GT doublet in the 5′ flanking region of the human interferon "α9" gene (*Figure 4.26*).

An example of a more complex indel that has occurred during vertebrate evolution is provided by the human chorionic gonadotropin β-subunit (*CGB*; 19q13.3) gene and is responsible for the introduction of a 24 amino acid C-terminal extension to the protein product (Talmadge *et al.*, 1984). The *CGB* gene emerged as a result of the duplication of an ancestral β-luteinizing hormone-like (*LHB*; 19q13.3) gene. A single base deletion (A1540) altered the translational reading frame of the ancestral gene allowing read-through into what was originally 3′ untranslated region (UTR). An insertion of a CG dinucleotide also occurred at nucleotide 1612 within the ancestral 3′ UTR region and served to extend the chorionic gonadotropin β-subunit protein by a further 8 amino acids. The *CGB* gene therefore evolved from its *LHB*-like ancestor by acquiring 8 amino acids through translation in a new reading frame and incorporating 24 novel amino acids from the 3′ UTR into its coding sequence.

8.4. Insertion of transposable elements in evolution

Transposable elements in the human genome are essentially of two kinds, those

Table 8.1. Retroelements in the human genome

Retroelement type	Retroelement family	Copy number	% of genome
C-type-related HERVs	HERV-ER1 superfamily		0.07%
	HERV-E (4–1, ERVA, MP-2)	35–50	
	HERV-E LTR	500–600	
	51–1	35–50	
	ERV1	10–15	
	HERV-R (ERV3)	10	
	RRHERV-1	20	
	S71	15–20	
	S71 LTR	50–100	
	ERV-FRD	5–7	
	ERV9	30–40	0.2%
	ERV9 LTR	3000–4000	
	HERV-P (HuERS-P1, HuERS- P2, HuERS-P3/HuERS-P)	50–90	0.01%
	HERV-I (RTV$_L$-I)	25–50	0.01%
	ERV-FTD	5–7	
C-type and HTLV-related HERVs	HERV-H (RTVL-H, RGH)	900–1000	0.2%
	HERV-H-LTR	1000	
	HRES1	2	
A-B- and D-type-related HERVs	HML families 1-6 HERV-K (HM, HLM, HML-1)	50	0.5%
	HERV-K LTR	10000–25000	
	ERV-MLN (HML-4)	20–25	
THE-1 elements	THE 1	10000	1%
	TTHE-1 LTR	30000	
Nonviral retroposons	LINE-1	100000	5%
	Alu	500000	5%
	SINE-R	5000	0.1%

that undergo transposition through a DNA intermediate, and those that undergo transposition through an RNA intermediate. Transposable elements with a DNA intermediate are termed *transposons* and these are characterized by terminal inverted repeats and duplication of the target site (visible as direct repeats flanking the element). However, the great majority of transposable elements in the human genome have undergone retrotransposition through an RNA intermediate. Such *retroelements* or *retroposons* may be of either viral or nonviral origin (*Table 8.1*). Whilst endogenous retroviral sequences comprise some 0.1–0.6% of the human genome (Leib-Mösch *et al.*, 1990), the nonviral *Alu* sequences and LINE elements may together make up as much as 10% of the genome.

8.4.1 Endogenous retroviral sequences and transposable elements

Retroposons. A number of retroposon families have been characterized in primate genomes (Leib-Mösch and Seifarth, 1996; McDonald, 1993; *Table 8.1*). Occasionally, these are human-specific (e.g. the HERV-K10-related SINE-R.C2; Zhu *et al.*, 1992; 1994) but usually they are found distributed through the genomes of other primate species (e.g. RTVL-H (Goodchild *et al.*, 1993), RTVL-I (Maeda and Kim, 1990), HERV-K (Steinhuber *et al.*, 1995), and HERV-L (Cordonnier *et al.*, 1995)). Type I and II HERV-H elements were amplified to ~1000 copies after the divergence of New World from Old World monkeys but before the divergence of apes from Old World monkeys (Leib-Mösch and Seifarth, 1996). By contrast, the family of type Ia HERV-H elements expanded to ~100 copies only after the divergence of apes from Old World monkeys. Analysis of the copy number, distribution and sequence characteristics of such endogenous elements promises to provide important clues as to the evolutionary history and phylogeny of the various mammalian orders, suborders, and species, and even the population genetics of human racial groups (Furano and Usdin, 1995).

Retroposons have sometimes become integrated into the vicinity of human genes. Thus, two copies of an RTV_L-I sequence are present in the human haptoglobin (*HP*; 16q22.1) gene cluster whilst an additional copy has been inserted in the same region in the orthologous chimpanzee gene (Maeda and Kim, 1990). Rather more dramatically, the endogenous retrovirus, HRES-1 lies within the coding sequence of the human transaldolase gene (*TALDO1*; 11p15; Banki *et al.*, 1994).

One view of endogenous retroviral elements is that they inserted themselves into the germline of our primate ancestors during the last 40 Myrs as a result of infection with exogenous retroviruses, persisting thereafter as proviruses, albeit rendered replication-defective by multiple mutational events (Shih *et al.*, 1991). Another (not incompatible) view is that retroviruses themselves originally arose from intracellular retroelements (the protovirus hypothesis), a view which is supported by the phylogenetic analysis of endogenous retroviral DNA sequences (*Figure 8.4*). Regardless of whether or not the horizontal transmission of retroviral elements has taken place, copy number amplification has certainly occurred.

Retrotransposition often generates retroposon sequence variants owing to its inherent imprecision: target site rearrangements combine with the infidelity of both reverse transcriptases and RNA-dependent RNA polymerases to ensure that the inserted sequences are highly variable thus providing new avenues for the highly opportunistic evolutionary process (Preston, 1996).

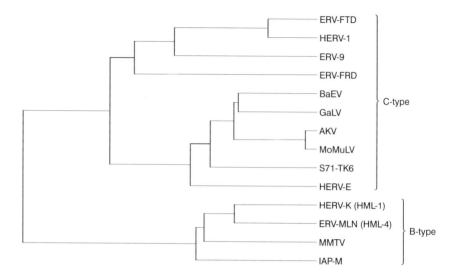

Figure 8.4. Phylogenetic analysis of mammalian endogenous retroviral *pol* sequences (after Leib-Mösch and Seiforth, 1996). HERV: Human endogenous retroviral elements ERV: Endogenous retrovirus BaEV: Baboon endogenous virus GaLV: Gibbon ape leukemia virus AKV: Endogenous murine leukemia virus MoMuLV: Moloney murine leukemia virus MMTV: Mouse mammary tumor virus IAP-M: Murine intracisternal A-type particles.

Transposons. Transposon-like THE-1 repeats, which lack any obvious homology to retroviral sequences, have been found in the deletion-prone intron 43 of the dystrophin (*DMD*; Xp21.2-p21.3) gene (Pizzuti *et al.*, 1992), the human blood group *GC* (4q12) gene (Witke *et al.*, 1993) and the 3′ untranslated region of the human calmodulin-related protein (*CALML1*; 7p13-pter) gene (Deka *et al.*, 1988b). A cluster of three THE-1 repeats located in a 26 kb region of intron 7 in the human *DMD* gene has arisen by three independent insertion events (McNaughton *et al.*, 1993, 1995). There is some evidence to support the hypothesis that the insertion of these elements has occurred at preferred target sites (Deka *et al.*, 1988a).

A pseudoautosomal gene sequence (*Tramp*) has recently been isolated which encodes within its single exon a protein with homology to transposases (enzymes that mediate transposition) of the Ac family (Esposito *et al.*, 1999). It is as yet unclear whether the Tramp protein has been involved in the transposition of other transposable elements or if it has instead become specialized for a novel cellular function. The centromeric protein CENP-B (*CENPB*; 20p13) may also represent an example of a transposase-encoded protein which has acquired a cellular function. This protein binds to the CENP-B box (TTCGNNNNANNCGGG) sequence in the alpha satellite DNA of human centromeres and has sequence similarity to the *pogo* family of transposases which includes the *Tigger* elements (Kipling and Warburton, 1997). Since CENP-B has nicking activity, it may promote homologous recombination and could have contributed to the species-specific patterns of evolution of satellite repeats.

Since many transposable elements contain enhancer sequences, their transposition may have served to alter the pattern of host gene expression at or around the integration site. Thus, once transposed, evolution may have recruited such enhancers to play a role in the transcriptional regulation of a gene in the vicinity of the integration site (see Chapter 5, section 5.1.12). One example of this is the human salivary amylase (*AMY1C*; 1p21) gene where the HERV-E-derived enhancer may be involved in tissue-specific expression (Ting *et al.*, 1992; Chapter 5, section 5.1.12, *Endogenous retroviral elements*).

Evolution may also recruit transposable elements as a means to alter mRNA processing. For example, a B2 (SINE) element has become inserted into the 3′ untranslated region of the murine (*lifr*) gene encoding the soluble form of the leukemia inhibitory factor receptor (LIFR; Michel *et al.*, 1997). Insertion of the B2 element has, by potentiating alternative 3′ mRNA processing and alternative splicing, given rise to a truncated mRNA species (relative to the mRNA encoding the membrane-anchored LIFR) which encodes soluble LIFR. In the rat, no such retrotranspositional event has occurred and the soluble form of LIFR is not found.

A very special case of the opportunistic recruitment of a transposable element may have been that of the recombination-activating gene (RAG) transposase postulated to have been inserted into an ancestral immunoglobulin/T-cell receptor gene soon after the divergence of jawed and jawless fishes (Chapter 9, section 9.4.2). The subsequent conversion of this transposon into a site-specific recombinase may have been the critical event in allowing the vertebrates to generate the genetic diversity so essential for the flexible adaptive response of their immune systems.

8.4.2 LINE elements

LINE elements have assumed considerable importance in the context both of gene pathology and gene evolution (Kazazian and Moran, 1998). They are present in a wide range of mammals (Furano and Usdin, 1995) and are represented by some 40 subfamilies (Smit *et al.*, 1995). They are nonrandomly distributed in the human genome, inserting preferentially into chromosomal G bands (Wichman *et al.*, 1992) and at the DNA level, into A-rich sequences (Vanin, 1984). The total numbers of LINE elements in four of the great apes have been estimated by Hwu *et al.* (1986): human, 107 000; chimpanzee, 51 000; gorilla, 64 000; and orangutan, 84 000. Since these figures differ markedly, it follows that numerous insertions and deletions of these sequences must have occurred during the evolution of the great apes.

In a pathological context, a number of examples of gene inactivation through insertion of LINE elements into gene coding sequences are known (Miki *et al.*, 1992; Narita *et al.*, 1992; reviewed by Cooper and Krawczak, 1993) and in some cases a preference for integration at AT-rich sequences is exhibited (Kariya *et al.*, 1987; Kazazian *et al.*, 1988). Further, the target sites of two LINE elements inserted into the factor VIII (*F8C*; Xq28) gene causing hemophilia A (Kazazian *et al.*, 1988) are 80% homologous to a 10 nucleotide motif (GAAGACATAC) present in one of the highly favored retroviral insertion target sequences reported by Shih *et al.* (1988).

Preferential target sites for LINE elements are also apparent in mammalian genes during evolution. For example, the interleukin-6 genes of rodents represent hotspots for LINE element retrotransposition (Qin *et al.*, 1991). During mammalian evolution, the introduction of LINE elements in the vicinity of genes has sometimes altered gene expression as a consequence of their being recruited to perform a regulatory function (examples of this phenomenon are given in Chapter 5, section 5.1.12, *Alu sequences*). LINE elements have also served to promote genetic rearrangements and indeed they may well have mediated both gene inversion (Chapter 9, section 9.1) and duplication (Section 8.5) events during evolution. Finally, and perhaps most importantly, LINE elements may have repeatedly transduced exons from one genomic location to another, thereby potentiating exon shuffling (Chapter 3, section 3.6.1), the transfer of exons encoding specific protein modules between genes.

8.4.3 Alu *sequences*

Evolution of Alu *sequences.* The fossil *Alu* monomer is thought to have arisen by the deletion of the central S domain of 7SL RNA (*RN7SL*) followed by the addition of a 3' poly(A) tract which may have facilitated reverse transcription of these RNA polymerase III transcripts (Mighell *et al.*, 1997; *Figure 8.5*). The free left arm monomer then arose by deletion of 42 bp from the fossil *Alu* monomer whilst the free right arm monomer arose by deletion of 11 bp from the fossil *Alu* monomer (Mighell *et al.*, 1997; *Figure 8.5*). The first *Alu* sequence may have been formed

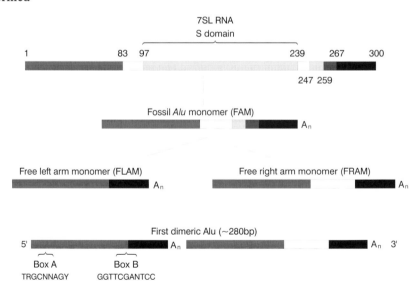

Figure 8.5. Proposed model of dimeric *Alu* formation via intermediate monomeric units derived from 7SL RNA which is neither capped nor polyadenylated. Important nucleotide positions are marked on the schematic 7SL RNA moiety. See text for explanation of the progression from 7SL RNA to the first dimeric *Alu* repeat. The approximate positions and consensus sequences of the RNA pol III promoter boxes A and B in the *Alu* sequence are marked. Redrawn from Mighell *et al.* (1997).

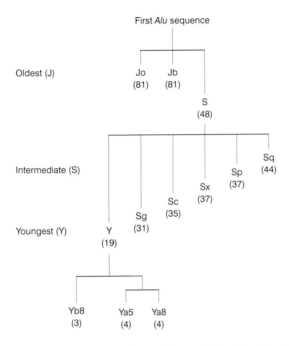

First *Alu* sequence

Figure 8.6. The proposed evolution of the 12 human *Alu* subfamilies. Numbers in parentheses represent approximate times (in Myrs) of insertion of different subfamilies into the human genome (redrawn from Mighell *et al.*, 1997).

by dimerization of a free left arm monomer with a free right arm monomer (*Figure 8.5*), an event which is thought to have occurred about 60 Myrs ago, before the divergence of prosimians (Zietkiewicz *et al.*, 1998). Subsequently, many rounds of sequential amplification took place to generate the 12 human *Alu* subfamilies seen today (Mighell *et al.*, 1997; *Figure 8.6*).

The total numbers of copies of *Alu* sequences in four of the great apes have been estimated by Hwu *et al.* (1986): human, 910 000; chimpanzee, 330 000; gorilla, 410 000; and orangutan, 580 000. As with the LINE elements, it would appear that numerous insertions and deletions of these sequences have occurred during the evolution of the great apes. At the chromosomal level, *Alu* sequences insert preferentially into R bands (Wichman *et al.*, 1992) whereas at the DNA level, they preferentially integrate into A-rich sequences (Batzer *et al.*, 1990; Daniels and Deininger, 1985; Matera *et al.*, 1990).

During mammalian evolution, the introduction of *Alu* sequences in the vicinity of genes has sometimes altered gene expression as a consequence of their being recruited to perform a regulatory function; examples of this phenomenon are given in Chapter 5, section 5.1.12, *Alu sequences*. *Alu* sequences may also have been involved in, or mediated, many other different types of gene rearrangement during gene evolution including gross deletions (Section 8.1), duplications (Section 8.5), transpositions (Chapter 9, section 9.2), gene fusions (Chapter 9, section 9.3), recombination (Chapter 9, section 9.4) and gene conversion events (Chapter 9, section 9.5).

Alu *sequence polymorphisms.* Once inserted, specific *Alu* sequences have often been relatively stable in terms of their location during primate evolution (Sawada *et al.*, 1985). This notwithstanding, some human *Alu* sequences are polymorphic in terms of their presence or absence (Batzer *et al.*, 1994; Edwards and Gibbs, 1992; Kass *et al.*, 1994; Meagher *et al.*, 1996; Milewicz *et al.*, 1996; Tishkoff *et al.*, 1996; Zucman-Rossi *et al.*, 1997), a situation which is sometimes also found in other primates (Bailey and Shen, 1993). Some of these polymorphisms may be of functional significance e.g. a common insertion/deletion polymorphism (0.40/0.60) within intron 16 of the human angiotensin I converting enzyme (*DCP1*; 17q23) gene, explicable in terms of the presence or absence of a 287 bp *Alu* repeat, is known to have an important influence on serum enzyme concentration (Rigat *et al.*, 1990). *Alu* sequence retrotransposition is also an occasional cause of genetic disease (e.g. Muratani *et al.*, 1991; Vidaud *et al.*, 1993; Wallace *et al.*, 1991).

Alu *sequence target sites.* Although *Alu* sequences occur on average every 3–6 kb, there are several examples of regions that appear to be preferential target sites for *Alu* sequence insertion in mammalian genes. Thus, a 40 kb region, spanning the spermatid-specific protamine genes *PRM1*, *PRM2* and the transition protein (*TNP2*) gene (16p13.2), contains a total of 42 *Alu* sequences (Nelson and Krawetz, 1994). Similarly, a 22 kb region telomeric to the HLA-B-associated transcript 2 (BAT2; *D6S51E*; 6p21.3) gene in the HLA class III locus contains 42 *Alu* repeats (Iris *et al.*, 1993), whilst a 2.2 kb segment 5′ to the human lysozyme (*LYZ*; 12) gene contains four such repeats (Riccio and Rossolini, 1993).

Alu *sequences within protein-coding sequences.* Many *Alu* sequences are found within introns and therefore this repeat is represented in heterogeneous nuclear RNA. *Alu* sequences are however also found at different locations within mRNA-homologous sequences, the majority occurring within the untranslated regions (UTRs). Thus, Yulug *et al.* (1995) reported that 5% of full-length human cDNAs contained an *Alu* sequence, with 82% and 14% of these being located in the 3′ UTR and 5′ UTR, respectively. In a few cases, however, *Alu* sequences have been incorporated into the coding sequences of human genes and have therefore altered the amino acid sequences of the encoded proteins. Thus, a 279 bp *Alu* sequence spans 103 bp of the coding region of a zinc finger protein (*ZNF91*; 19p12) gene and extends 166 bp into the 3′ UTR (Yulug *et al.*, 1995). Similarly, 110 bp of *Alu* sequence lies within the coding a region of the lectin-like type II integral membrane protein (*KLRC1*; 12) gene with 43 bp extending into the 3′ UTR (Yulug *et al.*, 1995). An *Alu* sequence is entirely contained within the coding region of the protein serine/threonine kinase stk2 (*STK2*; 3p21.1) (279 bp *Alu*; Yulug *et al.*, 1995) gene. Finally, and perhaps most dramatically, two *Alu* sequences (both with poly(A) tails) are entirely contained within the coding region of the regulator of mitotic spindle assembly 1 (*RMSA1*; 17p11.2-p12) gene accounting for 111 amino acids of its coding potential, some 40% of the total (Margalit *et al.*, 1994).

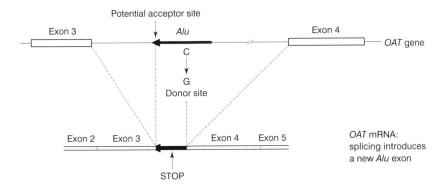

Figure 8.7. Ornithine δ-aminotransferase deficiency caused by a mutation in a resident intronic *Alu* element within the *OAT* gene. A point mutation within the right subunit of the inversely oriented intronic *Alu* repeat activates a donor splice site (from Labuda *et al.*, 1995).

Splice-mediated insertion of Alu *sequences. Alu* sequences have also been found to alter protein coding sequences through the *splice-mediated insertion* of the repeat and this probably represents the major mechanism by which *Alu* sequences have entered protein coding regions. *Figure 8.7* illustrates the principle involved by reference to the pathological example of ornithine δ-aminotransferase deficiency caused by a single base-pair substitution in an intronic *Alu* element in the human ornithine δ-aminotransferase (*OAT*; 10q26) gene; this lesion activates a cryptic donor splice site which results in the incorporation of the *Alu* sequence into the mRNA.

In an evolutionary context, there are several examples of the splice-mediated insertion of *Alu* repeats into human gene coding sequences. The splice-mediated insertion of a 95 bp *Alu* sequence has been reported in the lecithin: cholesterol acyltransferase (*LCAT*; 16q22.1) gene (Miller and Zeller, 1997). In humans, the alternate *Alu*-containing transcript represents between 5% and 20% of the *LCAT* mRNA. It is also present in *LCAT* mRNA from chimpanzee, gorilla and orang-utan; in the latter, the *Alu*-containing mRNA species constitutes 50% of the total *LCAT* mRNA pool (Miller and Zeller, 1997). It is not however present in the *LCAT* genes of gibbons, or Old World and New World monkeys (Miller and Zeller, 1997).

In the human biliary glycoprotein (*BGP*; 19q13.2) gene, three mRNA variants are produced as a result of the alternative splicing of an exon (IIa) with one of two virtually identical *Alu* cassettes derived from two intronic repeats (*Figure 8.8*). Other such examples are to be found in the human *REL* (2p12-p13) proto-onco-gene, and the complement decay-accelerating factor (*DAF*; 1q32) and complement C5 (*C5*; 9q33) genes (Makalowski *et al.*, 1994). Of the 17 *Alu* sequences found in mRNA coding regions by Makalowski *et al.* (1994), seven contained in-frame Stop codons and three others were predicted to cause frameshifts. Thus it is perhaps not surprising that in several cases of mRNAs containing *Alu* sequences, *allelic exclusion* is evident and the mRNA containing the *Alu* sequence is of low abundance compared to splice variants of the same gene that lack the repeat (Mighell *et al.*, 1997). This notwithstanding, it may well be that the splice-

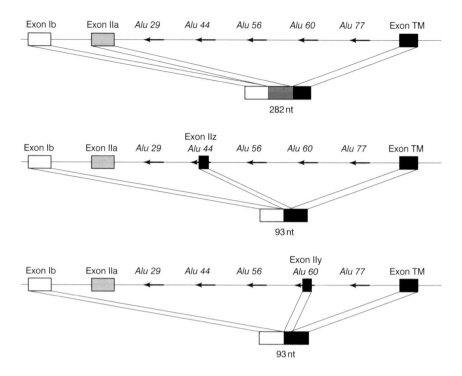

Figure 8.8. A scheme for alternative splicing in the human biliary glycoprotein (*BGP*) mRNA. Boxes represent exons and arrows the five intronic antisense *Alu* elements. TM, exon encoding the transmembrane domain. Dotted lines indicate the splicing patterns found in the three cDNAs (after Barnet *et al.*, 1993).

mediated insertion of *Alu* repeats has been an important evolutionary mechanism for creating diversity at the protein level.

Alu *sequence incorporation by intron sliding.* An alternative mechanism of *Alu* sequence incorporation into gene coding regions is intron sliding, and is illustrated by the example of the human *HLA-DRB1* (6p21.3) gene; an intronic *Alu* sequence has been incorporated into exon 4 of the HLA-DR-β1 mRNA (Labuda et al., 1995; Figure 8.9). Among three variants of the HLA-DR-β1 cDNA, detected by library screening, one was considered to be the usual form, whereas two others were alternatively spliced owing to a lack of splicing at the intron 5 donor site. As a result, exon 5 was extended into a nearby downstream Alu sequence in intron 5, either to include a stop codon within the *Alu* sequence, or to be spliced with exon 6 (the open reading frame in the extended exon 5 matches that of exon 6). These three cDNA clones may thus illustrate two phases of intron sliding: the inactivation of an existing splice site followed by the activation of a cryptic one.

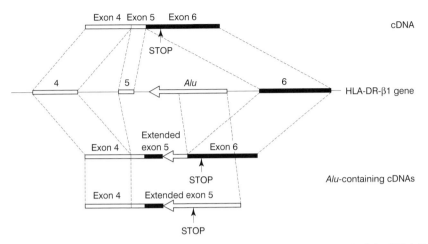

Figure. 8.9. A resident intronic *Alu* sequence is incorporated into exon 4 of the HLA-DR-β1 mRNA by intron sliding (from Labuda *et al.*, 1995). Among three variants of the HLA-DR-β1 cDNA, detected by library screening, the upper one is considered to be the usual form, whilst the two others are alternatively spliced, apparently due to a lack of splicing at the intron 5 donor site. As a result, exon 5 is extended into a nearby downstream *Alu* sequence in intron 5 either to include a stop codon within the *Alu* sequence or to be spliced with exon 6 (the open reading frame in the extended exon 5 matches that of exon 6). These three cDNA clones may illustrate two phases of 'intron sliding': inactivation of an existing splice site followed by activation of a cryptic one.

8.5 Gross gene duplications in evolution

Gene duplication (or partial duplication) events are a fairly uncommon cause of human genetic disease (reviewed by Hu and Worton, 1992; Mazzarella and Schlessinger, 1998). Two distinct mechanisms are currently envisaged: (i) *homologous unequal recombination* either between homologous chromosomes or sister chromatids and (ii) *nonhomologous recombination* at sites with minimal homology. Topoisomerase cleavage sites have been reported to be associated with pathological gene duplications (Kornreich *et al.*, 1990; Hu *et al.*, 1991) and potential sites for topoisomerases I and/or II have been found to coincide with the breakpoints of duplications in the human factor VIII (*F8C*; Xq28; Casula *et al.*, 1990) and dystrophin (*DMD*; Xp21.2-p21.3; Hu *et al.*, 1991) genes. These observations are potentially interesting since topoisomerase activity has been implicated in several cases of nonhomologous recombination (Bullock *et al.*, 1985).

One of the best studied gross duplications in human genome pathology is the 1.5 Mb duplication of the short arm of chromosome 17 associated with Charcot-Marie-Tooth disease type 1A. This recurring duplication is thought to be mediated by homologous unequal recombination between two misaligned ~30 kb CMT1A-REP repeat sequences flanking the CMT1A region in direct tandem orientation (Reiter *et al.*, 1996). In humans, these repeats are 98% identical. Chimpanzees have two copies of a CMT1A-REP-like sequence, whereas gorilla, orangutan and gibbon only have a single copy consistent with a duplication of the CMT1A-REP sequence after gorilla diverged from the human lineage but before the divergence of chimpanzee and human (Kiyosawa and Chance, 1996). Orthologous sequence comparison has provided evidence that the distal repeat was the progenitor copy.

8.5.1 Duplications and the emergence of paralogous genes

During vertebrate evolution, novel genes have arisen by genome duplication (tetraploidization; Chapter 2, section 2.1.1), intra-chromosomal regional duplication (Chapter 2, section 2.1.1), and localized individual gene duplication. All three mechanisms give rise to *paralogous* genes, genes that occur within the same species and which have a common ancestor. Paralogous genes therefore include the members of multigene families and superfamilies. Evidence for the common ancestry of paralogous genes may come from sequence homologies (e.g. as with the voltage-sensitive ion channel genes; Strong *et al.*, 1993) and/or from similar exon–intron organization, for example the cholesterol ester transfer protein (*CETP*; 16q21) and the phospholipid transfer protein (*PLTP*; 20q12-q13) (Tu *et al.*, 1995) genes, or the growth hormone receptor (*GHR*; 5p12-p14), prolactin receptor (*PRLR*; 5p13-p14) and interferon receptor α, β, and ω, 1 (*IFNAR1*; 21q22.1) genes (Lutfalla *et al.*, 1992).

8.5.2 Intra-chromosomal regional duplication

In the human genome, whole chromosomal segments have sometimes been duplicated (see Chapter 2, section 2.1) resulting in a series of paralogous genes retaining their syntenic arrangement (Endo *et al.*, 1997; Mazzarella and Schlessinger, 1997). Thus, a number of genes located at 6p21.3 have paralogous genes at 9q33-q34 (Endo *et al.*, 1997); the chromosome 6 loci include genes for type 11 collagen α2 subunit (*COL11A2*), *NOTCH4*, 70 kDa heat shock proteins (*HSPA1A*, *HSPA1B*, *HSPA1L*), valyl-tRNA synthetase 2 (*VARS2*), complement components (*C2*, *C4A*, *C4B*), pre-B cell leukemia transcription factor 2 (*PBX2*) and retinoid X receptor β (*RXRB*) whilst the chromosome 9 paralogues include *COL5A1*, *NOTCH1*, *HSPA5*, *VARS1*, *C5*, *PBX3*, and *RXRA*. Other extensive chromosomal duplications included genomic segments present at Xq28 and 16p11.1 involving the paralogous creatine transporter genes *SLC6A8* and *SLC6A10* respectively (Eichler *et al.*, 1996).

Some intra-chromosomal duplications may only involve one or a small number of genes, for example the duplication of the iduronate-2-sulphatase (*IDS*) locus at Xq28 (Timms *et al.*, 1995; Bondeson *et al.*, 1995). Another example is that of the inverted duplication at 5q13 which duplicated the spinal muscular atrophy (*SMA*) gene, the survival motor neuron (*SMN*) gene and the apoptosis inhibitory protein (*NAIP*) gene (Campbell *et al.*, 1997). Duplicated paralogous genes may however be translocated to quite different locations on the same chromosome, for example the adrenergic receptor (*ADRA1B* and *ADRB2*) genes on 5q23-q32 are quite distant from the evolutionarily related serotonin receptor (*HTR1A*) gene on 5cen-q11 (Oakey *et al.*, 1991).

8.5.3 Tandem duplications

Multigene families often form syntenic gene clusters as a result of the tandem duplication of an ancestral gene sequence. For example, the immunoglobulin genes are clustered at 14q32.33 (*IGHA, IGHD, IGHG*) and 22q11.2 (*IGLC, IGLL*) (see Chapter 4, section 4.2.4, *Immunoglobulin genes*), the T-cell receptor genes at 14q11.2 (*TCRA, TCRD*), 7q35 (*TCRB*) and 7p14-p15 (*TCRG*) (see Chapter 4, section 4.2.4, *T-cell receptor genes*), the pregnancy-specific glycoprotein (*PSG1, PSG2, PSG3, PSG4, PSG5, PSG6, PSG7, PSG8, PSG11, PSG12, PSG13*) genes at chromosome 19q13.2, the histocompatibility antigen (*HLA*) genes at chromosome 6p21.3 (see Chapter 4, section 4.2.1, *Genes of the major histocompatibility complex*) whilst the three alkaline phosphatase genes (*ALPP, ALPI, ALPPL2*) are clustered at chromosome 2q37.

Some genes appear especially prone to duplicate, probably by virtue of their already being clustered in multiple copies. Thus, the carcinoembryonic antigen (*PSG, CEA*; 19q13.2) gene family has undergone multiple, but independent, multiplication events in both the rodent and primate lineages (Rudert *et al.*, 1989). In similar vein, a disproportionate fraction of mapped zinc finger gene family members are located on chromosome 19 (Lichter *et al.*, 1992; Chapter 4, section 4.2.3, *Zinc finger genes*). Such clustering has probably arisen as a result of the serial duplication of a single ancestral gene on the same chromosome.

Members of gene families in different species may however be amplified differentially. For example, in the human genome, three members of the formylpeptide receptor gene family (*FPR1, FPRL1*, and *FPRL2*) cluster at chromosome 19q13.3, whereas in mouse, there are six *Fpr* genes on a syntenic region of chromosome 17 (Gao *et al.*, 1998). The human *FPRL2* gene and four murine *Fpr* genes arose after the divergence of human and mouse.

8.5.4 Translocation of duplicated genes

Gene family members have often become chromosomally separated through translocation (Chapter 9, section 9.2) and it would appear as if the older the multigene family, the more likely this is to happen. Thus, the chromosomally unlinked *CD36L2* (chromosome 4), *CD36L1* (chromosome 12) and *CD36* (7q11.2) membrane glycoprotein genes were duplicated and diverged from an ancestral gene prior to the separation of the arthropod and vertebrate lineages (Calvo *et al.*, 1995). Other examples of ancient duplicated genes that have become separated by translocation are the thrombospondin (*THBS1*, 15q15 and *THBS2*, 6q27) genes which arose ~900 Myrs ago (Lawler *et al.*, 1993) and the transforming growth factor-β (*TGFB1*, 19q13.2; *TGFB2*, 1q41; *TGFB3*, 14q24) genes which arose ~300 Myrs ago (Burt and Paton, 1992).

8.5.5 Duplication of translocated genes

Genomic duplication may be followed by further local tandem duplication as exemplified by the human homeobox gene clusters at 7p14-p15 (*HOXA*), 17q21-q22 (*HOXB*), 12q12-q13 (*HOXC*) and 2q31 (*HOXD*) (see Chapter 4, section 4.2.1, *Homeobox genes*). The same phenomenon is exhibited by the human sodium channel genes many of which reside in the same paralogous chromosome segments as the HOX gene clusters: *SCN1A*, *SCN2A*, *SCN3A*, *SCN6A*, *SCN7A* and *SCN9A* (2q23–q24), *SCN8A* (12q13), *SCN4A* (17q23–q25), *SCN5A* and *SCN10A* (3p21–p24) (Plummer and Meisler, 1999).

By contrast the localized duplication of translocated genes appears to be a general feature of the human olfactory receptor (*OLFR*) genes (Trask *et al.*, 1998b). Thus, several *OLFR* gene clusters on chromosome 11 (11p15, 11p13, 11q24; Buettner *et al.*, 1998) are more similar to each other than to *OLFR* genes on chromosome 17 (Ben-Arie *et al.*, 1994), implying that translocation was followed by regional tandem duplication. However, the *OLFR* gene cluster on chromosome 17p13.3 contains members that do not belong to the same subfamily, suggesting that it originated instead by duplication of an entire gene cluster followed by translocation of that cluster (Ben-Arie *et al.*, 1994).

8.5.6 Syntenic relationships and gene dispersal

Gene duplication sometimes creates gene families whose members are both syntenic and dispersed, for example the human purinoceptor gene family (*P2RY1*, 3; *P2RY2*, 11q13.5-q14.1; *P2RY4*, Xq13; *P2RY6*, 11q13.5, *P2RY7*, chromosome 14; Somers *et al.*, 1997). Seven human matrix metalloproteinase genes cluster at chromosome 11q22.3 (*MMP1*, *MMP3*, *MMP7*, *MMP8*, *MMP10*, *MMP12*, *MMP13*; Pendas *et al.*, 1996), but the other family members are dispersed between many other chromosomes (*MMP2*, 16q13; *MMP9*, 20q11.2-q13.1; *MMP11*, 22q11.2; *MMP14*, 14q11-q12; *MMP15*, 16q21; *MMP16*, 8q21; *MMP19*, 12q14). Other examples include the human fucosyltransferase (*FUT1*, *FUT2*, *FUT3*, *FUT5*, *FUT6*, 19p13.3; *FUT4*, 11q21; *FUT7*, 9q34; *FUT8*, 14q23; Costache *et al.*, 1997) and annexin genes (*ANX1*, 9q11-q22; *ANX2*, 15q21-q22; *ANX3*, 4q13-q22; *ANX4*, 2p13; *ANX5*, 4q28-q32; *ANX6*, 5q32-q34; *ANX7*, 10q21.1-q21.2; *ANX8*, 10q11.2; *ANX11*, 10q21.1-q21.2; *ANX13*, 8q24.1-q24.2; Morgan *et al.*, 1998).

In some cases, evolutionary conservation of post-duplicational clustering may be important for the coordinate regulation of individual genes by common control elements (see Chapter 5, section 5.1.14). Possible examples of this include the spermatid-specific protamine (*PRM1* and *PRM2*; 16p13.2) genes (Nelson and Krawetz, 1994), the platelet membrane glycoprotein (*ITGA2B* and *ITGA3*; 17q21-q22) genes (Bray *et al.*, 1988), the albumin gene family (*ALB*, *AFP*, *AFM*, *GC*; 4q11-q13; Nishio *et al.*, 1996), the pregnancy-specific glycoprotein (*PSG1*, *PSG2*, *PSG3*, *PSG4*, *PSG5*, *PSG6*, *PSG7*, *PSG8*, *PSG11*, *PSG12*, *PSG13*; 19q13.2) genes (Khan *et al.*, 1992) and the fibrinogen α, β and γ (*FGA*, *FGB* and *FGG*; 4q31) genes (Fu *et al.*, 1992; Roy *et al.*, 1994).

Individually duplicated genes may still exhibit synteny simply because recombination has not yet separated them. Thus, the paralogous pulmonary surfactant protein D (*SFTPD*) gene at 10q22-q23 lies in very close proximity to the pulmonary surfactant protein A (*SFTPA1*) gene (Kölble *et al.*, 1993). Similarly, the paralogous the interferon α/β receptor gene (*IFNAR1*) is closely linked to the cytokine receptor B4 (*CRFB4*) gene on chromosome 21q22.1 (Lutfalla *et al.*, 1995). Synteny may however be retained for very long periods of evolutionary time. Indeed, the human proteasome α2 subunit (*PSMA2*) and TATA box-binding protein (*TBP*) genes are linked on chromosome 6q27 and their orthologues are also syntenic in *Drosophila melanogaster* and *Caenorhabditis elegans* (Trachtulec, 1997). Similarly, conserved synteny is also apparent between the human genes *PIM1, RXRB, PBX2, NOTCH4* and *TNXA* at 6p21 and their orthologues in *D. melanogaster* and *C. elegans* (Trachtulec, 1997). The evolutionary conservation of close linkage over such long time periods is highly unusual and implies the existence of underlying functional reasons.

8.5.7 Functional redundancy and post-duplication diversification

Evolution is opportunistic and gene duplication provides the opportunity for structural and functional diversification. This is well exemplified by the origin of the α-lactalbumin gene (*LALBA*; 12q13). α-Lactalbumin may be regarded as essentially a mammalian 'invention'. It is a regulatory subunit of the enzyme lactose synthetase. Phylogenetic analysis has indicated that α-lactalbumin evolved from lysozyme (*LYZ*; chromosome 12) (a bacteriolytic enzyme present in both vertebrates and invertebrates) through a gene duplication event that occurred before the divergence of mammals and birds, and prior to the evolution of the mammary gland (Prager and Wilson 1988). The origin of the *LALBA* gene thus appears to have long preceded the acquisition of its modern function. Its rapid evolution in the mammalian lineage need not however have been solely due to its acquisition of a new function. Once its lysozyme activity was lost, many of the amino acids essential for the protein's original function were freed from selective pressure, allowing rapid change, not all of it necessarily adaptive (Nitta and Sugai, 1989; see also discussion on function switching in Chapter 7, section 7.1.3).

The protein products of a gene duplication may sometimes be targeted to different cellular locations. The human mitochondrial HMG CoA synthase (*HMGCS2*; 1p12-p13) gene encodes an enzyme which catalyses the first reaction of ketogenesis whilst the cytoplasmic form of the enzyme, encoded by the *HMGCS1* gene (5p13), catalyses the second step in cholesterol biosynthesis from acetyl CoA. The two genes are thought to have arisen from a gene duplication ~500 Myrs ago (Boukaftane *et al.*, 1994). The emergence of the two enzymes as distinct entities served to link the pathways of β-oxidation and leucine catabolism and created the HMG CoA pathway of ketogenesis thereby providing a lipid-derived energy source.

Gene duplication has also potentiated the emergence of *isozymes*, enzymes that catalyze the same biochemical reaction but which may differ from each other in terms of tissue specificity, developmental regulation or biochemical properties. Thus, in vertebrates, the two subunits of lactate dehydrogenase are encoded by two genes (*LDHA*; 11p14-p15 and *LDHB*; 12p12) and these subunits can be combined in such a way as to form five tetrameric isozymes (A_4, A_3B, A_2B_2, AB_3, and B_4) each with its own distinctive properties.

Iwabe *et al.* (1996) examined gene duplications during organismal evolution and concluded that most gene duplications giving rise to entirely novel functions predated the divergence of the vertebrate and arthropod lineages. Genes which encode proteins that are localized to cell compartments (compartmentalized isoforms) emerged by duplications which predated the separation of animals and fungi. By contrast, genes encoding products with virtually identical functions but differing tissue distribution (tissue-specific isoforms) have undergone duplications independently in vertebrates and arthropods after divergence of the vertebrate and arthropod lineages. Iwabe *et al.* (1996) concluded that there was a good correspondence between molecular evolution at the level of the gene, and tissue and organismal evolution.

Several gene duplications have occurred during primate evolution. For example, the tandem duplication responsible for the creation of the γ1- (*HBG1*) and γ2- (*HBG2*) globin genes occurred prior to the divergence of Old World from New World monkeys (Fitch *et al.*, 1991). The 5.5 kb duplicated segment is bounded by two related LINE elements suggesting that the duplication occurred via homologous unequal recombination (Fitch *et al.*, 1991). (The role of repetitive DNA sequences in mediating the recombinational events responsible for gene duplications is discussed in more detail in Chapter 9, section 9.4.1). Perhaps it was the functional redundancy initially introduced by the duplication which created an opportunity for the γ-globin genes to evolve a fetal function to replace their original embryonic function. Post-duplicational functional redundancy is sometimes still apparent in some systems. For instance, the MyoD family of transcription factors involved in myogenesis in skeletal muscle still exhibits functional redundancy (between the proteins encoded by the *MYOD1* (11p15.1), *MYF5* (chromosome 12), and *MYF6* (chromosome 12) genes) and this is probably indicative of overlapping functions (Atchley *et al.*, 1994).

The primate glycophorin genes (*GYPA, GYPB, GYPE*; 4q28-q31) are thought to have emerged by duplication, mediated perhaps by recombination between *Alu* sequences (Rearden *et al.*, 1993). *GYPA* probably represents the ancestral gene and is present in all primates studied. The *GYPB* gene is present in human, chimpanzee and gorilla but not orangutan and gibbon, whereas the *GYPE* gene is present in human, chimpanzee but intriguingly only in 7/16 gorillas tested (Rearden *et al.*, 1993). The complement C4 (*C4A, C4B*; 6p21.3) genes and the cytochrome *CYP21* (6p21.3) gene also emerged in the primate lineage as a result of a single duplication occurring prior to the divergence of the apes from the Old World monkeys (Horiuchi *et al.*, 1993).

There are several other examples of gene duplications occurring in only one mammalian order. For example, the bovine-specific coglutinin (*Cgn1*) gene is homologous to the human pulmonary surfactant protein D (*SFTPD*; 10q22-q23) gene and is thought to have evolved by gene duplication in the Bovidae after their divergence from the other mammals (Liou *et al.*, 1994). Sometimes the number and distribution of *Alu* sequences may help the reconstruction of the phylogeny of a gene duplication event as in the case of the duplication of the apolipoprotein CI (*APOC1*; 19q13.2) gene, timed at about 39 Myrs ago (Raissonier, 1991).

Some partially homologous genes may have undergone a process of duplication and divergence but one or other copy may have either acquired or lost DNA sequence at some stage thereby limiting the extent of observable homology between them. One example of this is the human α-*N*-acetylgalactosaminidase (*NAGA*; 22qter) and α-galactosidase A (*GLA*; Xq) genes (Wang and Desnick, 1991). Six of the seven *GLA* exons were identically positioned in the *NAGA* gene but there was no similarity between the predicted amino acid sequences of *GLA* exon 7 and *NAGA* exons 8 and 9.

8.5.8 Truncated gene copies

In some cases, only a portion of a gene may be involved in these intra-chromosomal duplication events [e.g. the polycystic kidney disease 1 (*PKD1*; 16p13.3) gene; Hughes *et al.*, 1995]. This often results, as in the *PKD1* example cited, in the generation of a linked pseudogene representing a partial copy of the parental source gene (for other examples, see Chapter 6, section 6.1.1). However, truncated copies need not always be inactive. Thus, the melanin-concentrating hormone (*PMCH*; 12q23-q24) gene has become partially duplicated during primate evolution to generate two truncated copies (*PMCHL1* and *PMCHL2*) which have been translocated to chromosome 5p14 and 5q12-q13 respectively (Viale *et al.*, 1998). These variant gene copies possess open reading frames, are expressed in a distinctive tissue-specific fashion and, in the authors' opinion, may represent 'genes in search of a function'.

8.5.9 Duplicational polymorphisms

Some gene duplications occur as polymorphic variants in the human population, for example the red (*RCP*; Xq28) and green (*GCP*; Xq28) visual pigment (*Figure 9.6*; Neitz *et al.*, 1995; Neitz and Neitz 1995) gene. Human males with trichromatic vision typically possess one *RCP* gene, one, two, or more *GCP* genes and an *RCP/GCP* hybrid gene (*Figure 9.6*; Drummond-Borg *et al.*, 1989; Nathans *et al.*, 1986a; 1986b; Neitz and Neitz, 1995; Neitz *et al.*, 1995). However, some males with apparently normal color vision can possess 4 *RCP* genes and 6 or 7 *GCP* genes (Neitz and Neitz, 1995). This notwithstanding, only one *GCP* gene is normally expressed, probably as a result of the activity of a locus control region upstream of the *RCP* gene (Winderickx *et al.*, 1992).

Other examples of duplicational polymorphisms include the α1-globin (*HBA1*, 16p13.3; Lie-Injo *et al.*, 1981), ζ-globin (*HBZ*, 16p13.3; Winichagoon *et al.*, 1982), Gγ-globin (*HBG2*; 11p15; Thein *et al.*, 1984), haptoglobin (*HP*; 16q22.1; Maeda *et al.*, 1986) and proline-rich protein (*PRB1*, *PRB2*, *PRB3*, *PRB4*; 12p13.2; Lyons *et al.*, 1988) genes. Duplicational polymorphisms may even manifest as gene cluster copy number variation, for example that involving the *CYP21*,

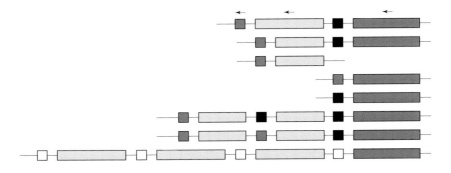

Figure 8.10. Common haplotypes of genes present on human chromosome 6p21.3 (after Figueroa, 1997). *CYP21* Open square. *CYP21P* Solid square. *C4B* Stippled square. *C4A* Hatched square.

CYP21P, *C4A* and *C4B* genes at human chromosome 6p21.3 (Collier *et al.*, 1989; Figueroa, 1997; *Figure 8.10*). Another example of a duplicational polymorphism is that of three members (one potentially functional) of the olfactory receptor (*OLFR*) gene family which occur in tandem within a block that is duplicated at 14 different subtelomeric locations in the human genome (Trask *et al.*, 1998a). This results in normal individuals possessing between 7 and 11 copies of this block in their genomes. Trask *et al.* (1998a) suggested that sub-telomeric regions could serve as 'nurseries' for the generation of diversity by promoting gene duplication.

Other gross duplicational polymorphisms are found in the immunoglobulin V_H gene cluster (*IGHV*; 14q32): one of 50 kb in length is present in 73% of individuals and results in the gain of 5 functional V_H segments (Walter *et al.*, 1993; Willems van Dijk *et al.*, 1992) whilst another of length ~80 kb is present in ~50% of individuals and involves the gain/loss of two functional V_H segments (Cook *et al.*, 1994). A third such polymorphism involving only a single additional V_H segment has been reported to occur in 27% of individuals (Cook and Tomlinson, 1995). A duplicational polymorphism is also apparent in the C_H gene cluster: the *IGHG4* gene (14q32) is duplicated in 44% of haplotypes (Brusco *et al.*, 1997). Finally, the partial duplication of the GABA A receptor α5 (*GABRA5*) gene is polymorphic in that individuals differ with respect to gene copy number (Ritchie *et al.*, 1998).

8.6 Intragenic gene duplications in evolution

Numerous examples of human genes have now come to light in which the encoded proteins have emerged through the introduction of individual exons or blocks of exons. Some of these exons encode specific protein domains (e.g. zinc finger, homeobox, immunoglobulin-like, epidermal growth factor-like, fibronectin, ABC cassette, Sushi, ankyrin, chymotrypsin etc; Doolittle, 1995; Henikoff *et al.*, 1997) which have come to be distributed between a large number of different proteins through *exon shuffling* (Chapter 3, section 3.6.1). Protein evolution may however also involve the internal duplication or amplification of individual exons, blocks of exons or alternatively repetitive sequence motifs within exons. Some typical examples of such intragenic duplication events are explored below.

8.6.1 Multi-exon duplications

Some genes encode proteins that comprise two homologous domains and are therefore likely to have originated through a gross internal duplication. Thus, the 17 exon human transferrin (*TF*; 3q21) gene is thought to have originated by an internal duplication which may have resulted from an unequal crossing over event (Park *et al.*, 1985; *Figure 8.11*). Similarly, the human angiotensin I converting enzyme is encoded by a gene (*DCP1*; 17q23) which comprises 26 exons that encode two homologous domains each containing an active site. Exon number and size in the *DCP1* gene are consistent with an ancient internal duplication, as are the codon phases at the exon-intron boundaries (Hubert *et al.*, 1991). An ancient

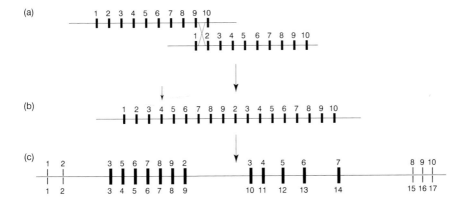

Figure 8.11. A possible scheme for the evolution of the human transferrin (*TF*) gene (after Park *et al.*, 1985). The ancestor of the transferrin gene was duplicated by an intragenic crossing over event (A) generating an internally duplicated gene which had lost one of its leader peptide coding exons and one of its terminal 3′ exons (B). During evolution, exon 4 (short arrow) was deleted. In C, the upper numbers correspond to the exon numbering in A and B whilst the lower ones correspond to the exons of the extant human transferrin gene.

internal duplication probably also occurred in the cystic fibrosis transmembrane conductance regulator (*CFTR*; 7q31.3) gene which encodes a protein with two transmembrane domains and two nucleotide binding fold domains (Hughes 1994). By contrast, successive duplications of ancestral domains appear to have occurred in the human kininogen (*KNG*; 3q27) gene (Kellermann *et al.*, 1986). Other examples of human proteins displaying internal domain duplication include calbindin (six 43 amino acid repeats), fibronectin (twelve 40 amino acid repeats), plasminogen (five 79 amino acid repeats) and α-tropomyosin (seven 42 amino acid repeats) (Li 1997). Many other genes are also likely to have evolved by internal gene duplication but the duplicated regions have probably diverged so much over evolutionary time that sequence homology between them is no longer discernible.

8.6.2 Exon duplication

It has been estimated that at least 6% of exons in human genes have arisen by the duplication of pre-existing exons (Fedorov *et al.*, 1998). One example is that of the human *CHC1* (1p36.1) gene, which encodes a protein involved in the coupling between DNA replication and mitosis. This gene comprises 14 exons, eight of which encode the 7 tandemly repeated domains of ~60 amino acids within the CHC1 protein; each repeat is encoded by a single exon except for repeat IV which is encoded by exons 10 and 11 separated by an inserted intron (Furuno *et al.*, 1991; *Figure 8.12*). The *CHC1* gene therefore appears to have arisen through the amplification of a primordial exon (Furuno *et al.*, 1991). Gene construction by exon amplification has also been employed in the macrophage mannose receptor (*MRC1*; 10p13; Kim *et al.*, 1992) gene; 26 of the 30 exons of the *MRC1* gene serve to encode the eight C-type carbohydrate recognition domains. Ceruloplasmin

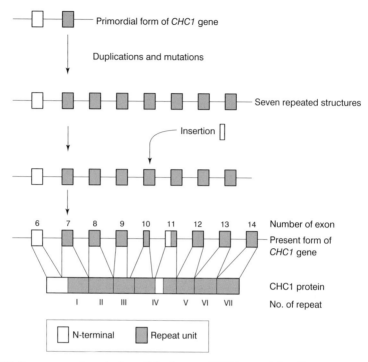

Figure 8.12. Structure and evolution of the human *CHC1* gene (after Furuno *et al.*, 1991).

(*CP*; 3q23-q25) consists of three homologous repeat units and appears to have evolved by successive duplication of a primordial copper-binding protein domain of ~350 amino acid residues (Ortel *et al.*, 1984). Finally, the human annexin II (*ANX2*; 15q21-q22) gene encodes a protein containing 8 copies of a conserved 70 amino acid repeat (Spano *et al.*, 1990).

Exon duplication/amplification is clearly a common mechanism of evolutionary change and its occurrence may often be inferred from studies of exon/intron distribution or from the repetitivity of protein domains (Chapter 3, section 3.6). Thus, the six paralogous genes of the human salivary proline-rich protein family, closely linked on chromosome 12p13.2, differ from each other in terms of the number of copies of a 63 bp tandem repeat in exons 3 of the gene: *PRH1* and *PRH2* (6 repeats), *PRB1* and *PRB3* (15 repeats), *PRB2* (16 repeats) and *PRB4* (11 repeats) (Azen *et al.*, 1987; Kim *et al.*, 1993). Examples of orthologous gene pairs in which only one orthologue manifests exon duplication/amplification are somewhat rarer. The oxidized low density lipoprotein receptor 1 (*OLR1*; 12p12.3-p13.1) gene serves to illustrate the principle; the rat *Olr1* gene encodes three 46 amino acid repeats between the transmembrane and lectin-like domains of the LOX-1 protein, whilst the human and bovine *OLR1* genes encode only one such repeat (Nagase *et al.*, 1998).

The primate semenogelin genes (*SEMG1*, *SEMG2*) which, in human, are closely linked on 20q12-q13.1, encode the major protein constituents of the seminal fluid. These genes differ however between species as a result of internal duplications of ~180 bp segments encoding 60 amino acid repeats. Human semenogelin II contains two fewer repeats than rhesus monkey semenogelin II which appears to contain two fewer repeats than baboon semenogelin II (Ulvsbäck and Lundwall, 1997). The primate semenogelin genes arose by duplication of an ancestral gene about 60 Myrs ago (Lundwall, 1996) and the human semenogelin I gene appears to have lost two 60 amino acid repeats by comparison with the paralogous semenogelin II gene (Ulvsbäck and Lundwall, 1997). The semenogelin I protein from cotton-top tamarin (*Saguinus oedipus*) possesses three additional repeats as compared to human and, intriguingly, possesses 14 potential glycosylation sites not present in the human protein (Lundwall, 1998).

Most known genes manifesting internal duplication exhibit strong conservation of intron positions between the duplicated domains (Chapter 3, sections 3.1 and 3.6.2). One notable exception, however, is the rabbit phosphofructokinase gene (Lee *et al.*, 1987). Sequence homologies between bacterial and rabbit phosphofructokinases and between the amino and carboxy terminal ends of the rabbit enzyme are consistent with the origin of this gene being via a process of internal duplication and divergence. However, intron positions are not conserved even between the two halves of the rabbit gene.

Exon duplication may potentiate alternative splicing. Thus, the human ketohexokinase (*KHK*; 2p23) gene contains two very similar 135 bp exons (termed 3a and 3c) which are mutually exclusively spliced into *KHK* mRNA (Hayward and Bonthron, 1998). Both the exon-intron structure and the pattern of alternative splicing are conserved between human, rat and mouse, consistent with the existence of two evolutionarily conserved KHK isoforms. This alternative splicing event is also tissue-specific since, in both rat and human, those tissues that express high levels of KHK incorporate exon 3c whereas other tissues incorporate exon 3a. Interestingly, a shift in splicing choice from exon 3a to 3c appears to occur during development between the human fetus and adult (Hayward and Bonthron, 1998).

8.6.3 Intra-exonic duplications

Internal gene duplications may sometimes be quite subtle. Thus, the 3′ untranslated regions of the human and monkey cytochrome *c* oxidase subunit II (*MTCO2*; mitochondrial genome) genes have been generated by duplication events involving a 13 bp region that occurred during primate evolution (Ramharack and Deeley, 1987). The RNA derived from the *MTCO2* gene has the potential to form stable stem-loop structures in a region immediately preceding the duplication site and these inverted repeat sequences may have played a role in promoting the duplicational events.

In some cases, homogenization of internal repeats has occurred as for example with the complement control protein repeats of the baboon and human complement receptor type 1 (*CR1*; 1q32; Clemenza *et al.*, 1997; Hourcade *et al.*, 1990) genes. This process has been termed *horizontal* or *concerted evolution* and the mechanism involved is likely to be either unequal crossing over or gene conversion. Whichever mechanism is responsible, the following example suggests that intragenic homogenization may be a less efficient process than intergenic homogenization. The human small proline-rich (SPRR) proteins, induced during keratinocyte differentiation, are encoded by two *SPRR1* genes (A and B), approximately seven *SPRR2* genes and a single *SPRR3* gene, closely linked on chromosome 1q21-q22 (Gibbs *et al.*, 1993). The central segments of the encoded polypeptides are composed of tandemly repeated units of either eight or nine amino acids. Thus the consensus octamer PKVPEPCH is found 6 times in *SPRR1A* and *SPRR1B*, the nonamer PKCPEPCPP three times in SPRR2 and the octamer TKVPEPGC 14 times in SPRR3. This is consistent with a process of internal duplication that began after the divergence of the *SPRR* genes into three distinct subfamilies. It is evident that, during the evolution of the *SPRR* gene family, there has been a bias toward either intragenic or intergenic duplications. Thus in SPRR2, intergenic recombination has occurred more frequently than intragenic recombination since there are seven *SPRR2* genes each with three repeats. By contrast, in SPRR3, no intergenic recombination has occurred but intragenic recombination has occurred frequently to generate 14 repeat copies. Interestingly, the percentage of amino acid conservation (relative to the consensus for each type of gene) is significantly higher for the SPRR2 repeats than for either SPRR1 or SPRR3 suggesting that intergenic homogenization may be more effective than intragenic homogenization.

Concerted evolution is not however an obligatory property of internally repetitive proteins. Take for example the case of the α- and β-spectrins. Spectrin is a red blood cell cytoskeletal component which consists of a tetramer of two antiparallel ab spectrin dimers. The α-spectrin (*SPTA1*; 1q21) and β-spectrin (*SPTB*; 14q22-q23) genes evolved by duplication of a common ancestral gene which existed before the divergence of the vertebrate and arthropod lineages ~600 Myrs ago. The structure of both protein subunits is consistent with successive intragenic duplications. The α- and β-subunits consist of tandemly repeated segments of 106 amino acids of which 20 occur in α-spectrin, 17 in β-spectrin. Although the α-spectrin segments appear to have evolved in homogeneous fashion, the β-spectrin segments exhibit considerable heterogeneity (Muse *et al.*, 1997; Thomas *et al.*, 1997). One explanation for this difference is that some segments with specific functions may have evolved differently from others. Indeed, on the basis of the similar locations of the heterogeneous α- and β-spectrin segments, Muse *et al.* (1997) suggested that the α- and β-spectrins have co-evolved, and those segments that are intimately involved in subunit dimerization have been evolutionarily constrained by the structures of their binding partners. Muse *et al.* (1997) found no evidence for interrepeat exchanges and therefore concluded that neither gene conversion nor recombination had operated. At some stage, probably before the divergence of arthropods and vertebrates, concerted evolution may have operated but this initial phase probably ceased as the individual segments began to diverge at the DNA level (Thomas *et al.*, 1997).

8.6.4 The emergence of primordial genes by oligomer duplication

It is possible that primordial coding sequences emerged by a process of sequential duplication of base oligomers to yield 'genes' encoding polypeptides with significant periodicity (Ohno, 1984; Trifonov and Bettecken, 1997). Ohno (1987) cites the example of a putative primordial rhodopsin gene, which gave rise to this ancient family of seven transmembrane domain-containing proteins including among others, bacterial rhodopsin and vertebrate retinal opsin, β$_2$-adrenergic receptor and muscarinic acetylcholine receptor. This primordial gene originally contained CCTGCTG, CCTGGCC and GCTGGCC heptameric repeats. Ohno (1987) showed that the gene encoding porcine muscarinic acetylcholine receptor still contains many of these oligomeric repeats. The original heptameric repeats appear to be more stringently conserved in those portions of the gene encoding the seven transmembrane domains whereas new repeat units comingle with old repeats in those portions that encode the extracellular and intra-cytoplasmic domains.

8.6.5 Intragenic duplicational polymorphisms

Partial internal duplications of genes can also occur as polymorphic variants. A particularly dramatic example is provided by the human apolipoprotein(a) (*LPA*; 6q27) gene. This gene possesses at least 34 different alleles containing a variable number (between 12 and 51) of tandemly repeated kringle IV-encoding domains. Each allele gives rise to a different apolipoprotein(a) isoform thereby explaining the size polymorphism of the protein: 300–800 kDa (Lackner *et al.*, 1993; reviewed by Scanu and Edelstein, 1995). The *LPA* gene has been described as a 'plasminogen gene gone awry'; an apt description in that the closely linked and evolutionarily related plasminogen (*PLG*; 6q27) gene encodes a protein with only 5 kringles (Ichinose, 1992; Magnaghi *et al.*, 1994). The *LPA* gene is thought to have arisen between 40 Myrs (McLean *et al.*, 1987) and 90 Myrs ago (Pesole *et al.*, 1994) during the adaptive radiation of the mammals. At least ten kringle IV-encoding domains are present in rhesus macaque suggesting that kringle number may have expanded progressively during primate evolution (Pesole *et al.*, 1994). Although the *LPA* gene was originally thought to be confined exclusively to primates, it has also been found in hedgehogs in which it possesses 31 kringle-encoding domains (Lawn *et al.*, 1997). Since these are kringle III repeats, we must surmise that an apolipoprotein(a)-like gene arose independently from a plasminogen-like gene in this insectivore (~80 Myrs ago) and experienced a similar expansion of a subset of its kringle repeats to that found in humans – a remarkable example of *convergent evolution*. In humans, the *LPA* kringle repeat number polymorphism may not be without clinical significance because (i) there is an inverse relationship between apolipoprotein(a) isoform size and plasma apolipoprotein(a) levels (van der Hoek *et al.*, 1993) and (ii) elevated levels of apolipoprotein(a) are associated with an increased risk of atherosclerosis and cardiovascular disease (Byrne and Lawn, 1994).

Several human mucin genes exhibit highly polymorphic tandemly repetitive regions, for example *MUC1* (1q21; Gengler *et al.*, 1990), *MUC2* (11p15; Toribara *et al.*, 1991), and *MUC4* (3q29; Nollet *et al.*, 1998). Intragenic repeat copy number polymorphisms are also evident in the human proline-rich protein (*PRH1*; 12p13.2) gene where alleles vary in terms of the number of copies of a 63 bp repeat in exon 3 (Azen *et al.*, 1987; Kim *et al.*, 1993) and the complement receptor type 1 (*CR1*; 1q32) gene where the two most common alleles differ in terms of the presence of a long homologous repeat of ~450 amino acids (Wong *et al.*, 1989). Finally, in the human filaggrin (*FLG*; 1q21) gene, a length polymorphism of 10, 11, or 12 copies of a 972 bp repeat occurs within its polyprotein precursor which is subsequently cleaved into individual functional filaggrin molecules (Gan *et al.*, 1991).

8.7 Coding sequence expansion and contraction resulting from the introduction or removal of initiation and termination codons

The mutation of initiation codons leading to the extension of the protein coding sequence of a gene is known to be a cause, albeit an infrequent one, of human genetic disease (Cooper and Krawczak, 1993). Such mutations have however also occurred on a number of occasions during gene evolution. Thus, an ATG→GTG substitution occurred in the Met initiator codon of the *ZNF80* (3q13.3) gene in the common ancestor of African green monkey and rhesus macaque which resulted in a change in the site of translational initiation to a location 20 codons amino terminal to the Met initiator codon used by humans, chimpanzees and gorillas (Di Cristofano *et al.*, 1995). However, in African green monkey, the ZNF80 protein is only 213 amino acid residues in length (as compared to 273 residues in humans and the great apes and 293 residues in rhesus macaque) owing to truncation of the protein due to an additional GAG→TAG substitution introducing a novel termination codon (Di Cristofano *et al.*, 1995).

Another example of mutation resulting in the species-specific use of alternative initiation codons is provided by an ATG→ATA transition in the human hepatic peroxisomal L-alanine: glyoxylate aminotransferase 1 (*AGXT*; 2q36-q37) gene (Takada *et al.*, 1990). This lesion removed the original initiation codon and an alternative downstream initiation codon is used in the translation of the human *AGXT* transcript. The rat *Agxt* orthologue encodes a protein which, by comparison with the human protein, possesses an extra 22 amino terminal amino acids specifying a leader sequence thought to contain a mitochondrial targeting signal that is absent in the human protein.

The mutational removal of a termination codon may also lead to the elongation of a protein product. One example of this is provided by the porcine P-glycoprotein genes whose coding regions may be seen to have been extended by such a mutation when compared to the human orthologous proteins (*PGY1*, *PGY3*; 7q21; Childs and Ling, 1996).

8.8 Minisatellites, microsatellites and telomeric repeats

> Repetition is the only form of permanence that nature can achieve.
> George Santayana (1922)

A detailed discussion of the evolution of the various families of satellite, minisatellite and microsatellite DNA would be somewhat tangential to the remit of this volume and would merely serve to recapitulate a previous volume in this series. However, a brief resumé will be given to provide the necessary background to guide the interested reader to the relevant literature.

8.8.1 Minisatellite DNA sequences

Minisatellite DNA sequences occur in all eukaryotes from yeast to human (Haber and Louis, 1998). Minisatellites frequently exhibit substantial allelic variability with respect to repeat number (Jeffreys *et al.*, 1990; reviewed by Armour, 1996) and allele length analysis has demonstrated germline mutation rates as high as 15% per gamete (Jeffreys *et al.*, 1988; Jeffreys, 1997). Minisatellite mutation may involve intra-allelic rearrangements whose frequency, unlike inter-allelic rearrangements, is influenced by the size of the tandem array (Buard *et al.*, 1998). Sequence similarities, manifested by a subset of minisatellites, to the Chi recombination promoting element of *E. coli* have led to the suggestion that this 'core sequence' might be recombinogenic and could serve to promote unequal crossing over. However, analysis of flanking polymorphisms has not indicated the exchange of markers predicted for the products of unequal exchange between alleles (Wolff *et al.*, 1988, 1989). This notwithstanding, recombination hotspots sometimes co-localize with minisatellites (Jeffreys *et al.*, 1998a) which has led to the suggestion that minisatellite instability may be a by-product of meiotic recombination (Jeffreys *et al.*, 1998b).

Monckton *et al.* (1994) have shown that minisatellite mutation can involve complex inter-allelic gene conversion events. These may exhibit polarity since the gain of a few repeat units appears to be confined to one end of the tandem repeat array (Jeffreys *et al.*, 1994). One alternative proposal is that minisatellite mutation may involve an array homogenization process which could operate by biased repair of intra-helical (slippage) or inter-helical (unequal sister chromatid exchange) heteroduplexes (Bouzekri *et al.*, 1998). Whatever the mechanism(s) underlining their allelic variability, the rapid evolution of minisatellites appears to have rendered them an important substrate for the opportunistic processes of molecular evolution. Indeed, several are known to have become recruited as gene regulatory elements (see Chapter 5, section 5.1.12, *Minisatellites and microsatellites*).

8.8.2 Microsatellite DNA sequences

Microsatellites typically mutate with a frequency of 10^{-3} to 10^{-4} although mutation rates vary between loci by several orders of magnitude (Brinkman *et al.*, 1998; Crawford and Cuthbertson 1996). Microsatellite mutation events in the male germline outnumber those in the female germline 5- to 6-fold (Brinkman *et al.*, 1998). Interestingly, alternative alleles at the same locus may differ dramatically in terms of their mutation rate (Jin *et al.*, 1996). By contrast, microsatellite location is often conserved at orthologous positions in the genomes of different species (Blanquer-Maumont and Crouau-Roy, 1995; Liao and Weiner, 1995; Meyer *et al.*, 1995; Moore *et al.*, 1991; Pausova *et al.*, 1995; Stallings *et al.*, 1991; Stallings, 1994, 1995; Sun and Kirkpatrick, 1996). Rubinsztein *et al.* (1995a,b) compared allele length distributions for a considerable number of human microsatellites with their orthologues in chimpanzees, gorillas, orangutans, baboons, and macaques and claimed a tendency for the loci to be longer in humans. Although some microsatellites are more variable in nonhuman primates than in humans (Kayser *et al.*, 1995), a tendency for the loci to be longer in humans would be consistent with the directionality of microsatellite evolution and compatible with evolution proceeding at different rates in different species. However, it could also be due to ascertainment bias in that it is precisely the longest, most mutable and therefore most informative of human microsatellites that have been selected as genetic markers (Ellegren *et al.*, 1995). In principle, the large size of the human population could have compounded this effect since it could support a higher level of genetic diversity; in smaller populations, much variability is lost as alleles go to fixation. In practice, however, it would appear as if ascertainment bias cannot be the sole explanation for inter-specific differences in microsatellite repeat length (Cooper *et al.*, 1998). Thus, the tendency for microsatellite loci to be longer in humans may not simply be an experimental artefact.

8.8.3 Telomeric and centromeric repetitive DNA

Telomeric TTAGGG repeat number varies not only between human chromosomal arms but also between individuals (Brown *et al.*, 1990; Martens *et al.*, 1998). Indeed, there is some evidence in humans for the exchange of telomeric and subtelomeric repeats between nonhomologous chromosomal ends (van Deutekom *et al.*, 1996). Since satellite DNA sequences at the telomeric junctions of chimpanzees do not show any similarity to their counterparts in human or orangutan, it is likely that telomeres became reorganized relatively recently during primate evolution (Baird and Royle, 1997; Royle *et al.*, 1994; Royle, 1996).

Alphoid satellite DNA is found as a tandem repeat in long chromosome-specific arrays at the centromeres of all primate chromosomes (Warburton and Willard, 1996). These arrays are highly variable within and between homologous chromosomes in the same species.

8.9 Expansion of unstable repeat sequences

DNA sequences that are internally repetitive are particularly prone to misalignment during DNA synthesis (Djian, 1998; Di Rienzo *et al.*, 1994; Levinson and Gutman, 1987; Schlötterer and Tautz, 1992; Valdes *et al.*, 1993). If such misalignment takes place, the nascent strand can slip back in multiples of the repeat unit and the newly synthesized DNA strand will be elongated by comparison with the parental strand. Since many coding sequences contain simple sequence repeats (Tautz *et al.*, 1986) (for example, the genes encoding the 28S and 18S ribosomal RNAs (*RNR1-5*; Hancock, 1995a; Hancock and Dover, 1988) and the TATA-binding protein TBP (*TBP*; Hancock, 1993)), repeat expansion may have been involved in the evolution of these sequences (Hancock 1995b; see Section 8.9.2). However, these simple repeats also have the potential to be involved in disease pathogenesis (see Section 8.9.1) and it is from studies of genetic disease that most of our knowledge of triplet repeat expansion is derived.

8.9.1 Triplet repeat expansion disorders

One manifestation of DNA slippage involving simple repetitive sequences is the instability of certain trinucleotide repeat sequences (reviewed by Djian, 1998; Monckton and Caskey, 1995; Richards and Sutherland, 1996; Sutherland and Richards, 1995; Timchenko and Caskey, 1996; Wells, 1996). This mutational mechanism underlies fragile X mental retardation syndrome, a condition associated with the presence of a fragile site on the X chromosome (FRAXA). The brain-expressed *FMR1* (Xq27.3) gene responsible contains an unusual (CGG)n repeat in its 5′ untranslated region. This repeat exhibits copy number variation of between 6 and 54 in normal healthy controls, between 52 and >200 in phenotypically normal transmitting males (the 'premutation') and between 300 and >1000 in affected males (the 'full mutation') (Verkerk *et al.*, 1991; Fu *et al.*, 1991). Thus a continuum exists between a copy number polymorphism present in the general population, the asymptomatic premutation which involves limited expansion of (CGG)n copy number, and the full mutation which appears to require copy number expansion beyond a certain threshold value. Expansion of premutations to full mutations is thought to be a prezygotic event (Moutou *et al.*, 1997) and occurs only during female meiotic transmission. Alleles with a repeat copy number of <46 do not exhibit elevated meiotic instability. By contrast, for alleles with 52–113 repeat copies, the premutation expands to the full mutation in 70% of transmissions whereas the corresponding figure for alleles with >90 repeat copies is 100%. The probability of repeat expansion thus correlates with the repeat copy number in the premutation allele, consistent with a mechanism of slipped mispairing during replication. Expansion of a sequence can thus itself lead to further expansion, a process termed 'dynamic mutation' by Richards *et al.* (1992). In FRAXA, triplet repeat expansion is thought to exert its pathological effects by down-regulation of *FMR1* gene expression through hypermethylation of the promoter region upstream of the CGG repeat and repression of translation of the *FMR1* transcript.

The discovery of this novel mutational mechanism has led to the recognition that the expansion of unstable triplet repeats is also responsible for a number of other human inherited diseases (*Table 8.2*). These often manifest a wide range of clinical severity and possess unusual features such as increasing severity and penetrance in successive generations ('anticipation') and a sex bias in the transmission of the disease which correlates with the degree of meiotic instability and allelic expansion. As can be seen from *Table 8.2*, the nature and location of the repeat sequence involved varies between disease states as does the extent of the expansion necessary to bring about symptoms of disease, and of course the mechanism of pathogenesis consequent to repeat expansion. It should however be noted that repeat expansion is not confined to triplet repeats; minisatellite expansion can also occur as in progressive myoclonus epilepsy and the clinically asymptomatic fragile sites FRA16A and FRA16B (*Table 8.2*).

Triplet repeat expansions have so far been found to be associated mainly with three types of sequence: CAG (with complement CTG), CGG (with complement CCG) and GAA (with complement TTC). Sequence specificity implies a role for DNA secondary structure in the expansion mechanism. Such repeats are known to form stable hairpin loop structures (Chen *et al.*, 1995; Gacy *et al.*, 1995; Mitas *et al.*, 1995) which become more stable with increasing repeat number. DNA polymerase progression appears to be blocked by CTG and CGG repeats and the resultant idling of the polymerase may serve to catalyze slippage leading to repeat expansion (Kang *et al.*, 1995).

A number of diseases are characterized by CAG repeat expansion within the

Table 8.2. Diseases/traits associated with expansions of unstable repeat sequences

Disease/trait	Gene/fragile site	Protein	Repeat motif	Location of repeat	Copy number (controls, patients)
Fragile X mental retardation (FRAXA)	FMR1 (Xq27)	RNA binding protein	CGG	5'UTR	5–52, >200
Fragile X E mental retardation (FRAXE)	FMR2 (Xq28)	Transcription factor?	CGG	5'UTR	4–39, >200
Fragile site, FRAXF	(Xq28)	–	(GCCGTC)n (GCC)n	?	12–26, >900
Fragile site, FRA16A	(16p13.11)	–	CGG	?	?
Fragile site, FRA16B	(16q22.1)	–	(ATATATTATATATTA TATCTAATAATATAT C/A TA)n	?	?
Myotonic dystrophy	DMPK (19q13)	Serine/threonine kinase	CTG	3'UTR	<30, 50–2000+
Friedreich ataxia	FRDA (9q)	Frataxin	GAA	Intronic	7–34, 120–1700
Progressive myoclonus epilepsy	CSTB (21q22)	Cystatin B	CCCCGCCCCGCG	5'UTR	2–3, 50–75
Spinobulbar muscular atrophy	AR (Xq)	Androgen receptor	CAG	Coding region	17–26, 40–52
Huntington disease	HD (4p16.3)	Huntingtin	CAG	Coding region	<35, 40–400
Dentatorubral pallidoluysian atrophy	DRPLA (12p)	Atrophin	CAG	Coding region	9–23, 40–100
Spinocerebellar ataxia type 1	SCA1 (6p23)	Ataxin 1	CAG	Coding region	25–36, 40–100
Spinocerebellar ataxia type 2	SCA2 (12q24)	Ataxin 2	CAG	Coding region	15–29, 35–59
Spinocerebellar ataxia type 3 (Machado-Joseph disease)	MJD (14q)	Ataxin 3	CAG	Coding region	14–40, 68–79
Spinocerebellar ataxia type 6	CACNA1A (19q13)	Calcium channel	CAG	Coding region	4–16, 21–28
Spinocerebellar ataxia type 7	SCA7 (3p)	Ataxin 7	CAG	Coding region	7–17, 38–130
Oculopharyngeal muscular dystrophy	PABP2 (14q11-q13)	Poly(A)-binding protein 2	GCG	Coding region	6, 7–13
Synpolydactyly	HOXD13 (2q)	Homeobox D13	(GCG)n(GCT)n/(GCA)n	Coding region	15, 22–29

gene coding region (*Table 8.2*). This is thought to constitute a gain-of-function mutation through the incorporation of a polyglutamine tract into the protein product (Houseman, 1995). Several factors are known to influence the stability of triplet repeats viz. the type of sequence, length of repeat, whether the repeat is interrupted or not, and the orientation of the repeat relative to the origin of replication (Andrew *et al.*, 1997). The removal of interrupting point mutations from repeat arrays is thought to be very important in promoting triplet repeat expansion (Eichler *et al.*, 1994) and is significant in an evolutionary context as well as in cases of disease. Removal of these point mutations could occur simply by single base-pair substitution or by gene conversion or unequal crossing over. Other inherited conditions result from expansion of a triplet repeat in the 3′ untranslated region (myotonic dystrophy; *DMPK*) and an intron (Friedreich ataxia; *FRDA*) (*Table 8.2*) and the resulting mechanisms of disease are consequently different.

8.9.2 Nature and distribution of triplet repeats in the human genome

Trinucleotide repeats are 10–100-fold less frequent than (AC)n repeats (Gastier *et al.*, 1995) and different types of trinucleotide repeat occur at different frequencies. Thus repeats (AAT)n and (AAC)n are the most frequent trinucleotide repeats found in the human genome (Gastier *et al.*, 1995; Stallings 1994), with (AAT)n exhibiting a high degree of copy number polymorphism (Gastier *et al.*, 1995). Both (CAG)n and (CCG)n repeats appear to be over-represented in the human genome (Han *et al.*, 1994), the former being polymorphic in number in the genomes of humans and non-human primates (Sirugo *et al.*, 1997). (CAG)n repeats are rare in intronic regions, possibly because of their similarity to the acceptor splice site consensus, CAGG (Stallings, 1994). Long homopeptides are present in 1.7% of human protein-coding sequences (Karlin and Burge, 1996).

A number of human genes contain polymorphic triplet repeats (e.g. cadherin 2 (*CDH*; 18q12), breakpoint cluster region (*BCR*; 22q11), glutathione-*S*-transferase (*GSTA1*; 6p12), Na$^+$/K$^+$ ATPase β1-subunit (*ATP1B1*; 1q22-q25) but these repeats are not known to exert any pathological effect (Li *et al.*, 1993; Riggins *et al.*, 1992). Other genes have been identified solely on account of their possession of polymorphic trinucleotide repeats and these genes represent candidate loci for involvement in complex diseases (Breschel *et al.*, 1997; Néri *et al.*, 1996). Trinucleotide repeat containing genes are also found in other species, including mouse (Kim *et al.*, 1997), but no nonhuman example of triplet repeat expansion as a cause of a genetic disease has yet been documented.

Replication slippage involving short GC-rich motifs ('expansion segments') has occurred during the evolution of the vertebrate genes encoding the 28S and 18S ribosomal RNAs (Hancock, 1995a; Hancock and Dover, 1988). Interestingly, different segments of the 28S rRNA subunit gene appear to have coevolved by 'compensatory slippage' allowing RNA secondary structure to be conserved as a consequence of runs of sequence motifs in one region being compensated for by complementary motifs in another (Hancock and Dover, 1990).

8.9.3 Origin of expanded triplet repeats

In myotonic dystrophy, all Caucasian and Japanese DM chromosomes possess a specific haplotype (Deka *et al.*, 1996; Imbert *et al.*, 1993; Tishkoff *et al.*, 1998;

Yamagata *et al.*, 1996), whilst a second disease-associated haplotype has been found in Africans (Krahe *et al.*, 1995). Similarly, haplotype analysis has pointed to a single origin for Japanese and Caucasian Machado-Joseph disease chromosomes (Takiyama *et al.*, 1995), a single origin for Japanese and Caucasian dentatorubral and pallidoluysian atrophy chromosomes (Yanagisawa *et al.*, 1996), a single origin for the Friedreich ataxia expansion (Cossée *et al.*, 1997), at least two origins for the Huntington disease expanded repeat (Squitieri *et al.*, 1994) and a small number of FRAXA progenitor chromosomes (Hirst *et al.*, 1994).

One of the best examples of the evolutionary emergence of an expanded triplet repeat is that found in the *SRP14* gene (15q22) encoding the 14 kDa *Alu* RNA-binding protein (Chang *et al.*, 1995). The human protein is larger than that of mouse and dog on account of an extra 28 residue alanine-rich C-terminal tail which is translated from a 3' GCA-rich trinucleotide repeat. In the prosimian *Galago*, the relevant sequence at the site of the human triplet repeat is GCA GCA, whereas the mouse possesses the sequence CCA GCA. By contrast, the African green monkey possesses 33 GCA repeats, whilst the owl monkey possesses 52 suggesting that a CCA→GCA substitution occurred in an ancestral prosimian thereby creating two consecutive GCA codons which then facilitated GCA expansion in higher primates. Interestingly, however, no intra-specific variability in repeat size was detected in primates.

8.9.4 Evolution of repeat number in the genes underlying disorders of triplet repeat expansion

The normal range for CAG repeat number in the *HD* gene is 8–35. Although modal CAG repeat length is fairly similar between different human populations, the degree of spread varies, being greatest among Africans and lowest among the Japanese (Rubinsztein *et al.*, 1994; Watkins *et al.*, 1995). The breadth of this normal range suggests that natural selection is acting weakly if at all on *HD* alleles below the disease threshold. However, the population distribution of CAG repeat number in the *HD* gene exhibits an apparent asymmetry in that more alleles lie above rather than below the modal length. Using computer simulations, Rubinsztein *et al.* (1994) have shown that the distribution of *HD* alleles in human populations is explicable in terms of a simple length-dependent mutational bias. The observed distribution of alleles is thus explicable merely in terms of mutation and genetic drift thereby obviating the need to invoke positive selection (Rubinsztein *et al.*, 1994). It may be, however, that coding sequences can tolerate CAG repeat-encoded polyglutamine tracts relatively well, thereby minimizing the effect of negative selection (Green and Wang, 1994). More controversially, in the case of myotonic dystrophy and Machado-Joseph disease, *meiotic drive* (also termed *segregation distortion*), the excess recovery of one of a pair of alleles in the gametes of an heterozygous parent, has been proposed as being responsible for maintaining the frequency in the population of chromosomes bearing triplet repeats capable of expansion into the disease range (Chakraborty *et al.*, 1996; Leeflang *et al.*, 1996; Takiyama *et al.*, 1997).

Djian *et al.* (1996) examined the disease-associated CAG repeats in the *HD*, *MJD*, *AR*, and *SCA1* orthologues of various nonhuman primate species. For the *HD* and *MJD* genes, CAG copy number was polymorphic in the nonhuman primates but there was essentially no overlap with the normal human range, findings also reported by Limprasert *et al.* (1996). The *AR* gene was also found to be polymorphic in nonhuman primates but with minimal overlap with the normal human range (Djian *et al.*, 1996). For the *SCA1* gene, the average repeat copy numbers exhibited by *Pan*, *Gorilla*, *Hylobates*, *Macaca*, and *Cercopithecus* were within the normal human range but were lower than the modal number in humans (Djian *et al.*, 1996). Thus, for these four loci, CAG copy number is lower in nonhuman primates than in humans. From what we know of the propensity of CAG repeats to expand, this is much more likely to be due to a higher rate of expansion in humans than the alternative: contraction in the other primate species. In the case of the *HD* gene, this interpretation is supported by the finding of a similar number of CAG repeats in the murine *Hd* gene to that found in the nonhuman primate *HD* genes (Lin *et al.*, 1994). The high number of CAG repeats in the human *HD* gene is therefore due primarily to expansion within the human lineage. This contrasts with the expansion of CAG number in the *AR* gene which began earlier in hominoid evolution; the great apes contain an expanded CAG repeat not found in rodents (Choong *et al.*, 1998; Djian *et al.*, 1996). Thus, the *AR* CAG repeat expansion began prior to ape-human divergence but continued in human after divergence from the great apes. The most dramatic example of CAG expansion is however that of the involucrin (*IVL*; 1q21) gene which is so rich in CAG repeats and codons derived from CAG that it is likely that it is descended from a simple poly (CAG) sequence (see Section 8.9.5 below).

The CGG repeat in the 5′ UTR of the *FMR1* gene has been studied in 44 mammalian species from 8 different orders (Eichler *et al.*, 1995). The presence of this repeat in all species examined indicates that the CGG repeat has been conserved for over 150 Myrs and is therefore likely to be have some functional significance (reviewed by Eichler and Nelson, 1998). Repeat length was found to be similar among the 24 nonprimate species, ranging from 4 to 12 units (mean 8.01 ± 0.8). By contrast, the mean length of the repeat among the 20 primate species examined was 20.1 ± 2.3. Copy number polymorphism was not found to be limited to human, with *Ornythorhyncus* (platypus), *Artibeus* (phyllostomid bat) and *Pan* (chimpanzee) all possessing polymorphic repeats. Parsimony analysis predicted that the early mammalian CGG repeat was short (4–9 units) and uninterrupted, whereas an increase in copy number beyond ~20 repeats appears to have occurred at least three times independently in the Catarrhini (*Figure 8.13*). These expansions in the hylobatid apes, great apes and the cercopithecoid monkeys were associated with the addition of specific interspersions, CGA, AGG and CGGG respectively (*Figure 8.13*). These interspersions are unlikely to have arisen through DNA polymerase slippage and may have been mediated by unequal crossing over or gene conversion.

A comparison of trinucleotide repeats in human versus rodent coding sequences has indicated that, by and large, orthologous repeats have not been conserved for long periods of evolutionary time, either in terms of their size or location (Stallings, 1994). As yet, there are no known pathological equivalents of human dynamic mutations in other species but this may merely reflect bias of ascertainment. Thus, nonhuman examples of disease-associated triplet repeat expansions may not yet have come to our attention and it may be that these expan-

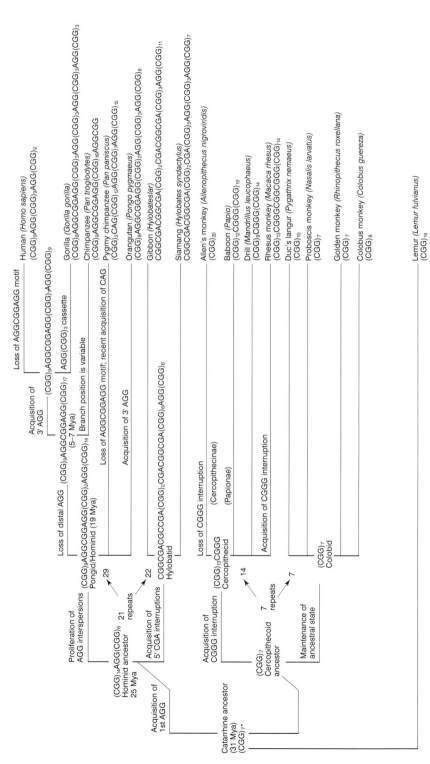

Figure 8.13. Hypothetical schema for the evolution of the *FMR1* CGG repeat in Old World monkeys (after Eichler *et al.,* 1995). Mya, million years ago.

sions have no counterpart in the human orthologues. Similarly, the observation that triplet repeats associated with human disease are sometimes found to be both polymorphic and expanded beyond the mammalian ancestral state in non-human primates may also be viewed in terms of a bias of ascertainment rather than any intrinsic propensity of triplet repeats to expand in primate genomes. Finally, noting that most of the expanded triplet repeats are found in genes expressed in the nervous system, Hancock (1996) proposed that triplet repeat expansion may have occurred in parallel with the increase in brain complexity so characteristic of the human lineage and may even have been instrumental in bringing about this increase in complexity.

8.9.5 The unique case of involucrin

Involucrin is a unique protein in that it has evolved very rapidly in the primate lineage through changes in the length and copy number of tandem repeats within the coding region with the result that a full two thirds of its coding region has been created within the anthropoid lineage (Green and Djian, 1992). The evolution of this gene/protein is therefore worth examining in some detail.

Involucrin is the most abundant protein component of the keratinocyte envelope and is cross-linked to membrane proteins by a transglutaminase. The coding

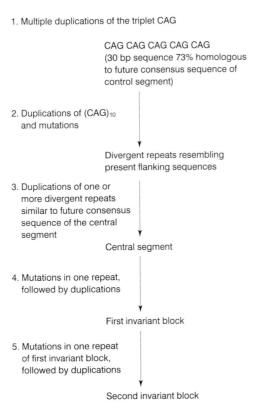

1. Multiple duplications of the triplet CAG

CAG CAG CAG CAG CAG
(30 bp sequence 73% homologous
to future consensus sequence of
control segment)

2. Duplications of (CAG)$_{10}$
   and mutations

Divergent repeats resembling
present flanking sequences

3. Duplications of one or
   more divergent repeats
   similar to future consensus
   sequence of the central
   segment

Central segment

4. Mutations in one repeat,
   followed by duplications

First invariant block

5. Mutations in one repeat
   of first invariant block,
   followed by duplications

Second invariant block

Figure 8.14. Proposed scheme for the evolution of the human involucrin (*IVL*) gene.

region of the human involucrin (*IVL*; 1q21) gene lies within a single exon and encodes a 585 amino acid glutamine-rich protein (Eckert and Green, 1986). The human *IVL* gene is extemely rich in CAG repeats (encoding glutamine) and codons derived from it (GAG and CTG). Thus, CAG codons, and codons one substitution removed from CAG, comprise 64% of the total codon number. It is possible therefore that this gene is descended from a simple poly(CAG) sequence which has been subsequently modified by nucleotide substitution (Tseng, 1997; *Figure 8.14*).

The time of origin of the involucrin gene is unclear but the gene is present in a wide variety of mammals and may be present in all terrestrial vertebrates. The length of the involucrin molecule has continued to grow by successive addition of repeats from prosimians (average 409 residues) through New World monkeys (average 488 residues) and Old World monkeys (average 514 residues) to hominoids (average 632 residues). The coding region outside the segment of repeats has by contrast changed very little in length. Since glutamine residues serve as amine acceptors in the transglutaminase-catalyzed cross-linking, the primordial

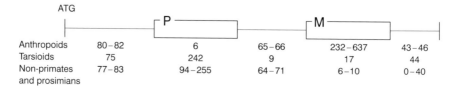

	ATG	P		M	
Anthropoids	80–82	6	65–66	232–637	43–46
Tarsioids	75	242	9	17	44
Non-primates and prosimians	77–83	94–255	64–71	6–10	0–40

Figure 8.15. Coding regions of primate involucrin genes illustrating the relative locations of repetitive regions P and M. Numbers given denote lengths (in codons) of segments of repeats and non-repetitive regions (redrawn from Green and Djian, 1992).

Repeat Number	Type	1	2	3	4	5	6	7	8	9	10	Number of altered codons per repeat
10	A	K	H	L	E	Q	Q	E	G	Q	L	5
9	B	E	L	P	E	Q	Q	V	G	Q	P	10
8	A	K	H	L	E	Q	E	E	K	Q	L	10
7	B	E	L	P	E	Q	Q	E	G	Q	L	3
6	A	K	H	L	E	K	Q	E	A	Q	L	2
5	B	E	L	P	E	Q	Q	V	G	Q	P	3
4	A	K	H	L	E	Q	Q	E	K	Q	L	2
3	B	E	H	P	E	Q	Q	E	G	Q	L	7
2	A	K	H	L	E	Q	Q	E	G	Q	L	6
1	A	K	D	L	E	Q	Q	E	G	Q	L	10

Mean 5.8

Figure 8.16. Consensus amino acid sequences (from 12 anthropoid ape species) for each of the 10 repeats of the early region of the M segment of involucrin. Alterations in one or more species are boxed or framed. Solid lines indicate deletions whilst dashed lines indicate amino acid substitutions. Over half of the amino acids have been deleted or replaced in one or more species. The doubly underlined Q in repeat 5 is the preferred residue for cross-linking in human involucrin (redrawn from Green and Djian, 1992).

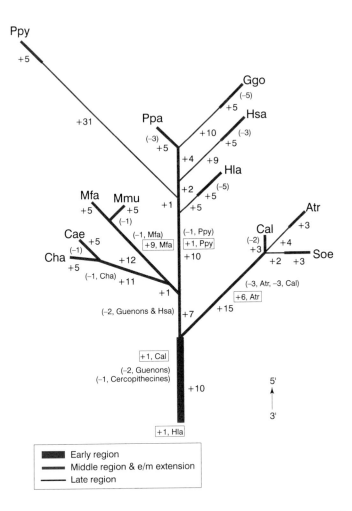

Figure 8.17. Evolutionary tree of the M region of the primate involucrin gene detailing repeat additions (after Green and Djian, 1992). The numbers along the branches of the tree indicate the number of repeats added to form each region. Four late additions within the middle or early regions and one 3' of the early region (boxed) are deviations from the usual vectorial pattern of repeat addition. The deletions assigned to the region, the lineage in which they occurred, and the number of repeats deleted are given in parentheses. Ppy: *Pongo pygmaeus*. Hsa: *Homo sapiens*. Mmu: *Macaca mulatta*. Cal: *Cebus albifrons*. Soe: *Saguinus oedipus*. Ppa: *Pan paniscus*. Hla: *Hylobates lar*. Cae: *Cercopithecus aethiops*. Atr: *Aotus trivirgatus*. Ggo: *Gorilla gorilla*. Mfa: *Macaca fascicularis*. Cha: *Cercopithecus hamlyni*.

glutamine-rich involucrin may have been able to function as a substrate for trans-glutaminase.

Involucrin gene sequences have now been determined in a range of primate and nonprimate species. Two different sites have been found within the coding region that contain variable numbers of repeat segments. One, at site P, lies ~80 codons from the ATG, contains a 16-amino acid repeat and is evident in nonprimate mammals, prosimians and tarsioids (*Figure 8.15*). The other, at site M, containing a 10-amino acid repeat, is located a further 68 codons C-terminal to site P and has increased in size in the anthropoid apes concomitant with a reduction in size of the P repeat segment (*Figure 8.15*).

The involucrin genes of the dog, pig and mouse have 6, 13, and 21 repeats, respectively at site P whilst the lemur (*Lemur catta*) and tarsier (*Tarsius bancanus*) possess 19 and 18, respectively. Within each species, however, P repeat segments differ; thus three codons have been deleted in eight of the 13 repeat copies in *Galago crassicaudatus* whilst two codons have been deleted in all repeat copies in *T. bancanus*. Consensus sequences also differ between nonanthropoid species in terms of nucleotide differences that occurred in any position in a codon. Changes in a consensus nucleotide must have arisen from the occurrence of a nucleotide substitution in one repeat followed by the correction of the analogous nucleotides at the corresponding positions in the other repeats, with adjacent repeats tending to be more homogeneous (Phillips *et al.*, 1990). Green and Djian (1992) have proposed that a form of intragenic gene conversion (see also Chapter 4, section 4.2.2, *Ribosomal RNA genes* and chapter 9, section 9.5) may have been responsible for this phenomenon. Although the P segment was retained by prosimians and tarsioids, it was modified by site-specific deletions, the addition of certain repeats and nucleotide substitutions which served to alter the consensus codons of the repeats by a process of correction operating between neighboring repeats (Djian and Green, 1991). It is possible that the retention of the P segment was essential for the function of the protein since repeats at site M were lacking.

In the anthropoid apes, the M segment increased in size concomitant with the dramatic reduction in size of the P repeat segment (*Figure 8.15*). Perhaps deletion of the P segment was only possible after a sufficient number of repeats had been generated within the M segment. This notwithstanding, the M segment continued to grow by successive addition of repeats from prosimians through New World monkeys and Old World monkeys to hominoids. However, the repeats have the same consensus sequence in all the anthropoid apes (*Figure 8.16*). The M segment has been divided into early, middle and late regions which have been added vectorially in a 3'→5' direction (Djian and Green, 1989) and the repeats from these regions are shared to differing extents by different anthropoid species (Teumer and Green, 1989). Thus, the early region is common to all anthropoids, the middle region is shared by different species but to a lesser extent, and the late region is species-specific and must have developed after the divergence of the different species. An evolutionary tree of the M region of involucrin in higher primates is shown in *Figure 8.17*. After its divergence from the cercopithecoids, the hominoid lineage added 10 repeats to the middle region. After divergence from the common lineage, *Hylobates*, *Pongo*, and *Homo* acquired species-specific repeats in the late region. Another shared repeat segment is the early/middle (e/m) extension 5' to the late region which was probably formed independently in Old World and New World monkeys.

The site of repeat addition, which in the common anthropoid lineage was located in the P segment, moved in a 5' direction during the evolution of both the New World and Old World monkeys after their divergence (Green and Djian, 1992). Deletions of repeats also took place within the early, middle and e/m

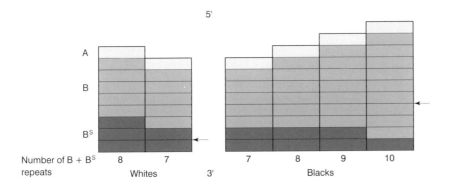

Figure 8.18. Polymorphism of the repeat pattern of the late region of the M segment of human involucrin. Each column of rectangular boxes indicates an allele, and each box denotes a 10-codon repeat. There are three different kinds of repeat: A (diagonal lines), B (shaded) and B^s (chequered) which differ with respect to the sequences of the first three codons (AAGCACCCG, GAGCTCCCA, and GAGCTCTCT respectively). The two columns on the left denote the variable number of B^s repeats in whites whilst the four columns on the right denote the variable numbers of B and B^s repeats found in blacks. Arrowheads indicate the position of the hotspots for repeat addition in the white and black populations (redrawn from Green and Djian, 1992).

regions of the M segment although these were relatively few in number as compared to the additions (Green and Djian, 1992). These deletions have occurred independently of the vectorial process of repeat addition both spatially and temporally (Green and Djian, 1992).

In addition to the inter-specific differences in the size of the involucrin molecule, the higher primates also exhibit size polymorphisms which result from variation in the number of repeats in the late region. In humans, the polymorphism comprises variable numbers of B and B^s repeats within the late region (*Figure 8.18*). The most common allele in white Caucasians contains nine repeats (3B^s, 5B, 1A) but a second 'Mormon' allele (2B^s, 5B, 1A) has also been described (Simon *et al.*, 1989; Urquhart and Gill, 1993). Blacks possess the 'Mormon' allele plus three other alleles containing 6, 7, or 9 B repeats (Simon *et al.*, 1991; Urquhart and Gill, 1993; *Figure 8.18*). Repeat number polymorphisms in the late region have also been reported in *Aotus trivirgatus*, *Macacca mulatta* and *Gorilla gorilla* (Green and Djian, 1992). The repeat pattern observed in anthropoid apes is explicable in terms of the presence in the involucrin gene of a hotspot at which repeats are generated by unequal recombination (Green and Djian, 1992). In the human lineage, the location of this hotspot varies between racial groups such that in whites it is located within the B^s repeat region whilst in blacks, it is within the B repeat region (*Figure 8.18*). Thus, although the process of vectorial repeat addition has operated in both blacks and whites, there has been a difference between these racial groups in the sites of repeat addition, a finding that is consistent with an endogenously controlled mechanism.

Attempts have been made to relate the evolution of a larger involucrin molecule with a different repeat structure to the trend in anthropoid apes towards relative

Table 8.3. Mechanism-based evolution of the repeat segments of involucrin in mammals (from Green and Djian, 1992)

Mechanism	Organism in which mechanism operates
Shortening or lengthening of pre-existing repeats by site-specific deletion or insertion.	Site P of prosimians, tarsioids and non-primate mammals. Not operative in anthropoid apes.
Change in consensus nucleotide at certain positions resulting from mutation in one repeat followed by correction of neighboring repeats (probably by gene conversion). Many of the corrections are silent.	Site P of prosimians, tarsioids and non-primate mammals. Not operative in anthropoid apes.
Addition of shorter repeats as incomplete copies of older ones.	Site P of tarsioids. Not operative in anthropoid apes.
Generation of a new duplication site (M) by vectorial addition of repeats at a controlled hotspot moving in a 3′→5′ direction. The site of addition differs in different human populations and is therefore under genetic control.	Tarsioids and anthropoid apes. Aborted in tarsioids. Not operative in prosimians and non-primate mammals.

hairlessness (Green and Djian, 1992). The P segment of repeats could conceivably have conferred some selective advantage during the early stages of mammalian and primate evolution as might the M segment when it emerged. At a biochemical level, natural selection might have favored the retention of repeat-bearing segments in involucrin so as to provide a substrate for transglutaminase. Once the M segment had been generated, the repeats at site P could then have been deleted without selective disadvantage. However, there is a very considerable degree of latitude between different species in terms of what constitutes an effective transglutaminase substrate. In a comparison of the involucrins of 19 mammalian species, only 3.1% of amino acid residues were uniformly conserved (Djian *et al.*, 1993). Thus, on balance, neither the very high rate of evolutionary change nor the observed patterns of mutational change in the involucrin gene provide convincing evidence for natural selection being the major motive force behind involucrin evolution in higher primates. Indeed, it is unlikely that natural selection could account on its own for either the highly variable pattern of repeat addition or the numbers of repeats added.

Natural selection and neutral mutation/genetic drift are both processes that govern how mutations, once they have arisen, spread through a population with the possibility of eventually becoming fixed. By contrast, the evolution of involucrin appears to be primarily mechanism-based, being dominated by endogenously controlled and spatially targeted mechanisms of mutagenesis. Such mechanisms are likely to be the major factor responsible for the *directed evolution* of involucrin, directed in the sense that the mechanisms promote change in a specific direction (i.e. increases in repeat number or gene conversion between homologous repeats; see *Table 8.3*) without the necessary assistance of natural selection.

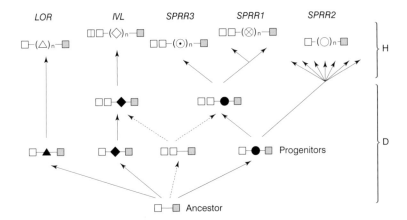

Figure 8.19. Hypothetical scheme for the evolution of the involucrin (*IVL*), loricrin (*LOR*) and small proline-rich protein (*SPRR*) genes from a single ancestral gene (after Backendorf and Hohl, 1992, Volz *et al.*, 1993). The repetitive central domains are represented by a triangle (loricrin), a diamond (involucrin) and circles (SPRR proteins). The number of repeats is represented by n (n=39 for involucrin, 6 for SPRR1, 3 for SPRR2, 14 for SPRR3). The broken arrows indicate an alternative evolutionary route. H denotes homogenization between subfamily members. D denotes divergence of different classes of genes.

Interestingly, several other human genes encoding epithelial proteins are located on the long arm of chromosome 1 [filaggrin (*FLG*), loricrin (*LOR*), epithelial mucin (*MUC1*), and the small proline-rich proteins 1A (*SPRR1A*), 1B (*SPRR1B*), 2A (*SPRR2A*), and 3 (*SPRR3*); Volz *et al.*, 1993]. These genes share a similar structure, contain multiple tandem repeats and are believed to be evolutionarily related both to involucrin and to each other (Backendorf and Hohl, 1992; Gibbs *et al.*, 1993; *Figure 8.19*). The *LOR* (Yoneda *et al.*, 1992), *FLG* (Gan *et al.*, 1990) and *MUC1* (Gendler *et al.*, 1990) genes also exhibit repeat copy number polymorphism. The apparent association between chromosomal location and copy number polymorphism almost certainly reflects the possession by evolutionarily related proteins of a repeat structure that is particularly prone to the process of vectorial addition of repeats.

References

Andrew S.E., Goldberg Y.P., Hayden M.R. (1997) Rethinking genotype and phenotype correlations in polyglutamine expansion disorders. *Hum. Molec. Genet.* **6**: 2005–2010.

Armour J.A.L. (1996) Tandemly repeated minisatellite: generating human genetic diversity via recombinational mechanisms. In: *Human Genome Evolution.* (eds. M Jackson, T Strachan, G Dover). BIOS Scientific Publishers, Oxford, pp 171–190.

Atchley W.R., Fitch W.M., Bronner-Fraser M. (1994) Molecular evolution of the MyoD family of transcription factors. *Proc. Natl. Acad. Sci. USA* **91**: 11 522–11 526.

Azen E.A., Kim H.-S., Goodman P., Flynn S., Maeda N. (1987) Alleles at the *PRH1* locus coding for the human salivary-acidic proline-rich proteins Pa, Db and PIF. *Am. J. Hum. Genet.* **41**: 1035–1047.

Backendorf C., Hohl D. (1992) A common origin for cornified envelope proteins? *Nature Genet.* **2**: 91.

Bailey A.D., Shen C.-K.J. (1993) Sequential insertion of *Alu* family repeats into specific genomic sites of higher primates. *Proc. Natl. Acad. Sci. USA* **90**: 7205–7209.

Baird D.M., Royle N.J. (1997) Sequences from higher primates orthologous to the human Xp/Yp telomore junction region reveal gross rearrangements and high levels of divergence. *Hum. Molec. Genet.* **6**: 2291–2299.

Banki K., Halladay D., Perl A. (1994) Cloning and expression of the human gene for transaldolase. A novel highly repetitive element constitutes an integral part of the coding sequence. *J. Biol. Chem.* **269**: 2847–2851.

Batzer M.A., Kilroy G.E., Richard P.E., Shaikh T.H., Desselle T.D., Hoppens C.L., Deininger P.L. (1990) Structure and variability of recently inserted *Alu* family members. *Nucleic Acids Res.* **18**: 6793–6799.

Batzer M.A., Stoneking M., Alegria-Hartman M., Bazan H., Kass D.H., Shaikh T.H., Novick G.E., Joannou P.A., Scheer W.D., Herrera R.J., Deininger P.L. (1994) African origin of human-specific polymorphic *Alu* insertions. *Proc. Natl. Acad. Sci. USA* **91**: 12 288–12 292.

Ben-Arie N., Lancet D., Taylor C., Khen M., Walker N., Ledbetter D.H., Carrozzo R., Paten K., Sheer D., Lehrach H., North M.A. (1993) Olfactory receptor gene cluster on human chromosome 17: possible duplication of an ancestral receptor repertoire. *Hum. Molec. Genet.* **3**: 229–235.

Blanquer-Maumont A., Crouau-Roy B. (1995) Polymorphisms, monomorphism, and sequences in conserved microsatellites in primate species. *J. Mol. Evol.* **41**: 492–497.

Board P., Coggan M., Johnston P., Ross V., Suzuki T., Webb G. (1990) Genetic heterogeneity of the human glutathione transferases: a complex of gene families. *Pharm. Therap.* **48**: 357–369.

Bondeson M.L., Malmgren H., Dahl N., Carlberg B.M., Petterson U. (1995) Presence of an IDS-related locus (*IDS2*) in Xq28 complicates the mutational analysis of Hunter syndrome. *Eur. J. Hum. Genet.* **3**: 219–227.

Boukaftane Y., Duncan A., Wang S., Labuda D., Robert M.-F., Sarrazin J., Schappert K., Mitchell G.A. (1994) Human mitochondrial HMG CoA synthase: liver cDNA and partial genomic cloning, chromosome mapping to 1p12-p13, and possible role in vertebrate evolution. *Genomics* **23**: 552–559.

Bouzekri N., Taylor P.G., Hammer M.F., Jobling M.A. (1998) Novel mutation processes in the evolution of a haploid minisatellite, MSY1: array homogenization without homogenization. *Hum. Molec. Genet.* **7**: 655–659.

Bray P.F., Barsh G., Rosa J.-P., Luo X.Y., Magenis E., Shuman M.A. (1988) Physical linkage of the genes for platelet membrane glycoproteins IIb and IIIa. *Proc. Natl. Acad. Sci. USA* **85**: 8683–8687.

Breschel T.S., McInnis M.G., Margolis R.L., Sirugo G., Corneliussen B., Simpson S.G., McMahon F.J., MacKinnon D.F., Xu J.F., Pleasant N., Huo Y., Ashworth R.G., Grundstrom C., Grundstrom T., Kidd K.K., DePaulo J.R., Ross C.A. (1997) A novel, heritable, expanding CTG repeat in an intron of the SEF2–1 gene on chromosome 18q21.1. *Hum. Molec. Genet.* **6**: 1855–1863.

Brinkman B., Klintschar M., Neuhuber F., Hühne J., Rolf B. (1998) Mutation rate in human microsatellites: influence of the structure and length of the tandem repeat. *Am. J. Hum. Genet.* **62**: 1408–1415.

Brockmoller J., Gross D., Kerb R., Drakoulis N., Roots I. (1992) Correlation between trans-stilbene oxide-glutathione conjugation activity and the deletion mutation in the glutathione S-transferase class mu gene detected by polymerase chain reaction. *Biochem. Pharmacol.* **43**: 647–650.

Brown W.R.A., MacKinnon P.J., Willasante A., Spurr N., Buckle V.J., Dobson M.J. (1990) Structure and polymorphism of human telomere-associated DNA. *Cell* **63**: 119–132.

Brusco A., Cinque F., Saviozzi S., Boccazzi C., DeMarchi M., Carbonara A.O. (1997) The G4 gene is duplicated in 44% of human immunoglobulin heavy chain constant region haplotypes. *Hum. Genet.* **100**: 84–89.

Buard J., Bourdet A., Yardley J., Dubrova Y., Jeffreys A.J. (1998) Influences of array size and homogeneity on minisatellite mutation. *EMBO J.* **17**: 3495–3502.

Buettner J.A., Glusman G., Ben-Arie N., Ramos P., Lancet D., Evans G.A. (1998) Organization and evolution of olfactory receptor genes on human chromosome 11. *Genomics* **53**: 56–68.

Bullock P., Champoux J.J., Botchan M. (1985) Association of crossover points with topoisomerase I cleavage sites: A model for non-homologous recombination. *Science* 230: 954–957.

Burt D.W., Paton I.R. (1992) Evolutionary origins of the transforming growth factor-β gene family. *DNA Cell Biol.* 11: 497–510.

Byrne C.D., Lawn R.M. (1994) Studies on the structure and function of the apolipoprotein (a) gene. *Clin. Genet.* 46: 34–41.

Calvo D., Dopazo J., Vega M.A. (1995) The CD36, CLA-1 (CD36L1), and LIMPII (CD36L2) gene family: cellular distribution, chromosomal location, and genetic evolution. *Genomics* 25: 100–106.

Campbell L., Potter A., Ignatius J., Dubowitz V., Davies K. (1997) Genomic variation and gene conversion in spinal muscular atrophy: implications for disease process and clinical phenotype. *Am. J. Hum. Genet.* 61: 40–50.

Casula L., Murru S., Pecorara M., Ristaldi M.S., Restagno G., Mancuso G., Morfini M., DeBiasi R., Baudo F., Carbonara A., Mori P.G., Cao A., Pirastu M. (1990) Recurrent mutations and three novel rearrangements in the factor VIII gene of hemophilia A patients of Italian descent. *Blood* 75: 662–670.

Chakraborty R., Stivers D.N., Deka R., Yu L.M., Shriver M.D., Ferrell R.E. (1996) Segregation distortion of the CTG repeats at the myotonic dystrophy locus. *Am. J. Hum. Genet.* 59: 109–118.

Chang D.-Y., Sasaki-Tozawa N., Green L.K., Maraia R.J. (1995) A trinucleotide repeat-associated increase in the level of *Alu* RNA-binding protein occurred during the same period as the major *Alu* amplification that accompanied anthropoid evolution. *Mol. Cell. Biol.* 15: 2109–2116.

Chen X., Santhana Mariappan S.V., Catasti P., Ratliff R., Moyzis R.K., Laayoun A., Smith S.S., Bradbury E.M., Gupta G. (1995) Hairpins are formed by the single DNA strands of the fragile X triplet repeats: structure and biological implications. *Proc. Natl. Acad. Sci. USA* 92: 5199–5203.

Childs S., Ling V. (1996) Duplication and evolution of the P-glycoprotein genes in pig. *Biochim. Biophys. Acta* 1307: 205–212.

Choong C.S., Kemppainen J.A., Wilson E.M. (1998) Evolution of the primate androgen receptor: a structural basis for disease. *J. Mol. Evol.* 47: 334–342.

Clemenza L., Subramanian B., Hourcade D., Nickells M., Atkinson J. (1997) Primary sequence of baboon CR1 demonstrates concerted evolution within the *CR1* gene. *Molec. Immunol.* 34: 297–304.

Christian S.L., Fantes J.A., Mewborn S.K., Huang B., Ledbetter D.H. (1999) Large genomic duplicons map to sites of instability in the Prader-Willi/Angelman syndrome chromosome region (15q11–q13). *Hum. Molec. Genet.* 8: 1025–1037.

Colin Y., Cherif-Zahar B., Le Van Kim C., Raynal V., Van Huffel V., Cartron J.-P. (1991) Genetic basis of the RhD-positive and RhD-negative blood group polymorphism as determined by Southern analysis. *Blood* 78: 2747–2752.

Collier S., Sinnott P.J., Dyer P.A., Price D.A., Harris R., Strachan T. (1989) Pulsed field gel electrophoresis identifies a high degree of variability in the number of tandem 21-hydroxylase and complement C4 genes repeats in the 21-hydroxlase deficiency haplotypes. *EMBO J.* 8: 1393–1402.

Cook G.P., Tomlinson I.M. (1995) The human immunoglobulin V$_H$ repertoire. *Immunology Today* 16: 237–242.

Cook G.P., Tomlinson I.M., Walter G., Riethman H., Carter N.P., Buluwela L., Winter G., Rabbitts T.H. (1994) A map of the immunoglobulin V$_H$ locus completed by analysis of the telomeric region of chromosome 14q. *Nature Genet.* 7: 162–168.

Cooper D.N., Krawczak M. (1993) *Human Gene Mutation.* BIOS Scientific Publishers, Oxford.

Cooper G., Rubinsztein D.C., Amos W. (1998) Ascertainment bias cannot entirely account for human microsatellites being longer than their chimpanzee homologues. *Hum. Molec. Genet.* 7: 1425–1429.

Cordonnier A., Casella J.F., Heidmann T. (1995) Isolation of novel human endogenous retrovirus-like elements with foamy virus-related *pol* sequence. *J. Virol.* 69: 5890–5897.

Cossee M., Schmirr M., Campuzano V., Reutenauer L., Moutou C., Mandel J.-L., Koenig M. (1997) Evolution of the Friedreich's ataxia trinucleotide repeat expansion: founder effect and premutations. *Proc. Natl. Acad. Sci. USA* 94: 7452–7457.

Costache M., Apoil P.-A., Cailleau A., Elmgren A., Larson G., Henry S., Blancher A., Iordachescu D., Oriol R., Mollicone R. (1997) Evolution of fucosyltransferase genes in vertebrates. *J. Biol. Chem.* 272: 29 721–29 728.

Crawford A.M., Cuthbertson R.P. (1996) Mutations in sheep microsatellites. *PCR Meth. Appl.* 6: 876–879.

Daniels G.R., Deininger P.L. (1985) Integration site preferences of the *Alu* family and similar repetitive sequences. *Nucleic Acids Res.* **13**: 8939–8946.

Deeb S.S., Alvarez A., Malkki M., Motolsky A.G. (1995) Molecular patterns and sequence polymorphisms in the red and green visual pigment genes of Japanese men. *Hum. Genet.* **95**: 501–506.

De Jong W.W., Rydén L. (1981) Causes of more frequent deletions than insertions in mutations and protein evolution. *Nature* **290**: 157–159.

DeJong J.L., Mohandas T., Tu C.P. (1991) The human Hb (mu) class glutathione S-transferases are encoded by a dispersed gene family. *Biochem. Biophys. Res. Commun.* **180**: 15–22.

Deka N., Willard C.R., Wong E., Schmid C.W. (1988a) Human transposon-like elements insert at a preferred target site: evidence for a retrovirally mediated process. *Nucleic Acids Res.* **16**: 1143–1151.

Deka N., Wong E., Matera A.G., Kraft R., Leinwand L.A., Schmid C.W. (1988b) Repetitive nucleotide sequence insertions into a novel calmodulin-related gene and its processed pseudogene. *Gene* **71**: 123–134.

Deka R., Majumder P.P., Shriver M.D., Stivers D.N., Zhong Y., Yu L.M., Barrantes R., Yin S.J., Miki T., Hundreiser J., Bunker C.H., McGarvey S.T., Sakallah S., Ferrell R.E., Chakraborty R. (1996) Distribution and evolution of CTG repeats at the myotonin protein kinase gene in human populations. *Genome Res.* **6**: 142–154.

Deutekom J.C.T., Bakker E., Lemmers R.J.L.F., van der Wielen M.J.R., Bik E., Hofker M.H., Padberg G.W., Frants R.R. (1996) Evidence for subtelomeric exchange of 3.3kb tandemly repeated unit between chromosomes 4q35 and 10q26: implications for genetic counselling and etiology of FSHD1. *Hum. Molec. Genet.* **5**: 1997–2003.

Di Cristofano A., Strazzullo M., Parisi T., La Mantia G. (1995) Mobilization of an ERV9 human endogenous retroviral element during primate evolution. *Virology* **213**: 271–275.

Di Rienzo A., Peterson A.C., Garza J.C., Valdes A.M., Slatkin M., Freimer N.B. (1994) Mutational processes of simple sequence repeat loci in human populations. *Proc. Natl. Acad. Sci. USA* **91**: 3166–3170.

Djian P. (1998) Evolution of simple repeats in DNA and their relation to human disease. *Cell* **94**: 155–160.

Djian P., Green H. (1989) Vectorial expansion of the involucrin gene and the relatedness of the hominoids. *Proc. Natl. Acad. Sci. USA* **86**: 8447–8451.

Djian P., Green H. (1991) Involucrin gene of tarsioids and other primates: alternatives in evolution of the segment of repeats. *Proc. Natl. Acad. Sci. USA* **88**: 5321–5325.

Djian P., Phillips M., Easley K., Huang E., Simon M., Rice R.H., Green H. (1993) The involucrin genes of the mouse and the rat: study of their shared repeats. *Mol. Biol. Evol.* **10**: 1136–1149.

Dijan P., Hancock J.M., Chana H.S. (1996) Codon repeats in genes associated with human diseases: fewer repeats in the genes of non-human primates and nucleotide substitutions concentrated at the sites of reiteration. *Proc. Natl. Acad. Sci. USA* **93**: 417–421.

Doolittle R.F. (1995) The multiplicity of domains in proteins. *Annu. Rev. Biochem.* **64**: 287–314.

Drummond-Borg M., Deeb S.S., Motulsky A.G. (1989) Molecular patterns of X chromosome-linked color vision genes among 134 men of European ancestry. *Proc. Natl. Acad. Sci. USA* **86**: 983–987.

Eckert R.L., Green H. (1986) Structure and evolution of the human involucrin gene. *Cell* **46**: 583–589.

Edwards M.C., Gibbs R.A. (1992) A human dimorphism resulting from loss of an *Alu. Genomics* **14**: 590–596.

Efstratiadis A., Posakony J.W., Maniatis T., Lawn R.M., O'Connor C., Spritz R.A., DeRiel J.K., Forget B.G., Weissman S.M., Slightom J.L., Blechl A.E., Smithies O., Baralle F.E., Shoulders C.C., Proudfoot N.J. (1980) The structure and evolution of the human β-globin gene family. *Cell* **21**: 653–668.

Eichler E.E., Nelson D.L. (1998) The FRAXA fragile site and fragile X syndrome. In: *Analysis of Triplet Repeat Disorders*, (eds. DC Rubinsztein, MR Hayden). BIOS Scientific Publishers, Oxford, pp 13–50.

Eichler E.E., Holden J.J.A., Popovich B.W., Reiss A.L., Snow K., Thibodeau S.N., Richards C.S., Ward P.A., Nelson D.L. (1994) Length of uninterrupted CGG repeats determines stability in the *FMR1* gene. *Nature Genet.* **8**: 88–94.

Eichler E.E., Kunst C.B., Lugenbeel K.A., Ryder O.A., Davison D., Warren S.T., Nelson D.L. (1995) Evolution of the cryptic *FMR1* CGG repeat. *Nature Genet.* **11**: 301–308.

Eichler E.E., Lu F., Shen Y., Antonacci R., Jurecic V., Doggett N.A., Moyzis R.K., Baldini A., Gibbs R.A., Nelson D.L. (1996) Duplication of a gene-rich cluster between 16p11.1 and Xq28: a novel pericentromeric-directed mechanism for paralogous genome evolution. *Hum. Mol. Genet.* 5: 899–912.

Ellegren H., Primmer C.R., Sheldon B.C. (1995) Microsatellite 'evolution': directionality or bias? *Nature Genet.* 11: 360–362.

Endo T., Imanishi T., Gojobori T., Inoko H. (1997) Evolutionary significance of intra-genome duplications on human chromosomes. *Gene* 205: 19–27.

Esposito T., Gianfrancesco F., Ciccodicola A., Montanini L., Mumm S., D'Urso M., Forabosco A. (1999) A novel pseudoautosomal human gene encodes a putative protein similar to Ac-like transposases. *Hum. Molec. Genet.* 8: 61–67.

Fedorov A., Fedorova L., Starshenko V., Filatov V., Grigorev E. (1998) Influence of exon duplication on intron and exon phase distribution. *J. Mol. Evol.* 46: 263–271.

Felice A.E., Cleek M.P., Marino E.M., McKie K.M., McKie V.C., Chang B.K., Huisman T.H.J. (1986) Different ζ-globin gene deletions among black Americans. *Hum. Genet.* 73: 221–224.

Figueroa F. (1997) The primate class III MHC region encoding complement components and other genes. In: *Molecular Biology and Evolution of Blood Group and MHC Antigens in Primates,* (eds A Blancher, J Klein, WW Socha). Springer-Verlag, Berlin. pp 433–448.

Fitch D.H., Bailey W.J., Tagle D.A., Goodman M., Sien L., Slightom J.L. (1991) Duplication of the γ-globin gene mediated by L1 long interspersed repetitive elements in an early ancestor of simian primates. *Proc. Natl. Acad. Sci. USA* 88: 7396–7400.

Fracasso C., Patarnello T. (1998) Evolution of the dystrophin muscular promoter and 5′ flanking region in primates. *J. Mol. Evol.* 46: 168–179.

Fu Y.-H., Kuhl D.P.A., Pizzuti A., Pieretti M., Sutcliffe J.S., Richards S., Verkerk A.J.M.H., Holden J.J.A., Fenwick R.G., Warren S.T., Oostra B.A., Nelson D.L., Caskey C.T. (1991) Variation of the CGG repeat at the fragile X site results in genetic instability: resolution of Sherman paradox. *Cell* 67: 1047–1058.

Fu Y., Weissbach L., Plant P.W., Oddoux C., Cao Y., Liang T.J., Roy S.N., Redman C.M., Grieninger G. (1992) Carboxy-terminal-extended variant of the human fibrinogen a subunit: α novel exon conferring marked homology to β and γ subunits. *Biochemistry* 31: 11 968–11 972.

Funkhouser W., Koop B.F., Charmley P., Matindale D., Slightom J., Hood L. (1997) Evolution and selection of primate T cell antigen receptor BV8 gene subfamily. *Mol. Phylogenet. Evol.* 8: 51–64.

Furano A.V., Usdin K. (1995) DNA 'fossils' and phylogenetic analysis. *J. Biol. Chem.* 270: 25 301–25 304.

Furuno N., Makagawa K., Eguchi U., Ohtubo M., Nishimoto T., Soeda E. (1991) Complete nucleotide sequence of the human *RCC1* gene involved in coupling between DNA replication and mitosis. *Genomics* 11: 459–461.

Gacy A.M., Giellner G., Juranic N., Macura S., McMurray C.T. (1995) Trinucleotide repeats that expand in human disease form hairpin structures *in vitro. Cell* 81: 533–540.

Galili U., Swanson K. (1991) Gene sequences suggest inactivation of α-1,3-galactosyltransferase in catarrhines after the divergence of apes from monkeys. *Proc. Natl. Acad. Sci. USA* 88: 7401–7404.

Gan S.-Q., McBride O.W., Idler W.W., Markova N., Steinert P.M. (1990) Organization, structure and polymorphisms of the human profilaggrin gene. *Biochemistry* 29: 9432–9440.

Gao J.-L., Chen H., Filie J.D., Kozak C.A., Murphy P.M. (1998) Differential expansion of the N-formylpeptide receptor gene cluster in human and mouse. *Genomics* 51: 270–276.

Gastier J.M., Pulido J.C., Sunden S., Brody T., Neutow K.H., Murray J.C., Weber J.L., Hudson T.J., Sheffield V.C., Duyk G.M. (1995) Survey of trinucleotide repeats in the human genome: assessment of their utility as genetic markers. *Hum. Molec. Genet.* 4: 1829–1836.

Gendler S.J., Lancaster C.A., Taylor-Papadimitriou J., Duhig T., Peat N., Burchell J., Pemberton L., Lalani E.-N., Wilson D. (1990) Molecular cloning and expression of human tumor-associated polymorphic epithelial mucin. *J. Biol. Chem.* 265: 15 286–15 293.

Ghanem N., Buresi C., Moisan J.-P., Bensmana M., Chuchana P., Huck S., Lefranc G., Lefranc M.-P. (1989) Deletion, insertion, and restriction site polymorphism of the T-cell receptor gamma variable locus in French, Lebanes, Tunisian, and Black African populations. *Immunogenetics* 30: 350–360.

Gibbs S., Fijneman R., Wiegant J., Geurts van Kessel A., van de Putte P., Backendorf C. (1993) Molecular characterization and evolution of the SPRR family of keratinocyte differentiation markers encoding small proline-rich proteins. *Genomics* **16**: 630–637.

Goodchild N.L., Wilkinson D.A., Mager D.L. (1993) Recent evolutionary expansion of a subfamily of RTV$_L$-H human endogenous retrovirus-like elements. *Virology* **196**: 778–788.

Goodman B.K., Klein D., Tabor D.E., Vockley J.G., Cedarbaum S.D., Grody W.W. (1994) Functional and molecular analysis of liver arginase promoter sequences from man and *Macaca fascicularis*. *Somat. Cell Molec. Genet.* **20**: 313–325.

Graur D., Shuali Y., Li W.-H. (1989) Deletions in processed pseudogenes accumulate faster in rodents than in humans. *J. Mol. Evol.* **28**: 279–285.

Green H., Djian P. (1992) Consecutive actions of different gene-altering mechanisms in the evolution of involucrin. *Mol. Biol. Evol.* **9**: 977–1017.

Green H., Wang N. (1994) Codon reiteration and the evolution of proteins. *Proc. Natl. Acad. Sci. USA* **91**: 4298–4302.

Gritzmacher C.A. (1989) Molecular aspects of heavy-chain class switching. *Crit. Rev. Immunol.* **9**: 173–200.

Groot P.C., Bleeker M.J., Pronk J.C., Arwert F., Mager W.H., Planta R.J., Eriksson A.W., Frants R.R. (1989) The human α-amylase multigene family consists of haplotypes with variable numbers of genes. *Genomics* **5**: 29–42.

Haber J.E., Louis E.J. (1998) Minisatellite origins in yeast and humans. *Genomics* **48**: 132–135.

Han J., Hsu C., Zhu Z., Longshore J.W., Finley W.H. (1994) Over-representation of the disease associated (CAG) and (CGG) repeats in the human genome. *Nucleic Acids Res.* **22**: 1735–1740.

Hancock J.M. (1993) Evolution of sequence repetition and gene duplications in the TATA-binding protein TPD (TFIID). *Nucleic Acids Res*. **21**: 2823–2830.

Hancock J.M. (1995a) The contribution of DNA slippage to eukaryotic nuclear 18S rRNA evolution. *J. Mol. Evol.* **40**: 629–639.

Hancock J.M. (1995b) The contribution of slippage-like processes to genome evolution. *J. Mol. Evol.* **41**: 1038–1047.

Hancock J.M. (1996) Microsatellites and other simple sequences in the evolution of the human genome. In: *Human Genome Evolution*, (eds. M. Jackson, T. Strachan, G. Dover). BIOS Scientific Publishers, Oxford, pp191–210.

Hancock J.M., Dover G.A. (1988) Molecular coevolution among cryptically simple expansion segments of eukaryotic 26S/28S rRNAs. *Mol. Biol. Evol.* **5**: 377–391.

Hancock J.M., Dover G.A. (1990) 'Compensatory slippage' in the evolution of ribosomal RNA genes. *Nucleic Acids Res*. **18**: 5949–5954.

Hayward B.E., Bonthron D.T. (1998) Structure and alternative splicing of the ketohexokinase gene. *Eur. J. Biochem.* **257**: 85–91.

Henikoff S., Greene E.A., Pietrokovski S., Bork P., Attwood T.K., Hood L. (1997) Gene families: the taxonomy of protein paralogs and chimeras. *Science* **278**: 609–614.

Hirst M.C., Grewal P.K., Davies K.E. (1994) Precursor arrays for triplet repeat expansion at the fragile X locus. *Hum. Molec. Genet.* **3**: 1553–1560.

Horiuchi Y., Kawaguchi H., Figueroa F., O'hUigin C., Klein J. (1993) Dating the primigenial *C4-CYP21* duplication in primates. *Genetics* **134**: 331–339.

Hourcade D., Miesner D.R., Bee C., Zeldes W., Atkinson J.P. (1990) Duplication and divergence of the amino-terminal coding region of the complement receptor 1 (CR1) gene. An example of concerted (horizontal) evolution within a gene. *J. Biol. Chem.* **265**: 974–980.

Housman D. (1995) Gain of glutamines, gain of function? *Nature Genet.* **10**: 3–4.

Hu X., Worton R.G. (1992) Partial gene duplication as a cause of human disease. *Hum. Mutation* **1**: 3–12.

Hu X., Ray P.N., Worton R.G. (1991) Mechanisms of tandem duplication in the Duchenne muscular dystrophy gene include both homologous and nonhomologous intrachromosomal recombination. *EMBO J*. **10**: 2471–2477.

Hubert C., Houot A.-M., Corvol P., Soubrier F. (1991) Structure of the angiotensin I-converting enzyme gene. *J. Biol. Chem.* **266**: 15 377–15 383.

Hughes A.L. (1994) Evolution of the ATP-binding cassette transmembrane transporters of vertebrates. *Mol. Biol. Evol.* **11**: 899–910.

Hughes J., Ward C.J., Peral B., Aspinwall R., Clark K., San Millan J.L., Harris P.C. (1995) The polycystic kidney disease 1 (*PKD1*) gene encodes a novel protein with multiple cell recognition domains. *Nature Genet.* **10**: 151–160.

Hunt D.M., Cowing J.A., Patel R., Appukuttan B., Bowmaker J.K., Mollon J.D. (1995) Sequence and evolution of the blue cone pigment gene in Old and New World primates. *Genomics* **27**: 535–538.

Hwu H.R., Roberts J.W., Davidson E.H., Britten R.J. (1986) Insertion and /or deletion of many repeated DNA sequences in human and higher ape evolution. *Proc. Natl. Acad. Sci. USA* **83**: 3875–3879.

Ichinose A. (1992) Multiple members of the plasminogen-apolipoprotein(a) gene family associated with thrombosis. *Biochemistry* **31**: 3113–3118.

Imbert G., Kretz C., Johnson K., Mandel J.-L. (1993) Origin of the expansion mutation in myotonic dystrophy. *Nature Genet.* **4**: 72–76.

Iris F.J., Bouguéleret L., Prieur S., Caterina D., Primas G., Perrot V., Jurka J., Rodriguez-Tome P., Claverie J.M., Dausset J., Cohen D. (1993) Dense *Alu* clustering and a potential new member of the NFκB family within a 90 kilobase HLA class III segment. *Nature Genet.* **3**: 137–145.

Iwabe N., Kuma K., Miyata T. (1996) Evolution of gene families and relationship with organismal evolution: rapid divergence of tissue-specific genes in the early evolution of chordates. *Mol. Biol. Evol.* **13**: 483–493.

Jeffreys A.J. (1997) Spontaneous and induced minisatellite instability in the human genome. *Clin. Sci.* **93**: 383–390.

Jeffreys A.J., Royle N.J., Wilson V., Wong Z. (1988) Spontaneous mutation rates to new length alleles at tandem-repetitive hypervariable loci in human DNA. *Nature* **332**: 278–281.

Jeffreys A.J., Neumann R., Wilson V. (1990) Repeat unit sequence variation in minisatellites: a novel source of DNA polymorphism for studying variation and mutation by single molecule analysis. *Cell* **60**: 473–485.

Jeffreys A.J., Tamaki K., MacLeod A., Monckton D.G., Neil D.L., Armour J.A.L. (1994) Complex gene conversion events in germline mutation at human minisatellites. *Nature Genet.* **6**: 136–145.

Jeffreys A.J., Murray J., Neumann R. (1998a) High-resolution mapping of crossovers in human sperm defines a minisatellite-associated recombination hotspot. *Molec. Cell* **2**: 267–273.

Jeffreys A.J., Neil D.L., Neumann R. (1998b) Repeat instability at human minisatellites arising from recombination. *EMBO J.* **17**: 4147–4157.

Jin L., Macaubas C., Hallmeyer J., Kimura A., Mignot E. (1996) Mutation rate varies among alleles at a microsatellite locus: phylogenetic evidence. *Proc. Natl. Acad. Sci. USA* **93**: 15 285–15 288.

Kang S., Ohshima K., Simizu M., Amirhaeri S., Wells R.D. (1995) Pausing of DNA synthesis *in vitro* at specific loci in CTG and CGG triplet repeats from human hereditary disease genes. *J. Biol. Chem.* **270**: 27 014–27 021.

Kariya Y., Kato K., Hayashizaki Y., Himeno S., Tarui S., Matsubara K. (1987) Revision of consensus sequence of human *Alu* repeat – a review. *Gene* **53**: 1–10.

Karlin S., Burge C. (1996) Trinucleotide repeats and long homopeptides in genes and proteins associated with nervous system disease and development. *Proc. Natl. Acid. Res. USA* **93**: 1560–1565.

Kass D.H., Aleman C., Batzer M.A., Deininger P.L. (1994) Identification of a human specific *Alu* insertion in the factor XIIIB gene. *Genetica* **94**: 1–8.

Kayser M., Nürnberg P., Bercovitch F., Nagy M., Roewer L. (1995) Increased microsatellite variability in *Macaca mulatta* compared to humans due to a large scale deletion/insertion event during primate evolution. *Electrophoresis* **16**: 1607–1611.

Kazazian H.H., Wong C., Youssoufian H., Scott A.F., Phillips D.G., Antonarakis S.E. (1988) Haemophilia A resulting from *de novo* insertion of L1 sequences represents a novel mechanism for mutation in man. *Nature* **332**: 164–166.

Kazazian H.H., Moran J.V. (1998) The impact of L1 retrotransposons on the human genome. *Nature Genet.* **19**: 19–24.

Kellermann J., Lottspeich F., Henschen A., Müller-Esterl W. (1986) Completion of the primary structure of human high-molecular-mass kininogen. *Eur. J. Biochem.* **154**: 471–478.

Khan W.N., Teglund S., Bremer K., Hammarström S. (1992) The pregnancy-specific glycoprotein family of the immunoglobulin superfamily: identification of new members and estimation of family size. *Genomics* **12**: 780–787.

Kim S.J., Ruiz N., Bezouska K., Drickamer K. (1992) Organization of the gene encoding the human macrophage mannose receptor (*MRC1*). *Genomics* **14**: 721–727.

Kim H.S., Lyons K.M., Saitoh E., Azen E.A., Smithies O., Maeda N. (1993) The structure and evolution of the human salivary proline-rich protein gene family. *Mamm. Genome* **4**: 3–14.

Kim S.J., Shon B.H., Kang J.H., Hahm K.-S., Yoo O.J., Park Y.S., Lee K.-K. (1997) Cloning of novel trinucleotide-repeat (CAG) containing genes in mouse brain. *Biochem. Biophys. Res. Commun.* **240**: 239–243.

Kipling D., Warburton P.E. (1997) Centromeres, CENP-B and *Tigger* too. *Trends Genet.* **13**: 141–145.

Kiyosawa H., Chance P.F. (1996) Primate origin of the CMT1A-REP repeat and analysis of a putative transposon-associated recombinational hotspot. *Hum. Molec. Genet.* **5**: 745–753.

Kölble K., Lu J., Mole S.E., Kaluz S., Reid K.B.M. (1993) Assignment of the human pulmonary surfactant protein D gene (*SFTP4*) to 10q22-q23 close to the surfactant protein A gene cluster. *Genomics* **17**: 294–298.

Kornreich R., Bishop D.F., Desnick R.J. (1990) α-Galactosidase A gene rearrangements causing Fabry disease. *J. Biol. Chem.* **265**: 9319–9326.

Krahe R., Eckhart M., Ogunniyi A.O., Osuntokun B.O., Siciliano M.J., Ashizawa T. (1995) *De novo* myotonic dystrophy mutation in a Nigerian kindred. *Am. J. Hum. Genet.* **56**: 1067–1074.

Krawczak M., Cooper D.N. (1991) Gene deletions causing human genetic disease: mechanisms of mutagenesis and the role of the local DNA sequence environment. *Hum. Genet.* **86**: 425–441.

Labuda D., Zietkiewicz E., Mitchell G.A. (1995) *Alu* elements as a source of genomic variation: deleterious effects and evolutionary novelties. In: *The Impact of Short Interspersed Elements (SINEs) on the Host Genome*, (ed. RJ Maraia). RG Landes Co, Austin, Texas.

Lackner C., Cohen J.C., Hobbs H.H. (1993) Molecular definition of the extreme size polymorphism in apolipoprotein(a). *Hum. Molec. Genet.* **2**: 933–940.

Larsen R.D., Rivera-Marrero C.A., Ernst L.K., Cummings R.D., Lowe J.B. (1990) Frameshift and nonsense mutations in a genomic sequence homologous to a murine UDP-Gal: beta-D-Gal(1,4)-D-GlcNAc alpha (1,3)-galactosyltransferase cDNA. *J. Biol. Chem.* **265**: 7055–7061.

Lawler J., Duquette M., Urry L., McHenry K., Smith T.F. (1993) The evolution of the thrombospondin gene family. *J. Mol. Evol.* **36**: 509–516.

Lawn R.M., Schwartz K., Patthy L. (1997) Convergent evolution of apolipoprotein(a) in primates and hedgehog. *Proc. Natl. Acad. Sci. USA* **94**: 11 992–11 997.

Lee C.-P., Kao M.-C., French B.A., Putney S.D., Chang S.H. (1987) The rabbit muscle phosphofructokinase gene. *J. Biol. Chem.* **262**: 4195–4199.

Leeflang E.P., McPeek M.S., Arnheim N. (1996) Analysis of meiotic segregation, using single sperm typing: meiotic drive at the myotonic dystrophy locus. *Am. J. Hum. Genet.* **59**: 896–904.

Leib-Mösch C., Brack-Werner R., Werner T., Bachmann M., Faff O., Erfle V., Hehlmann R. (1990) Endogenous retroviral elements in human DNA. *Cancer Res.* **50**: 5636S-5642S.

Leib-Mösch C., Seifarth W. (1996) Evolution and biological significance of human retroelements. *Virus Genes* **11**: 133–145.

Levinson G., Gutman G.A. (1987) Slipped-strand mispairing: a major mechanism for DNA sequence evolution. *Mol. Biol. Evol.* **4**: 203–221.

Li S.-H., McInnis M.G., Margolis R.L., Antonarakis S.E., Ross C.A. (1993) Novel triplet repeat containing genes in human brain: cloning, expression and length polymorphisms. *Genomics* **16**: 572–579.

Li W.-H. (1997) *Molecular Evolution.* Sinauer Associates, Sunderland, Mass.

Liao D., Weiner A.M. (1995) Concerted evolution of the tandemly repeated genes encoding primate U2 small nuclear RNA (the RNU2 locus) does not prevent rapid diversification of the (CT)n (GA)n microsatellite embedded within the U2 repeat unit. *Genomics* **30**: 583–593.

Lichter P., Bray P., Ried T., Dawid I.B., Ward D.C. (1992) Clustering of C2-H2 zinc finger motif sequences within telomeric and fragile site regions of human chromosomes. *Genomics* **13**: 999–1007.

Limrasert P., Nouri N., Heyman R.A., Nopparatana C., Kamonsilp M., Deininger P.L., Keats B.J.B. (1996) Analysis of CAG repeat of the Machado-Joseph gene in human, chimpanzee and monkey populations: a variant nucleotide is associated with the number of CAG repeats. *Hum. Molec. Genet.* **5**: 207–213.

Lin B., Nasir J., MacDonald H., Hutchinson G., Graham R.K., Rommens J.M., Hayden M.R. (1994) Sequence of the murine Huntington disease gene: evidence for conservation, alternate splicing and polymorphism in a triplet (CCG) repeat. *Hum. Molec. Genet.* **3**: 85–92.

Liou L.S., Sastry R., Hartshorn K.L., Lee Y.M., Okarma T.B., Tauber A.I., Sastry K.N. (1994) Bovine conglutinin gene exon structure reveals its evolutionary relationship to surfactant protein-D. *J. Immunol.* **153**: 173–180.

Lundwall A. (1996) The structure of the semenogelin gene locus. Nucleotide sequence of the intergenic and the flanking DNA. *Eur. J. Biochem.* **235**: 466–470.

Lundwall A. (1998) The cotton-top tamarin carries an extended semenogelin I gene but no semenogelin II gene. *Eur. J. Biochem.* **255**: 45–51.

Lutfalla G., Gardiner K., Proudhon D., Vielh E., Uzé G. (1992) The structure of the human interferon α/β receptor gene. *J. Biol. Chem.* **267**: 2802–2809.

Lutfalla G., McInnis M.G., Antonarakis S.E., Uzé G. (1995) Structure of the human *CRFB4* gene: comparison with its *IFNAR* neighbor. *J. Mol. Evol.* **41**: 318–344.

McLean J.W., Tomlinson J.E., Kuang W.J., Eaton D.L., Chen E.Y., Fless G.M., Scanu A.M., Lawn R.M. (1987) cDNA sequence of human apolipoprotein(a) is homologous to plasminogen. *Nature* **330**: 132–137.

McDonald J.F. (1993) Evolution and consequences of transposable elements. *Curr. Opin. Genet. Devel.* **3**: 855–864.

McEvoy S.M., Maeda N. (1988) Complex events in the evolution of the haptoglobin gene cluster in primates. *J. Biol. Chem.* **263**: 15 740–15 747.

McNaughton J.C., Broom J.E., Hill D.F., Jones W.A., Marshall C.J., Renwick N.M., Stockwell P.A., Petersen G.B. (1993) A cluster of transposon-like repetitive sequences in intron 7 of the human dystrophin gene. *J. Mol. Biol.* **232**: 314–321.

Maeda N., Kim H.S. (1990) Three independent insertions of retrovirus-like sequences in the haptoglobin gene cluster of primates. *Genomics* **8**: 671–683.

Magnaghi P., Citterio E., Malgaretti N., Acquati F., Ottolenghi S., Taramelli R. (1994) Molecular characterization of the human apo(a)-plasminogen gene family clustered on the telomeric region of chromosome 6 (6q26–27). *Hum. Molec. Genet.* **3**: 437–442.

Makalowski W., Mitchell G.A., Labuda D. (1994) *Alu* sequences in the coding regions of mRNA: a source of protein variability. *Trends Genet.* **10**: 188–192.

Margalit H., Nadir E., Ben-Sasson S.A. (1994) A complete *Alu* element within the coding sequence of a central gene. *Cell* **78**: 173–174.

Martens U.M., Zijlmans M.J.-M., Poon S.S.S., Dragowska W., Yui J., Chavez E.A., Ward R.K., Lansdorp P.M. (1998) Short telomeres on human chromosome 17p. *Nature Genet.* **18**: 76–80.

Matera A.G., Hellmann U., Hintz M.F., Schmid C.W. (1990) Recently transposed *Alu* repeats result from multiple source genes. *Nucleic Acids Res.* **18**: 6019–6026.

Mazzarella R., Schlessinger D. (1997) Duplication and distribution of repetitive elements and non-unique regions in the human genome. *Gene* **205**: 29–38.

Mazzarella R., Schlessinger (1998) Pathological consequences of sequence duplications in the human genome. *Genome Res.* **8**: 1007–1021.

Meagher M.J., Jorgensen A.L., Deeb S.S. (1996) Sequence and evolutionary history of the length polymorphism in intron 1 of the human red photopigment gene. *J. Mol. Evol.* **43**: 622–630.

Meireles C.M., Schneider M.P., Sampaio M.I., Schneider H., Slightom J.L., Chiu C.H., Neiswanger K., Gumucio D.L., Czelusniak J., Goodman M. (1995) Fate of a redundant gamma-globin gene in the atelid clade of New World monkeys: implications concerning fetal globin gene expression. *Proc. Natl. Acad. Sci. USA* **92**: 2607–2611.

Meyer E., Wiegand P.R. and S.P., Kuhlmann D., Brack M., Brinkmann B. (1995) Microsatellite polymorphisms reveal phylogenetic relationships in primates. *J. Mol. Evol.* **41**: 10–14.

Michel D., Chatelain G., Mauduit C., Benahmed M., Brun G. (1997) Recent evolutionary acquisition of alternative pre-mRNA splicing and 3′ processing regulations induced by intronic B2 SINE insertion. *Nucleic Acids Res.* **25**: 3228–3234.

Mighell A.J., Markham A.F., Robinson P.A. (1997) *Alu* sequences. *FEBS Letts.* **417**: 1–5.

Miki Y., Nishisho I., Horii A., Miyoshi Y., Utsunomiya J., Kinzler K.W., Vogelstein B., Nakamura Y. (1992) Disruption of the *APC* gene by a retrotransposal insertion of L1 sequence in a colon cancer. *Cancer Res.* **52**: 643–645.

Milewicz D.M., Byers P.H., Reveille J., Hughes A.L., Duvic M. (1996) A dimorphic *Alu* Sb-like insertion in *COL3A1* is ethnic specific. *J. Mol. Evol.* **42**: 117–123.

Millar M., Zeller K. (1997) Alternative splicing in lecithin: cholesterol acyltransferase mRNA: an evolutionary paradigm in humans and great apes. *Gene* **190**: 309–313.

Mitas M., Yu A., Dill J., Kamp T.J., Chambers E.J., Howorth I.S. (1995) Hairpin properties of single-stranded DNA containing a GC-rich triplet repeat: $(CTG)_{15}$. *Nucleic Acids Res.* **23**: 1050–1059.

Monckton D.G., Neumann R., Guram T., Fretwell N., Tamaki K., MacLeod A., Jeffreys A.J. (1994) Minisatellite mutation rate variation associated with a flanking DNA sequence polymorphism. *Nature Genet.* **8**: 162–166.

Monckton D.G., Caskey C.T. (1995) Unstable triplet repeat diseases. *Circulation* **91**: 513–520.

Monnat R.J., Hackmann A.F.M., Chiaverotti T.A. (1992) Nucleotide sequence analysis of human hypoxanthine phosphoribosyltransferase (*HPRT*) gene deletions. *Genomics* **13**: 777–787.

Moore S., Sargeant L., King T., Mattik J., Georges M., Hetzel D. (1991) The conservation of dinucleotide microsatellites among mammalian genomes allows the use of heterologous PCR primer pairs in closely related species. *Genomics* **10**: 654–660.

Morgan R.O., Bell D.W., Testa J.R., Fernandez M.P. (1998) Genomic locations of *ANX11* and *ANX13* and the evolutionary genetics of human annexins. *Genomics* **48**: 100–110.

Moutou C., Vincet M.-C., Biancalana V., Mandel J.-L. (1997) Transition from premutation to full mutation in fragile X syndrome is likely to be prezygotic. *Hum. Molec. Genet.* **6**: 971–979.

Muratani K., Hada Y., Yamamoto T., Kaneko Y., Shigeto T., Ohue J., Furuyama J., Higashino K. (1991) Inactivation of the cholinesterase gene by *Alu* insertion: possible mechanism for human gene transposition. *Proc. Natl. Acad. Sci. USA* **88**: 11 315–11 319.

Muse S.V., Clark A.G., Thomas G.H. (1997) Comparisons of the nucleotide substitution process among repetitive segments of the α- and β-spectrin genes. *J. Mol. Evol.* **44**: 492–500.

Nagase M., Hirose S., Fujita T. (1998) Unique repetitive sequence and unexpected regulation of expression of rat endothelial receptor for oxidized low-density lipoprotein (LOX-1). *Biochem. J.* **330**: 1417–1422.

Narita N., Nishio H., Kitoh Y., Ishikawa Y., Ishikawa R., Minami H., Nakamura Y., Matsuo M. (1992) Insertion of a truncated 5′ L1 element into the 3′ end of exon 44 of the dystrophin gene resulted in skipping of the exon during splicing in a case of Duchenne muscular dystrophy. *J. Clin. Invest.* **91**: 1862–1867.

Nathans J., Thomas D., Hogness D.S. (1986a) Molecular genetics of human color vision: the genes encoding blue, green and red pigments. *Science* **232**: 193–201.

Nathans J., Piantanida T.P., Eddy T., Shows T.B., Hogness D.S. (1986b) Molecular genetics of inherited variation in human color vision. *Science* **232**: 203–210.

Neitz M., Neitz J. (1995) Numbers and ratios of visual pigment genes for normal red-green color vision. *Science* **267**: 1013–1016.

Neitz M., Neitz J., Grishok A. (1995) Polymorphism in the numbers of genes encoding long-wavelength-sensitive cone pigments among males with normal color vision. *Vision Res.* **35**: 2395–2407.

Nelson H.H., Wiencke J.K., Christiani D.C., Cheng T.J., Zuo Z.F., Schwartz B.S., Lee B.K., Spitz M.R., Wang M., Xu X. (1995) Ethnic differences in the prevalence of the homozygous deleted genotype of glutathione S-transferase theta. *Carcinogenesis* **16**: 1243–1245.

Nelson J.E., Krawetz S.A. (1994) Characterization of a human locus in transition. *J. Biol. Chem.* **269**: 31 067–31 073.

Neri C., Albanese V., Lebre-A.-S., Holbert S., Saada C., Bouguleret L., Meier-Ewert S., Le Gall I., Millasseau P., Bui H., Giudicelli C., Massaart C., Guillou S., Gervy P., Poullier E., Rigault P., Weissenbach J., Lennon G., Chumakov I., Dausset J., Lehrach H., Cohen D., Cann H.M. (1996) Survey of CAG/CTG repeats in human cDNAs representing new genes: candidates for inherited neurological disorders. *Hum. Molec. Genet.* **5**: 1001–1009.

Nishio H., Heiskanen M., Palotie A., Bélanger L., Dugaiczyk A. (1996) Tandem arrangement of the human serum albumin multigene family in the sub-centromeric region of 4q: evolution and chromosomal direction of transcription. *J. Mol. Biol.* **259**: 113–119.

Nitta K., Sugai S. (1989) The evolution of lysozyme and α-lactalbumin. *Eur. J. Biochem.* **182**: 111–118.

Nollet S., Moniaux N., Maury J., Petitprez D., Degand P., Laine A., Porchet N., Aubert J.-P. (1998) Human mucin gene *MUC4*: organization of its 5′-region and polymorphism of its central tandem repeat array. *Biochem. J.* **332**: 739–748.

Oakey R.J., Caron M.G., Lefkowitz R.J., Seldin M.F. (1991) Genomic organization of adrenergic and serotonin receptors in the mouse: linkage mapping of sequence-related genes provides a method for examining mammalian chromosome evolution. *Genomics* **10**: 338–344.

Ohno S. (1984) Repeats of base oligomers as the primordial coding sequences of the primeval earth and their vestiges in modern genes. *J. Mol. Evol.* **20**: 313–321.

Ohno S. (1987) Early genes that were oligomeric repeats generated a number of divergent domains on their own. *Proc. Natl. Acad. Sci. USA* **84**: 6486–6490.

Olsson M.L., Chester M.A. (1996) Evidence for a new type of O allele at the ABO locus, due to a combination of the A2 nucleotide deletion and the Ael nucleotide insertion. *Vox. Sang.* **71**: 113–117.

Ortel T.L., Takahashi N., Putnam F.W. (1984) Structural model of human ceruloplasmin based on internal triplication, hydrophilic/hydrophobic character, and secondary structure of domains. *Proc. Natl. Acad. Sci. USA* **81**: 4761–4765.

Park I., Schaeffer E., Sidoli A., Baralle F.E., Cohen G.N., Zakin M.M. (1985) Organization of the human transferrin gene: direct evidence that it originated by gene duplication. *Proc. Natl. Acad. Sci. USA* **82**: 3149–3153.

Pascarella S., Argos P. (1992) Analysis of insertions/deletions in protein structures. *J. Mol. Biol.* **224**: 461–471.

Pausova Z., Morgan K., Fujiwara R.M., Hendy G.N. (1995) Evolution of a repeat sequence in the parathyroid hormone-related peptide gene in primates. *Mamm. Genome* **6**: 408–414.

Pearson W.R., Vorachek W.R., Xu S., Berger R., Hart I., Vannais D., Patterson D. (1993) Identification of class-mu glutathionetransferase genes *GSTM1-GSTM5* on human chromosome 1p13. *Am. J. Hum. Genet.* **53**: 220–233.

Pemble S., Schroeder K.R., Spenser S.R., Meyer D.J., Hallier E., Bolt H.M., Ketterer B., Taylor J.B. (1994) Human glutathione S-transferase theta (GSTT1): cDNA cloning and the characterization of a genetic polymorphism. *Biochem. J.* **300**: 271–276.

Pendas A.M., Santamaria I., Alvarez M.V., Pritchard M., Lopez-Otin C. (1996) Fine physical mapping of the human metalloproteinase genes clustered on chromosome 11q22.3. *Genomics* **37**: 266–269.

Pesole G., Gerardi A., di Jeso F., Saccone C. (1994) The peculiar evolution of apolipoprotein(a) in human and rhesus macaque. *Genetics* **136**: 255–260.

Pizzuti A., Pieretti M., Fenwick R.G., Gibbs R.A., Caskey C.T. (1992) A transposon-like element in the deletion-prone region of the dystrophin gene. *Genomics* **13**: 594–600.

Plummer N.W., Meisler M.H. (1999) Evolution and diversity of mammalian sodium channel genes. *Genomics* **57**: 323–331.

Prager E.M., Wilson A.C. (1988) Ancient origin of lactalbumin from lysozyme: analysis of DNA and amino acid sequences. *J. Mol. Evol.* **27**: 326–335.

Preston B.D. (1996) Error-prone retrotransposition: rime of the ancient mutators. *Proc. Natl. Acad. Sci. USA* **93**: 7427–7431.

Qin Z.H., Schuller I., Richter G., Diamantstein T., Blankenstein T. (1991) The interleukin-6 gene locus seems to be a preferred target site for retrotransposon integration. *Immunogenetics* **33**: 260–266.

Rabbani H., Pan Q., Kondo N., Smith C.I.E., Hammarström L. (1996) Duplications and deletions of the human *IGHC* locus: evolutionary implications. *Immunogenetics* **45**: 136–141.

Raisonnier A. (1991) Duplication of the apolipoprotein C-1 gene occurred about forty million years ago. *J. Mol. Evol.* **32**: 211–219.

Ramharack R., Deeley R.G. (1987) Structure and evolution of primate cytochrome *c* oxidase subunit II gene. *J. Biol. Chem.* **262**: 14 014–14 021.

Rearden A., Magnet A., Kudo S., Fukuda M. (1993) Glycophorin B and glycophorin E genes arose from the glycophorin A ancestral gene via two duplications during primate evolution. *J. Biol. Chem.* **268**: 2260–2267.

Reiter L.T., Murakami T., Koeuth T., Pentao L., Muzny D.M., Gibbs R.A., Lupski J.R. (1996) A recombination hotspot responsible for two inherited peripheral neuropathies is located near a mariner transposon-like element. *Nature Genet.* **12**: 288–297.

Riccio M.L., Rossolini G.M. (1993) Unusual clustering of *Alu* repeats within the 5′-flanking region of the human lysozyme gene. *DNA Sequence* **4**: 129–134.

Richards R.I., Sutherland G.R. (1996) Repeat offenders: simple repeat sequences and complex genetic problems. *Hum. Mutation* **8**: 1–7.

Rigat B., Hubert C., Alhenc-Gelas F., Cambien F., Corvol P., Soubrier F. (1990) An insertion/deletion polymorphism in the angiotensin I-converting enzyme gene accounting for half the variance of serum enzyme levels. *J. Clin. Invest.* **86**: 1343–1346.

Riggins G.J., Lokey L.K., Chastain J.L., Leiner H.A., Sherman S.l., Wilkinson K.D., Warren S.T. (1992) Human genes containing polymorphic trinucleotide repeats. *Nature Genet.* **2**: 186–191.

Ritchie R.J., Mattei M.G., Lalande M. (1998) A large polymorphic repeat in the pericentromeric region of human chromosome 15q contains three partial gene duplications. *Hum. Molec. Genet.* **7**: 1253–1260.

Rousseau F., Heitz D., Biancalana V., Blumenfeld S., Kretz C., Boué J., Tommerup N., Van der Hagen C., DeLozier-Blanchet C., Croquette M.-F., Gilgenkrantz S., Jalbert P., Voelckel M.-A., Oberlé I., Mandel J.-L. (1991) Direct diagnosis by DNA analysis of the fragile X syndrome of mental retardation. *New Engl. J. Med.* **325**: 1673–1681.

Rowen L., Koop B.F., Hood L. (1996) The complete 685-kilobase DNA sequence of the human β T cell receptor locus. *Science* **272**: 1755–1762.

Roy S., Overton O., Redman C. (1994) Overexpression of any fibrinogen chain by HepG2 cells specifically elevates the expression of the other two chains. *J. Biol. Chem.* **269**: 691–695.

Royle N.J. (1996) Telomeres, subterminal sequences, variation and turnover. In: *Human Genome Evolution*, (eds. M Jackson, T Strachan, G Dover). BIOS Scientific Publishers, Oxford, pp 147–169.

Royle N.J., Baird D.M., Jeffreys A.J. (1994) A subterminal satellite located adjacent to telomeres in chimpanzees is absent from the human genome. *Nature Genet.* **6**: 52–55.

Rubinsztein D.C., Amos W., Leggo J., Goodburn S., Ramesar R.S., Old J., Bontrop R., McMahon R., Barton D.E., Ferguson-Smith M.A. (1994) Mutational bias provides a model for the evolution of Huntington's disease and predicts a general increase in disease prevalence. *Nature Genet.* **7**: 525–530.

Rubinsztein D.C., Amos W., Leggo J., Goodburn S., Jain S., Li S.-H., Margolis R.L., Ross C.A., Ferguson-Smith M.A. (1995a) Microsatellite evolution – evidence for directionality and variation in rate between species. *Nature Genet.* **10**: 337–343.

Rubinsztein D.C., Leggo J., Amos W. (1995b) Microsatellites evolve more rapidly in humans than in chimpanzees. *Genomics* **30**: 610–612.

Rudert F., Zimmermann W., Thompson J.A. (1989) Intra- and interspecies analyses of the carcinoembryonic antigen (*CEA*) gene family reveal independent evolution in primates and rodents. *J. Mol. Evol.* **29**: 126–134.

Saitou N., Ueda S. (1994) Evolutionary rates of insertion and deletion in noncoding nucleotide sequences of primates. *Mol. Biol. Evol.* **11**: 504–512.

Scanu A.M., Edelstein C. (1995) Kringle-dependent structural and functional polymorphism of apolipoprotein(a). *Biochim. Biophys. Acta* **1256**: 1–12.

Schlötterer C., Tautz D. (1992) Slippage synthesis of simple sequence DNA. *Nucleic Acids Res.* **20**: 211–215.

Seidegard E., Vorachek W.R., Pero R.W., Pearson W.R. (1988) Hereditary differences in the expression of the human glutathione transferase active on *trans*-stilbene oxide are due to a gene deletion. *Proc. Natl. Acad. Sci. USA* **85**: 7293–7297.

Shih C.-C., Stoye J.P., Coffin J.M. (1988) Highly preferred targets for retrovirus integration. *Cell* **53**: 531–537.

Shih A., Coutavas E.E., Rush M.G. (1991) Evolutionary implications of primate endogenous retroviruses. *Virology* **182**: 495–502.

Simon M., Phillips M., Green H., Stroh H., Glatt G., Bruns G., Latt S.A. (1989) Absence of a single repeat from the coding region of the human involucrin gene leading to RFLP. *Am. J. Hum. Genet.* **45**: 910–916.

Simon M., Phillips M., Green H. (1991) Polymorphism due to variable number of repeats in the human involucrin gene. *Genomics* **9**: 576–580.

Sirugo G., Deinard A.S., Kidd J.R., Kidd K.K. (1997) Survey of maximum CTG/CAG repeat lengths in humans and non-human primates: total genome scan in populations using the Repeat Expansion Detection method. *Hum. Molec. Genet.* **6**: 403–408.

Smit A.F.A., Toth G., Riggs A.D., Jurka J. (1995) Ancestral, mammalian-wide subfamilies of LINE-1 repetitive sequences. *J. Mol. Biol.* **246**: 401–417.

Somers G.R., Hammet F., Woollatt E., Richards R.I., Southey M.C., Venter D.J. (1997) Chromosomal localization of the human P2y6 purinoceptor gene and phylogenetic analysis of the P2y purinoceptor family. *Genomics* **44**: 127–130.

Spano F., Raugei G., Palla E., Colella C., Melli M. (1990) Characterization of the human lipocortin-2-encoding multigene family: its structure suggests the existence of a short amino acid unit undergoing duplication. *Gene* **95**: 243–251.

Squitieri F., Andrew S.E., Goldberg Y.P., Kremer B., Spense N., Zeisler J., Nichol K., Theilmann J., Greenberg J., Goto J., Kanazawa I., Vesa V., Peltonen L., Almqvist E., Anvret M., Telenius H., Lin B., Napolitano G., Morgan K., Hayden M.R. (1994) DNA haplotype analysis of Huntington disease reveals clues to the origins and mechanisms of CAG expansion and reasons for geographic variations of prevalence. *Hum. Molec. Genet.* **3**: 2103–2114.

Stallings R.L. (1994) Distribution of trinucleotide microsatellites in different categories of mammalian genomic sequence: implications for human genetic diseases. *Genomics* **21**: 116–121.

Stallings R.L. (1995) Conservation and evolution of $(CT)_n/(GA)_n$ microsatellite sequences at orthologous positions in diverse mammalian genomes. *Genomics* **25**: 107–113.

Stallings RL., Ford A.F., Nelson D., Torney D.C., Hildebrand C.E., Moyzis R.K. (1991) Evolution and distribution of $(GT)_n$ repetitive sequences in mammalian genomes. *Genomics* **10**: 807–815.

Steinhuber S., Brack M., Hunsmann G., Schwelberger H., Dierich M.P., Vogetseder W. (1995) Distribution of human endogenous retrovirus HERV-K genomes in humans and different primates. *Hum. Genet.* **96**: 188–192.

Strong M., Chandy K.G., Gutman G.A. (1993) Molecular evolution of voltage-sensitive ion channel genes: on the origins of electrical excitability. *Molec. Biol. Evol.* **10**: 221–224.

Sun H.S., Kirkpatrick B.W. (1996) Exploiting dinucleotide microsatellites conserved among mammalian species. *Mamm. Genome* **7**: 128–132.

Sutherland G.R., Richards R.I. (1995) Simple tandem DNA repeats and human genetic disease. *Proc. Natl. Acad. Sci. USA* **92**: 3636–3641.

Swada I., Willard C., Shen C.K., Chapman B., Wilson A.C., Schmid C.W. (1985) Evolution of *Alu* family repeats since the divergence of human and chimpanzee. *J. Mol. Evol.* **22**: 316–322.

Taggart R.T., Mohandas T.K., Bell G.I. (1987) Parasexual analysis of human pepsinogen molecular heterogeneity. *Somat. Cell Molec. Genet.* **13**: 167–172.

Takada Y., Kaneko N., Esumi H., Purdue P.E., Danpure C.J. (1990) Human peroxisomal L-alanine: glyoxylate aminotransferase. Evolutionary loss of a mitochondrial targeting signal by point mutation of the initiation codon. *Biochem. J.* **268**: 517–520.

Takiyama Y., Igarashi S., Rogaeva E.A. et al. (1995) Evidence for inter-generational instability in the CAG repeat in the *MJD1* gene and for conserved haplotypes at flanking markers amongst Japanese and Caucasian subjects with Machado-Joseph disease. *Hum. Molec. Genet.* **4**: 1485–1491.

Takiyama Y., Sakoe K., Soutome M., Namekawa M., Ogawa T., Nakano I., Igarashi S., Oyake M., Tanaka H., Tsuji S., Nishizawa M. (1997) Single sperm analysis of the CAG repeats in the gene for Machado-Joseph disease (*MJD1*): evidence for non-mendelian transmission of the *MJD1* gene and for the effect of the intragenic CGG/GGG polymorphism on the intergenerational instability. *Hum. Mol. Genet.* **6**: 1063–1068.

Talmadge K., Vamvakopoulos N.C., Fiddes J.C. (1984) Evolution of the genes of the β subunits of human chorionic gonadotrophin and luteinizing hormone. *Nature* **307**: 37–39.

Tautz D., Trick M., Dover G.A. (1986) Cryptic simplicity in DNA is a major source of genetic variation. *Nature* **322**: 652–656.

Teisberg P., Jonassen R., Mevag B., Gedde-Dahl T., Olaisen B. (1988) Restriction fragment length polymorphisms of the complement component C4 loci on chromosome 6: studies with emphasis on the determination of gene number. *Ann. Hum. Genet.* **52**: 77–84.

Teumer J., Green H. (1989) Divergent evolution of part of the involucrin gene in the hominoids: unique intragenic duplications in the gorilla and human. *Proc. Natl. Acad. Sci. USA* **86**: 1283–1286.

Thein S.L., Hill F.G.H., Weatherall D.J. (1984) Haematological phenotype of the triplicated γ-globin gene rearrangement. *Br. J. Haematol.* **57**: 349–351.

Thiel C., Bontrop R.E., Lanchbury J.S. (1995) Structure and diversity of the T-cell receptor α chain in rhesus macaque and chimpanzee. *Hum. Immunol.* **43**: 85–94.

Thomas G.H., Newbern E.C., Korte C.C., Bales M.A., Muse S.V., Clark A.G., Kiehart D.P. (1997) Intragenic duplication and divergence in the spectrin superfamily of proteins. *Mol. Biol. Evol.* **14**: 1285–1295.

Timchenko L.T., Caskey C.T. (1996) Trinucleotide repeat disorders in humans: discussions of mechanisms and medical issues. *FASEB J.* **10**: 1589–1597.

Timms K.M., Lu F., Shen Y., Pierson C.A., Muzny D.M., Gu Y., Nelson D.L., Gibbs R.A. (1995) 130 kb of DNA sequence reveals two new genes and a regional duplication distal to the human iduronate-2-sulfate sulfatase locus. *Genome Res.* **5**: 71–78.

Ting C.N., Rosenberg M.P., Snow C.M., Samuelson L.C., Meisler M.H. (1992) Endogenous retroviral sequences are required for tissue-specific expression of a human salivary amylase gene. *Genes Devel.* **6**: 1457–1465.

Tishkoff S.A., Ruano G., Kidd J.R., Kidd K.K. (1996) Distribution and frequency of a polymorphic *Alu* insertion at the plasminogen activator locus in humans. *Hum. Genet.* **97**: 759–764.

Tishkoff S.A., Goldman A., Calafell F., Speed W.C., Deinard A.S., Bonné-Tamir B., Kidd J.R., Pakstis A.J., Jenkins T., Kidd K.K. (1998) A global haplotype analysis of the myotonic dystrophy locus: implications for the evolution of modern humans and for the origin of myotonic dystrophy mutations. *Am. J. Hum. Genet.* **62**: 1389–1402.

Toribara N.W., Gum J.R., Culhane P.J., Lagace R.E., Hicks J.W., Petersen G.M., Kim Y.S. (1991) MUC-2 human small intestinal mucin gene structure: repeated arrays and polymorphism. *J. Clin. Invest.* **88**: 1005–1013.

Toyosawa S., O'hUigin C., Figueroa F., Tichy H., Klein J. (1998) Identification and characterization of amelogenin genes in monotremes, reptiles, and amphibians. *Proc. Natl. Acad. Sci. USA* **95**: 13 056–13 061.

Trachtulec Z., Hamvas R.M.J., Forejt J., Lehrach H.R., Vincek V., Klein J. (1997) Linkage of TATA-binding protein and proteasome subunit C5 genes in mice and humans reveals synteny conserved between mammals and invertebrates. *Genomics* **44**: 1–7.

Trask B.J., Friedman C., Martin-Gallardo A. *et al.* (1998a) Members of the olfactory receptor gene family are contained in large blocks of DNA duplicated polymorphically near the ends of human chromosomes. *Hum. Molec. Genet.* **7**: 13–26.

Trask B.J., Massa M., Brand-Arpon V., Chan K., Friedman C., Nguyen O.T.H., Eichler E., van den Engh G., Rouquier S., Shizuya H., Giorgi D. (1998b) Large multi-chromosomal duplications encompass many members of the olfactory receptor gene family in the human genome. *Hum. Molec. Genet.* **7**: 2007–2020.

Trifonov E.N., Bettecken T. (1997) Sequence fossils, triplet expansion, and reconstruction of earliest codons. *Gene* **205**: 1–6.

Tseng H. (1997) Complementary oligonucleotides and the origin of the mammalian involucrin gene. *Gene* **194**: 87–95.

Tu A.Y., Deeb S.S., Iwasaki L., Day J.R., Albers J.J. (1995) Organization of human phospholipid transfer protein gene. *Biochem. Biophys. Res. Commun.* **207**: 552–558.

Ulvsbäck M., Lundwall A. (1997) Cloning of the semenogelin II gene of the rhesus monkey. Duplications of 360 bp extend the coding region in man, rhesus monkey and baboon. *Eur. J. Biochem.* **245**: 25–31.

Urquhart A., Gill P. (1993) Tandem-repeat internal mapping (TRIM) of the involucrin gene: repeat number and repeat-pattern polymorphism within a coding region in human populations. *Am. J. Hum. Genet.* **53**: 279–286.

Valdes A.M., Slatkin N., Freimer N.B. (1993) Allele frequencies at microsatellite loci: the stepwise mutation model revisited. *Genetics* **133**: 737–749.

van der Hoek Y.Y., Wittekoek M.E., Beisiegel U., Kastelein J.J., Koschinsky M.L. (1993) The apolipoprotein(a) kringle IV repeats which differ from the major repeat kringle are present in variably sized isoforms. *Hum. Molec. Genet.* **2**: 361–366.

Vanin E. (1984) Processed pseudogenes: characteristics and evolution. *Biochim. Biophys. Acta* **782**: 231–246.

Verkerk A.J.M.H., Pieretti M., Sutcliffe J.S. *et al.* (1991) Identification of a gene (*FMR1*) containing a CGG repeat coincident with a breakpoint cluster region exhibiting length variation in fragile X syndrome. *Cell* **65**: 905–914.

Viale A., Ortola C., Richard F., Vernier P., Presse F., Schilling S., Dutrillaux B., Nahon J.-L. (1998) Emergence of a brain-expressed variant melanin-concentrating hormone gene during higher primate evolution: a gene 'in search of a function'. *Mol. Biol. Evol.* **15**: 196–214.

Vidaud D., Vidaud M., Bahnak B.R., Siguret V., Sanchez S.G., Laurian Y., Meyer D., Goossens M., Lavergne J.M. (1993) Haemophilia B due to a *de novo* insertion of a human-specific *Alu* subfamily member within the coding region of the factor IX gene. *Eur. J. Hum. Genet.* **1**: 30–36.

Vincent K.A., Wilson A.C. (1989) Evolution and transcription of Old World monkey globin genes. *J. Mol. Biol.* **207**: 465–479.

Volz A., Korge B.P., Compton J.G., Ziegler A., Steinert P.M., Mischke D. (1993) Physical mapping of a functional cluster of epidermal differentiation genes on chromosome 1q21. *Genomics* **18**: 92–99.

Wallace M.R., Andersen L.B., Saulino A.M., Gregory T.W., Collins F.S. (1991) A *de novo Alu* insertion results in neurofibromatosis type 1. *Nature* **353**: 864–866.

Walter G., Tomlinson I.M., Cook G.P., Winter G., Rabbitts T.H., Dear P.H. (1993) HAPPY mapping of a YAC reveals alternative haplotypes in the human immunoglobulin V_H locus. *Nucleic Acids Res.* **21**: 4524–4529.

Wang A.M., Desnick R.J. (1991) Structural organization and complete sequence of the human α-N-acetylgalactosaminidase gene: homology with the α-galactosidase A gene provides evidence for evolution from a common ancestral gene. *Genomics* **10**: 133–142.

Warburton P.E., Willard H.F. (1996) Evolution of centromeric alpha satellite DNA: molecular organization within and between human and primate chromosomes. In: *Human Genome Evolution*, (eds. M Jackson, T Strachan, G Dover). BIOS Scientific Publishers, Oxford, pp 121–145.

Watkins W.S., Bamshad M., Jorde L.B. (1995) Population genetics of trinucleotide repeat polymorphisms. *Hum. Molec. Genet.* **4**: 1485–1491.

Weaver D.T., DePamphilis M.L. (1982) Specific sequences in native DNA that arrest synthesis by DNA polymerase α. *J. Biol. Chem.* **257**: 2075–2086.

Weber W.W. (1997) *Pharmacogenetics*. Oxford University Press, New York.

Wells R.D. (1996) Molecular basis of genetic instability of triplet repeats. *J. Biol. Chem.* **271**: 2875–2878.

Wichman H.A., Van den Bussche R.A., Hamilton M.J., Baker R.J. (1992) Transposable elements and the evolution of genome organization in mammals. *Genetica* **86**: 287–293.

Willems van Dijk K., Milner L.A., Sasso E.H., Milner E.C.B. (1992) Chromosomal organization of the heavy chain variable region gene segments comprising the human fetal antibody repertoire. *Proc. Natl. Acad. Sci. USA* **89**: 10 430–10 434.

Winderickx J., Battisti L., Motulsky A.G., Deeb S.S. (1992) Selective expression of human X chromosome-linked green opsin genes. *Proc. Natl. Acad. Sci. USA* **89**: 9710–9714.

Witke W.F., Gibbs P.E.M., Zielinski R., Yang F., Bowman B.H., Dugaiczyk A. (1993) Complete structure of the human Gc gene: differences and similarities between members of the albumin gene family. *Genomics* **16**: 751–754.

Wolffe R., Nakamura Y., White R. (1988) Molecular characterization of a spontaneously generated new allele at a VNTR locus: no exchange of flanking DNA sequences. *Genomics* **3**: 347–351.

Wolffe R.K., Plaetke R., Jeffreys A.J., White R. (1989) Unequal crossing over between homologous chromosomes is not the major mechanism involved in the generation of new alleles at VNTR loci. *Genomics* **5**: 382–384.

Wong W.W., Cahill J.M., Rosen M.D., Kennedy C.A., Bonaccio E.T., Morris M.J., Wilson J.G., Klickstein L.B., Fearon D.T. (1989) Structure of the human CR1 gene. Molecular basis of the structural and quantitative polymorphisms and identification of a new CR1-like allele. *J. Exp. Med.* **169**: 847–863.

Wu X.W., Muzny D.M., Lee C.C., Caskey C.T. (1992) Two independent mutational events in the loss of urate oxidase during hominoid evolution. *J. Mol. Evol.* **34**: 78–84.

Xu S.-J., Wang Y.-P., Roe B., Pearson W.R. (1998) Characterization of the human class mu glutathione *S*-transferase gene cluster and the *GSTM1* deletion. *J. Biol. Chem.* **273**: 3517–3527.

Yamada K., Yuan X., Ishiyama S., Ichiyama S., Ichikawa F., Kohno S., Shoji S., Hayashi H., Nonaka K. (1998) Identification of a single nucleotide insertion polymorphism in the upstream region of the insulin promoter factor-1 gene: an association study with diabetes mellitus. *Diabetologia* **41**: 603–605.

Yamagata H., Miki T., Nakagawa M., Johnson K., Deka R., Ogihara T. (1996) Association of CTG repeats and the 1-kb *Alu* insertion/deletion polymorphism at the myotonin protein kinase gene in the Japanese population suggests a common Eurasian origin of the myotonic dystrophy mutation. *Hum. Genet.* **97**: 145–147.

Yanagisawa J., Fujii K., Nagafuchi S. *et al.* (1996) A unique origin and multistep process for the generation of expanded DRPLA triplet repeats. *Hum. Molec. Genet.* **5**: 373–379.

Yokoi T., Kosaka Y., Chida M., Chiba K., Nakamura H., Ishizaki T., Kinoshita M., Sato K., Gonzalez F.J., Kamataki T. (1996) A new *CYP2D6* allele with a nine base insertion in exon 9 in a Japanese population associated with poor metabolizer phenotype. *Pharmacogenetics* **6**: 395–401.

Yoneda K., Hohl D., McBride O.W., Wang M., Cehrs K.U., Idler W.W., Steinert P.M. (1992) The human loricrin gene. *J. Biol. Chem.* **267**: 18 060–18 066.

Yulug I.G., Yulug A., Fisher E.M.C. (1995) The frequency and position of *Alu* repeats in cDNAs, as determined by database searching. *Genomics* **27**: 544–548.

Zhu Z.B., Hsieh S.L., Bentley R., Campbell R.D., Volanakis J.E. (1992) A variable number of tandem repeats locus within the human complement C2 gene is associated with a retroposon derived from a human endogenous retrovirus. *J. Exp. Med.* **175**: 1783–1787.

Zhu Z.B., Jian B., Volanakis J.E. (1994) Ancestry of SINE-R.C2, a human-specific retroposon. *Hum. Genet.* **93**: 545–551.

Zietkiewicz E., Richer C., Sinnett D., Labuda D. (1998) Monophyletic origin of *Alu* elements in primates. *J. Mol. Evol.* **47**: 172–182.

Zucman-Rossi J., Batzer M.A., Stoneking M., Delattre O., Thomas G. (1997) Interethnic polymorphism of EWS intron 6: genome plasticity mediated by *Alu* retroposition and recombination. *Hum. Genet.* **99**: 357–363.

9

Gross gene rearrangements

Cases are known of the doubling of the entire chromosome outfit, the doubling of single chromosomes, and of parts of chromosomes; in other cases a part of a chromosome appears to be translocated from its habitual site and attached to some other chromosome. The grosser forms of mutation may indeed play a special evolutionary role in supplying a mechanism of reproductive incompatibility.

R.A. Fisher (1930) *The Genetical Theory of Natural Selection.*

Gross gene rearrangements involving the inversion, translocation or fusion of DNA sequences are relatively rare in inherited disease although much more common in cancer. Clearly, such mutations can lead to dramatic changes in gene structure, function and/or expression. At first glance, it may therefore seem somewhat surprising that there are numerous examples of such mutations that have occurred during evolution. However, genes do not necessarily always change by a slow incremental process of single base-pair substitution and sudden gross changes have also played an important role in fashioning our genes.

9.1 Inversions

Most examples of inversions that have occurred during mammalian evolution have involved large portions of chromosomes and are usually *pericentric* in that the breakpoints are located on opposite chromosomal arms with the inversion spanning the centromere. This contrasts with *paracentric* inversions in which the inversion involves breaks on the same chromosomal arm, with a segment from only one chromosome arm being inverted.

9.1.1 Pericentric inversions

A good example of a pericentric inversion that has occurred during evolution is that of two clusters of zinc finger protein genes on human chromosome 10 (Tunnacliffe *et al.*, 1993). Cluster A, comprising *ZNF11A*, *ZNF25*, *ZNF33A*, and *ZNF37A*, is located on chromosome 10p11.2 whilst cluster B, which is a partial duplicate of cluster A, comprises *ZNF11B*, *ZNF33B*, and *ZNF37B* and is located on chromosome 10q11.2. Duplicated gene clusters are usually contiguous but these ZNF clusters are located on opposite sides of the centromere consistent with their involvement in a pericentric inversion subsequent to the duplication. Tunnacliffe *et al.* (1993) proposed that this event took place during primate evolution.

The human genes encoding NADP-dependent malate dehydrogenase (*ME1*; 6q12), glutathione *S*-transferase 2 (*GSTA2*; 6p12) and phosphoglucomutase 3 (*PGM3*; 6p12) form part of a chromosome 9 syntenic group in mouse (Kasahara *et al.*, 1990). The extension of this syntenic group across the centromere of chromosome 6 in human is indicative of a pericentric inversion which must have occurred since the divergence of rodents and humans. Other examples of inferred pericentric inversions that have occurred during primate evolution involve the β-casein (*CSN2*; 4q13-q21; McConkey *et al.*, 1996) gene, the *GLI* oncogene on chromosome 12q13 (Conte *et al.*, 1998), the fibroblast growth factor genes *FGF3* and *FGF4* on chromosome 11q13 (Conte *et al.*, 1998) and the steroid sulfatase pseudogene (*STSP*; Yq11; Yen *et al.*, 1988). Finally, a Y-specific pericentric inversion occurs as a polymorphism among Gujarati Indians (Spurdle and Jenkins, 1992).

Pericentromeric regions appear to be particularly prone to rearrangement (Eichler, 1998) and this genetic instability could be related to the presence of specific repetitive sequences (Wöhr *et al.*, 1996).

9.1.2 Paracentric inversions

An example of a human-specific paracentric inversion is that involving the short arm of the Y chromosome (Schwartz *et al.*, 1998). A single contiguous segment of Xq21 is homologous to two noncontiguous segments of Yp and it is likely that the transposition of a ~4 Mb segment from the X to the Y chromosome ~3–4 Myrs ago after the divergence of human from chimpanzee was followed by an inversion of the sequence. Schwartz *et al.* (1998) suggested that this inversion could have been mediated by recombination between two LINE elements misaligned at homologous CATTATTCT motifs. Since humans from different racial groups (Caucasian, African, and Asian) have all been found to possess this Yp inversion, the rearrangement must have occurred prior to the radiation of the human racial groups.

Another rather different type of paracentric inversion is exemplified by the Xq28 inversion polymorphism involving the Emery-Dreyfuss muscular dystrophy (*EMD*) and filamin (*FLN1*) genes. Recombination between two large 11.3 kb inverted repeats (with >99% sequence homology) flanking these genes has been responsible for a frequent and clinically asymptomatic inversion of the 48 kb *FLN1/EMD* region (Small *et al.*, 1997; see *Figure 9.1*). The inversion polymorphism may have resulted either from inter- or intra-chromatid exchange between misaligned repeats. Whichever, it is very common in the general population occurring in the heterozygous state in 33% of females and 19% of males (18% of human X chromosomes surveyed).

Other regionally localized paracentric inversions may be inferred from the presence of long inverted repeat regions containing homologous but divergently transcribed genes. Examples of this include the α-amylase (*AMY1A* and *AMY1B*;

Figure 9.1. Structure of the Xq28 region containing the human emerin (*EMD*) and filamin (*FLN1*) gene loci (from Small *et al.*, 1997). The exons of the *EMD* and *FLN1* genes are indicated by black boxes, some of which are numbered for orientation. The direction of transcription for each gene is indicated by small arrows below. The thick black arrows represent the 11.3 kb inverted repeats. The black circle represents the centromere and the position of the telomere is indicated.

1p21; Groot *et al.*, 1990) and the α- and β-fibrinogen (*FGB* and *FGG*; 4q31; Kant *et al.*, 1985) genes. Genes which are divergently transcribed, evolutionarily related and which share bidirectional promoters (see Chapter 5, section 5.1.5) are likely to have arisen by a process of duplication and inversion. Other genes have clearly evolved by this same process but are almost certainly too far apart to share promoter elements, for example the serum amyloid A1 and A2 genes (*SAA1, SAA2*; 15–20 kb apart on chromosome 11p15) and the GABA receptor β3 and α5 genes (*GABRB3, GABRA5*; 100 kb apart on chromosome 15q11-q12).

9.1.3 Physiological and pathological inversions

Somatic inversions, both relatively small and rather larger involving megabase-sized DNA fragments, occur physiologically during rearrangement of DNA at the immunoglobulin κ (*IGKV*; 2p12) and T cell receptor β (*TCRB*; 7q35) loci in human (Weichhold *et al.*, 1990). In human pathology, sporadic chromosomal inversions are not uncommon although each type of inversion is likely to be individually rare. By contrast, intragenic DNA sequence inversions, occurring as a result of recombination between inverted repeats in the germline, are highly unusual. The best known example is that found in the factor VIII (*F8C*) gene causing hemophilia A: this rearrangement occurs in about 40% of severely affected patients and recurs at high frequency (Lakich *et al.*, 1993; Naylor *et al.*, 1993). The mechanism responsible is thought to be homologous intrachromosomal recombination between a gene (*F8A*) located in intron 22 of the *F8C* gene and one of two additional homologues of the *F8A* gene situated 500 kb upstream of the *F8C* gene.

9.1.4 Intragenic inversions

Not surprisingly, intragenic inversions occurring during gene evolution are extremely uncommon. One example is however provided by the family of inter-α-trypsin inhibitors encoded by four genes in the human genome (*ITIH1, ITIH3, ITIH4*, 3p21; *ITIH2*, 10p14-p15). The ancestral *ITIH* gene was first duplicated with one copy being translocated to chromosome 10 about 300 Myrs ago (Diarra-Mehrpour *et al.*, 1998). Prior to further gene duplication and divergence, the primordial *ITIH1/ITIH3/ITIH4* gene experienced an inversion of exons 3–13.

9.1.5 Common sites for inversions in pathology and evolution?

It has been known for some time that chromosomal inversions in human pathology are nonrandomly distributed although it remains unclear whether this is due to interchromosomal differences or to bias of ascertainment (Dutrillaux *et al.*, 1986; Madan, 1995). Some of these inversions have nevertheless been characterized with sufficient precision to allow comparison with inversions that have occurred in the karyotypic evolution of the hominids (Dutrillaux, 1988; Yunis and Prakash, 1982). Miro *et al.* (1992) performed such a comparison and demonstrated that 10/20 pericentric inversions and 1/4 paracentric inversions which had occurred during human chromosome evolution coincided (albeit at low level resolution) with known sites of pathological inversion. These (evolutionary) inversions were inv(1) (p12q21.22), inv(2) (p11.2q13), inv(4) (p14q21.1), inv(5) (p13.3q13.3), inv(7) (q11.23q22.2), inv(8) (p21.1q22.1), inv(9) (p24.2q12), inv(11) (p15.5q13), inv(16) (p11.2q12.1), inv(18) (p11.32q11.2), and inv(Y) (p11.2q11.23). If the sites that have been involved in chromosome inversions during primate evolution were also to be involved in cases of human chromosome pathology, this would argue for certain chromosomal regions possessing sequence characteristics (possibly long highly conserved inverted repeats) that could predispose to this type of lesion. Thus higher resolution studies such as that on human chromosome 12q15 (which contains breakpoints associated with both benign solid tumors and a pericentric inversion that occurred during hominoid evolution; Nickerson and Nelson, 1997) could yield valuable insights into the mechanisms underlying chromosome rearrangement in both pathology and evolution.

9.2 Translocations and transpositions

The human genome is replete with examples of the transposition and translocation of gene sequences during evolution. A selection of some of the most notable examples are given here while others have been discussed in the context of gene duplication (see Chapter 8, section 8.5). The occurrence of most translocations has been inferred from the localization of evolutionarily related genes on different chromosomes or different chromosomal arms. One example is provided by the human aminoacyl-tRNA synthetase gene family (Brenner and Corrochano, 1996) whose evolution is depicted in *Figure 9.2*. Another is the human γ-aminobutyric acid receptor (GABA$_A$R) which comprises several different types of subunit that combine to form a pentameric channel complex and which are encoded by a small but dispersed gene family. The *GABRA1*, *GABRB2* and *GABRG2*, genes have been localized to chromosome 5q34-q35 (Russek and Farb, 1994) whilst similar clusters comprising *GABRA2*, *GABRB1*, and *GABRG1* (4p13-p12) and *GABRA5*, *GABRB3* and *GABRG3* (15q11.2-q12) are present on two other chromosomes. This organization is compatible with the duplication of an ancestral gene cluster and its translocation to other chromosomes.

Tryptophan hydroxylase, tyrosine hydroxylase and phenylalanine hydroxylase are members of the family of pterin-dependent aromatic amino acid hydroxylases and are encoded by the *TPH* (11p15.3-p14), *TH* (11p15.5) and *PAH* (12q22-q24.1) genes, respectively. *PAH* and *TPH* are estimated to have arisen by a process of

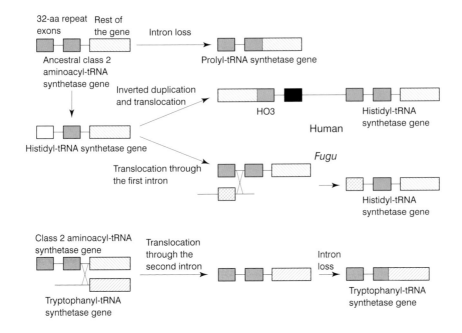

Figure 9.2. Hypothetical translocation events in the evolution of the aminoacyl-tRNA synthetase genes (after Brenner and Corrochano, 1996). The two exons encoding the 32-amino acid repeat and the rest of the gene are shown by stippled and hatched boxes respectively. An ancestral class 2 aminoacyl-tRNA synthetase gene containing the two repeat-containing exons gave rise to the prolyl-tRNA synthetase and histidyl-tRNA synthetase (*HARS*) genes. The prolyl-tRNA synthetase gene of extant animals contains several copies of the repeat and is fused to the gene encoding glutamyl-tRNA synthetase (*EPRS*). The *HARS* gene underwent an inverted duplication and a translocation in the human lineage resulting in two genes in opposite orientation. The translocation may have occurred through the first intron, resulting in the capture of a new exon in the histidyl-tRNA synthetase-homologous (HO3) gene. In the *Fugu* lineage, a translocation through the first intron also allowed the capture of a new exon. The human tryptophanyl-tRNA synthetase (*WARS*) gene, encoding a class 1 enzyme, has captured the two exon repeat by a translocation into a class 2 gene. The loss of a further intron resulted in the structure of the extant *WARS* gene in humans.

duplication and divergence some 750 Myrs ago whilst *PAH* and *TH* diverged subsequently about 600 Myrs ago (Craig *et al.*, 1986; Ledley *et al.*, 1987; *Figure 9.3*). The synteny of the extant *TPH* and *TH* genes implies that the *PAH* gene must have been translocated to chromosome 12 only after the second duplicational event 600 Myrs ago (*Figure 9.3*). A similar explanation may pertain for the insulin (*INS*; 11p15.5), insulin-like growth factor 1 (*IGF1*; 12q22-q24.1) and insulin-like growth factor 2 (*IGF2*; 11p15.5) genes (Bell *et al.*, 1985; Tricoli *et al.*, 1984).

In humans and orangutans, two tRNA[Asn] gene clusters are located on the short and long arms of chromosome 1, respectively (*TRN*, 1p36.1; *TRNL*, 1q21) (Buckland *et al.*, 1992). By contrast, Old World monkeys possess only one tRNA[Asn] gene cluster on chromosome 1p whilst the capuchin (a New World monkey)

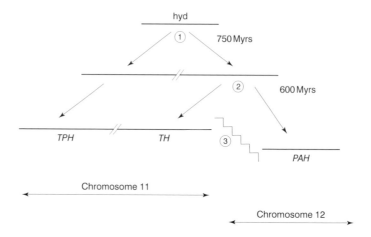

Figure 9.3. Genetic events in the evolution of the tryptophan hydroxylase (*TPH*), tyrosine hydroxylase (*TH*) and phenylalanine hydroxylase (*PAH*) genes from an ancestral hydroxylase gene (hyd). 1. Duplication giving rise to *TPH* gene. 2. Duplication giving rise to *TH* and *PAH* genes. 3. Translocation resulting in *TPH* and *TH* remaining on the same chromosome whilst *PAH* is separated from them.

possesses a single cluster on 1q. These data are consistent with the interpretation that the tRNA^Asn gene cluster split before the divergence of Old World monkeys and hominoids about 30 Myrs ago. No simple inversion can account for this situation and, since this chromosome is well conserved in primate evolution, Buckland *et al.* (1992) speculated that one tRNA^Asn gene cluster could have been relocated from one chromosome arm to the other by some form of replicative transposition.

Another example of a gene translocation occurring during the evolution of the primates is provided by the adenine nucleotide translocator 3 (*ANT3*) and steroid sulfatase (*STS*) genes. The *ANT3* and *STS* genes are pseudoautosomal (Xp22.32) in humans and other higher primates but both localize to an autosome in lemurs suggesting that there was an autosome-to-X/Y translocation after the simians diverged from the prosimians (Toder *et al.*, 1995).

9.2.1 Pericentromeric-directed transposition

Several examples of pericentromeric-directed transposition have been described in primate genomes and since these have been well characterized, they are worthy of discussion in some detail here. Comparative FISH analysis of chromosomes from various primates has shown that a 27 kb region of Xq28 containing the creatine transporter (*SLC6A8*) gene and five exons of the 'CDM' gene (DXS1357E) has been duplicated in its entirety and translocated to 16p11.1, probably within the last 10 Myrs (Eichler *et al.*, 1996; *Figure 9.4*). An inverted cluster of *Alu* repeats and a number of immunoglobulin-like CAGGG repeat motifs lie in the vicinity of the Xq28 breakpoint. Either type of repeat could have mediated the transposition event but the CAGGG repeats may be the more likely candidate for involvement since they have been found to flank a number of other recently duplicated gene sequences (Eichler *et al.*, 1998a). The chromosome 16 *SLC6A8* paralogue (*SLC6A10*) retained both its putative promoter and transcriptional activity (Iyer *et al.*, 1996) but the chromosome 16 CDM paralogue clearly represents a truncated pseudogene. A very similar Xq28–16p11 transposition, reported by Eichler *et al.* (1997), involves the independent transposition of a 9.7 kb segment encompassing exons 7–10 of the adrenoleukodystrophy (*ALD*; Xq28) gene to the pericentromeric regions of chromosomes 2p11, 10p11, 16p11, and 22q11. This *ALD* paralogy domain lies only 27 kb telomeric to the *SLC6A8*/CDM paralogy domain (*Figure 9.4*). The autosomal sequences represent truncated non-processed pseudogenes with sequence divergence being consistent with the duplication/transposition events having occurred between 5 Myrs and 10 Myrs ago. As far as the *ALD* transposition events are concerned, an initial event directed to chromosome 2p11 appears to have served as a 'seed' for further pericentromeric-directed transposition events to occur (*Figure 9.4*). An inverted cluster of *Alu* repeats lies in the vicinity of the Xq28 breakpoint. In addition, Eichler *et al.* (1997) noted the presence of a GCTTTTTGC repeat flanking the duplicated region which they speculated might serve as a sequence-specific integration site for transposition. Such a sequence may provide a hyper-recombinogenic signal which could account for the propensity of this locus to be involved in transpositional events.

Other pericentromeric-directed duplications/transpositions in the human genome (reviewed by Eichler 1998) involve the immunoglobulin Vκ light chain

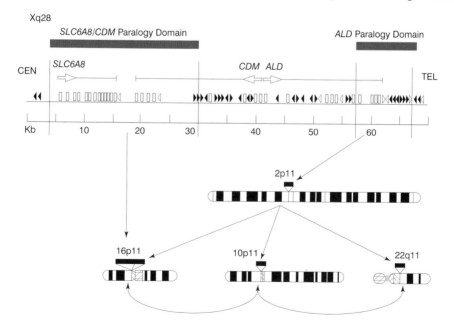

Figure 9.4. Pericentromeric-directed transposition events from the *ALD/SLC6A8* locus at Xq28 to other chromosomes. An initial duplication/transposition targeted chromosome 2p11 from where subsequent transposition events were directed to 16p11, 10p11, and 22q11. The paralogy domains encompassing, (i) the *SLC6A8* and *CDM* genes, and (ii) the *ALD* gene are boxed. CEN, centromere; TEL, telomere.

locus (Borden *et al.*, 1990; Zimmer *et al.*, 1990), the immunoglobulin heavy chain V_H region (Tomlinson *et al.*, 1994), the GABA A receptor α5 (*GABRA5*) gene (Ritchie *et al.*, 1998) and the neurofibromatosis type 1 (*NF1*) gene (Regnier *et al.*, 1997). The inter-chromosomal duplication and transposition of *NF1*-related sequences may be related to the presence in the vicinity of α-satellite sequences that could have served to promote their dispersal. The transposition associated with the immunoglobulin Vκ light chain locus (*IGKV*; chromosome 2p11.1→chromosome 1; Arnold *et al.*, 1995) was directed to a site containing an *ALD* like GCTTTTTGC repeat suggesting that a similar mechanism may have been responsible for these transpositional events.

The pericentromeric zinc finger gene cluster on chromosome 19p12 which harbors the *ZNF208* gene is flanked by large blocks of β-satellite repeat sequences (Eichler *et al.*, 1998b). This gene cluster is thought to have arisen early in primate evolution (~50 Myrs ago) by a process of pericentromeric-directed transposition. Eichler *et al.* (1998b) proposed a model in which an ancestral *ZNF* gene became associated with β-satellite repeat sequences at 19p12. Such repeats are capable of rapid expansion, possibly by unequal crossing over, and may have served to promote the rapid amplification of the associated *ZNF* gene.

The above examples serve to indicate that the pericentromeric regions of human chromosomes have frequently acquired sequences from remote genomic locations. Indeed, these regions appear to be very dynamic, being subject to amplification, duplication, deletion and inversion events as well as translocations (Eichler, 1999; Jackson *et al.*, 1999). Many of these pericentromeric rearrangements have occurred relatively recently during primate evolution leading to marked inter-specific differences even among the great apes. The potential evolutionary importance of pericentromeric regions can perhaps be gauged from Eichler's (1999) description of them as 'recruitment stations for repeats' and 'reservoirs for the accumulation of transposed genic segments.'

9.2.2 Sub-telomeric transposition

Sub-telomeric regions may be similarly dynamic. Thus the rapid proliferation of multigene families clustered near telomeres may have occurred by repeat-mediated bursts of duplication/transposition of short stretches of genomic DNA (e.g. the olfactory receptor genes; Trask *et al.*, 1998). Duplicational transposition also resulted in the telomeric localization of a number of pseudogenes derived from the *Chl1*-related helicase (*DDX11*, 12p11; *DDX12*, 12p13; Amann *et al.*, 1996) and interleukin 9 receptor (*IL9R*; Xq28/Yq12; Kermouni *et al.*, 1995) genes. The 'spreading' of a sub-telomeric region has also been described by Monfouilloux *et al.* (1998); originally localized to 17qter in chimpanzee and orangutan, a specific sub-telomeric domain has been translocated in humans and has colonized several other chromosome ends. Finally, there is emerging evidence for sequence exchange between human telomeric and centromeric regions (Eichler *et al.*, 1997; Jackson *et al.*, 1999; Vocero-Akbani *et al.*, 1996). This may have had consequences for the growth and distribution of multigene families as in the case of the human olfactory receptor genes which, although clustered predominantly in subtelomeric regions, are also located in pericentromeric regions (Rouquier *et al.*, 1998).

Some sequence exchanges that have been reported as translocations may, how-ever, upon closer scrutiny be explicable by other mechanisms. The 'inter-chro-mosomal polymorphism' involving a sub-telomeric exchange between regions on chromosomes 4q35 and 10q26, which has been shown to be present in at least 20% of the human population, is a case in point (van Deutekom *et al.*, 1996). These regions are thought to be 100–400 kb in length and contain 3.3 kb tandem repeat units (each containing two homeodomain sequences and two different classes of GC-rich repetitive DNA) that are 98% homologous to each other (Cacurri *et al.*, 1998). An explanation invoking frequent inter-chromosomal translocation would be virtually unprecedented. The most parsimonious explanation would appear to be inter-chromosomal gene conversion followed by nonhomologous recombina-tion leading to repeat homogenization.

9.2.3 Translocations and chromosome associations

Do some chromosomal rearrangements occur as a consequence of the spatial asso-ciation of particular chromosomes on the metaphase plate? Nagele *et al.* (1995) have suggested that human chromosomes are arranged in antiparallel fashion as a rosette and that this chromosome arrangement is both consistent and cell-type independent in human cells. The high frequency of translocations involving cer-tain chromosomes, whether in pathology (somatic/germline; Rabbitts, 1994; Mitelman *et al.*, 1997) or evolution, may therefore be due at least in part to certain chromosomes being positioned in close proximity to each other during the cell cycle.

9.2.4 Translocations in pathology and evolution

As with inversions (Section 9.1), the chromosomal sites that have been involved in translocations during mammalian evolution sometimes crop up in cases of human chromosome pathology. Thus the breakpoint of X;1 papillary renal cell carcinoma-associated translocations has been mapped to a <200 kb region between the *SPTA1* and *CD1C* genes (Weterman *et al.*, 1996) at the same 1q21 location that contains the boundary between human and mouse syntenic regions (Oakey *et al.*, 1992). Another example of such a correspondence is provided by the holoprosencephaly critical region at 2p21, a site which is associated with a translocation breakpoint in a gibbon karyotype (Arnold *et al.*, 1996). If such cor-respondences are not simply coincidental, these chromosomal regions could pos-sess characteristics that would predispose to translocations. Their study could therefore yield valuable insights into the mechanisms underlying chromosome rearrangement in both pathology and evolution.

9.3 Gene fusion

Gene fusions resulting from somatic chromosomal translocations are a common cause of tumorigenesis (Barr, 1998). By contrast, gene fusions occurring in the germline and causing genetic disease (e.g. hemoglobin Lepore) are a highly unusual cause of human genome pathology (reviewed by Cooper and Krawczak 1993). Such fusions are however thought to have had an important role in evolution through the creation of novel genes encoding novel combinations of functional protein domains; some examples of this phenomenon are described below. As we have seen in Chapter 3, section 3.6, exon shuffling has been an important process in evolution, serving to bring together coding sequence blocks from different sources to generate new proteins capable of novel interactions and therefore, potentially, with novel functions. In a sense, therefore, all proteins have evolved by a series of gene fusion events. In this section, however, we concentrate solely on the generation of novel genes through the fusion of once distinct and independently functional gene sequences by recombination.

9.3.1 Gene fusion during evolution

Carboxyesterase E1, an enzyme responsible for the detoxification of ingested xenobiotics, exhibits sequence homology both with acetylcholinesterase and the carboxy terminal end of thyroglobulin, a precursor of thyroid hormone (Takagi *et al.*, 1991). Phylogenetic analysis has suggested that both acetylcholinesterase (human *ACHE* gene on 7q22) and thyroglobulin (human *TG* gene on 8q24.2-q24.3) evolved from a common ancestral gene that encoded a carboxyesterase and that the emergence of thyroglobulin must have preceded the divergence of the vertebrates and invertebrates (Takagi *et al.*, 1991). The evolutionary origin of the amino terminal portion of extant thyroglobulin is however unknown.

The human multidrug resistance (*PGY1*; 7q21.1) gene encodes a membrane-associated pump protein called P-glycoprotein. Sequence homology between the amino and carboxy terminal halves of the protein was initially held to be consistent with the view that the protein evolved by duplication of a primordial gene. However, once the structure of the 29 exon *PGY1* gene had been determined, it became clear that only two intron pairs (both within nucleotide binding domains) were located in conserved positions in the two halves of the protein. Thus, rather than a primordial duplication, Chen *et al.* (1990) proposed that primordial proteins corresponding to the left and right halves of P-glycoprotein were formed independently by the fusion of closely related genes encoding the nucleotide-binding domain with genes for different transmembrane domains. Subsequent fusion of these two independently derived genes then resulted in the formation of the *PGY1* gene. Pauly *et al.* (1995) have speculated that a highly conserved poly(CA).poly(TG) sequence in intron 15 of the *PGY1* gene could have, with *Alu* sequences in introns 14 and 17, mediated such a fusion event by recombination.

The human glutaminyl-tRNA synthetase (*EPRS*) gene serves to encode two distinct aminoacyl-tRNA synthetase activities joined as part of a multi-enzyme synthetase complex. The *EPRS* gene, located at chromosome 1q41-q42, comprises 29 exons spanning some 90 kb (Kaiser *et al.*, 1994). Exons 4 to 10 encode a glutaminyl-tRNA synthetase (Kaiser *et al.*, 1992) whilst exons 19 to 29 encode a prolyl-tRNA synthetase (Kaiser *et al.*, 1994). The function of the intervening region (exons 11 to 18) is unclear (actin binding?; Kaiser *et al.*, 1994). Since the glutaminyl- and prolyl-tRNA synthetases belong to different enzyme classes which are believed to have evolved by separate pathways (Nagel and Doolittle,

1991), the composite *EPRS* gene would appear to represent the consequence of an ancient gene fusion.

The human quiescin Q6 (*QSCN6*; 1q24) gene also appears to represent the product of an ancient gene fusion that occurred during metazoan evolution (Coppock *et al.*, 1998). The N-terminal portion of the quiescin Q6 protein is related to the thioredoxin family (human members include thioredoxin (*TXN*; 9q31) and prolyl-4 hydroxylase β polypeptide (*P4HB*; 17q25)) whilst the C-terminal is related to the yeast *Erv1* growth regulatory gene (human homologue, *GFER*, 16p13).

The human lamin B receptor is an integral protein of the inner nuclear membrane encoded by a gene (*LBR*) on chromosome 1q42.1. The *LBR* gene comprises 13 protein coding exons, the first four of which encode the N-terminal domain and the remainder the C-terminal sterol reductase-like domain (Holmer *et al.*, 1998). A relatively large intron separating exons 4 and 5 is suggestive of an ancient recombination between two genes, one encoding a basic nuclear protein, the other a sterol reductase. Consistent with this interpretation, the 3' end of the *LBR* gene is structurally homologous with two other human genes viz. the transmembrane protein TM7SF2 (*TM7SF2*; 11q13) and 7-dehydrocholesterol reductase (*DHCR7*; 11q13) (Holmer *et al.*, 1998).

A highly unusual example of gene fusion is provided by the human *HLA-DRB6* (6p21.3) gene. This gene had represented something of a paradox in that although both exon 1 and the promoter region are absent in the orthologous human, chimpanzee and rhesus macaque genes, these genes are still capable of being transcribed. This apparent paradox has now been resolved by the elucidation of the molecular basis of the change: the insertion of a retroviral (mouse mammary tumor virus) LTR into intron 1 of the primate *HLA-DRB6* gene >23 Myrs ago, prior to the divergence of the Old World monkeys from the human-ape lineage (Mayer *et al.*, 1993; see Chapter 5, section 5.1.12, *Endogenous retroviral elements*). Whether the exon/promoter deletion accompanied or instead followed the LTR insertion is unclear. What is clear is that an open reading frame for a new exon was created by the insertion which serendipitously encoded a hydrophobic sequence that was able to function as a leader for the truncated *HLA-DRB6* protein. The new exon provided a functional donor splice site at its 3' end which potentiated in-register splicing with exon 2 of the *HLA-DRB6* gene. The LTR also provided a substitute promoter region essential for the transcription of the downstream gene. Thus, this case not only provides an example of the *de novo* creation of an exon but also illustrates a potential evolutionary mechanism to retarget proteins within the cell. The incorporation of a novel leader peptide into, for example, a cytoplasmic protein that previously lacked one could result either in an alteration in the cellular localization of the protein or its conversion into a secreted protein.

The mammalian defensins constitute a family of microbicidal and cytotoxic peptides made by neutrophils. In humans, they are encoded by a gene cluster (*DEFA1*, *DEFA3*, *DEFA4*, *DEFA5*, *DEFA6*, *DEFB1*) at chromosome 8p23 (Bevins *et al.*, 1996; Liu *et al.*, 1997). After a primordial defensin gene had duplicated to yield two genes ancestral to extant *DEFA5* and *DEFA6*, an unequal crossing over event generated a novel hybrid defensin gene which was the ancestor of the present day hematopoietic defensin genes *DEFA1*, *DEFA3*, *DEFA4* (*Figure 9.5*).

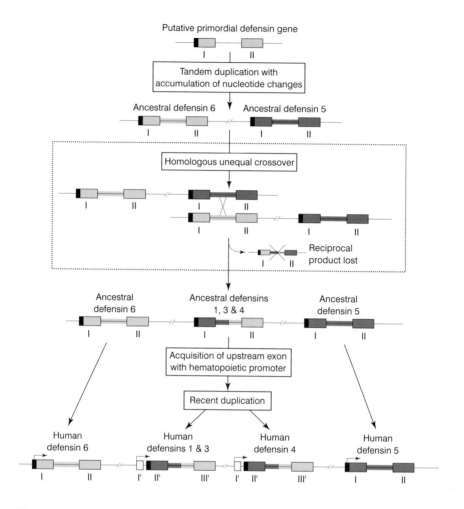

Figure 9.5. Model for the involvement of an homologous unequal crossover in the evolution of the human defensin gene family (after Bevins *et al.*, 1996).
Shaded enclosed boxes denote exons, with exons I and II representing the two exons of the epithelial defensin genes *DEFA5* and *DEFA6*. Exon I′ is the characteristic upstream exon of hematopoietic defensins whilst exons II′ and III′ are homologous to exons I and II of the epithelial defensins. Solid boxes denote regions of striking conservation between all defensins. Small arrows denote the locations of the transcription start sites.

The *SCYA18* (17q11.2) gene, which encodes a member of the small inducible cytokine family, appears to have been generated by the fusion of two macrophage inflammatory protein-1α-like (*SCYA3*) genes with subsequent deletion and selective use of exons (Tasaki *et al.*, 1999). Since there are several related genes (*SCYA3*, *SCYA3L1* and *SCYA3L2*) in the vicinity of *SCYA18*, the authors suggested that the *SCYA3* gene might represent a 'hot spring' that continually generates new genes by duplication and fusion.

The human immunoglobulin γFc receptor IIC (*FCGR2C*; 1q23) gene is also thought to have resulted from an unequal crossover, this time between the

FCGR2A and *FCGR2B* genes (1q23; Warmerdam *et al.*, 1993); the 5' end of the *FCGR2C* gene exhibits significant homology to the *FCGR2B* gene whilst its 3' end is homologous to the *FCGR2A* gene. Finally, two human ubiquitin genes represent the products of an in-frame fusion event between a ubiquitin gene and a ribosomal protein gene encoding a protein of either 52 (*UBA52*; 19p13; Baker and Board, 1991, 1992) or 76–80 (*UBA80*) amino acids in length (Lund *et al.*, 1985).

9.3.2 Internal methionines as evidence for ancient gene fusion events

Berman *et al.* (1994) demonstrated that some 20% of eukaryotic proteins are multiples of 123 ± 3 amino acids. This modular structure has been held to reflect the combinatorial fusion of genes, initially of the same elementary size, during

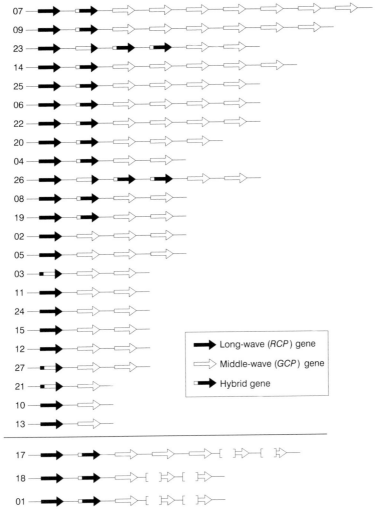

Figure 9.6. Inferred visual pigment gene (*RCP*, *GCP*) arrangements in 26 individuals with normal color vision (redrawn from Neitz *et al.*, 1995).

evolution. One consequence of this fusion process would be the frequent occurrence of methionine-encoding ATG triplets at the borders between unit length sequence segments, a prediction which has been bourne out by statistical analysis (Kolker and Trifonov, 1995). This positional preference of methionines testifies to the excision-reinsertion mechanism of protein construction and means that these internal methionines can justifiably be termed the 'fossils of gene fusion.' Further, it illustrates the probable importance of gene fusion events in the evolutionary construction of extant gene sequences.

9.3.3 Fusion gene polymorphisms

Some fusion genes occur in the human genome as polymorphic variants. One example is the red and green visual pigment genes (*RCP*, *GCP*; Xq28). Normal human males with trichromatic vision typically possess one *RCP* gene, one, two or more *GCP* genes plus an *RCP/GCP* hybrid gene (*Figure 9.6*; Neitz *et al.*, 1995; see Chapter 7 section 7.5.2, *The visual pigments*). In some individuals, the *RCP* genes contain a substantial amount of *GCP* sequence e.g. exon 4 of the *GCP* gene (individuals 23, 26, 03, 27, and 21 in *Figure 9.6*). Other individuals possess *RCP* genes that contain 5′ sequences derived from *GCP* genes (individuals 07, 09, 14, 25, 06, 22, 20, 04, 08, 19, 18, 01 in *Figure 9.6*). The degree of polymorphism is dramatic in that ~70% of individuals with normal trichromatic color vision possess one or other type of fusion gene (*Figure 9.6*). Other gene fusions have occurred as a result of unequal homologous crossing over to generate polymorphic variants in the MNS (*GYPA*; 4q31; Huang and Blumenfeld, 1991) and ABO blood group systems (*ABO*; 9q; Olsson *et al.*, 1997).

9.3.4 Fusion splicing

Although gene fusion can arise through deletion or translocation, it need not invariably occur as a result of DNA rearrangement. Gene fusion can, in functional terms, also be brought about by the *fusion splicing* of mRNA transcripts derived from two closely linked but unrelated genes. Evidence for such a mechanism has come from the cotranscription of two human genes encoding galactose-1-phosphate uridylyl-transferase (*GALT*) and interleukin-11 receptor α-chain (*IL11RA*) which are closely linked (separated by only 4 kb) on chromosome 9p13 (Magrangeas *et al.*, 1998). GALT is a 43 kDa enzyme required for the conversion of galactose to glucose and is encoded by a gene which comprises 11 exons spanning 4 kb and which specifies a 1.4 kb mRNA. The *IL11RA* gene, a member of the hematopoietin receptor superfamily, comprises 13 exons spanning 8 kb and encodes a 48 kDa protein. Magrangeas *et al.* (1998) demonstrated that the two genes are sometimes cotranscribed in normal human cells and that the 3 kb fusion mRNA encodes an 85 kDa protein of unknown function and biological significance. The in-frame fusion transcript probably results from a combination of inefficient RNA polymerase II termination of *GALT* gene transcription and an alternative splicing event between exon 10 of the *GALT* gene and exon 2 of the *IL11RA* gene.

Magrangeas *et al.* (1998) speculated that fusion splicing could represent an 'exploratory event for evolution'; fusion proteins thus formed might initially be produced at low cellular levels but would represent a reservoir of novel functions which could be called upon to confer a selective advantage under particular conditions. In such a situation, it may be envisaged that a point mutation at the acceptor splice site or polyadenylation site could then lead to the generation of much larger amounts of the fusion protein. Such a mutation is evident in the guinea pig gene that encodes seminal vesicle secretory proteins 1, 3, and 4 (Hagstrom *et al.*, 1996). The 5' half of this guinea pig gene is homologous to the human semenogelin II (*SEMG2*; 20q12-q13) gene. Indeed, sequences related to the human *SEMG2* gene are also found in the first intron of the guinea pig gene (Hagstrom *et al.*, 1996). However, the 3' half of the guinea pig gene shares homology with the closely linked skin-derived antileukoproteinase/elafin (*PI3*; 20q12-q13) gene. It would appear that, as a result of an A→G transition, the guinea pig gene contains a novel AG dinucleotide 7 bp upstream of the 3' splice site used in the *SEMG2* gene. If used as a splice acceptor site, this AG dinucleotide would lead to a spliced product that is out of frame with the product of the *SEMG2* gene, and the use of a splice acceptor site in the downstream *PI3* gene would have been favored (Hagstrom *et al.*, 1996). Thus an alternative splicing pathway has led to the creation of a novel guinea pig gene that presumably encodes a protein with a new function.

In principle, fusion splicing could also lead to the creation of novel genes through the germline retrotransposition of the fusion mRNA. Such novel fusion genes would however be expected to be characterized by the absence of introns (unlike the case of the guinea pig seminal vesicle secretory protein gene cited above) and to be chromosomally distant from the parental genes. There are at least two other possible ways of generating novel genes by fusion splicing. The first would be via *trans*-splicing (involving the joining of independently transcribed coding sequences from genetically unlinked loci) followed by retrotransposition back into the genome. As yet, however, the available evidence for *trans*-splicing in mammalian cells must be regarded as somewhat tentative (Eul *et al.*, 1995; Fujieda *et al.*, 1996). The second mechanism would involve the retrotransposition of abnormally spliced gene transcripts comprising exons joined in an order different from that in which they are normally found in the genome (*scrambled exons*). Exon scrambling has been described for at least two human genes: *DCC* (18q21; Nigro *et al.*, 1991) and *MLL* (11q23; Caldas *et al.*, 1998).

9.4 Recombination

> Mankind...will not willingly admit that its destiny can be revealed by the breeding of flies or the counting of chiasmata.
> C. D. Darlington (1960)

9.4.1 Homologous recombination

Numerous recombination events have now been documented as having occurred during the evolution of mammalian genes. Indeed, most gene duplication and amplification events have been mediated by recombination (see Chapter 8, section 8.5). Examples of the emergence of novel human gene sequences by recombination-mediated duplication include those encoding saposin C (*PSAP*; 10q; Rorman *et al.*, 1992), pepsinogen A3 (*PGA3*; 11q13; Evers *et al.*, 1989), mannose-binding protein (*MBP*; 10q11-q21; Sastry *et al.*, 1989) and the α-amylase genes (*AMY1A, AMY1B, AMY1C, AMY2A, AMY2B*; 1p21; Groot *et al.*, 1990). Recombination events have also been invoked to explain cases of gene fusion (see Section 9.3).

Multiple independent recombination events are thought to have occurred during the evolution of the two human haptoglobin genes (*HP* and *HPR*; 16q22.1) and their counterparts in the other primates (Erickson *et al.*, 1992; Erickson and Maeda 1994; Maeda *et al.*, 1986; Maeda, 1985; McEvoy and Maeda, 1988). Whilst the great apes and Old World monkeys possess three haptoglobin genes (*Hp, Hpr,* and *Hpp*), New World monkeys only have one (*Hp*). This is consistent with a gene triplication after the divergence of Old World from New World monkeys followed by a *Hpp* gene deletion in the human lineage. Breakpoint analysis suggests that both duplications and deletions of gene copies may have been mediated by homologous unequal recombination between *Alu* repeats. That this region is prone to recombination is also evidenced by the haptoglobin gene copy number polymorphism found in the black population (Maeda *et al.*, 1986). Other examples of homologous recombination events thought to be responsible for human gene copy number polymorphisms include the α-globin (*HBA1*; Lie-Injo *et al.*, 1981) genes, the ζ-globin (*HBZ*; Winichagoon *et al.*, 1982) gene, the α-amylase genes (Groot *et al.*, 1989), the proline-rich protein genes (*PRB1, PRB2, PRB3, PRB4*; 12p13.2; Lyons *et al.*, 1988) and the pepsinogen A genes (*PGA3, PGA4, PGA5*; 11q13; Zelle *et al.*, 1988); none of these copy number polymorphisms are known to have any clinical significance.

Recombination has also resulted in the alteration of the expression level of a gene. Thus, the galago δ-globin gene is expressed at an unusually high level in the adult (18% of β-like globin chains, cf. 0–6% in other primates) as a result of a recombination event which replaced 2.4 kb of δ-globin gene sequence by β-globin gene sequence containing 800 bp of the promoter region (Tagle *et al.*, 1991).

Considerable effort has been made to localize and characterize the DNA sequences responsible for mediating the recombinational events known to have occurred during gene evolution. Thus, the duplication events involved in the evolution of the human glycophorin (*GYPA, GYPB,* and *GYPE*; 4q28.2-q31.1) genes (Labuda *et al.*, 1995; *Figure 9.7*), the human growth hormone gene (*GH1, GH2, CSH1, CSH2*; 17q22-q24; Figure 4.13) cluster (Chen *et al.*, 1989) and the mouse lysozyme genes (Cross and Renkawitz, 1990) are considered to have been mediated by recombination between *Alu* repeats. A correspondence between illegitimate recombination junctions and the sites of *Alu* sequence insertion has also been proposed for the primate α-globin (*HBA1, HBA2*; 16p13.3) genes (Bailey *et al.*, 1997; Shaw *et al.*, 1991). The duplication of the primate γ-globin (*HBG1*; 11p15.5) gene may have been mediated by homologous recombination between LINE elements flanking a fetal globin progenitor gene (Fitch *et al.*, 1991). High G+C content may promote recombination (Eyre-Walker, 1993) whilst Ashley *et al.* (1993) have suggested that telomeric repeats could promote meiotic recombination.

1. Duplication

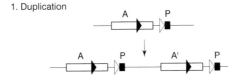

2. Unequal homologous *Alu-Alu* recombination

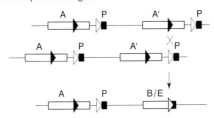

3. Duplication

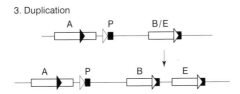

Figure 9.7. Three-step scenario in the evolution of the glycophorin gene family. A: *GYPA*, B: *GYPB*, E: *GYPE*. In the second step, unequal *Alu-Alu* recombination created the subsequently duplicated B/E gene locus. Vertical triangles denote *Alu* repeats, open boxes the genes A, A′, B/E, B, and E, whereas a filled box – 'precursor' sequence (P) that became an integral part of the contemporary genes B and E (redrawn from Labuda *et al.*, 1995).

High resolution genetic mapping studies have provided evidence for both 'hotspots' and 'coldspots' of recombination (Nagaraja *et al.*, 1997; Shiroishi *et al.*, 1993). A frequent disease-associated recombinational hotspot in the human genome occurs at 17p11.2-p12 where a CMT1A-REP tandem repeat mediates meiotic crossing over events through chromosome misalignment. Humans and chimpanzees have two copies of this repeat whilst gorilla, orangutan and gibbon have only a single copy (Kiyosawa and Chance, 1996). The CMT1A-REP repeat must therefore have appeared before the divergence of chimpanzee and human. The repeat contains a *mariner*-like element in the vicinity of the recombination hotspot and, since this element occurs in association with the CMT1A-REP repeat in all primates, it must have predated the emergence of the proximal and distal copies of the repeat in the human-chimpanzee common ancestor (Kiyosawa and Chance, 1996). Interestingly, the *mariner* element of *Drosophila* exhibits sequence homologies to transposons from *Caenorhabditis elegans* as well as heptamer and nonamer signal sequences of the vertebrate immunoglobulin somatic recombination pathway (Dreyfus 1992). Whether these homologies that transcend the invertebrate-vertebrate divide are merely coincidental, or whether they reflect either a common evolutionary origin or alternatively convergent evolution, is as yet unclear.

Recombination may have been involved in at least some instances of 'exon shuffling', the process of intron-mediated recombination by which functional domains encoded by one or more exons are dispersed to a variety of different proteins (reviewed in more detail in Chapter 3, section 3.6). For example, a combination of X-ray crystallographic data and studies of exon organization have led to the conclusion that the two exons encoding the nucleotide-binding domain have been independently transferred to the human phosphoglycerate kinase (*PGK1*; Xq13.3) gene as well as to the maize alcohol dehydrogenase and chicken glyceraldehyde-3-phosphate dehydrogenase genes (Michelson *et al.*, 1985). Whilst the almost ubiquitous presence of introns may permit exon shuffling by intron-mediated recombination (see Chapter 3, section 3.6), it has been suggested (Matsuo *et al.*, 1994) that the presence of short unconserved introns with different types of splice junction may serve to prevent recombination between exon clusters encoding highly conserved functional domains without inhibiting the potential for domain (multiexon) shuffling.

9.4.2 V(D)J recombination

> It is natural selection that gives direction to changes, orients chance, and slowly, progressively produces more complex structures, new organs, and new species. Novelties come from previously unseen association of old material. To create is to recombine.
> F. Jacob (1977)

In the lymphocyte, functional immunoglobulin and T-cell receptor genes are assembled from their constituent gene coding segments by V(D)J recombination (Gellert, 1997; see Chapter 4, section 4.2.4, *Immunoglobulin genes* and *T-cell receptor genes*). Recombination is directed by recombination signal sequences (RSS) adjacent to each coding sequence. The RSS comprises conserved heptamer and nonamer motifs separated by a non-conserved spacer region of either 12 bp or 23 bp (termed 12 signals and 23 signals); recombination only occurs between a 12 signal and a 23 signal. V(D)J recombination is initiated by the products of two lymphoid-specific intronless recombination-activating genes, *RAG1* and *RAG2*, closely linked to each other on human chromosome 11p13. The RAG1 and RAG2 proteins bind the two recombination signals, bringing them into close juxtaposition and cleaving the DNA molecule thereby separating the recombination signals from the flanking coding segments. The *RAG1* and *RAG2* genes are present in all vertebrates but have not been reported either in jawless vertebrates or invertebrates.

Together RAG1 and RAG2 constitute a transposase that is capable of excising a DNA segment containing recombination signals from a donor site and inserting it into a target DNA molecule (Agrawal *et al.*, 1998; Hiom *et al.*, 1998). The product of transposition contains a short target site duplication immediately flanking the transposed fragment that is reminiscent both of retroviral integration and the insertional mechanisms employed by other transposases. It is therefore reasonable to speculate that the *RAG1* and *RAG2* genes may once have been contained

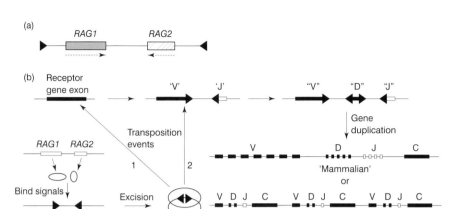

Figure 9.8. Model for the evolution of the immunoglobulin and T-cell receptor (TCR) genes and the role of RAG-mediated transposition (from Agrawal *et al.*, 1998). (a) Putative structure of the recombination activating gene (*RAG*) transposon that integrated into the germline of a vertebrate ancestor. Arrows denote direction of transcript. (b) The split organization of extant immunoglobulin and TCR genes may have arisen by RAG-mediated transposition through the introduction of signal end/signal end elements into a primordial receptor gene exon, thereby dividing the exon into two or three gene segments, each flanked by one or two recombination signals (triangles). These gene segments would be the evolutionary precursors of current V, D, and J gene segments. Different patterns of gene duplication could have given rise to both the mammalian heavy chain locus and the 'cluster' configuration in cartilaginous fishes.

within a transposable element that also possessed flanking RSS sequences (*Figure 9.8*). Since in extant genomes, the *RAG* genes and the RSS elements are unlinked, transposition now serves to relocate pairs of RSS elements plus the intervening sequence without the necessity of the *RAG* genes themselves being displaced.

The split nature of the vertebrate immunoglobulin light chain and T-cell receptor α and γ genes could have originated from the germline insertion of the RAG transposon into an ancestral receptor gene soon after the divergence of jawed and jawless fishes (Agrawal *et al.*, 1998; Hiom *et al.*, 1998; Thompson, 1995; *Figure 9.8*). This gene could then only have been expressed if the inserted transposon were excised by the RAG proteins and the two ends of the exon rejoined and repaired. The tripartite structure characteristic of the immunoglobulin heavy chain and T-cell receptor β and δ genes could have arisen as a result of the further insertion of a second RAG transposon into the same exon. Subsequent duplication of the individual gene segments or of the entire gene would then have served to generate the 'mammalian-type' or 'cluster-type' organization of mammals and cartilaginous fishes, respectively (*Figure 9.8*; Litman *et al.*, 1993). It therefore appears likely that our early vertebrate ancestors were successful in taming a transposon by transforming it into a site-specific recombinase that was then harnessed in the cause of generating the genetic diversity so vital for the flexible adaptive response of our immune system.

9.5 Gene conversion

Gene conversion is the 'modification of one of two alleles by the other' (Vogel and Motulsky, 1997). The end result is very similar to that resulting from a double unequal crossing-over event and in humans, it is difficult to distinguish the two processes (Cooper and Krawczak, 1993). However, in practical terms, the difference is that the correction of an 'acceptor' gene or DNA sequence by gene conversion is nonreciprocal leaving the 'donor' sequence physically unchanged.

The process of gene conversion may involve the whole or only a part of a gene (Dover, 1993) and usually occurs between highly homologous, but nonallelic genes. Gene conversion may be suspected in cases where the degree of sequence homogeneity exhibited by related genes is much greater than expected from what is known of their evolutionary history. Thus, two or more gene sequences may exhibit strong homology over all or parts of their coding or promoter sequences yet they are known from phylogenetic studies to have diverged a considerable time ago and might reasonably have been expected to have accumulated significant numbers of both synonymous and non-synonymous substitutions.

There are now numerous examples of gene conversion causing human gene pathology (Cooper and Krawczak, 1993; see Chapter 6, section 6.1.6) but gene conversion has also had an important influence in evolution. Examples of gene conversion in human gene evolution often involve duplicated gene sequences lying in close physical proximity: the G-γ (*HBG2*; 11p15.5) and A-γ (*HBG1*; 11p15.5) globin genes (Shen *et al.*, 1981; Scott *et al.*, 1984; Stoeckert *et al.*, 1984; Powers and Smithies, 1986; Fitch *et al.*, 1990; see *Figure 9.9*), the β- and δ-globin genes (*HBB, HBD*; 11p15.5; Koop *et al.*, 1989), the α1-(*HBA1*) and α2-(*HBA2*) globin genes (Liebhaber *et al.*, 1981), the *HLA-DQB1* loci (6p21.3; Wu *et al.*, 1986), the visual pigment (*GCP* and *RCP*; Xq28; Kuma *et al.*, 1988; Shyue *et al.*, 1994; 1995; Winderickx *et al.*, 1993; Zhou and Lee, 1996) genes, the rhesus blood group (*RHD, RHCE*; 1p34-p36.2; Carritt *et al.*, 1997) genes, the salivary (*AMY1A, AMY1B, AMY1C*; 1p21) and pancreatic (*AMY2A* and *AMY2B*; 1p21) amylase genes (Gumucio *et al.*, 1988), the α-interferon genes *IFNA1* and *IFNA13* (9p22; Todokoro *et al.*, 1984), the haptoglobin genes (*HP, HPR*; 16q22.1; Erickson and Maeda, 1994; Maeda, 1985), the p15 and p16 cyclin-dependent kinase inhibitor (*CDKN2B* and *CDKN2A*; 9p21) genes (Jiang *et al.*, 1995), the glycophorin genes *GYPA* and *GYPE* (4q28.2-q31.1; Kudo and Fukuda 1994; Onda and Fukuda 1995), the P glycoprotein 1 and 3 genes (*PGY1, PGY3*; 7q21.1; van der Bliek *et al.*, 1988), the salivary proline-rich protein genes (*PRB1, PRB2, PRB3, PRB4*; 12p13.2; Kim *et al.*, 1993), the cardiac α- and β-myosin heavy chain (*MYH6, MYH7*; 14q12; Epp *et al.*, 1995) genes, the chorionic somatomammotropin (*CSH1, CSH2*; 17q22-q24) genes (Hirt *et al.*, 1987), the immunoglobulin $C_H\alpha$ (*IGHA1* and *IGHA2*, 14q32; Kawamura *et al.*, 1992) and V_H genes (*IGHV*, 14q32; Haino *et al.*, 1994), and the α1-acid glycoprotein (*ORM1, ORM2*; 9q34.1–34.3) genes (Merritt *et al.*, 1990).

In the context of gene conversion, if there is any middle ground between pathology and evolution, it is probably occupied by the cytochrome P450 *CYP2A6* gene (19q13.2). A null *CYP2A6* variant, formed by gene conversion between the *CYP2A6* gene and the closely linked but inactive *CYP2A7* gene

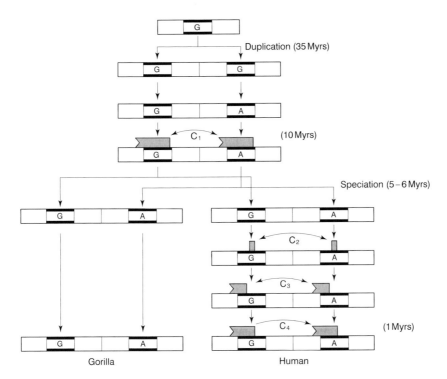

Figure 9.9. History of γ-globin gene evolution and conversion events in the great apes (from Scott *et al.*, 1984). A duplication of a γ-globin gene encoding glycine at position 136 occurred in the early catarrhine primates about 35 Myrs ago. Since a glycine is encoded at this position in both γ-globin genes of *Pongo* and Old World monkeys whereas a replacement with alanine in the 3′ gene is found in *Homo* and *Gorilla*, this change must have occurred after the divergence of *Pongo* (about 8 Myrs ago) but before *Homo* and *Gorilla* branched off (about 6–7 Myrs ago). This replacement may have occurred before the first conversion (C1), or it may have happened after it but still before the separation of *Homo* and *Gorilla*. If the Gly→Ala replacement occurred before C1, the 3′ boundary of C1 can be placed in exon 3 at codon position 135 or nucleotide position 1543. If the replacement occurred after C1, the 3′ boundary can be placed at the start of the 3′ untranslated region. C1 is common to both humans and gorillas and was estimated to have occurred about 10 Myrs ago, using a replacement rate change of 1% for every 10 Myrs.

No further conversions have been identified in the gorilla lineage, but three have been identified in the human lineage. C2 and C3 are estimated to have occurred about 2–3 Myrs ago either in a common ancestor of human chromosome types A and B or in an early version of chromosome B itself. Conversion C2 is evident in the BAγ-gene from positions 901 to 1128 and extending into the 'hot spot,' and C3 is also evident in the BAγ-gene, but from positions 42 to 777. C4 extending over some 1500 bp is estimated to have occurred no earlier than 1 Myrs ago on human chromosome type A. Its effects are evident in the AAγ-gene from positions 42 to 1128 with its 3′ boundary being located in the hot spot region. Taken from Scott *et al.* (1994) with kind permission.

occurs at polymorphic frequencies in the Japanese (28%) and African-American (2.5%) populations (Fernandez-Salguero *et al.*, 1995).

Presumably, the greater the number of homologous repetitive elements, the greater the opportunity for gene conversion. Consistent with this assertion, multigene families frequently exhibit extensive sequence homogeneity compatible with the consequences of gene conversion. Thus, the human ribosomal RNA gene family, which is composed of ~400 members arranged in tandem repeats on five different chromosomes (*RNR1*, 13p12; *RNR2*, 14p12; *RNR3*, 15p12; *RNR4*, 21p12; *RNR5*, 22p12) are much more similar to one another than they are to members of the rDNA family in other primates (Arnheim *et al.*, 1980; Li, 1997). This genomic organization also results in a greater degree of homogeneity within rDNA clusters than between them (Gonzalez *et al.*, 1992; Seperack *et al.*, 1988), a phenomenon also noted for the polyubiquitin genes (*UBA52*, 19p13.1; *UBB*, 17p11.1-p12; *UBC*, 12q24.3) where gene conversion appears to occur exclusively within rather than between gene clusters (Sharp and Li, 1987). Other examples of repetitive gene families being subject to the homogenizing effects of gene conversion are those of the human immunoglobulin Cα (*IGHA1*; 14q32-q33; Kawamura *et al.*, 1992; McCormack *et al.*, 1993), Vκ light chain (*IGKV*; 2p12; Huber *et al.*, 1993), γ heavy chain (*IGHG1*; 14q32.33; Lefranc *et al.*, 1986) and λ light chain (*IGLC1*; 22q11.12; Udey and Blomberg, 1988) genes, U2 snRNA (*RNU2*; 17q21-q22; Liao *et al.*, 1997) genes and the T-lymphocyte antigen receptor β (*TCRB*; 14q11.2; Funkhouser *et al.*, 1997; Tunnacliffe *et al.*, 1985) genes.

The pituitary-expressed growth hormone (*GH1*; 17q22-q24) gene promoter region has been found to exhibit a very high level of sequence polymorphism with 8 variant nucleotides within a 134 bp stretch (Giordano *et al.*, 1997). These eight variable positions have been ascribed to 12 different haplotypes ranging in frequency from 2% to 31% in the general population. Since they occur in the same positions in which the *GH1* gene differs from the other placentally expressed genes of the growth hormone cluster [two chorionic somatomammotropin (*CSH1* and *CSH2*) genes, a chorionic somatomammotropin 'pseudogene' (*CSHL1*) and a second growth hormone gene (*GH2*)], the mechanism responsible is likely to be gene conversion with the placentally expressed genes serving as donors of the converted sequences. Various examples of gene conversion of functional genes templated by pseudogenes have been documented as a cause of human pathology (discussed in Chapter 6 section 6.1.6).

Studies of gene conversion in fungi have shown that gene conversion occurs not only between alleles but also intrachromosomally between duplicated sequences on either the same chromatid or the sister chromatid and even between sequences on nonhomologous chromosomes. Does gene conversion in the human genome occur predominantly within a single chromosome (i.e. through sister chromatid exchanges) or instead between either homologous or nonhomologous chromosomes? Data from both the human ribosomal RNA (Seperack *et al.*, 1988) and U2 snRNA gene families (Liao *et al.*, 1997) have suggested that intra-chromosomal events are more frequent than inter-chromosomal events. Inter-chromosomal gene conversion events have been reported in human gene pathology (see Chapter 6, section 6.1.6) but these are comparatively rare. In an evolutionary context, possible examples of gene conversion operating between homologous loci on nonhomologous chromosomes are provided by non-reciprocal exchanges between the X-linked and autosomal phosphoglycerate kinase (*PGK1*, Xq13; *PGK2*; chromosome 19) genes and the X-linked and autosomal pyruvate dehydrogenase (*PDHA1*, Xp22.1-p22.2; *PDHA2*; 4q22-q23) genes (Fitzgerald *et al.*, 1996).

Although gene conversion is generally considered to involve only short stretches of DNA sequence, it has often not been possible to determine the precise length of the converted sequence owing to the high degree of homology manifested by the genes involved. Papadakis and Patrinos (1999) reviewed data from the human Gγ- (*HBG2*; 11p15.5) and Aγ- (*HBG1*; 11p15.5) globin genes and concluded that the length of the converted fragments is usually less than 400 bp but can vary from as little as 113 bp to as much as 2266 bp.

The mechanism underlying gene conversion remains elusive but must presumably entail the close physical interaction between homologous DNA sequences. It may involve heteroduplex formation followed by mismatch repair. Both Amor *et al.* (1988) and Matsuno *et al.* (1992) have noted the presence of Chi-like sequences (GCTGGGG; known to promote recombination both in *E. coli* and in mouse immunoglobulin genes; Smith, 1983) in the vicinity of regions of the *CYP21/CYP21P* and *HBB* genes. These authors speculated that the Chi-like sequences might play a role in gene conversion events. Various other sequences e.g. *Alu* repeat sequences (Merritt *et al.*, 1990), a retroviral Long Terminal Repeat (Pavelitz *et al.*, 1995), alternating purine.pyrimidine tracts (Papadakis and Patrinos, 1999) and a $(CT)_n \cdot (GA)_n$ microsatellite (Liao and Weiner, 1995) have also been postulated to be involved in promoting gene conversion. Many gene conversion events involving the *HBG1* and *HBG2* genes appear to terminate near a polypyrimidine stretch (Fitch *et al.*, 1990). Papadakis and Patrinos (1999) suggested that palindromic sequences may be associated with the termination of gene conversion, possibly through secondary structure formation blocking branch migration. The insertion of some transposable elements into members of multigene families can however reduce the rate, and limit the extent of, gene conversion by reducing the degree of homology between potential donor and acceptor sequences (Hess *et al.*, 1983; Schimenti and Duncan, 1984).

Gene conversion has thus had an important influence on the evolution of multigene families by homogenizing the sequences of duplicated or repeated genes. Sequence homogenization can of course hinder divergence and hence potential adaptation. New substitutions occurring in a sequence will tend to be lost because they are likely to be converted back to the sequence of the more common allele in the population. However, gene conversion may also promote diversity by introducing multiple sequence changes simultaneously into different members of large gene families as with the genes encoding the proteins of the major histocompatibility complex (*HLA-B*, *HLA-DRA*; 6p21.3) (Belich *et al.*, 1992; Gorski and Mach, 1986; Kuhner and Peterson, 1992; Kuhner *et al.*, 1991; Ohta, 1991; Parham and Lawlor, 1991; Seemann *et al.*, 1986) and the kallikreins (Ohta and Basten, 1992).

9.6 Gene creation by retrotransposition

A mutation that produces a new elementary species is due to the sudden appearance or creation of a new element – a new gene. Put in another way, we

> witness at mutation the birth of a new gene or at least its activation. The number of active genes in the world has been increased by one.
>
> T.H. Morgan (1926) *The Theory of the Gene*

During the evolution of the mammalian genome, new genes have often emerged by duplication and divergence (Chapter 4 and Section 9.5). Another, albeit less common route has been via retrotransposition (Brosius, 1991; Nouvel, 1994). At least five functional human genes have a structure consistent with their having arisen by retrotransposition: the phosphoglycerate kinase 2 (*PGK2*; 19) gene (Boer *et al.*, 1987; McCarrey and Thomas, 1987; McCarrey, 1990), the pyruvate dehydrogenase α2 (*PDHA2*; 4q22-q23) gene (Dahl *et al.*, 1990), the cAMP-dependent protein kinase Cγ subunit (*PRKACG*; 9q13) gene (Reinton *et al.*, 1998), the SR splicing factor SRp46 (*Srp46*; 11q22; Soret *et al.*, 1998) gene and the Y-encoded zinc finger protein (*ZFY*: Yp11.3) gene (Ashworth *et al.*, 1990). All are characterized by a lack of introns and a chromosomal location different and distinct from their presumed parent genes (*PGK1*, Xq13; *PDHA1*, Xp22.1-p22.2; *PRKACA*, 19p13.1; *ZFX*, Xp21.3-p22.2 respectively). In the case of the *PRKACG* gene, there are also remnants of flanking direct repeats and a poly(A) tail providing further evidence of the retrotranspositional origin of the gene (Reinton *et al.*, 1998). This type of sequence has sometimes been referred to as a 'functional retropseudogene' but this is surely a misnomer. Functional it may be, but pseudogene it is not, and so a better term would be 'retrotransposed gene'.

The retrotransposed *PGK2* gene originated prior to the divergence of eutherian and metatherian mammals some 125 Myrs ago and has been characterized in some detail (McCarrey, 1990). It is expressed only in spermatogenic cells whereas its putative parent (*PGK1*) is ubiquitously expressed. Sequence comparison studies are consistent with the newly retrotransposed *PGK2* gene having possessed regulatory sequences derived from the *PGK1* gene which facilitated the initial expression of the intronless gene. Inclusion of promoter sequence could have come about if the *PGK1*-derived *PGK2* mRNA intermediate had initiated aberrantly at an upstream start site as McCarrey (1990) suggested, or instead from an alternative upstream promoter. This promoter sequence might then be envisaged to have evolved cell type specificity, a process presumably involving loss of the CpG island still apparent in the *PGK1* gene. Interestingly, the *PDHA2* and *PRKACG* genes are also expressed exclusively in the spermatogenic cells of the testis. In the case of the X-linked *PGK1* and *PDHA1* genes, transposition to an autosome might have been selectively advantageous in that it would have ensured expression of these important enzyme encoding genes in Y-bearing as well as X-bearing spermatozoa.

References

Agrawal A., Eastman Q.M., Schatz D.G. (1998) Transposition mediated by RAG1 and RAG2 and its implications for the evolution of the immune system. *Nature* **394**: 744–751.

Amann J., Valentine M., Kidd V.J., Lahti J.M. (1996) Localization of *Chl*1-related helicase genes to human chromosome regions 12p11 and 12p13: similarity between parts of these genes and conserved human telomeric-associated DNA. *Genomics* **32**: 260–265.

Amor M., Parker K.L., Globerman H., New M.I., White P.C. (1988) Mutation in the *CYP21B* gene (Ile→Asn) causes steroid 21-hydroxylase deficiency. *Proc. Natl. Acad. Sci. USA.* **85**: 1600–1604.

Arnheim N., Krystal M., Schmickel R., Wilson G., Ryder O., Zimmer E. (1980) Molecular evidence for genetic exchanges among ribosomal genes on non-homologous chromosomes in man and apes. *Proc. Natl. Acad. Sci. USA* **77**: 7323–7327.

Arnold N., Wienberg J., Emert K., Zachau H. (1995) Comparative mapping of DNA probes derived from the Vκ immunoglobulin gene regions on human and great ape chromosomes by fluorescence *in situ* hybridization. *Genomics* **26**: 147–156.

Arnold N., Stanyon R., Jauch A., O'Brien P., Wienberg J. (1996) Identification of complex chromosome rearrangements in the gibbon (*Hylobates klossii*) by fluorescent *in situ* hybridization (FISH) of a human chromosome 2q specific microlibrary, yeast artificial chromosomes and reciprocal chromosome painting. *Cytogenet. Cell Genet.* **74**: 80–85.

Ashley T., Cacheiro N.L., Russell L.B., Ward D.C. (1993) Molecular characterization of a pericentric inversion in mouse chromosome 8 implicates telomeres as promoters of meiotic recombination. *Chromosoma* **102**: 112–120.

Ashworth A., Skene B., Swift S., Lovell B.R. (1990) *Zfa* is an expressed retroposon derived from an alternative transcript of the *Zfx* gene. *EMBO J.* **9**: 1529–1534.

Bailey A.D., Shen C.C., Shen C.-K.J. (1997) Molecular origin of the mosaic sequence arrangements of higher primate α-globin duplication units. *Proc. Natl. Acad. Sci. USA* **94**: 5177–5182.

Baker R.T., Board P.G. (1991) The human ubiquitin 52 amino acid fusion protein gene shares several structural features with mammalian ribosomal protein genes. *Nucleic Acids Res.* **19**: 1035–1040.

Baker R.T., Board P.G. (1992) The human ubiquitin/52-residue ribosomal protein fusion gene subfamily (UbA$_{52}$) is composed primarily of processed pseudogenes. *Genomics* **14**: 520–522.

Barr F.G. (1998) Translocations, cancer and the puzzle of specificity. *Nature Genet.* **19**: 121–124.

Belich M.P., Madrigal J.A., Hildebrand W.H., Zemmour J., Williams R.C., Luz R., Petzl E.M., Parham P. (1992) Unusual HLA-B alleles in two tribes of Brazilian Indians. *Nature* **357**: 326–329.

Bell G.I., Gerhard D.S., Fong N.M., Sanchez-Pescador R., Rall L.B. (1985) Isolation of the human insulin-like growth factor genes: insulin-like growth factor II and insulin genes are contiguous. *Proc. Natl. Acad. Sci. USA* **82**: 6450–6454.

Berman A.L., Kolker E., Trifonov E.N. (1994) Underlying order in protein sequence organization. *Proc. Natl. Acad. Sci. USA* **91**: 4044–4047.

Bevins C.L., Jones D.E., Dutra A., Schaffzin J., Muenke M. (1996) Human enteric defensin genes: chromosomal map position and a model for possible evolutionary relationships. *Genomics* **31**: 95–106.

Boer P.H., Adra C.N., Lau Y.F., McBurney M.W. (1987) The testis-specific phosphoglycerate kinase gene *pgk-2* is a recruited retroposon. *Mol. Cell. Biol.* **7**: 3107–3112.

Borden P., Jaenichen R., Zachau H. (1990) Structural features of transposed human Vκ genes and implications for the mechanism of their transpositions. *Nucleic Acids Res.* **18**: 2101–2107.

Brenner S., Corrochano L.M. (1996) Translocation events in the evolution of aminoacyl-tRNA synthetases. *Proc. Natl. Acad. Sci. USA* **93**: 8485–8489.

Brosius J. (1991) Retroposons – seeds of evolution. *Science* **251**: 753.

Buckland R.A. (1992) A primate transfer RNA cluster and the evolution of human chromosome 1. *Cytogenet. Cell Genet.* **61**: 1–4.

Cacurri S., Piazzo N., Deidda G., Vigneti E., Galluzzi G., Colantoni L., Merico B., Ricci E., Felicetti L. (1998) Sequence homology between 4qter and 10qter loci facilitates the instability of subtelomeric *Kpn*I repeat units implicated in facioscapulohumeral muscular dystrophy. *Am. J. Hum. Genet.* **63**: 181–190.

Caldas C., So C.W., MacGregor A., Ford A.M., McDonald B., Chan L.C., Wiedemann L.M. (1998) Exon scrambling of MLL transcripts occur commonly and mimic partial genomic duplication of the gene. *Gene* **208**: 167–176.

Carritt B., Kemp T.J., Poulter M. (1997) Evolution of the human RH (rhesus) blood group genes: a 50 year old prediction (partially) fulfilled. *Hum. Molec. Genet.* **6**: 843–850.

Chen E.Y., Liao Y.C., Smith D.H., Barrera-Saldana H.A., Gelinas R.E., Seeburg P.H. (1989) The human growth hormone locus: nucleotide sequence, biology and evolution. *Genomics* **4**: 479–497.

Chen C., Clark D., Ueda K., Pastan I., Gottesman M.M., Roninson I.B. (1990) Genomic organization of the human multidrug resistance (*MDR1*) gene and origin of P-glycoproteins. *J. Biol. Chem.* **265**: 506–514.

Conte R.A., Samonte R.V., Verma R.S. (1998) Evolutionary divergence of the oncogenes GLI, HST and INT2. *Heredity* **81**: 10–13.

Cooper D.N., Krawczak M. (1993) *Human Gene Mutation.* BIOS Scientific Publishers, Oxford.

Coppock D.L., Cina-Poppe D., Gilleran S. (1998) The quiescin Q6 gene (*QSCN6*) is a fusion of two ancient gene families: thioredoxin and ERV1. *Genomics* **54**: 460–468.

Craig S.P., Buckle V.J., Lamouroux A., Mallet J., Craig I. (1986) Localization of the human tyrosine hydroxylase gene to 11p15: gene duplication and evolution of metabolic pathways. *Cytogenet. Cell Genet.* **42**: 29–32.

Cross M., Renkawitz R. (1990) Repetitive sequence involvement in the duplication and divergence of mouse lysozyme genes. *EMBO J.* **9**: 1283–1288.

Dahl H.H., Brown R.M., Hutchison W.M., Maragos C., Brown G.K. (1990) A testis-specific form of the human pyruvate dehydrogenase E1 alpha subunit is coded for by an intronless gene on chromosome 4. *Genomics* **8**: 225–232.

Diarra-Mehrpour M., Sarafan N., Bourguignon J., Bonnet F., Bost F., Martin J.-P. (1998) Human inter-α-trypsin inhibitor heavy chain H3 gene. *J. Biol. Chem.* **273**: 26 809–26 819.

Dover G.A. (1993) Evolution of genetic redundancy for advanced players. *Curr. Op. Genet. Devel.* **3**: 902–910.

Dreyfus D.H. (1992) Evidence suggesting an evolutionary relationship between transposable elements and immune system recombination sequences. *Molec. Immunol.* **29**: 807–810.

Dutrillaux B. (1988) Chromosome evolution in primates. *Fol. Primatol.* **50**: 134.

Dutrillaux B., Prieur M., Aurias A. (1986) Theoretical study of inversions affecting human chromosomes. *Ann. Génét.* **29**: 184–188.

Eichler E.E. (1998) Masquerading repeats: paralogous pitfalls of the human genome. *Genome Res.* **8**: 758–762.

Eichler E.E. (1999) Repetitive conundrums of centromere structure and function. *Hum. Molec. Genet.* **8**: 151–155.

Eichler E.E., Lu F., Antonacci R., Jurecic V., Doggett N.A., Moyzis R.K., Baldini A., Gibbs R.A., Nelson D.L. (1996) Duplication of a gene-rich cluster between 16p11.1 and Xq28: a novel pericentromeric-directed mechanism for paralogous genome evolution. *Hum. Molec. Genet.* **5**: 899–912.

Eichler E.E., Budarf M.L., Rocchi M., Deaven L.L., Doggett N.A., Baldini A., Nelson D.L., Mohrenweiser H.W. (1997) Interchromosomal duplications of the adrenoleukodystrophy locus: a phenomenon of pericentromeric plasticity. *Hum. Molec. Genet.* **6**: 991–1002.

Eichler E.E., O'Keefe C.L., Rocchi M., Archidiacono N., Matera A.G. (1998a) Pericentromeric CAGGG repeat elements: markers of genomic instability and gene duplication. *Am. J. Hum. Genet.* **63**: A41.

Eichler E.E., Hoffman S.M., Adamson A.A., Gordon L.A., McCready P., Lamerdin J.E., Mohrenweiser H.W. (1998b) Complex β-satellite repeat structures and the expansion of the zinc finger cluster in 19p12. *Genome Res.* **8**: 791–808.

Epp T.A., Wang R., Sole M.J., Liew C.-C. (1995) Concerted evolution of mammalian cardiac myosin heavy chain genes. *J. Mol. Evol.* **41**: 284–292.

Erickson L.M., Maeda N. (1994) Parallel evolutionary events in the haptoglobin gene clusters of rhesus monkey and human. *Genomics* **22**: 579–589.

Erickson L.M., Kim H.S., Maeda N. (1992) Junctions between genes in the haptoglobin gene cluster of primates. *Genomics* **14**: 948–958.

Eul J., Graessmann M., Graessmann A. (1995) Experimental evidence for RNA trans-splicing in mammalian cells. *EMBO J.* **14**: 3226–3235.

Evers M.P.J., Zelle B., Bebelman J.P., van Beusechem V., Kraakman L., Hoffer M.J.V., Pronk J.C., Mager W.H., Planta R.J., Eriksson A.W., Frants R.R. (1989) Nucleotide sequence comparison of five human pepsinogen A (PGA) genes: evolution of the PGA multigene family. *Genomics* **4**: 232–239.

Eyre-Walker A. (1993) Recombination and mammalian genome evolution. *Proc. R. Soc. Lond. B* **252**: 237–243.

Fernandez-Salguero P., Hoffman S.M.G., Cholerton S., Mohrenweiser H., Raunio H., Rautio A., Pelkonen O., Huang J., Evans W.E., Idle J.R., Gonzalez F.J. (1995) A genetic polymorphism in coumarin 7-hydroxylation: sequence of the human *CYP2A* genes and identification of variant *CYP2A6* alleles. *Am. J. Hum. Genet.* **57**: 651–660.

Fitch D., Mainone C., Goodman M., Slightom J. (1990) Molecular history of gene conversion in the primate fetal γ-globin genes. *J. Biol. Chem.* **265**: 781–793.

Fitch D.H.A., Bailey W.J., Tagle D.A., Goodman M., Sieu L., Slightom J.L. (1991) Duplication of the gamma-globin gene mediated by L1 long interspersed repetitive elements in an early ancestor of simian primates. *Proc. Natl. Acad. Sci. USA* **88**: 7396–7400.

Fitzgerald J., Dahl H.-H.M., Jakobsen I.B., Easteal S. (1996) Evolution of mammalian X-linked and autosomal *Pgk* and *Pdh E1α* subunit genes. *Mol. Biol. Evol.* **13**: 1023–1031.

Fujieda S., Lin Y.O., Saxon A., Zhang K. (1996) Multiple types of chimeric germ-line Ig heavy chain transcripts in human B cells: evidence for trans-splicing of human Ig RNA. *J. Immunol.* **157**: 3450–3459.

Funkhouser W., Koop B.F., Charmley P., Martindale D., Slightom J., Hood L. (1997) Evolution and selection of primate T cell antigen receptor BV8 gene subfamily. *Molec. Phylogenet. Evol.* **8**: 51–64.

Gellert M. (1997) Recent advances in understanding V(D)J recombination. *Adv. Immunol.* **64**: 39–64.

Giordano M., Marchetti C., Chiorboli E., Bona G., Richiardi P.M. (1997) Evidence for gene conversion in the generation of extensive polymorphism in the promoter of the growth hormone gene. *Hum. Genet.* **100**: 249–255.

Gonzalez I.L., Wu S., Li W.-M., Kuo B.A., Sylvester J.E. (1992) Human ribosomal RNA intergenic spacer sequence. *Nucleic Acids Res.* **20**: 5846.

Gorski J., Mach B. (1986) Polymorphism of human Ia antigens: gene conversion between two DR beta loci results in a new HLA-D/DR specificity. *Nature* **322**: 67–70.

Groot P.C., Bleeker M.J., Pronk J.C., Arwert F., Mager W.H., Planta R.J., Eriksson A.W., Frants R.R. (1989) The human α-amylase multigene family consists of haplotypes with variable numbers of genes. *Genomics* **5**: 29–42.

Groot P.C., Mager W.H., Henriquez N.V., Pronk J.C., Arwert F., Planta R.J., Eriksson A.W., Frants R.R. (1990) Evolution of the human α-amylase multigene family through unequal, homologous and inter- and intrachromosomal crossovers. *Genomics* **8**: 97–105.

Gumucio D.L., Wiebauer K., Caldwell R.M., Samuelson L.C., Meisler M.H. (1988) Concerted evolution of human amylase genes. *Molec. Cell. Biol.* **8**: 1197–1205.

Hagstrom J.E., Fautsch M.P., Perdok M., Vrabel A., Wieben E.D. (1996) Exons lost and found. Unusual evolution of a seminal vesicle transglutaminase substrate. *J. Biol. Chem.* **271**: 21 114–21 119.

Haino M., Hayashida H., Miyata T., Shin E.K., Matsuda F., Nagaoka H., Matsumura R., Taka-ishi S., Fukita Y., Fujikura J., Honjo T. (1994) Comparison and evolution of human immunoglobulin V$_H$ segments located in the 3′ 0.8-megabase region. *J. Biol. Chem.* **269**: 2619–2626.

Hess J.F., Fox M., Schmid C., Shen C.-K.J. (1983) Molecular evolution of the adult α-globin-like gene region: insertion and deletion of *Alu* family repeats and non-*Alu* DNA sequences. *Proc. Natl. Acad. Sci. USA* **80**: 5970–5974.

Hiom K., Melek M., Gellert M. (1998) DNA transposition by the RAG1 and RAG2 proteins: a possible source of oncogenic translocations. *Cell* **94**: 463–470.

Hirt H., Kimelman J., Birnbaum M.J., Chen E.Y., Seeburg P.H., Eberhardt N.L., Barta A. (1987) The human growth hormone gene locus: structure, evolution, and allelic variations. *DNA* **6**: 59–70.

Holmer L., Pezhman A., Worman H.J. (1998) The human lamin B receptor/sterol reductase multigene family. *Genomics* **54**: 469–476.

Huang C.H., Blumenfeld O.O. (1991) Identification of recombination events resulting in three hybrid genes encoding human MiV, MiV(J.L.) and Sta glycophorins. *Blood* **77**: 1813–1820.

Huber C., Schable K.F., Huber E., Klein R., Meindl A., Thiebe R., Lamm R., Zachau H.G. (1993) The V kappa genes of the L regions and the repertoire of V kappa gene sequences in the human germ line. *Eur. J. Immunol.* **23**: 2868–2875.

Iyer G., Krahe R., Goodwin L., Doggett N., Siciliano M., Funanage V., Proujansky R. (1996) Identification of a testis-expressed creatine transporter gene at 16p11.2 and confirmation of the X-linked locus to Xq28. *Genomics* **34**: 143–146.

Jackson M.S., Rocchi M., Thompson G., Hearn T., Crosier M., Guy J., Kirk, Mulligan L., Ricco A., Piccininni S., Marzella R., Viggiano L., Archidiacono N. (1999) Sequences flanking the centromere of human chromosome 10 are a complex patchwork of arm-specific sequences, stable duplications and unstable sequences with homologies to telomeric and other centromeric locations. *Hum. Molec. Genet.* **8**: 205–215.

Jiang P., Stone S., Wagner R., Wang S., Dayananth P., Kozak C.A., Wold B., Kamb A. (1995) Comparative analysis of *Homo sapiens* and *Mus musculis* cyclin-dependent kinase (CDK) inhibitor genes P16 (MTS1) and P15 (MTS2). *J. Mol. Evol.* **41**: 795–802.

Kaiser E., Eberhard D., Knippers R. (1992) Exons encoding the highly conserved part of human glutaminyl-tRNA synthetase. *J. Mol. Evol.* **34**: 45–53.

Kaiser E., Hu B., Becher S., Eberhard D., Schray B., Baack M., Hameister H., Knippers R. (1994) The human *EPRS* locus (formerly the QARS locus): a gene encoding a class I and a class II aminoacyl-tRNA synthetase. *Genomics* **19**: 280–290.

Kant J.A., Fornace A.J., Saxe D., Simon M.I., McBride O.W., Crabtree G.R. (1985) Evolution of the fibrinogen locus on chromosome 4: gene duplication accompanied by transposition and inversion. *Proc. Natl. Acad. Sci. USA* **82**: 2344–2348.

Kasahara M., Matsumara E., Webb G., Board P.G., Figueroa F., Klein J. (1990) Mapping of the class alpha glutathione *S*-transferase 2 (*Gst-2*) genes to the vicinity of the *d* locus on mouse chromosome 9. *Genomics* **8**: 90–96.

Kawamura S., Saitou N., Ueda S. (1992) Concerted evolution of the primate immunoglobulin α-gene through gene conversion. *J. Biol. Chem.* **267**: 7359–7367.

Kermouni A., van Roost E., Arden K.C., Vermeesch J.R., Weiss S., Godelaine D., Flint J., Lurquin C., Szikora J.-P., Higgs D.R., Marynen P., Renauld J.-C. (1995) The IL-9 receptor gene (*IL9R*): genomic structure, chromosomal localization in the pseudoautosomal region of the long arm of the sex chromosomes, and identification of IL9R pseudogenes at 9qter, 10pter, 16pter and 18pter. *Genomics* **29**: 371–382.

Kim H.S., Lyons K.M., Saitoh E., Azen E.A., Smithies O., Maeda N. (1993) The structure and evolution of the human salivary proline-rich protein gene family. *Mamm. Genome* **4**: 3–14.

Kiyosawa H., Chance P.F. (1996) Primate origin of the CMT1A-REP repeat and analysis of a putative transposon-associated recombinational hotspot. *Hum. Molec. Genet.* **5**: 745–753.

Kolker E., Trifonov E.N. (1995) Periodic recurrence of methionines: fossil of gene fusion? *Proc. Natl. Acad. Sci. USA* **92**: 557–560.

Koop B.F., Siemieniak D., Slightom J.L., Goodman M., Dunbar J., Wright P.C., Simons E.L. (1989) Tarsius δ-and β-globin genes: conversions, evolution and systematic implications. *J. Biol. Chem.* **264**: 68–79.

Kudo S.R., Fukuda M. (1994) Contribution of gene conversion to the retention of the sequence for M blood group type determinant in glycophorin E gene. *J. Biol. Chem.* **269**: 22 969–22 974.

Kuhner M.K., Lawlor D.A., Ennis P.D., Parham P. (1991) Gene conversion in the evolution of the human and chimpanzee MHC class I loci. *Tissue Antigens* **38**: 152–164.

Kuhner M.K., Peterson M.J. (1992) Genetic exchange in the evolution of the human MHC class II loci. *Tissue Antigens* **39**: 209–215.

Kuma K., Hayashida H., Miyata T. (1988) Recent gene conversion between genes encoding human red and green visual pigments. *Idengaku Zasshi* **63**: 367–371.

Labuda D., Zietkiewicz E., Mitchell G.A. (1995) *Alu* elements as a source of genomic variation: deleterious effects and evolutionary novelties. In: *The Impact of Short Interspersed Elements (SINEs) on the Host Genome*, (ed. RJ Maraia). RG Landes Co, Austin, Texas.

Ledley F.D., Grenett H.E., Bartos D.P., van Tuinen P., Ledbetter D.H., Woo S.L.C. (1987) Assignment of human tryptophan hydroxylase locus to chromosome 11: gene duplication and translocation in evolution of aromatic amino acid hydroxylases. *Somat. Cell Molec. Genet.* **13**: 575–580.

Lefranc M.P., Helal A.N., de Lange G., Chaabani H., van Loghem E., Lefranc G. (1986) Gene conversion in human immunoglobulin gamma locus shown by unusual location of IgG allotypes. *FEBS Letts.* **196**: 96–102.

Li W.-H. (1997) *Molecular Evolution*. Sinauer Associates, Sunderland, Mass.

Liao D., Weiner A.M. (1995) Concerted evolution of the tandemly repeated genes encoding primate U2 small nuclear RNA (the *RNU2* locus) does not prevent rapid diversification of the $(CT)_n \cdot (GA)_n$ microsatellite embedded within the U2 repeat unit. *Genomics* **30**: 583–593.

Liao D., Paveliz T., Kidd J.R., Kidd K.K., Weiner A.M. (1997) Concerted evolution of the tandemly repeated genes encoding human U2 snRNA (the *RNU2* locus) involves rapid intrachromosomal homogenization and rare interchromosomal gene conversion. *EMBO J.* **16**: 588–598.

Lie-Injo L.E., Herrera A.R., Kan Y.W. (1981) Two types of triplicated α-globin loci in humans. *Nucleic Acids Res.* **9**: 3707–3717.

Liebhaber S.A., Goossens M., Kan Y.W. (1981) Homology and concerted evolution at the α1 and α2 loci of human α-globin. *Nature* **290**: 26–29.

Litman G.W., Rast J.P., Shamblott M.J., Haire R.N., Hulst M., Roess W., Litman R.T., Hinds-Frey K.R., Zilch A., Amemiya C.T. (1993) Phylogenetic diversification of immunoglobulin genes and the antibody repertoire. *Mol. Biol. Evol.* **10**: 60–72.

Liu L., Zhao C., Heng H.H., Ganz T. (1997) The human beta-defensin-1 and alpha-defensins are encoded by adjacent genes: two peptide families with differing disulphide topology share a common ancestry. *Genomics* **43**: 316–320.

Lund P.K., Moats-Staats B.M., Simmons J.G., Hoyt E., D'Ercole A.J., Martin F., Van Wyk J.J. (1985) Nucleotide sequence analysis of a cDNA encoding human ubiquitin reveals that ubiquitin is synthesized as a precursor. *J. Biol. Chem.* **260**: 7609–7613.

Lyons K.M., Stein J.H., Smithies O. (1988) Length polymorphisms in human proline-rich protein genes generated by intragenic unequal crossing over. *Genetics* **120**: 267–278.

McCarrey J.R. (1990) Molecular evolution of the human Pgk-2 retroposon. *Nucleic Acids Res.* **18**: 949–955.

McCarrey J.R., Thomas K. (1987) Human testis-specific *PGK* gene lacks introns and possesses characteristics of a processed gene. *Nature* **326**: 501–505.

McConkey E.H., Menon R., Williams G., Baker E., Sutherland G.R. (1996) Assignment of the gene for β-casein (*CSN2*) to 4q13 q21 in humans and 3p13→p12 in chimpanzees. *Cytogenet. Cell Genet.* **72**: 60–62.

McCormack W.T., Tjoelker L.W., Thompson C.B. (1993) Immunoglobulin gene diversification by gene conversion. *Prog. Nucleic Acid Res. Molec. Biol.* **45**: 27–45.

McEvoy S.M., Maeda N. (1988) Complex events in the evolution of the haptoglobin gene cluster in primates. *J. Biol. Chem.* **263**: 15 740–15 747.

Madan K. (1995) Paracentric inversions: a review. *Hum. Genet.* **96**: 503–515.

Maeda N. (1985) Nucleotide sequence of the haptoglobin (*Hp*) and haptoglobin-related (*Hpr*) gene pair: *Hpr* contains a retrovirus-like element. *J. Biol. Chem.* **260**: 6698–6709.

Maeda N., McEvoy S.M., Harris H.F., Huisman T.H.J., Smithies O. (1986) Polymorphisms in the human haptoglobin gene cluster: chromosomes with multiple haptoglobin-related genes. *Proc. Natl. Acad. Sci. USA* **83**: 7395–7399.

Magrangeas F., Pitiot G., Dubois S., Bragado-Nilsson E., Chérel M., Jobert S., Lebeau B., Boisteau O., Lethé B., Mallet J., Jacques Y., Minivielle S. (1998) Cotranscription and intergenic splicing of human galactose-1-phosphate uridylyltransferase and interleukin-11 receptor α-chain genes generate a fusion mRNA in normal cells. *J. Biol. Chem.* **273**: 16 005–16 010.

Matsuno Y., Yamashiro Y., Yamamoto K., Hattori Y., Yamamoto K., Ohba Y., Miyaji T. (1992) A possible example of gene conversion with a common β-thalassaemia mutation and Chi sequence present in the β-globin gene. *Hum. Genet.* **88**: 357–358.

Matsuo K., Clay O., Kunzler P., Georgiev O., Urbanek P., Schaffner W. (1994) Short introns interrupting the Oct-2 POU domain may prevent recombination between POU family genes without interfering with potential POU domain 'shuffling' in evolution. *Biol. Chem. Hoppe-Seyler* **375**: 675–683.

Mayer W.E., O'hUigin C., Klein J. (1993) Resolution of the *HLA-DRB6* puzzle: a case of grafting a *de novo*-generated exon on an existing gene. *Proc. Natl. Acad. Sci. USA* **90**: 10 720–10 724.

Merritt C.M., Easteal S., Board P.G. (1990) Evolution of human α1-acid glycoprotein genes and surrounding *Alu* repeats. *Genomics* **6**: 659–665.

Michelson A.M., Blake C.C.F., Evans S.T., Orkin S.H. (1985) Structure of the human phosphoglycerate kinase gene and the intron-mediated evolution and dispersal of the nucleotide-binding domain. *Proc. Natl. Acad. Sci. USA* **82**: 6965–6969.

Miró R., Fuster C., Clemente I.C., Caballin M.R., Egozcue J. (1992) Chromosome inversions involved in the chromosome evolution of the *Hominidae* and in human constitutional chromosome abnormalities. *J. Hum. Evol.* **22**: 19–22.

Mitelman F., Mertens F., Johansson B. (1997) A breakpoint map of recurrent chromosomal rearrangements in human neoplasia. *Nature Genet.* Special issue, April: 417–430.

Monfouilloux S., Avet-Loiseau H., Amarger V., Balazs I., Pourcel C., Vergnaud G. (1998) Recent human-specific spreading of a subtelomeric domain. *Genomics* **51**: 165–176.

Nagaraja R., MacMillan S., Kere J. *et al.* (1997) X chromosome map at 75-kb STS resolution, revealing extremes of recombination and GC content. *Genome Res.* **7**: 210–222.

Nagel G.M., Doolittle R.F. (1991) Evolution and relatedness in two aminoacyl-tRNA synthetase families. *Proc. Natl. Acad. Sci. USA* **88**: 8121–8125.

Nagele R., Freeman T., McMorrow L., Lee H. (1995) Precise spatial positioning of chromosomes during prometaphase: evidence for chromosomal order. *Science* **270**: 1831–1834.

Neitz M., Neitz J., Grishok A. (1995) Polymorphism in the numbers of genes encoding long-wavelength-sensitive cone pigments among males with normal color vision. *Vision Res.* **35**: 2395–2407.

Nickerson E., Nelson DL. (1997) Precise localization of a pericentric inversion breakpoint at human 12q15 occurring in hominoid evolution. *Am. J. Hum. Genet.* **61**: A136.

Nigro J.M., Cho K.R., Fearon E.R., Kern S.E., Ruppert J.M., Loiner J.D., Kinzler K.W., Vogelstein B. (1991) Scrambled exons. *Cell* **64**: 607–613.

Nouvel P. (1994) The mammalian genome shaping activity of reverse transcriptase. *Genetica* **93**: 191–201.

Oakey R.J., Watson M.L., Seldin M.F. (1992) Construction of a physical map on mouse and human chromosome 1: comparison of 13 Mb of mouse and 11 Mb of human DNA. *Hum. Molec. Genet.* **1**: 613–620.

Ohta T. (1991) Role of the diversifying selection and gene conversion in evolution of major histocompatibility complex loci. *Proc. Natl. Acad. Sci. USA* **88**: 6716–6720.

Ohta T., Basten C.J. (1992) Gene conversion generates hypervariability at the variable regions of kallikreins and their inhibitors. *Mol. Phyl. Evol.* **1**: 87–90.

Olsson M.L., Guerreiro J.F., Zago M.A., Chester M.A. (1997) Molecular analysis of the O alleles at the blood group ABO locus in populations of different ethnic origin reveals novel crossing-over events and point mutations. *Biochem Biophys. Res. Commun.* **234**: 779–782.

Onda M., Fukuda M. (1995) Detailed physical mapping of the genes encoding glycophorins A, B and E, as revealed by P1 plasmids containing human genomic DNA. *Gene* **159**: 225–230.

Parham P., Lawlor D.A. (1991) Evolution of class I major histocompatibility complex genes and molecules in humans and apes. *Hum. Immunol.* **30**: 119–128.

Papadakis M.N., Patrinos G.P. (1999) Contribution of gene conversion in the evolution of the human β-like globin gene family. *Hum. Genet.* **104**: 117–125.

Pauly M., Kayser I., Schmitz M., Ries F., Hentges F., Dicato M. (1995) Repetitive DNA sequences located in the central region of the human *mdr1* (multidrug resistance) gene may account for a gene fusion event during its evolution. *J. Mol. Evol.* **41**: 974–978.

Pavelitz T., Rusché L., Matera A.G., Scharf J.M., Weiner A.M. (1995) Concerted evolution of the tandem array encoding primate U2 snRNA occurs *in situ*, without changing the cytological context of the *RNU2* locus. *EMBO J.* **14**: 169–177.

Powers P.A., Smithies O. (1986) Short gene conversion in the human fetal globin gene region: a by-product of chromosome pairing during during meiosis? *Genetics* **112**: 343–358.

Rabbitts T.H. (1994) Chromosomal translocations in human cancer. *Nature* **372**: 143–149.

Regnier V., Medeb M., Lecointre G., Richard F., Duverger A., Nguyen V., Dutrillaux B., Bernheim A., Danglot G. (1997) Emergence and scattering of multiple neurofibromatosis (NF1)-related sequences during hominoid evolution suggest a process of pericentromeric interchromosomal transposition. *Hum. Molec. Genet.* **6**: 9–16.

Reinton N., Haugen T.B., Ørstavik S., Skalhegg B.S., Hansson V., Jahnsen T., Taskén K. (1998) The gene encoding the Cγ catalytic subunit of cAMP-dependent protein kinase is a transcribed retroposon. *Genomics* **49**: 290–297.

Ritchie R.J., Mattei M.G., Lalande M. (1998) A large polymorphic repeat in the pericentromeric region of human chromosome 15q contains three partial gene duplications. *Hum. Molec. Genet.* **7**: 1253–1260.

Rorman E., Scheinker V., Grabowski G.A. (1992) Structure and evolution of the human prosaposin chromosomal gene. *Genomics* **13**: 312–318.

Rouquier S., Taviaux S., Trask B.J., Brand-Arpon V., van den Engh G., Demaille J., Giogi D. (1998) Distribution of olfactory receptor genes in the human genome. *Nature Genet.* **18**: 243–250.

Russek S.J., Farb D.H. (1994) Mapping of the β_2 subunit gene (*GABRB2*) to microdissected human chromosome 5q34-q35 defines a gene cluster for the most abundant $GABA_A$ receptor isoform. *Genomics* **23**: 528–533.

Sastry K., Herman G.A., Day L., Deignan E., Bruns G., Morton C.C., Ezekowitz R.A.B. (1989) The human mannose-binding protein gene. *J. Exp. Med.* **170**: 1175–1189.

Schimenti J.C., Duncan C.H. (1984) Ruminant globin gene structures suggest an evolutionary role for *Alu*-type repeats. *Nucleic Acids Res.* **12**: 1641–1655.

Schwartz A., Chan D.C., Brown L.G., Alagappan R., Pettay D., Disteche C., McGillivray B., de la Chapelle A., Page D.C. (1998) Reconstructing hominid evolution: X-homologous block, created by X-Y transposition, was disrupted by Yp inversion through LINE-LINE recombination. *Hum. Molec. Genet.* **7**: 1–11.

Seeman G.H., Rein R.S., Brown C.S., Ploegh H.L. (1986) Gene conversion-like mechanisms may generate polymorphism in human class I genes. *EMBO J.* **5**: 547–552.

Seperack P., Slatkin M., Arnheim N. (1988) Linkage equilibrium in human ribosomal genes: implications for multigene family evolution. *Genetics* **119**: 943–949.

Sharp P.M., Li W.-H. (1987) Ubiquitin genes as a paradigm of concerted evolution of tandem repeats. *J. Mol. Evol.* **25**: 58–64.

Shaw J.-P., Marks J., Shen C.-K.J. (1991) The adult α-globin locus of Old World monkeys: an abrupt breakdown of sequence similarity to human is defined by an *Alu* family repeat insertion site. *J. Mol. Evol.* **33**: 506–513.

Shen S., Slightom J.L., Smithies O. (1981) A history of the human fetal globin gene duplication. *Cell* **26**: 191–203.

Shiroishi T., Sagai T., Moriwaki K. (1993) Hotspots of meiotic recombination in the mouse major histocompatibility complex. *Genetica* **88**: 187–196.

Shyue S.K., Li L., Chang B.H., Li W.H. (1994) Intronic gene conversion in the evolution of human X-linked color vision genes. *Mol. Biol. Evol.* **11**: 548–551.

Shyue S.K., Hewett-Emmett D., Sperling H.G., Hunt D.M., Bowmaker J.K., Mollon J.D., Li W.H. (1995) Adaptive evolution of color vision genes in higher primates. *Science* **269**: 1265–1267.

Small K., Iber J., Warren S.T. (1997) Emerin deletion reveals a common X-chromosome inversion mediated by inverted repeats. *Nature Genet.* **16**: 96–99.

Smith G.R. (1983) Chi hotspots of generalized recombination. *Cell* **34**: 709–710.

Soret J., Gattoni R., Guyon C., Sureau A., Popielarz M., Le Rouzic E., Dumon S., Apiou F., Dutrillaux B., Voss H., Ansorge W., Stevenin J., Perbal B. (1998) Characterization of SRp46, a novel human SR splicing factor encoded by a PR264/SC35 retropseudogene. *Mol. Cell. Biol.* **18**: 4924–4934.

Spurdle A., Jenkins T. (1992) The inverted Y-chromosome polymorphism in the Gujarati muslim Indian population of South Africa has a single origin. *Hum. Hered.* **42**: 330–332.

Stoeckert C.J., Collins F.S., Weissman S. (1984) Human fetal globin DNA sequences suggest novel conversion event. *Nucleic Acids Res.* **12**: 4469–4479.

Tagle D.A., Slightom J.L., Jones R.T., Goodman M. (1991) Concerted evolution led to high expression of a prosimian primate δ globin gene locus. *J. Biol. Chem.* **266**: 7469–7480.

Takagi Y., Omura T., Go M. (1991) Evolutionary origin of thyroglobulin by duplication of esterase gene. *FEBS Letts* **282**: 17–22.

Tasaki Y., Fukuda S., Lio M., Miwa R., Imai T., Sugano S., Yoshie O., Hughes A.L., Nomiyama H. (1999) Chemokine PARC gene (*SCYA18*) generated by fusion of two MIP–1 alpha/LD78 alpha-like genes. *Genomics* **55**: 353–357.

Thompson C.B. (1995) New insights into V(D)J recombination and its role in the evolution of the immune system. *Immunity* **3**: 531–539.

Toder R., Rappold G.A., Schiebel K., Schempp W. (1995) *ANT3* and *STS* are autosomal in prosimian lemurs: implications for the evolution of the pseudoautosomal region. *Hum. Genet.* **95**: 22–28.

Todokoro K., Kioussis D., Weissmann C. (1984) Two non-allelic human interferon alpha genes with identical coding regions. *EMBO J.* **3**: 1809–1812.

Tomlinson I., Cook G., Carter N., Elaswarapu R., Smith S., Walter G., Buluwela L., Rabbitts T., Winter G. (1994) Human immunoglobulin V_H and D segments on chromosomes 15q11.2 and 16p11.2. *Hum. Molec. Genet.* **3**: 853–860.

Trask B.C., Friedman C., Martin-Gallardo A., Rowen L., Akinbami C., Blankenship J., Collins C., Giorgi D., Iadonato S., Johnson F. (1998) Members of the olfactory receptor gene family are contained in large blocks of DNA duplicated polymorphically near the ends of human chromosomes. *Hum. Molec. Genet.* **7**: 13–26.

Tricoli J.V., Rall L.B., Scott J., Bell G.I., Shows T.B. (1984) Localization of insulin-like growth factor genes to human chromosomes 11 and 12. *Nature* **310**: 784–786.

Tunnacliffe A., Kefford R., Milstein C., Forster A., Rabbitts T.H. (1985) Sequence and evolution of the human T-cell antigen receptor beta-chain genes. *Proc. Natl. Acad. Sci. USA* **82**: 5068–5072.

Tunnacliffe A., Liu L., Moore J.K., Leversha M.A., Jackson M.S., Papi L., Ferguson-Smith M.A., Thiesen H.-J., Ponder B.A.J. (1993) Duplicated KOX zinc finger gene clusters flank the centromere of human chromosome 10: evidence for a pericentric inversion during primate evolution. *Nucleic Acids Res.* **21**: 1409–1417.

Udey J.A., Blomberg B.B. (1988) Intergenic exchange maintains identity between two human lambda light chain immunoglobulin gene intron sequences. *Nucleic Acids Res.* **16**: 2959–2969.

van der Bliek A.M., Kooiman P.M., Schneider C., Borst P. (1988) Sequence of mdr3 cDNA encoding a human P-glycoprotein. *Gene* **71**: 401–411.

van Deutekom J.C.T., Bakker E., Lemmers R.J.L.F., van der Wielen M.J.R., Bik E., Hofker M.H., Padberg G.W., Frants R.R. (1996) Evidence for subtelomeric exchange of 3.3 kb tandemly repeated units between chromosomes 4q35 and 10q26: implications for genetic counselling and etiology of FSHD1. *Hum. Molec. Genet.* **5**: 1997–2003.

Vocero-Akbani A., Helms C., Wang J.-C., Sanjuro F.J., Korte-Sarfaty J., Veile R.A., Liu L., Jauch A., Burgess A.K., Hing A.V., Holt M.S., Ramachandra S., Whelan A.J., Anker R., Ahrent L., Chen M., Gavin M.R., Iannantuoni K., Morton S.M., Pandit S.D., Read C.M., Steinbrück T., Warlick C., Smoller D.A., Donnis-Keller H. (1996) Mapping human telomere regions with YAC and P1 clones: chromosome-specific markers for 27 telomeres including 149 STSs and 24 polymorphisms for 14 proterminal regions. *Genomics* **36**: 492–506.

Vogel F., Motulsky A.G. (1997) *Human Genetics: Problems and Approaches.* 3rd Edn. Springer, Berlin.

Warmerdam P.A., Nabben N.M., van de Graaf S.A., van de Winkel J.G., Capel P.J. (1993) The human low affinity immunoglobulin G Fc receptor IIC gene is a result of an unequal crossover event. *J. Biol. Chem.* **268**: 7346–7349.

Weichhold G.M., Klobeck H.-G., Ohnheiser R., Combriato G., Zachau H.G. (1990) Megabase inversions in the human genome as physiological events. *Nature* **347**: 90–92.

Weterman M.A., Wilbrink M., Dijkhuizen E., van den Berg E., Geurts van Kessel A. (1996) Fine mapping of the 1q21 breakpoint of the papillary renal cell carcinoma-associated (X;1) translocation. *Hum. Genet.* **98**: 16–21.

Winderickx J., Battisti L., Hibiya Y., Motulsky A.G., Deeb S.S. (1993) Haplotype diversity in the human red and green opsin genes: evidence for frequent sequence exchange in exon 3. *Hum. Molec. Genet.* **2**: 1413–1421.

Winichagoon P., Higgs D.R., Goodbourn S.E.Y., Lamb J., Clegg J.B., Weatherall D.J. (1982) Multiple arrangements of the human embryonic zeta globin genes. *Nucleic Acids Res.* **10**: 5853–5868.

Wöhr G., Fink T., Assum G. (1996) A palindromic structure in the pericentromeric region of various human chromosomes. *Genome Res.* **6**: 267–279.

Wu S., Sounders T.L., Bach F.H. (1986) Polymorphism of human Ia antigens generated by reciprocal exchange between two DRB loci. *Nature* **324**: 676–679.

Yen P.H., Marsh B., Allen E., Tsai S.P., Ellison J., Connolly L., Neiswanger K., Shapiro L.J. (1988) The human X-linked steroid sulfatase gene and a Y-encoded pseudogene: evidence for an inversion of the Y chromosome during primate evolution. *Cell* **55**: 1123–1135.

Yunis J.J., Prakash O. (1982) The origin of man: a chromosomal pictorial legacy. *Science* **215**: 1525–1530.

Zelle B., Evers M.P.J., Groot P.C., Bebelman J.P., Mager W.H., Planta R.J., Pronk J.C., Meuwissen S.G.M., Hofker M.H., Eriksson A.W., Frants R.R. (1988) Genomic structure and evolution of the human pepsinogen A multigene family. *Hum. Genet.* **78**: 79–82.

Zhou Y.H., Li W.H. (1996) Gene conversion and natural selection in the evolution of X-linked color vision genes in higher primates. *Mol. Biol. Evol.* **13**: 780–783.

Zimmer F., Hameister H., Schek H., Zachau H. (1990) Transposition of human immunoglobulin V kappa genes within the same chromosome and the mechanism of their amplification. *EMBO J.* **9**: 1535–1542.

Molecular reconstruction of ancient genes/proteins

10.1 Introduction

Proteins from extinct organisms can be studied by the analysis of DNA recovered from preserved organic specimens. This approach, however, requires biological material that has been completely protected from the oxidative effects of oxygen and water (Audic and Beraud-Colomb, 1997; DeSalle *et al.*, 1992; Pääbo, 1993; reviewed by Li, 1997) and significant doubts may remain as to the authenticity of the DNA recovered (Austin *et al.*, 1997; Stoneking, 1995). Alternatively, an attempt can be made to reconstruct the amino acid sequence of an ancient protein from the sequences of its extant descendants (Malcolm *et al.*, 1990; Shih *et al.*, 1993; Stackhouse *et al.*, 1990). This methodology relies upon the principle of maximum parsimony (i.e. the assumption that extant proteins have evolved from that of an extinct ancestor by the smallest number of mutational changes). DNA sequence data can also be treated in the same way (Hillis *et al.*, 1993), one example being the investigation of ribosomal DNA phylogeny (Friedrich and Tautz, 1995).

Jermann *et al.* (1995) examined the evolutionary history of artiodactyl RNase A, a pancreatic digestive enzyme, derived the ancestral enzymes phylogenetically and reconstructed them by site-directed mutagenesis. The kinetic properties of the reconstructed enzymes were found to be similar to those of extant RNases but less stable to thermal denaturation, more susceptible to proteolysis and five-fold more active toward double-stranded RNA. A Gly38→Asp substitution, which occurred ~40 Myrs ago around the time when foregut rumination evolved, was found to be associated with the reduction in activity toward double-stranded RNA.

A similar analysis has been attempted with the chymases (mast cell proteases) which hydrolyse angiotensin I to generate angiotensin II, a potent vasoconstrictor hormone. Primate α-chymases are highly specific and only hydrolyse the Phe8-His9 peptide bond. By contrast, rat β-chymase is less specific and further hydrolyses angiotensin I by cleavage of the Tyr4-Ile5 bond. Chandrasekharan *et al.* (1996) determined the phylogeny of four mammalian α-chymases and six mammalian β-chymases and reconstructed the putative ancestral enzyme by chemical synthesis. This enzyme proved capable of cleaving angiotensin I at the Phe8-His9

Human Gene Evolution, David N. Cooper.
© 1999 BIOS Scientific Publishers Ltd, Oxford.

bond but not at the Tyr4-Ile5 bond. Thus the relatively narrow specificity of the primate α-chymase represents the ancestral state of the enzyme while the broader specificity of the rat β-chymase must have been secondarily derived.

10.2 Molecular reconstruction and homology modelling of the catalytic domain of the common ancestor of the hemostatic vitamin K-dependent serine proteases

Adopting the maximum parsimony principle and employing a novel cDNA-based strategy, Krawczak *et al.* (1996) reconstructed the catalytic domains of the early mammalian ancestors of the vitamin K-dependent factors and then went on to reconstruct the putative common ancestor of all five proteins from an earlier stage of vertebrate evolution. A tertiary model of the ancestral vitamin K-dependent serine protease was built that was both energetically satisfactory and possessed a credible fold, and its topological and biophysical properties were examined. Wacey *et al.* (1997) then traced the evolution of specific structural features in protein C from the common ancestor of the vitamin K-dependent serine proteases toward extant human protein C. These studies will now be described in some detail since they serve to illustrate the potential value of homology modelling in evolutionary studies.

10.2.1 The evolution of the vitamin K-dependent coagulation factors

The vitamin K-dependent serine proteases of coagulation (factors VII, IX, and X, prothrombin and protein C) exhibit substantial sequence and structural homology (Greer, 1990). Factors VII, IX, X and protein C all contain an N-terminal domain of glutamic acid (Gla) residues, two epidermal growth factor-like (EGF) domains and a catalytic domain (Blake *et al.*, 1987); prothrombin differs slightly in that it possesses two kringle domains instead of the EGF domains (Gojobori and Ikeo, 1994). With the exception of prothrombin, the genes encoding the vitamin K-dependent coagulation factors have a very similar exon/intron organization (Tuddenham and Cooper, 1994), suggesting that they have arisen from a relatively recent common ancestor through a process of gene duplication and divergence (Neurath, 1984; Patthy, 1985; Patthy, 1990). The organization of these genes also reflects the functional modular assembly of the respective proteins and is thought to have emerged through exon shuffling (Chapter 3 section 3.6). Evolution of these genes has proceeded by repeated insertions, duplications, exchanges and deletions of modules. The presence/absence of modules such as the calcium-binding Gla domains, the EGF-like domains and kringles was used by Patthy (1990) to reconstruct the evolutionary past of the genes (*Figure 10.1*). More recently, it has been realised, however, that each module or domain has its own distinctive evolutionary history as a result of different evolutionary pressures (Ikeo *et al.*, 1995).

Doolittle (1993) also proposed a tentative scheme for the evolution of the vitamin K-dependent coagulation factors (*Figure 10.1*). First, an ancestral prothrombin emerged by serine protease gene duplication and the acquisition of Gla and

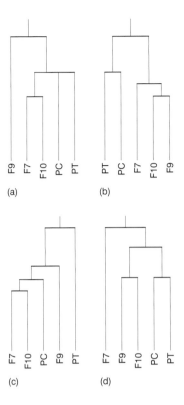

Figure 10.1. Proposed phylogenies for the vitamin K-dependent factors of coagulation. A: Krawczak *et al.* (1996), B: Doolittle (1993); C: Doolittle and Feng (1987), D: Patthy (1990).

EGF domains. Prothrombin's EGF domain(s) served as a site for the binding of tissue factor which at this time possessed the ability to activate it to yield thrombin. After the emergence of fibrinogen, factor X appeared as a result of a prothrombin gene duplication. The ability of factor X to activate prothrombin released the latter from its dependence on tissue factor. Factor VII, duplicated from factor X, was able to bind tissue factor and to activate factor X. Factor IX emerged last, again duplicated from factor X. Prothrombin then acquired kringle domains via exon shuffling (Rogers, 1985) allowing it to bind fibrin. The concomitant loss of prothrombin's EGF domains abolished its now redundant interaction with tissue factor. The plausibility of this scheme was tested by Krawczak *et al.* (1996) whose approach is described in subsequent sections.

10.2.2 Reconstruction of mammalian ancestral cDNAs

Krawczak *et al.* (1996) reconstructed the catalytic domains of the early mammalian ancestors of the vitamin K-dependent factors and the common ancestor of all five proteins. The study of Krawczak *et al.* (1996) relied upon mammalian phylogenies from different sources (Nei, 1987; Novacek, 1992; Vogel and Motulsky,

1986) and included cDNA sequences from human, macaque, sheep, pig, rabbit, rat, mouse, dog, and cow. Conserved cDNA blocks were used as anchors to align intervening non-conserved cDNA blocks for a particular protein, and since protein function must have been retained during evolution, frameshift mutations were precluded in all alignments. For each protein, the authors then deduced from the alignments the most likely cDNA sequence at each node of the mammalian phylogenies employed, including the roots representing the respective mammalian ancestors. A small number of ambiguities remained as to the ancestral mammalian cDNA sequences. These were resolved by reference to relative nearest neighbor-dependent mutation rates in humans. Use of human-derived parameters in this context was justified on the basis of the long-term evolutionary stability of relative single base-pair substitution rates (Krawczak and Cooper, 1996).

The evolution of a gene family whose members acquire different functions is invariably accompanied by rapid amino acid sequence divergence in functionally important regions (Ohta, 1991, 1994). Indeed, specific examples of this phenomenon include the 'accelerated evolution' (hypervariability) of the reactive center regions of serine protease inhibitors (Hill and Hastie, 1987) and the active site regions of some serine proteases (Creighton and Darby, 1989). For humans, the mutation rates derived during the reconstruction process were found by Krawczak et al. (1996) to be significantly higher in the factor VII, factor X and protein C lineages than in the factor IX and prothrombin lineages. Similarly in the dog, the factor VII lineage exhibited a higher mutation rate than that of factor IX. That factor IX exhibited a lower mutation rate than the other proteins was indicative of its early emergence. Once adapted to its functions, factor IX would have had to change rather less over evolutionary time than the other proteins of more recent origin which still had to adapt to their new-found roles. For factor IX, protein C and prothrombin, mutation rate differences were also apparent between species and exhibited an inverse correlation with generation time. Consistent with previous results (Britten, 1986; Collins and Jukes, 1994), the mutation rate in humans was much less than that found in rodents.

10.2.3 Evolutionary divergence of the vitamin K-dependent coagulation factors

Krawczak et al. (1996) identified a number of highly conserved regions in their reconstructed ancestral mammalian cDNA sequences. Mismatches in these regions were classified on the basis of whether they corresponded to either a silent or a missense mutation during the process of evolutionary divergence. Interestingly, the numbers of silent and missense mutations did not correlate with each other. This finding was interpreted in terms of the existence of two distinct molecular clocks: one would be based upon silent mutations and would run constantly after the divergence of any two sequences. The other would be based upon missense mutations which would continue to run immediately after gene duplication but would stop, or at least be dramatically slowed, once the new protein product had acquired, and then become adapted to, its new biological function.

Since by far the smallest number of silent mutations in conserved codons had occurred since the divergence of factors VII and X, Krawczak et al. (1996)

concluded that these two proteins must have their most recent root in common (*Figure 10.1*). This assertion is not inconsistent with the fact that the human factor VII (*F7*) and factor X (*F10*) genes are not only syntenic but also very closely linked on chromosome 13q34, as a result of their recent emergence through a process of duplication and divergence. Since the conserved cDNA blocks of factor IX exhibited the largest number of neutral differences with respect to other proteins, this was held to imply that factor IX was the first to diverge from the other proteins. However, a potential pitfall with this conclusion could have been a faster rate of substitution at the X-linked factor IX (*F9*) locus as compared to the other, autosomal genes. However, at least for human, substitution rates in the *F9* gene were significantly lower than in the factor *F7*, *F10* and protein C (*PROC*) genes, which also appeared to be the case in dog. Thus, divergence time rather than a higher propensity to mutate was held to be responsible for the large number of silent substitutions which separate the *F9* cDNA sequence from the other cDNAs. The precise order of divergence events for protein C, prothrombin and the common ancestor of factors VII and X could not be clarified unequivocally by Krawczak *et al.* (1996) on the basis of their data.

The cDNA-based phylogeny of the five vitamin K-dependent factors presented in *Figure 10.1* emphasizes the previously recognized relatedness of (i) protein C and prothrombin and (ii) factors VII and X. However, it is markedly different from other proposed phylogenies (Doolittle and Feng, 1987; Patthy, 1985; Patthy, 1990; Doolittle, 1993; *Figure 10.1*). This may have been because earlier attempts employed alignments of amino acid rather than cDNA sequences. The phylogeny of Krawczak *et al.* (1996) was closest to that presented by Doolittle and Feng (1987): both phylogenies claimed a comparatively recent common root for the catalytic domains of factors VII and X. However, the major difference lies in the much earlier branching out of the catalytic domain of factor IX, postulated by Krawczak *et al.* (1996).

Gene duplication has played a very important role in the evolution of the genomes of higher organisms (Ohno, 1970). Indeed, there is now considerable evidence for saltatory increases in the number of genes around the time of the emergence of the vertebrates 500 Myrs ago (Bird, 1995). These increases appear to have been caused by the duplication and subsequent divergence of many different gene sequences (see Chapter 2, section 2.1). It is still impossible to say for certain, however, when the duplication and divergence of the vitamin K-dependent serine proteases of coagulation occurred during vertebrate evolution. Prothrombin is present in bony fish (trout), cartilaginous fish (dogfish) and in the hagfish, one of the modern representatives of the jawless *Agnatha* (Banfield and MacGillivray, 1992; Doolittle, 1993). Thus, although thrombin is found in the most primitive of modern vertebrates, there is no evidence for its existence in either protochordates or echinoderms. Whether the other four vitamin K-dependent factors of coagulation are present in these types of fish is as yet unclear (Doolittle, 1993). If they are, the adaptive radiation of the vitamin K-dependent factors of hemostasis must have occurred during the space of some 50 Myrs between the divergence of the protochordates and the appearance of the *Agnatha*, some 450 Myrs ago (Doolittle, 1993). The processes of gene duplication and divergence that have led to the emergence of the five present day vitamin K-dependent factors of coagulation provide further evidence of what must have been a very active phase in the evolution of the modern vertebrate genome.

10.2.4 Reconstruction of a common ancestor of the vitamin K-dependent coagulation factors

The evolutionary scenario depicted in *Figure 10.1* was based solely upon the highly conserved regions of the reconstructed ancestral mammalian cDNA sequences. However, this phylogeny was used by Krawczak *et al.* (1996) in an attempt to reconstruct the common ancestor of all five vitamin K-dependent factors by means of molecular modelling. To this end, the less well conserved regions of the mammalian ancestral sequences were first aligned in the order stipulated by the phylogeny. The alignment was then used to determine the nucleotides present at each node of the phylogeny, moving upwards through the tree to its root. The resulting cDNA sequence, representing the putative common ancestor of all five vitamin K-dependent factors, contained several gaps which yielded ambiguities in the sequence of the ancestral protein. However, since the protein core had to contain a full complement of residues for meaningful molecular modelling to be possible, all such amino acids were replaced with the analogous residue found in extant human thrombin as a 'best guess'.

One way to examine the plausibility of the deduced amino acid sequence of the vitamin K-dependent factor ancestral protein would have been to express it *in vitro* and to characterize it biochemically (Malcolm *et al.*, 1990; Shih *et al.*, 1993). An alternative strategy was to construct a model of the tertiary structure of the protein by comparative methods and examine its topology and biophysical properties. Such a model was created by Krawczak *et al.* (1996) using the X-ray crystallographic coordinates of the heavy chain of α-thrombin as a template. This template was aligned against the high resolution structure of seven other serine proteases and the amino acid sequence of the reconstructed ancestral protein (Greer, 1990). Sequence modifications became necessary in this process at two variable regions, which were longer than the analogous loops present in extant serine proteases, and a single unpaired Cys residue (Cys22), which was replaced by Ala as in extant human thrombin. Extant human thrombin was chosen for comparison in this and other instances since, of all of the modern hemostatic serine proteases, it bore the strongest sequence homology to the ancestral protein. Moreover, it was assumed (Doolittle, 1993) that the vitamin K-dependent factor ancestral protein would have been capable of performing the end effector function of thrombin in the hemostatic cascade, that is the cleavage of soluble fibrinogen to generate insoluble fibrin clot.

The putative ancestral protein was found to contain 86 charged groups at pH 7.0, very similar to the number (89) found in extant human thrombin. In both cases, the great majority of these charges were noted to be accessible to water. The global electrostatic distribution across the ancestral protein's surface was relatively uniform by comparison with the dipolar distribution evident in extant human thrombin. Both structures still possessed an unbroken equatorial belt of negative charge but the extent of the electrostatic field strength in the ancient protein was not as strong as that of extant thrombin. It was speculated that the increase in charge intensity and dispersal over evolutionary time reflected a trend toward increasing protein binding specificity.

The fibrinogen-binding exosite (P1) was found to have greatly increased in size during the evolution of thrombin. From the electostatic contour map, only five Arg (126, 165, 233) and three Lys residues (230, 243, 245) of the putative ancestral protein appeared to contribute to a small patch of positive charge (*Figure 10.2*). Moreover, several thrombin residues known to be important in the binding of fibrinogen (Tsiang *et al.*, 1995) were not present in the ancient protein suggesting that the ancestral protease bound fibrinogen only weakly (*Figure 10.3*). A number of residues in thrombin have been shown to be involved in the binding of protein C (Tsiang *et al.*, 1995) and these are also components of the fibrinogen binding site (Lys36, Trp60D, Lys70, His71, Arg73, Tyr76, Arg77A, Lys81, Lys109, Lys110, Glu217, Arg221A; *Figure 10.3*) . However, only a fraction of these residues were present in the ancestral protein (Trp60D, His71, Arg77A, Lys110, Glu217, Arg221A). This was not surprising since, at the dawn of vertebrate evolution, protein C had yet to evolve from the vitamin K-dependent factor ancestral protein

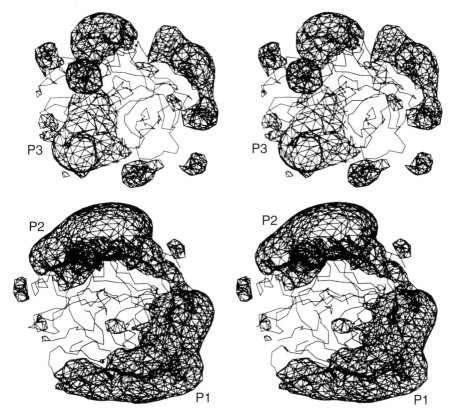

Figure 10.2. Stereo views of the electrostatic profiles of the putative vitamin K-dependent factor ancestral protein (above) and extant human thrombin (below). The view is towards the active site canyon. The positions of the fibrinogen-binding site (P1) and heparin-binding site (P2) are indicated. A large positively charged patch (P3) is also present on the vitamin-K dependent factor ancestral protein (after Krawczak *et al.*, 1996).

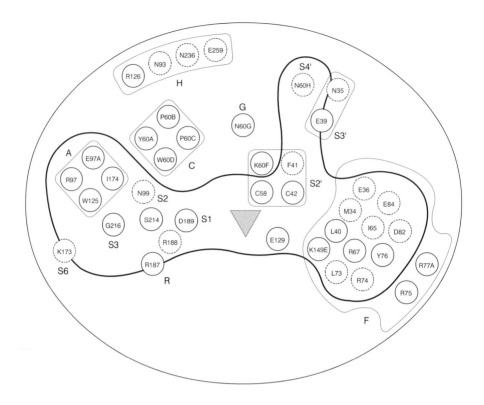

Figure 10.3. Schematic view of the active site canyon (bold contour) of the vitamin-K-dependent factor ancestral protein (Krawczak *et al.*, 1996). The active site triad (His57, Asp102, Ser195, chymotrypsin numbering) is denoted by a triangle.
H: heparin-binding site, G: glycosylation site, C: chemotactic region, R: RGD sequence, A: aryl-binding site, F: fibrinogen-binding exosite, S: specificity sites, N-terminal to cleavage, S′: specificity sites C-terminal to cleavage. Residues that are conserved between the vitamin-K-dependent factor ancestral protein and extant human thrombin are circled. Those residues that have not been conserved are circled with a broken line.

(Doolittle, 1993). By contrast, both thrombin residues implicated in binding thrombomodulin (Gln38, Arg75; Tsiang *et al.*, 1995), the endothelial cell surface thrombin receptor, were present in the ancestral protein. Thus an ancient thrombomodulin-like molecule might have been able to interact with the vitamin K-dependent factor ancestral protein.

When the residues responsible for binding thrombomodulin and protein C were excluded from the fibrinogen binding patch, five of the remaining seven residues were found to be conserved between the vitamin K-dependent factor ancestral protein and extant human thrombin (Lys60F, Asn60G, Asp186A, Lys186D, Glu192) (*Figure 10.3*). Both of the altered residues (Thr60I, Asp222) exhibit conservative changes (to Ser and Glu, respectively). These seven residues may have constituted the original fibrinogen binding patch that was to increase in size and binding affinity as well as diversifying functionally over evolutionary time.

The extensive positive electrostatic charge associated with the heparin binding site in extant thrombin was much smaller in size and field strength in the model of the putative ancient protein (*Figure 10.2*). In prothrombin, the second kringle domain interacts with the heparin-binding site to slow down antithrombin III/heparin-mediated inhibition (Arni *et al.*, 1993) prior to proteolytic prothrombin activation. The primitive heparin binding site evident in the ancient protease would have been unable to bind the kringle 2 domain as strongly as its extant descendant. This is consistent with the view that the ancient protein contained a light chain of Gla and EGF domains, the latter only being replaced by kringles at a later stage in the evolution of the protease (Patthy, 1985). The heparin-binding site may then have evolved in such a way as to balance the dual requirements of antithrombin III/heparin-mediated inhibition of thrombin and the kringle 2-mediated protection of prothrombin from premature inhibition.

10.2.5 The evolution of human protein C

One of the best characterized vitamin K-dependent serine proteases of coagulation is protein C. Activated protein C (APC) exerts a negative feedback regulatory effect on both the intrinsic and extrinsic pathways of coagulation through the proteolytic inactivation of factors Va and VIIIa in the presence of protein S and a negatively charged phospholipid surface (Tuddenham and Cooper, 1994). Protein C is activated by thrombin through cleavage of the Arg169-Leu170 bond with release of a dodecapeptide from the heavy chain. This reaction is enhanced some 20 000-fold by thrombomodulin, an endothelial cell surface glycoprotein which binds thrombin with high affinity (reviewed by Esmon, 1995).

The substrate specificity of serine proteases is largely dependent upon the structure and properties of the substrate-binding pocket adjacent to the active site. By analogy with other trypsin-like serine proteases, Asp189 (chymotrypsin numbering) of APC forms the bottom of this binding pocket (Segal *et al.*, 1971; Shekter and Berger, 1967). In principle, the presence of Ser198 at the S2 substrate-binding site of APC would allow the binding pocket to accommodate larger amino acid residues such as Phe or Leu (Stone and Hofsteege, 1985). However, this theoretical diversity appears not to be capitalised upon by the native structure since the substrate specificity of APC is in reality confined to factors Va and VIIIa (Esmon, 1987).

A molecular model of the catalytic domain of early mammalian protein C was also derived and was used to examine the functional architecture of that ancient domain and to explore its evolutionary progression from the putative common ancestor of the vitamin K-dependent serine proteases toward extant human protein C (Wacey *et al.*, 1997). This application of homology modeling to a reconstructed amino acid sequence made it possible to trace the evolution of structural features in a human protein and, in so doing, to make certain inferences regarding the development of its functional specificity. Since its first appearance in evolution as a result of a gene duplication, the catalytic domain of human protein C has undergone 41 amino acid changes, at least some of which have presumably served to 'fine-tune' interactions with its substrates. Thus, those sites of protein-protein interaction already present in early mammalian protein C should contain the amino acid residues most essential for interaction and hence for the biological functions of the protein.

10.2.6 Comparison of the functional architecture of early mammalian protein C with the putative ancestor of all vitamin K-dependent factors

Six of the 13 variable regions (VR's) in the model of early mammalian protein C were derived by reference to the crystal structure of factor Xa. The similarity in VR loop length between early mammalian protein C and extant factor X is consistent with both proteins sharing a considerable period of evolutionary history. It would seem reasonable to suppose that factor X (an activator of prothrombin and therefore a procoagulant) and protein C (an anticoagulant by virtue of its inhibitory action in inactivating factors Va and VIIIa) have coevolved in such a way as to ensure that thrombin generation is appropriately promoted yet adequately limited in response to hemostatic challenge.

Doolittle (1993), Patthy (1990), and Krawczak *et al.* (1996) all concluded that protein C and prothrombin probably emerged at a similar stage of vertebrate evolution. In the primordial vertebrate hemostatic system, one role for protein C could have been to act as part of a negative feedback control mechanism. Since both the catalytic efficiency of thrombin and the efficiency of activation of prothrombin were presumably undergoing a continual process of optimization at that time, there would have been an increasing requirement for an efficient negative regulator of thrombin.

Any change of function from the common ancestor of the vitamin K-dependent factors to early mammalian protein C would have necessarily required changes in the active site of the protease, thereby altering its substrate specificity from fibrinogen to the emerging co-factors VIIIa and Va. The 'accelerated evolution' of active sites is thought to have been important in the diversification of the substrate specificity of serine proteases, subsequent to gene duplication (Creighton and Derby, 1989; Ohta, 1994). From inspection of our models, changes in the active site region of protein C appear to have involved the elimination of the aryl binding site (with the exception of residue W215 which is highly conserved between all serine proteases) and the chemotactic binding site (*Figure 10.4*). Both of these modifications resulted from the loss of two pincer-like highly mobile insertion loops, Leu59-Asn62 and (to a lesser extent) Leu144-Gly150, originally present in the vitamin K-dependent factor ancestral protein. Since both loops are required for the catalytic activity of extant thrombin (Stubbs and Bode, 1993), it may be inferred that the early mammalian ancestor of protein C had already lost its ability to cleave fibrinogen. Moreover, the absence of those residues which in thrombin bind protein C [with the exception of Leu73, Arg75, and Asp186A (Tsiang *et al.*, 1995) and see below] implies that early mammalian protein C was probably not auto-catalytic.

In the early mammalian protein C molecule, the distribution of residues analogous to the fibrinogen-binding site of extant thrombin was very different from that predicted in the putative vitamin K-dependent factor ancestral protein (*Figure 10.4*). In terms of evolutionary conservation, the fibrinogen-binding patch

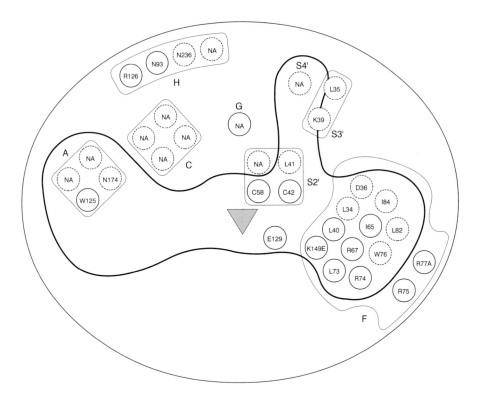

Figure 10.4. Schematic view of the active site canyon (bold contour) of early mammalian protein C (Wacey *et al.*, 1997). The location of the active site triad residues (His 57, Asp 102, Ser195, chymotrypsin numbering) is denoted by a shaded triangle. Residues which are conserved between the vitamin K-dependent factor ancestral protein and early mammalian protein are circled. Nonconserved residues are circled with a broken line. H: heparin-binding site, G: glycosylation site, C: chemotactic region, A: aryl-binding site, F: fibrinogen-binding exosite, S′: specificity sites C-terminal to cleavage.

of early mammalian protein C was split along its equatorial axis. The 'North' patch had acquired three non-conservative mutations, all of which were to neutral Ile and Leu residues; the resulting changes in both polarity and geometry probably reflected adaptation to new substrates viz. factors VIIIa and Va. By contrast, the 'South' patch had not experienced any non-conservative substitutions. With the exception of Leu73, Arg75, and Asp186A, residues analogous to the fibrinogen-binding patch of extant thrombin (Stubbs and Bode, 1993) which are located outside of the active site were absent from early mammalian protein C. It may be, therefore, that early mammalian protein C was unable to bind fibrinogen.

An electrostatic view of the early mammalian protein C molecule (*Figure 10.5*, bottom right) reveals an anionic patch (P4) in the 'South-West' corner of the molecule. This patch appears to have become moderately expanded in extant human protein C and comprises (i) residues analogous to the fibrinogen-binding residues of the 'South' active site, (ii) Lys38, a fibrinogen-binding residue external to the active site, and (iii) the thrombomodulin-binding residues Lys36 and Arg75 of extant thrombin. P4 also contained residues analogous to Asn36, Gln37, Glu38 (chymotrypsin numbering) of the vitamin K-dependent factor ancestral protein (Glu38 belongs to the S3′ site of extant thrombin). In early mammalian protein C, these three residues had become lysines which would have served to increase the net positive charge in this area. Since this region binds thrombomodulin in extant protein C (Wacey *et al.*, 1993; Greengard *et al.*, 1994; Vinceno *et al.*, 1995; Grinnell *et al.*, 1994), the P4 site may represent the nascent thrombomodulin-binding patch in early mammalian protein C. *Figure 10.5* depicts the envisaged evolutionary transformation of the primitive original fibrinogen/ thrombomodulin-binding patch into a region specialised for thrombomodulin binding. In extant protein C, substitution of Trp76 by Arg has served to increase further the anionic potential, and thus specificity, of this patch. Finally, it should be noted that the anionic residues of P4 (K36, K37, K38, R74, K148, R149, R151) in early mammalian protein C appear to have undergone relatively few subsequent substitutions in order to 'fine-tune' binding to thrombomodulin. Thus the P4 patch in early mammalian protein C was probably of the minimum size required to bind thrombomodulin, and was subsequently extended and refined by evolution.

The binding patch P3 (comprising anionic residues Lys20, Arg23, Lys159, and Arg188) of the vitamin K-dependent factor ancestral protein is also apparent in early mammalian protein C. However, in the latter molecule, this patch was less expansive and confined to a higher latitude in the molecule. The apparent migration of the patch away from the equatorial belt appears to have continued during

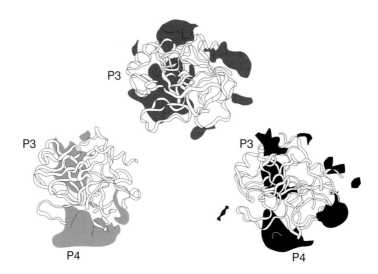

Figure 10.5. The electrostatic profiles of the vitamin K-dependent factor ancestral protein (top, alpha carbon backbone is shown as a ribbon, electrostatic surface is solid), early mammalian protein C (bottom right, alpha carbon backbone is shown as a ribbon, electrostatic surface is solid) and extant human protein C (bottom left, alpha carbon backbone is shown as a ribbon, electrostatic surface is solid) are shown (Wacey *et al.*, 1997). The electrostatic equipotential surfaces are contoured at +1 kcalmol⁻¹. The view is towards the active site canyon. The locations of the anionic patches P3 and P4 are indicated.

evolution toward extant human protein C. The persistence of the anionic potential of this area through evolutionary time, together with the report of a type II (dysfunctional) substitution (R352W, Reitsma *et al.*, 1995) in this region, implies a functional role in both ancient and extant proteins.

10.2.7 Comparative geometry of the active sites of early mammalian and extant protein C

Extant human protein C and early mammalian protein C exhibit 83.6% sequence identity at the amino acid sequence level, and exhibit a backbone RMS divergence of 2.6 Å (computed against a model of the serine protease domain of extant human protein C; Wacey *et al.*, 1993). Recently, a relatively low resolution crystal structure of human APC (des Gla domain) has been solved (Mather *et al.*, 1996). Comparison of our model with this structure served to demonstrate that the majority of functionally important residues identified in extant protein C were already present in its early mammalian predecessor. The Ca^{2+} binding loop (residues 70–80 with the exception of N78), the active site loop (residues 146–152) and the insertion helix (residue 129) are all to be found intact in early mammalian protein C. The asymmetric surface charge distribution of extant protein C is also apparent in early mammalian protein C. Finally, residues E192 and Y143 (the major determinants of substrate specificity) which span the active site, are present in both proteins.

Since the regions close to the catalytic triad of extant human and early mammalian protein C are likely to possess a similar functional architecture (owing to the minimal divergence in the template structures underlying the two models), the active site geometries of the two proteins may be reliably contrasted at relatively high resolution. Within the catalytic site pocket of extant human protein C, the S2 residue (Ser 198) is predicted to have arisen by substitution of a Phe residue present in early mammalian protein C. This would imply that an initially tightly fitting catalytic pocket became more capacious during evolution. Why, however, the substrate specificity of extant human protein C is restricted to factors Va and VIIIa, although its active site pocket can theoretically accommodate larger substrates, remains unclear. With this exception, the amino acid sequence of the active site pocket of early mammalian protein C is identical to that of extant human protein C. Thus the structural features noted in the crystal structure of human APC would probably also have been found in the early mammalian protein.

A prominent hydrophobic and solvent-accessible region is found on the surface of the serine protease domain of early mammalian protein C. This region has been shown to bind the second (C-terminal) EGF-like domain of the light chain in factors Xa, IXa, and protein C (residues 100–116 and 45–50 respectively). Early mammalian protein C may therefore have possessed a light chain, an hypothesis consistent with Patthy's (1985) view of serine protease evolution.

In summary, the application of homology modeling to a reconstructed amino acid sequence allowed Wacey *et al.* (1997) to trace the evolution of specific structural features in protein C. This approach provided new insights into the evolution of protein C, allowed an assessment of the nature of the minimal thrombomodulin binding site and permitted inferences to be made as to the possible catalytic mechanism in early mammalian protein C. Such an approach lends itself readily to the study of the common ancestors of other multigene families and may therefore allow further 'structure-function studies' of phylogenetically reconstructed proteins to be performed.

References

Arni R.K., Padmanabhan K., Padmanabhan K.P., Wu T.P., Tulinsky A. (1993) Structures of the non-covalent complexes of human and bovine prothrombin fragment 2 with human PPACK-thrombin. *Biochemistry* **32**: 4727–4737.

Audic S., Beraud-Colomb E. (1997) Ancient DNA is thirteen years old. *Nature Biotechnol.* **15**: 855–858.

Austin J.J., Ross A.J., Smith A.B., Fortey R.A., Thomas R.H. (1997) Problems of reproducibility – does geologically ancient DNA survive in amber-preserved insects? *Proc. Roy. Soc. Lond. Series B: Biological Sciences* **264**: 467–474.

Banfield D.K., MacGillivray R.T.A. (1992) Partial characterization of vertebrate prothrombin cDNAs: amplification and sequence analysis of the B chain of thrombin from nine different species. *Proc. Natl. Acad. Sci. USA* **89**: 2779–2783.

Bird A.P. (1995) Gene number, noise reduction and biological complexity. *Trends Genet.* **11**: 94–100.

Blake C.C.F., Harlos K., Holland S.K. (1987) Exon and domain evolution in the proenzymes of blood coagulation and fibrinolysis. *Cold Spring Harb. Symp. Quant. Biol.* **52**: 925–931.

Britten R.J. (1986) Rates of DNA sequence evolution differ between taxonomic groups. *Science* **231**: 1393–1398.

Chandrasekharan U.M., Sanker S., Glynias M.J., Karnik S.S., Husain A. (1996) Angiotensin II-forming activity in a reconstructed ancestral chymase. *Science* **271**: 502–505.

Collins D.W., Jukes T.H. (1994) Rates of transition and transversion in coding sequences since the human-rodent divergence. *Genomics* **20**: 386–396.

Creighton T.E., Darby N.J. (1989) Functional evolutionary divergence of proteolytic enzymes and their inhibitors. *Trends Biochem. Sci.* **14**: 319–324.

DeSalle R., Gatesy J., Wheeler W., Grimaldi D. (1992) DNA sequences from a fossil termite in Oligo-Miocene amber and their phylogenetic implications. *Science* **257**: 1933–1936.

Doolittle R.F., Feng D.F. (1987) Reconstructing the evolution of vertebrate blood coagulation from a consideration of the amino acid sequences of clotting proteins. *Cold Spring Harb. Symp. Quant. Biol.* **52**: 869–874.

Doolittle R.F. (1993) The evolution of vertebrate blood coagulation: a case of Yin and Yan. *Thromb. Haemost.* **70**: 24–28.

Esmon C.T. (1987) The regulation of natural anticoagulant pathways. *Science* **235**: 1348–1352.

Esmon C.T. (1995) Thrombomodulin as a model of molecular mechanisms that modulate protease specificity and function at the vessel surface. *FASEB J.* **9**: 946–955.

Friedrich M., Tautz D. (1995) Ribosomal DNA phylogeny of the major extant arthropod classes and the evolution of myriapods. *Nature* **376**: 165–167.

Gojobori T., Ikeo K. (1994) Molecular evolution of serine protease and its inhibitor with special reference to domain evolution. *Phil. Trans. R. Soc. Lond. B* **344**: 411–415.

Greengard J.S., Fisjer C.L., Villoutreix B., Griffin J.H. (1994) Structural basis for type I and type II deficiencies of thrombotic plasma protein C: patterns revealed by three dimensional molecular modeling of mutations of the protease domain. *Proteins Struct. Funct. Genet.* **18**: 367–380.

Greer J. (1990) Comparative modeling methods: application to the family of the mammalian serine proteases. *Proteins Struct. Funct. Genet.* **7**: 317–334.

Grinnell B.W., Gerlitz B., Berg D.T. (1994) Identification of a region in protein C involved in thrombomodulin-stimulated activation by thrombin: potential repulsion at anion-binding site I in thrombin. *Biochem. J.* **303**: 929–933.

Hill R.E., Hastie N.D. (1987) Accelerated evolution in the reactive centre regions of serine protease inhibitors. *Nature* **326**: 96–99.

Hillis D.M., Allard M.W., Miyamoto M.M. (1993) Analysis of DNA sequence data: phylogenetic inference. *Meths. Enzymol.* **224**: 456–187.

Ikeo K., Takahashi K., Gojobori T. (1995) Different evolutionary histories of kringle and protease domains in serine proteases: a typical example of domain evolution. *J. Mol. Evol.* **40**: 331–336.

Jermann T.M., Opitz J.G., Stackhouse J., Benner S.A. (1995) Reconstructing the evolutionary history of the artiodactyl ribonuclease superfamily. *Nature* **374**: 57–59.

Krawczak M., Cooper D.N. (1996) Single base-pair substitutions in pathology and evolution: two sides to the same coin. *Hum. Mutation* **8**: 23–31.

Krawczak M., Wacey A., Cooper D.N. (1996) Molecular reconstruction and homology modelling of the catalytic domain of the common ancestor of the haemostatic vitamin K-dependent serine proteinases. *Hum. Genet.* **98**: 351–370.

Li W.-H. (1997) *Molecular Evolution.* Sinauer Associates, Sunderland.

Malcolm B.A., Wilson K.P., Matthews B.W., Kirsch J.F., Wilson A.C. (1990) Ancestral lysozymes reconstructed, neutrality tested, and thermostability linked to hydrocarbon packing. *Nature* **345**: 86–89.

Mather T., Oganessyan V., Hof P., Huber R., Foundling S., Esmon C., Bode W. (1996) The 2.8Å crystal structure of Gla-domainless activated protein C. *EMBO J.* **15**: 6822–6831.

Nei M. (1987) *Molecular Evolutionary Genetics.* Columbia University Press, New York.

Neurath H. (1984) Evolution of proteolytic enzymes. *Science* **224**: 350–357.

Novacek M.J. (1992) Mammalian phylogeny: shaking the tree. *Nature* **356**: 121–125.

Ohno S. (1970) *Evolution by Gene Duplication.* George Allen & Unwin, London.

Ohta T. (1991) Multigene families and the evolution of complexity. *J. Mol. Evol.* **33**: 34–41.

Ohta T. (1994) On hypervariability at the reactive center of proteolytic enzymes and their inhibitors. *J. Mol. Evol.* **39**: 614–619.

Pääbo S. (1993) Ancient DNA. *Sci. Am.* Nov. **269**(5): 86–92.

Patthy L. (1985) Evolution of the proteases of blood coagulation and fibrinolysis by assembly from modules. *Cell* **41**: 657–663.

Patthy L. (1990) Evolutionary assembly of blood coagulation proteins. *Sem. Thromb. Haemost.* **16**: 245–259.

Reitsma P.H., Bernardi F., Doig R.G., Gandrille S., Greengard J.S., Ireland H., Krawczak M., Lind B., Long G.L., Poort S.R., Saito H., Sala N., Witt I., Cooper D.N. (1995) Protein C deficiency: a database of mutations, 1995 update. *Thromb. Haemost.* **73**: 876–889.

Rogers J. (1985) Exon shuffling and intron insertion in serine protease genes. *Nature* **315**: 458–459.

Segal D.M., Powers J.C., Cohen G.H., Davies D.R., Wilcox P.E. (1971) Substrate binding site in bovine chymotrypsin. A crystallographic study using chloromethylketones as site-specific inhibitors. *Biochemistry* **10**: 3728–3738.

Shekter I., Berger A. (1967) On the size of the active site in protease I. *Biochem. Biophys. Res. Commun.* **27**: 157–162.

Shih P., Malcolm B.A., Rosenberg S., Kirsch J.F., Wilson A.C. (1993) Reconstruction and testing of ancestral proteins. *Meths. Enzymol.* **224**: 576–591.

Stackhouse J., Presnell S.R., McGeehan G.M., Nambiar K.P., Benner S.A. (1990) The ribonuclease from an extinct bovid ruminant. *FEBS Letts.* **262**: 104–106.

Stone S.R., Hofsteenge J. (1985) Specificity of activated protein C. *J. Biochem.* **230**: 497–502.

Stoneking M. (1995) Ancient DNA: how do you know when you have it and what can you do with it? *Am. J. Hum. Genet.* **57**: 1259–1262.

Stubbs M.T., Bode W. (1993) A player of many parts: the spotlight falls on thrombin's structure. *Thromb. Res.* **69**: 1–58.

Tsiang M., Jain A.K., Dunn K.E., Rojas M.E., Leung L.L.K., Gibbs C.S. (1995) Functional mapping of the surface residues of human thrombin. *J. Biol. Chem.* **270**: 16 854–16 863.

Tuddenham E.G.D., Cooper D.N. (1994) *The Molecular Genetics of Haemostasis and its Inherited Disorders.* Oxford University Press, Oxford.

Tugendreich S., Bassett D.E., McKusick V., Boguski M.S., Hieter P. (1994) Genes conserved in yeast and humans. *Hum. Molec. Genet.* **3**: 1509–1517.

Vinceno A., Gaussem P., Pittet J.L., Debot S., Aiach M. (1995) Amino acids 225–235 of protein C serine protease domain are important for the interaction with the thrombin-thrombomodulin complex. *FEBS Letts.* **367**: 153–157.

Vogel F., Motulsky A. (1979) *Human Genetics: Problems and Approaches.* Springer, New York. pp 442–462.

Wacey A.I., Pemberton S., Cooper D.N., Kakkar V.V., Tuddenham E.G.D. (1993) A molecular model of the serine protease domain of activated protein C: application to the study of missense mutations causing protein C deficiency. *Br. J. Haematol.* **84**: 290–300.

Wacey A.I., Krawczak M., Kakkar V.V., Cooper D.N. (1994) Determinants of the factor IX mutational spectrum in haemophilia B: an analysis of missense mutations using a multi-domain molecular model of the activated protein. *Hum. Genet.* **94**: 594–608.

Wacey A.I., Krawczak M., Kemball-Cook G., Cooper D.N. (1997) Homology modelling of the catalytic domain of early mammalian protein C: evolution of structural features. *Hum. Genet.* **101**: 37–42.

Index to Human Gene Symbols used in text

(arranged alphabetically by gene symbol)

Gene symbols are as recommended by the *Human Gene Nomenclature Committee –* (**http://www.gene.ucl.ac.uk/nomenclature/**) at the time of going to press. Genes described by symbols in lower case letters have not yet been allocated a permanent symbol.

Pseudogene symbols are not included. Symbols are sometimes abbreviated in the text when a cluster of genes is being described e.g. *HOXA* for the genes which comprise the homeobox A cluster.

Index to Human Gene Symbols used in text

(arranged alphabetically by gene)

Gene symbols are as recommended by the *Human Gene Nomenclature Committee* – (**http://www.gene.ucl.ac.uk/nomenclature/**) at the time of going to press. Genes described by symbols in lower case letters have not yet been allocated a permanent symbol.

Pseudogene symbols are not included. Symbols are sometimes abbreviated in the text when a cluster of genes is being described e.g. *HOXA* for the genes which comprise the homeobox A cluster.

Index